W9-BQZ-824

ANNUAL REVIEW OF PHYTOPATHOLOGY

EDITORIAL COMMITTEE

D. GOTTLIEB

G. H. HEPTING

J. G. HORSFALL

R. R. NELSON

W. F. ROCHOW

G. A. ZENTMYER

Responsible for organization of Volume 7
(Editorial Committee, 1967)

G. H. HEPTING

J. G. HORSFALL

A. KELMAN

R. A. LUDWIG

W. F. ROCHOW

G. A. ZENTMYER

ANNUAL REVIEW OF PHYTOPATHOLOGY

JAMES G. HORSFALL, *Editor*
The Connecticut Agricultural Experiment Station

KENNETH F. BAKER, *Associate Editor*
University of California, Berkeley

D. C. HILDEBRAND, *Acting Associate Editor*
University of California, Berkeley

VOLUME 7

1969

ANNUAL REVIEWS, INC.
4139 EL CAMINO WAY
PALO ALTO, CALIFORNIA, U.S.A.

ANNUAL REVIEWS, INC.
PALO ALTO, CALIFORNIA, U.S.A.

© 1969 by Annual Reviews, Inc.
All Rights Reserved

Standard Book Number: 8243-1308-9
Library of Congress Catalogue Card Number: 63-8847

FOREIGN AGENCY

Maruzen Company, Limited
6, Tori-Nichome, Nihonbashi
Tokyo

PRINTED AND BOUND IN THE UNITED STATES OF AMERICA BY
GEORGE BANTA COMPANY, INC.

PREFACE

In offering Volume 7 of the Annual Review of Phytopathology, the Editorial Committee wishes to thank Dr. R. G. Grogan of the University of California at Davis and Dr. M. N. Schroth of the University of California at Berkeley who served as guest editors in planning the volume.

During the year Dr. R. A. Ludwig of the Canada Department of Agriculture completed his term as a member of the Editorial Committee. His enthusiastic support of the Review is hereby gratefully acknowledged. We welcome aboard his replacement, Dr. R. R. Nelson of Pennsylvania State University.

We have had excellent help from Mrs. Mary E. King and Mrs. Janet H. Greene of the Annual Reviews staff. Especially do we thank Dr. D. C. Hildebrand of the University of California at Berkeley, for his fine help as acting associate editor during the absence of Kenneth F. Baker on sabbatic leave. In addition we thank Dr. Hildebrand for his usual skillful index prepared for this volume.

K. F. Baker
D. C. Hildebrand
D. Gottlieb
G. H. Hepting
J. G. Horsfall
R. R. Nelson
W. F. Rochow
G. A. Zentmyer

ERRATA

Volume 6 (1968)

"Trends in the Development of Plant Virology" by Francis O. Holmes, p. 44, line 32 of text: "are as" should read "as are."

"Mechanisms of Biological Control of Soil-Borne Pathogens" by Ralph Baker, p. 283, line 6 of text: "Figure 3" should read "Figure 2."

CONTENTS

Annual Reviews, Inc., and the Editors of its publications assume no responsibility for the statements expressed by the contributors of this *Review*.

Copyright 1969. All rights reserved

A PERSPECTIVE ON PLANT PATHOLOGY

By S. E. A. McCallan

Boyce Thompson Institute, Yonkers, New York

When I was invited by the editor, a good friend for over 40 years, to write the prefatory chapter for this volume, I was honored but also a little overwhelmed. My predecessors include some of the most noted philosophers in our science, and all, I believe, are teachers. I am not a teacher. Perhaps, therefore, my perspective will be different. Truly I agree with Professor Muskett (24) that "I struggled to find a proper response," and with Professor Bailey (12) that such honored persons "should be assigned texts." Not having been assigned a text, I shall simply do what the editor asked, give you my personal perspective of plant pathology as I have seen it labor to respond to pressures both from within and without our profession.

The changes during the first 60 or 70 years of plant pathology before my day are admirably covered in the historical accounts of Whetzel (26), Large (15), and Keitt (14). While most historians consider deBary the father of scientific plant pathology, it is interesting to note that the Englishman, Marsh (22) cites the Reverend M. J. Berkeley, a mycologist of England, and Whetzel of German ancestry, promotes Julius Kühn, a practical plant pathologist of Germany. By any account, however, the science would be 110 to 120 years old. Of the last 50 years, I have some personal knowledge. I shall discuss that.

Why did I become a plant pathologist?—Perhaps I must answer first the question, why did I become a plant pathologist? I think that in general, one becomes a plant pathologist (*a*), because he falls under the influence of an enthusiastic person who stimulates his interest in the field, (*b*) because he is a farm boy who has a natural awareness of plant diseases, and (*c*), of course, just plain chance. But everyone must answer specifically for himself. Here is my answer.

I was born on a tiny coral island in the mid-Atlantic, a 12th generation Bermudian. Fifty years ago it was agreed in my family that I would follow my father's footsteps to the Ontario Agricultural College in Canada. About this time and shortly before, various plant pathologists, mostly from the United States Department of Agriculture, had visited Bermuda to observe the recently recognized leaf roll disease of potatoes. In fact, in 1918 one of the first articles to be published on virus diseases in *Phytopathology* was that on potato leaf roll by Paul A. Murphy of Ireland and E. J. Wortley, the Director of Agriculture in Bermuda. The visiting plant pathologists were the first scientists I had ever seen.

The journal, *Phytopathology*, with its unpronounceable name, created in me an impression of great awe. I would become a plant pathologist and publish in *Phytopathology*.

During the summer of 1921, between my sophomore and junior years in college, I returned for the first time to Bermuda for a vacation. By then my father had succeeded Wortley as Director of Agriculture and had invited Professor H. H. Whetzel of Cornell University to come to Bermuda on sabbatical leave.

That was a memorable summer. I worked with "Prof." Whetzel as we came to call him so affectionately. Whetzel was one of the most enthusiastic and best teachers of his day. Had he been an entomologist, a plant breeder, or even a marine biologist, I probably would have followed him. In any case, this I did upon my graduation from Guelph in 1923. Because of my Bermuda background I might have become a potato pathologist or perhaps even a virologist. In fact my summers at Cornell were spent as a potato inspector. However, during the winters Whetzel asked me to do a little testing of fungicides for a few industrial sponsors and so by chance I went into the field of fungicides.

In due course, armed with a brand new Ph.D., a new model A Ford, and a new wife, I headed southeast for a new job at the Boyce Thompson Institute to work on the nature of fungicidal action. The group leader was Frank Wilcoxon, a physical chemist, and our work together lasted until World War II. Frank was just as stimulating and instructive as Whetzel, and in the environment of this new Institute I profited much. Perhaps the most gratifying remark I have ever heard was made to Frank and me by the late John W. Roberts of the U.S.D.A. who commented, "You made research on fungicides respectable."

The moral, if any, to be drawn from this little personal story of mine is that we must expose young people who are potential plant pathologists to stimulating teachers if we wish to keep our profession growing in stature and esteem. Naturally, there is also the need for good teachers: in sum, to provide the right environment.

We move now from why I became a plant pathologist to an examination of how plant pathology has transferred its emphasis with the needs of the times. From there we shall examine briefly the size and quality of our profession in the U.S.A.

From mycology to biochemistry.—Perhaps I can best illustrate the adaptability of plant pathology by the enormous change in emphasis in my day from mycology to biochemistry. The plant pathologists that came to Bermuda and fired my enthusiasm were primarily mycologists. They were interested in the Fungi That Cause Plant Disease—the title of a text popular at the time. Fitzpatrick was a giant in the land and so were Arthur and Clinton and Stevens and Shear.

Professors nowadays must attempt to stir interest in their prospective students with the mysteries of biochemical plant pathology and how they can attain fame there.

The shift in emphasis reflects our efforts to follow ever farther back along the chain of events in the causation of plant disease. First we asked what organisms produce disease. Then we asked what enzymes are produced by the organisms that produce disease. And now we ask what produces the enzymes that are produced by the organisms that produce disease.

The change from mycology to biochemistry is a pretty obvious trend in plant pathology, but we need to dig data from the records to uncover other, perhaps less obvious trends in our field. Two rather ready sources of data are at hand—the abstracts of papers at the annual meetings and the Directories published by The American Phytopathological Society beginning in 1953.

The abstracts of papers given at the annual meeting of the Society provide data for earlier years. I have analyzed these at 10 year intervals from 1948 back to 1918 (3, 6, 8, 10), which quite by coincidence is the year of my conversion to plant pathology. The data are displayed in Table I. Since the war-time meeting of 1918 was a small one, I have combined it for the meetings for 1917 and 1919 (2, 4).

There were approximately 100 abstracts for each of the four decennial meetings. This is admittedly a small sample, but I believe adequate to establish major trends. A few papers had to be classified into two fields of interest, e.g., spraying experiments to control potato blight.

It was my honor and pleasure to initiate and edit the publication of the first Directory of the Society during my terminal year as Secretary of the Society in 1953 (16). Subsequent directories have become available at five-year intervals (17, 18, 23).

While editing the first three Directories, I pursued my avocation by asking the members of the Society to list their "Fields of interest." This they did in their own words and without restrictions. Their statements for the 1958 Directory were tabulated in a President's Column (20), and for 1963 in the Directory for that year. A tabulation for 1953 has been made for this chapter and data for all Directories are included in Table II.

In 1968 the new editor, Dr. Mirocha (23), wisely requested each member to check a list of specific diseases and other fields of interest, or to use his more definite expression, "Areas of Expertise." While this was somewhat more restrictive for a check list, it encouraged at the same time the checking of some 50 per cent more areas per member than had been indicated previously. The areas or fields listed in Table II follow those of 1968 with several modifications, mainly the addition of diseases of field crops, of forage crops, or other crops, and of pathological anatomy and histology to give a better integration with the previous Directories. In all, a total of 47 fields of interest appear in Table II.

In the following discussions it should be understood that while the absolute number of abstracts or workers in a given field, e.g., cereal diseases, may be increasing or decreasing, it is the interest relative to the total effort in all fields of plant pathology that is important.

TABLE I

Comparative Fields of Interest
1918–1948

Fields of Interest	No. per 100 Abstracts in:			
	1918	1928	1938	1948
Diseases of Cereal Crops	19	16	5	7
Deciduous Fruits & Nuts	11	8	9	8
Fiber Crops	1	1	1	0
Forage Crops	0	0	2	2
Forest & Shade Trees	8	1	9	4
Grasses & Turf	1	0	0	0
Legume Crops	3	3	4	3
Market & Storage	0	1	0	3
Ornamentals & Nursery Crops	4	4	5	4
Potatoes	8	2	4	5
Small Fruits	0	0	4	1
Sugar Crops	0	2	2	1
Tobacco	3	4	3	2
Tropical & Subtropical Crops	0	1	1	1
Vegetable Crops	26	8	12	9
Other Crops	0	0	0	1
Diseases of Crops—Subtotal	84	51	61	51
Bacteriology & Bacterial Diseases	0	12	5	1
Control—Breeding for Resistance	2	5	1	7
Fungicides	5	15	17	21
Nematocides	0	0	1	2
Antibiotics	0	0	0	2
Deterioration of Plant Products	2	0		1
Epidemiology	0	0	1	1
Extension	1	0	0	0
Genetics of Microorganisms	0	4	3	4
Mycology & Fungus Diseases	13	7	4	3
Nematology & Nematode Diseases	3	2	2	2
Parasitic Seed Plants	0	1	0	0
Pathological Anatomy & Histology	0	3	1	0
Physiology of Microorganisms	0	0	5	9
Physiology of Parasitism	0	3	1	0
Plant Disease Survey	2	0	0	0
Plant Pathology General	1	0	0	1
Soil Borne Pathogens & Diseases	0	0	1	4
Virology & Virus Diseases	1	11	14	19
Total Fields of Interest	114	114	117	128
Number of Abstracts	104	91	111	112

TABLE II

COMPARATIVE FIELDS OF INTEREST
1953–1968

Fields of Interest	No. per 1000 Members Reporting in:			
	1953	1958	1963	1968
Diseases of Cereal Crops	111	119	116	113
Deciduous Fruits & Nuts	90	84	72	70
Fiber Crops	17	20	14	23
Field Crops	18	27	20	—
Forage Crops	30	47	27	—
Forest & Shade Trees	76	84	85	100
Grasses & Turf	8	7	15	34
Legume Crops	24	19	27	45
Market & Storage	11	13	18	24
Ornamentals & Nursery Crops	62	70	54	47
Potatoes	35	35	30	45
Small Fruits	18	27	17	30
Sugar Crops	17	15	14	22
Tobacco	19	17	15	23
Tropical & Subtropical	37	58	43	52
Vegetables	135	141	91	104
Other Crops	8	11	7	10
Disease of—Subtotal	716	794	665	742
Aero & Space Biology	2	0	3	7
Agricultural Chemicals	8	9	22	50
Air Pollution	5	8	11	22
Bacteriology & Bacterial Diseases	25	16	30	63
Control—Biological	1	2	5	39
Breeding & Resistance	79	60	75	117
General	18	10	20	127
Fungicides	182	175	156	131
Antibiotics & Bactericides	39	25	18	34
Nematocides	12	24	14	52
Regulatory	9	10	42	37
Deterioration of Plant Products	14	12	11	15
Ecology of Organisms	2	2	14	67
Epidemiology	12	17	18	74
Extension	5	5	28	52
Genetics of Microorganisms	19	21	16	39
Industrial Microbiology	12	5	8	15
Mycology & Fungus Diseases	92	81	47	32
Mycotoxicology	0	0	0	20
Nematology & Nematode Diseases	24	76	91	82
Parasitic Seed Plants	0	0	1	6
Pathological Anatomy & Histology	10	6	11	—
Physiology of Microorganisms	68	31	37	89
Physiology of Parasitism	27	39	112	155
Plant Disease Survey	11	12	6	35
Plant Pathology General	51	74	75	117
Soil Borne Pathogens & Diseases	28	55	80	162
Soil Microbiology	4	15	30	36
Teaching	15	21	81	124
Virology & Virus Diseases	137	158	197	181
Total Fields of Interest	1,627	1,763	1,924	2,722
Total Members Reporting	1,310	1,682	2,075	1,895

The trends are dramatically clear. I have spoken of mycology. It fell steadily in popularity during the first years (Table I) and has continued to fall in the last years (Table II). I have spoken, too, of the physiology and chemistry of the disease process. This subject was essentially absent from the abstracts from 1918 to 1948, but now is climbing rapidly. By 1968 it (physiology of parasitism) had reached third in the list of the "areas of expertise." A related field, physiology of microorganisms (actually for the most part fungi) has had a somewhat erratic growth in interest but is currently relatively high.

Other changing trends.—Another great shift is underway. Table I shows clearly that most of the interest in plant pathology in the early years was concentrated on the diseases of *crops* not on diseases *per se*. Diseases of crops occupied 80 per cent of the abstracts in 1918. The plant pathologists of the day were crop oriented. They were concerned more with plants than with pathology. Although the relative interest fell off during the 30 years after 1918, the crop diseases still held the center of the stage at nearly 55 per cent in 1948. The textbooks reflected this—"Diseases of Vegetables," "Diseases of Citrus," etc.

The interest in crop diseases has fallen rapidly in the last 15 years, however. It is now down to 27 per cent of the total.

My field, of course, is fungicides (21) which show a rise and then a fall. Interest in fungicides rose steadily during the 30 years after 1918 coming up from about 4 per cent to about 16 per cent of the total interest. This reflects the rise of the "fixed coppers" in the thirties and the organic fungicides in the forties. Now the trend is as sharply downward.

There are several reasons for this, not the least of which is the book "Silent Spring." There is also the general effectiveness of modern fungicides for many kinds of applications, and the increasing difficulties and expense of finding and developing newer and better ones. A really successful, hence practical fungicide for the rust diseases, powdery mildews, and soil fungi, and especially a systemic or chemotherapeutic fungicide would greatly stimulate renewed interest.

Virology and virus diseases are areas of considerable prestige and justifiably so, especially the former. After all, it brought a Nobel prize to one member of our Society. As pointed out earlier (18), twice as many members in the 1963 Directory preferred virology as their area of concern over virus diseases. The opposite ratio among plant pathologists would seem more realistic. You would think they would be more interested in diseases than in viruses but they claim not. Rather than encourage the members to compromise themselves, Dr. Mirocha, the editor, in 1968 (23) at my suggestion combined the two. This broad field has grown steadily from only one per cent of the abstracts in 1918 to the leading position in both 1963 and 1968. The results of 1968 suggest that the relative interest may have reached a plateau or perhaps even to have begun a downward trend.

Epidemiology and the ecology of organisms have been given a considerable increase in attention in 1968, as has also been the case with biological control, bactericides, plant disease survey, and soil borne pathogens and diseases.

A dramatic new area of plant pathology is mycotoxicology. It was wholly unrepresented as late as 1963, but 20 people claimed expertise in 1968.

Another discipline included in the broad field of plant pathology is nematology. This area, from the beginning of the tabulations in 1918, has received only modest attention but this has risen markedly since 1958.

The field of antibiotics (separately recorded before 1968) shows an interesting pattern of change in concern. The emergence of interest in antibiotics following the discovery of penicillin and streptomycin is very evident. In 1948 there were two abstracts. The year 1953 was the high point with 36 members reporting an interest. In 1958 this interest had dropped to 23, and in 1963 to 15. Antibiotics do not control plant disease very well and we have lost interest.

Several of the areas of interest require a special explanation since I think that the apparent marked increase in 1968 is due in part at least to the changed method for reporting. Some of the teachers and extension workers who reported in the earlier years possibly felt that only subject areas of interest were to be reported. However, in 1968 these two areas were listed and could be checked. Hence the marked increase. The elimination of a free choice in the areas to be listed in 1968 probably caused nonplant pathologists, e.g., chemists, entomologists, administrators, etc. to check the broad categories of general control (general chemical control), plant pathology general, or both, with the resultant marked increase.

How broad is the interest in plant pathology?—We have discussed the breadth of interest within the usually defined area of our subject. We can examine the width of our field by examining the ancillary specialties that impinge sufficiently on plant pathology to warrant membership in The American Phytopathological Society.

Data are available in this for the 1953 and 1958 Directories (16, 19) and they have been obtained for 1968. This survey is made on the assumption that from the title and field of interest (when recorded) the scientific discipline can be determined. For example, a person who lists himself as professor of plant pathology is a plant pathologist and not a nematologist or a virologist and a professor of biology is a biologist and not a plant pathologist. This is not necessarily always the case, but I believe that the assumption is reasonably accurate.

People calling themselves plant pathologists, including graduate students in departments of plant pathology, constituted 73 per cent of the membership in 1953 and also in 1958. In 1968 this dominant group had fallen slightly to 65 per cent. The ten other leading professions represented in the

Society over this period, accounting for approximately 15 per cent of the membership, were, in the numerical order of 1968: biologist, botanist, nematologist, entomologist, chemist or biochemist, microbiologist, agronomist, mycologist, geneticist or plant breeder, and horticulturist. The remaining percentage covers a wide variety of occupations which, as may be seen in the Directory for 1953 (16) and for 1958 in a President's Column (19), ranges from university president to housewife. Among the 10 other professions recorded above, botanist, entomologist, agronomist, and mycologist appear to be falling in relative numbers. However, biologist, nematologist, and chemist or biochemist appear to be rising. The founding of the Society of Nematologists in 1961 so far appears not to have materially affected membership in the plant pathology society since the number of nematologists per 1000 increased from 10 to 21 to 24 in the three years reported. However, the publication of their own journal which is expected shortly, may cause a decrease.

Somewhat surprisingly the professions of bacteriology, plant physiology, and virology are not in the first ten above, at least as far as they can be identified. There were only 5 plant physiologists per 1000 in 1968. This discipline is as closely related to plant pathology as any other. Probably the plant physiologists consider plant pathology too applied. In 1953 there were no identifiable virologists, in 1958 only 1 per 1000, and in 1968 only 6 per 1000.

The general conclusion from this appraisal is that in a modest way more scientists from other disciplines are being attracted to plant pathology. This of course is borne out by the increasing emphasis on the areas in the lower half of Table II.

As is well known, plant pathology developed mainly from mycology and the term "applied mycology" was very appropriate. This is no longer the case. Plant Pathology can now be called a conglomerate science; at least it is broad and appears to be broadening. In some countries the term plant pathology covers all diseases and disorders of plant including attacks by insect pests. The inclusion of plant pathological entomology is most logical especially from the applied and extension point of view. Whether this organic union will ever come about in the U.S.A. is doubtful, especially since there are about twice as many entomologists as there are plant pathologists. We would be outvoted. However, one can be optimistic. Thus when our small committee of three was working on a design for the official seal of the Society (11), I had in the back of my mind that with a little stretching a small insect could be inserted in the center of the shield as befits the most important cause of plant disease and disorders!

Who employs plant pathologists?—The Directory returns tell us who employs plant pathologists, at least in the U.S.A. Answers were gathered from the 1968 returns to add to those already available for 1953 and 1958 (16, 20). These answers are summarized per 1000 individuals (see next page).

Type of employer	1953	1958	1968
State	582	611	615
Federal (incl. cooperation with State)	187	169	170
Industrial	135	142	123
Endowed colleges, institutions, foundations, etc.	52	46	55
Retired	30	28	37
Self-employed	14	4	0

Fifty years ago employment of plant pathologists by industry was practically unheard of, but a decade later an upward trend had begun. Possibly the peak was reached about 1958 and has fallen since. More evidence would be required to substantiate this. The general impression that the government, specifically the U.S.D.A., has been expanding its opportunities for employment at a higher rate than others appears not to be borne out. Certainly the state and endowed institutions are keeping pace.

Apparently, self-employment of plant pathologists is becoming less attractive. Actually, only one self-employed person was reported in 1968, another old friend from Cornell days, Dr. Cynthia Westcott, the Plant Doctor. However, since the above returns are rounded to the nearest whole number, active Cynthia at 0.1 becomes a lost cypher!

The costs of research in plant pathology.—We are all aware of the trend in the costs which follows the motto of New York State—"Excelsior." Excellence is expensive. Much of the rise in costs is due to inflation which over the last 40 years is about fourfold. However, this is not the whole story. Forty years ago, when arriving at the Boyce Thompson Institute, I was assigned space in a well built laboratory; adequate glassware and chemicals were available. Also, I was assigned a secondhand microscope and balance, and that was my individual equipment. Today one cannot hire a young Ph.D. fresh out of graduate school without expecting to buy some $10,000 of new equipment immediately. The complication of present research with its increasing emphasis on biochemistry demands this.

Rather than quote some multimillion dollars of national effort in phytopathological research (which I do not have) some figures from my relatively small Institute might be more meaningful. Since the Institute is a completely self-contained unit, all costs are accounted for including such indirect things as administration, the library, other service departments, and maintenance of buildings, grounds, and a small farm. Items such as these do not appear in the budgets of most university departments of plant pathology; nevertheless, they are a part of the over-all costs.

In the broad field of plant pathology at the Institute, the equivalent of 20 full-time senior scientists (Ph.D.'s) plus a technical staff of 35 are engaged. A third or less of these consider themselves plant pathologists; the others are plant physiologists, chemists or biochemists, virologists, and nematologists. Nevertheless, this is still plant pathology. The costs in this broad area in 1968 for direct expenses were $768,000. This figure includes salaries, supplies, specific charges from various service departments, and travel, but not capitalized equipment. The indirect charges were $378,000 or 73.7 per cent of research salaries. These indirect costs include a prorated share of administration costs, employee benefits, net costs of the service departments, and depreciation costs of buildings and equipment, an area not too well understood by research workers and some administrators as pointed out by Faiman (13). The total cost for plant pathology is $1,146,000 or $57,300 per senior research scientist. During the past decade these groups of workers have been furnished, for their own use, equipment costing $490,000.

The status of plant pathology.—Travel has become a status symbol in modern science. When I was a graduate student we were lucky to get to one scientific meeting in two years and that at our own expense. In fact a graduate student friend of mine from Cornell hitchhiked his way to one of these meetings in the dead of winter. This so impressed the elderly Dr. Erwin F. Smith that he asked to meet the young man.

Today most of us, including graduate students, expect to attend several meetings each year, on expense accounts which are often grant funds. This increase in travel, both national and international, while necessary and desirable for an exchange of information has become of concern to some administrators. As asked by Pound (25), when do they have time to stay home and get some work done?

Although I cannot produce any numbers to substantiate it, I suspect that the consensus would be that the science of plant pathology should be better recognized by top management and especially by the public at large. By the same token, I suspect that such a consensus could be derived for most other sciences and most other professions as well.

We need no numbers to show that plant pathology is submerged and diluted in the various crop departments of the U.S.D.A. It is my understanding that Erwin F. Smith in the early days had an opportunity to organize a named unit of plant pathology in the U.S.D.A. comparable to that in Entomology but that he was not interested. This seems too bad.

On the other hand, the trend in the Land Grant Colleges is upward according to Pound (25). The number of independent departments of plant pathology is upward. This is good for our profession. Pound notes one distressing trend—that our society shows the lowest growth rate of seven societies listed among the "ten other" disciplines closely related to plant pathology as noted above. Pound ascribes this to an inadequate publication program.

Nevertheless the percentage of foreign members in the society has been increasing from 1917 to 1968. The numbers at intervals of more or less a decade are: 4, 12, 14, 14, 13, and 17 per cent of the total membership (1, 5, 7, 9, 17, 23). Since the foreign members in general would join primarily to receive and to publish in our professional journal, this would appear to be an endorsement of the increasing interest and quality of *Phytopathology*.

Finis.—And this is my personal perspective of plant pathology. I have enjoyed 50 years of associating with plant pathologists. I look forward to many more. I am confident that we can provide the adjustments needed in a rapidly changing world—that we can continue to do our bit to help feed a hungry world.

LITERATURE CITED

1. American Phytopathological Society. *Constitution and list of members of The American Phytopathological Society* (Am. Phytopath. Soc., 26 pp., 1917)
2. American Phytopathological Society. Abstr. Am. Phytopath. Soc., 9th Meeting, Pittsburgh, December 28, 1917 to January 1, 1918. *Phytopathology,* **8,** 68–82 (1918)
3. American Phytopathological Society. Abstr. Am. Phytopath. Soc., 10th Meeting, Baltimore, December 26 to 28, 1918. *Phytopathology,* **9,** 49–55 (1919)
4. American Phytopathological Society. Abstr. Am. Phytopath. Soc., 11th Meeting, St. Louis, December 30, 1919 to January 1, 1920. *Phytopathology,* **10,** 51–68 (1920)
5. American Phytopathological Society. List of patrons and members. *Phytopathology,* **15,** 625–649 (1925)
6. American Phytopathological Society. Abstr. Am. Phytopath. Soc., 20th Meeting, New York, N.Y., December 28, 1928 to January 1, 1929. *Phytopathology,* **19,** 79–108 (1929)
7. American Phytopathological Society. List of members of The American Phytopathological Society, 1935. *Phytopathology,* **25,** Suppl., 1–22 (1935)
8. American Phytopathological Society. Abstr. Am. Phytopath. Soc., 30th Meeting, Richmond, Virginia, December 37 to 30, 1938, inclusive. *Phytopathology,* **30,** 1–25 (1939)
9. American Phytopathological Society. List of members of The American Phytopathological Society. *Phytopathology,* **38,** No. 8, Sect. 2, 1–40 (1948)
10. American Phytopathological Society, Abstr. Am. Phytopath. Soc., 40th Meeting, Pittsburgh, Pennsylvania, December 6 to 8, 1948. *Phytopathology,* **39,** 1–27 (1949)
11. Anonymous. The birth of our seal. *Phytopathology News,* **1,** No. 4, 1 (1967)
12. Bailey, D. L. Whither pathology. *Ann. Rev. Phytopathol.,* **4,** 1–8 (1966)
13. Faiman, R. N. Regulation of indirect costs. *Science,* **162,** 1433 (1968)
14. Keitt, G. W. History of plant pathology. In *Plant Pathology,* Vol. 1, 61–97 (Horsfall, J. G., Dimond, A. E., Eds., Academic Press, New York, 674 pp., 1959)
15. Large, E. C. *The Advance of the Fungi* (Henry Holt & Co., New York, 488 pp., 1940)
16. McCallan, S. E. A. Directory. *Phytopathology,* **43,** D1–D58 (1953)
17. McCallan, S. E. A. Directory. *Phytopathology,* **48,** D1–D68 (1958)
18. McCallan, S. E. A. Directory of The American Phytopathological Society. *Phytopathology,* **53,** D1-D110 (1963)
19. McCallan, S. E. A. The President's Column. *Phytopathology,* **51,** 207 (1961)
20. McCallan, S. E. A. The President's Column. *Phytopathology,* **51,** 267 (1961)
21. McCallan, S. E. A. History of fungicides. In *Fungicides.* Vol. I, 1–37 (Torgeson, D. C., Ed., Academic Press, New York, 697 pp., 1967)
22. Marsh, R. W. Moulds, mildews and men. *Nature,* **218,** 1017–1019 (1968)
23. Mirocha, C. J. *Directory of Members* (The American Phytopathological Society, St. Paul, Minnesota, 105 pp., 1968)

24. Muskett, A. E. Plant pathology and the plant pathologist. *Ann. Rev. Phytopath.*, **5**, 1–16 (1967)
25. Pound, Glenn S. A Midstream View of Plant Pathology (A Paper presented at the 60th anniversary of the founding of The American Phytopathological Society, Columbus, Ohio, September 5, 1968)
26. Whetzel, H. H. *An Outline of the History of Phytopathology* (Saunders, Philadelphia, 130 pp., 1918)

Copyright 1969. All rights reserved

HISTORY OF PLANT PATHOLOGY IN GREAT BRITAIN

By G. C. Ainsworth

Commonwealth Mycological Institute, Kew, Surrey, England

Shield the young Harvest from devouring blight,
The Smut's dark poison, and the Mildew white,
Deep-rooted Mould, the Ergot's horn uncouth,
And break the Canker's desolating tooth.
First in one point the festering wound confin'd
Mines unperceived beneath the shrivel'd rind;
Then climbs the branches with increasing strength,
Spreads as they spread, and lengthens with their length. . . .

Erasmus Darwin, *The Botanic Garden* (Part I, *The Economy of Vegetation*), pp. 202–3, 1791

To trace the origin of a scientific concept or the first observation of a critical phenomenon is frequently difficult. Posterity, with the advantage of hindsight, is liable to read into earlier writings senses and overtones that were not intended and many fundamental observations were first made and recorded without their significance being realized. In writing about the history of science it is customary, and in brief historical reviews almost essential, to dramatize developments by giving prominence to certain dates and to associate discoveries and advances with outstanding workers, thus falsifying the record by oversimplification. For example, the concept of evolution is almost universally associated with the name of Charles Darwin but the appearance of *The Origin of Species* in 1859 was precipitated by the independent formulation of the theory of natural selection by Alfred Wallace the previous year, while the idea of evolution had been very much in the air since the time of Lamarck and Erasmus Darwin, Charles Darwin's versatile grandfather, at the end of the 18th century. Again, Fleming's discovery of penicillin in 1928 was from one aspect merely another example of antagonism between microorganisms, a phenomenon well known to plant pathologists. During the 1930's Weindling, uninfluenced by Fleming, attempted the control of plant pathogenic soil fungi by the use of gliotoxin produced by *Gliocladium viride,* and the topic can be traced back to the Manchester physician, William Roberts, who, in 1870, noted the dissolution of bacteria in cultures contaminated by a mould, interpreted what he saw in the Darwinian terms of a struggle for existence, and was the first to use antagonism in this sense (59).

1846

In this review, tradition will be followed in claiming the Reverend Miles

Joseph Berkeley [1803–89] as the "father of plant pathology" in Great Britain and by pinpointing the opening of the modern era in this country with a sentence from his classic paper, "Observations, botanical and physiological, on the potato murrain," which appeared in the *Journal of the Horticultural Society of London* for January 1846: "The decay is the consequence of the presence of the mould, and not the mould of the decay" (7). This conclusion regarding the cause of potato blight was not original. As Berkeley records, it was held by the American, J. E. Teschemacher, the Belgian, Dr. C. J. E. Morren, and in France by Monsieur Antoine de Payen and also by Dr. J. F. C. Montagne who, on 30 August 1845, was the first to name the potato blight fungus which he described and designated *Botrytis infestans*. It was, however, a minority opinion. Most authors, including the French mycologist, J. B. H. J. Desmazières (who also described and named the potato blight fungus), and Dr. John Lindley [1799–1865], editor of the *Gardeners' Chronicle* and professor of botany at University College, London, considered the mould to be the result of the decay and attributed the primary disease to diverse environmental and physiological factors.

Berkeley belonged to a profession whose members have done much to advance biology, and from 1833 he lived first at King's Cliffe and then at Sibbertoft, both in Northamptonshire, the county of his birth. While an undergraduate at Cambridge he developed a deep interest in natural history and early established himself as the leading expert on fungi in the British Isles by compiling, at the invitation of William J. Hooker [1785–1865], then professor of botany at Glasgow (he subsequently became director of the Royal Botanic Gardens, Kew), the fungi for J. E. Smith's *English Flora* in a volume published in 1836. Berkeley was incredibly industrious and, stimulated by the need to provide for a large family, kept a school for boys and wrote prolifically books, scientific papers (in which he described for the first time more than 5000 fungi), and semipopular articles. He was also a collector and in 1879 gave to Kew the mycological herbarium representing 10,000 species which he had amassed.

It was thus only natural that Berkeley's expert opinion on potato blight should be sought. The controversy which raged during the early and mid-1840's on the aetiology of this most significant outbreak of plant disease is well covered by Large (39), whose account may now be supplemented by Mrs. Woodham-Smith's (71) graphic description of the social effects of this devastating epidemic (with particular reference to the Irish famine) and Bourke's (12) assessment of its origin and spread.

Beginnings

There are many references to plant disease in the English agricultural and general literature. According to the *Oxford Dictionary,* "mildew" and "rust" were first applied to plant disease in the 14th century, "canker" dates from the 15th century, "smut," "blight," and "ergot" from the 17th century. The verb "to wilt" was another 17th century introduction but "wilt" was not used as a substantive until the 19th century. With the exception of "ergot," all these terms were at first used very imprecisely (some still are) and frequently covered depredations by insects. They were also used figuratively by Shakespeare and Milton, among others.

To Robert Hooke [1635–1703] goes the credit of being the first to illustrate in detail a plant pathogenic microfungus when, in his *Micrographia,* 1665, he figured the rose rust fungus, *Phragmidium mucronatum,* which he considered to be derived from putrified host tissues (36). The Reverend Stephen Hales [1677–1761], in *Vegetable Statics,* 1727, held a similar view regarding the origin of hop mildew when he wrote: "in a rainy moist state of air . . . too much moisture hovers about the hops, so as to hinder in good measure the kindly perspiration of the leaves, whereby the stagnating sap corrupts, and breeds moldy fen. . . ." He surmised that the rapid spread of the disease was "because the small seeds of the quick growing mold, which soon come to maturity, are blown over the whole ground," and asked, "might it not then be adviseable to burn the fenny hop-vines as soon as the hops are picked, in the hopes thereby to destroy some of the mold?" (32). In his *Horse-Hoeing Husbandry* (65), 1733, Jethro Tull [1674–1741] had two technical chapters, "Of Smuttiness" and "Of Blight," and by the end of the century, Erasmus Darwin's appeal to the "Sylphs," quoted above, reflects his clear idea of the range of disease symptoms.

Of particular interest among early writings is the slim quarto pamphlet by Sir Joseph Banks [1743–1820], *A Short Account of the Cause of the Disease in Corn* (4), 1805, with its two attractive coloured plates by Francis Bauer [1758–1840], "Botanical Painter to His Majesty" [King George III ("Farmer George")]. Even if the teliospores of the black rust (*Puccinia graminis*) are shown bursting and shedding their "seeds," Banks took it for granted that the rust was a parasitic plant, acknowledged that Felice Fontana in Italy in 1767 was the first to give "an elaborate account of this mischievious weed" (26) (by describing both the uredinial and telial states) and wrote: "It has long been admitted by farmers, though scarcely credited by botanists, that wheat in the neighbourhood of a barberry bush seldom escapes the Blight. The village of Rollesby in Norfolk, where barberry bushes abound, is called by the opprobious appellation of Mildew Rollesby." He goes on that it is "notorious to all botanical observers, that the leaves of the barberry are very subject to the attack of a yellow parasitic fungus, larger, but otherwise much resembling rust in corn," and asks "Is it not more than possible that the parasitic fungus of the barberry and that of the wheat are one and the same species, and that seed is transferred from barberry to the corn?"—a view for which experimental support had been obtained only the year before by Thomas Andrew Knight [1759–1838], F.L.S., F.R.S., and president of the Horticultural Society for 27 years, who, in 1806, wrote to Banks reporting the development of rust in wheat plants ten days after being brushed with a branch of rust-diseased barberry damped with water (37, 55).

In 1807, Prévost (54) in France experimentally demonstrated that *Tilletia caries* caused bunt of wheat, while in 1835, the Italian, Bassi (6) showed the fungus now known as *Beauveria bassiana* to be the causal agent of the muscardine disease of silkworms, and shortly afterwards, Gruby (73) in Paris demonstrated the mycotic nature of human ringworm. These findings were known to the experts consulted regarding potato blight, but the main factor mitigating against their acceptance and the generalization that fungi could function as pathogens was undoubtedly the difficulty so frequently experienced by man in accepting novelty when it conflicts with orthodoxy.

Textbooks

Standard texts, even if invariably more or less out of date on publication, do reflect the general level of knowledge in any branch of science. The first British plant pathological "textbook," which was the Rev. Berkeley's outstanding contribution to plant pathology, was never published in book form. It comprised a series of 173 short articles on "Vegetable Pathology" which Berkeley contributed to *The Gardeners' Chronicle and Agricultural Gazette* between January 1854 and October 1857 (8). His approach was comprehensive and detailed. After a series of articles on the plant in a state of health and a discussion of definitions of disease, the different causes of disease are enumerated and then all types of abnormality in plants are considered under the two main headings "Internal or Constitutional" and "External or Accidental," phanerogamic and fungal parasites, mosses, lichens, algae, and insects being dealt with under the latter. Finally, "Remedies of Disease" are considered. Berkeley thus treated "vegetable pathology" as an equivalent to human and veterinary medicine, two disciplines which, unlike plant pathology, have retained their comprehensive approach.

The first British book devoted to plant disease was Worthington G. Smith's [1835–1917] little *Diseases of Field and Garden Crops* (62), 1884, which was based on a series of addresses given at the request of the officers of the Institute of Agriculture at the British Museum in South Kensington. "Plants in a state of health" is omitted (the addresses were preceded by a course of twenty lectures on vegetable physiology in relation to farm crops) as is insect damage, but one nematode infection—corn cockle [*Anguina tritici*]—is included among the thirty or so fungal diseases which are treated in more or less detail. Worthington Smith was an artist and wood engraver on the staff of the *Gardeners' Chronicle* and an amateur mycologist of note. His clear, if rather mechanical, drawings reflect his early training as an architect but his finished wood engravings are distinguished examples of botanical art. Although published some twenty years after de Bary's classical publications conclusively proving heteroecism in rusts, Smith did not concede that *Puccinia graminis* has an aecial state on barberry and his book also includes polemical writing against de Bary in which he (Smith) attempts to justify his mistake, made a decade earlier, in claiming oogonia of a *Pythium* as the much sought resting state of *Phytophthora infestans,* a "discovery" for which he was awarded the Knightian Gold Medal of the Royal Horticultural Society.

In 1889, another and even slighter book, *Diseases of Plants* (68), appeared under the imprint of the Society for the Promotion of Christian Knowledge. In this volume a much more modern note is struck in straightforward descriptions of nine common diseases by H. Marshall Ward [1854–1906], professor of botany in the Forestry School of the Royal Indian Engineering College, Cooper's Hill, Surrey. In 1902, Ward, by then professor of botany in the University of Cambridge, published *Disease in Plants* (69) in which the principles of plant pathology are discussed. This was two years before the first edition of Massee's *Text-book of Plant Diseases* (44) had appeared.

George Edward Massee [1850–1917], (57), a Yorkshireman and another prolific writer on fungi, took over from Mordecai Cubitt Cooke [1825–1914] (56) in 1893 the charge of the fungal collections at the Royal Botanic Gardens, Kew, where, among his other duties, he acted as "mycologist" to the Board of Agriculture. His *Text-book* (of which a second edition appeared in 1903) and its replacement, *Diseases of Cultivated Plants and Trees* (1910; 2nd ed., 1915) (45), were the standard British texts until the publication in 1928 of *Plant Diseases* (14) by F. T. Brooks [1882–1952], lecturer in botany at Cambridge University and one of Marshall Ward's former students.

Berkeley was a man of his time, and Marshall Ward was imbued with the "new botany" and was an advocate of the experimental approach but it must be admitted that British plant pathological texts were of a lower standard than their German contemporaries, while in the present century American books on general plant pathology have been widely used by British students. During the second half of the 19th century little attention was paid to plant disease, and advances made on the continent of Europe were neglected in Britain. Apart from de Bary's famous textbook of mycology (5), only two foreign language works on general plant pathology were translated into English during the century—the second edition of Ré, *Saggio teorico-practico sulle malattie delle piante,* 1817, and in 1897, *Pflanzenkrankheiten durch kryptogame Parasiten versursacht* by Karl von Tubeuf (64). The first appeared in the *Gardeners' Chronicle* for 1849–50 (58), when long past any practical usefulness, presumably because Ré's generic classification of disease had influenced Berkeley.

Monographs.—The specialized publications related to plant pathology were of higher quality. The rusts and smuts were comprehensively treated (52) in 1889 by Charles B. Plowright [1848–1910], a medical man whose book, which incorporated much original observation, gave the standard accounts for British rusts and smuts until 1913 (30) and 1950 (2), respectively. E. S. Salmon's [1871–1959] monograph on the powdery mildews (60) was of international significance and later W. B. Grove's [1848–1938] coelomycetes (31) and Wilson [1882–1960] & Henderson's rust monograph

(70) set regional standards. The reprint in 1906 of M. C. Cooke's series of articles on "fungoid pests of cultivated plants" provided a handbook (20) which is still consulted, while W. F. Bewley's account of diseases of glass-house crops (9) was influential in the 1920's. A revised edition of *Diseases of Fruit and Hops* (72) by H. Wormald [1879–1955], first published on the eve of the Second World War, is still in current use.

Forest pathology.—Forest pathology was one of Marshall Ward's early preoccupations and in 1894 he edited and revised an English translation of Hartig's textbook (33). W. E. Hiley's little monograph on fungus diseases of the larch (34) in 1919, was well received but, by and large, diseases of forest trees have been neglected in Great Britain. Only during the last few decades, stimulated by the extensive planting under the auspices of the Forestry Commission which has issued a series of bulletins and leaflets on disease topics, has much attention been paid to this aspect of plant pathology. The most recent addition to its literature is the notable book (51) by T. R. Peace [d. 1963].

Fungi were the first plant pathogenic organisms to be recognized and they are still the most important (1). One does, however, gain the impression that mycological aspects were given undue prominence in Great Britain where emphasis on the pathogen, the description of plant pathogenic fungi as an end in itself, persisted longer than elsewhere. This bias is correlated with the practice, which continued until the National Agricultural Advisory Service was instituted in 1946, of calling plant pathologists "mycologists" or "advisory mycologists"—in the 1930's even a specialist on virus diseases was a "mycologist" while the designation of the Commonwealth Mycological Institute gave the Institute's activities an advantageous ambivalence. Pathogens other than fungi did, however, receive attention.

Bacterial Diseases

Massee drew attention to bacterial diseases of plants in 1899 and by 1915 fifteen pages of his textbook were devoted to such diseases (fungal diseases were given more than four hundred), and bacterial diseases were reviewed in a short chapter by Brooks (15) in 1928. It was not until the next decade that the sporadic interest in phytopathogenic bacteria became intensified. During that period, investigations of bacterial diseases of stone fruits are associated with the pioneering work of Wormald at the East Malling Research Station. Dr. Margaret Lacey elucidated the role of *Corynebacterium fascians* in leafy gall of strawberry, chrysanthemum, and other plants, while at Professor Brooks' instigation, W. J. Dowson [1887–1963] at Cambridge, undertook research and teaching which culminated in his *Manual of Bacterial Plant Diseases* in 1949, which may be consulted for references to the work mentioned above (24).

When, in 1947, the National Collection of Type Cultures decided to limit the bacterial collection then maintained at the Lister Institute to bacteria of medical interest it handed over the plant pathogenic bacteria to Dowson at the Botany School at Cambridge. In 1956, these and Dowson's personal collection were transferred to the Ministry of Agriculture's Plant Pathology Laboratory at Harpenden and recognized as the National Collection of Plant Pathogenic Bacteria.

Virus Diseases

Although virus diseases of plants were recognized more recently than bacterial diseases, records of what were clearly virus infections antedate records of the latter. Grafting has long been a horticultural practice and there are three, apparently independent, English 18th century records in which grafting resulting in virus transmission. The Reverend John Laurence in his *Clergy-Man's Recreation,* 1714, describes how a normal jasmine may be variegated by "inoculating" (bud grafting) it from a striped plant. He observed "that if the bud live one or two months, and after that happen to die, . . . it will have communicated its Virtue to the sap, and the tree will become entirely strip'd. This Discovery undoubtedly proves the Circulation of the sap. Q.E.D." (40). In 1720, Mr. Henry Cane reported (18) that in 1692 he had transmitted the mottle of jasmine by grafting, and in the last decade of the century Erasmus Darwin (21) noted a case of a yellow spotting of a passion tree being transmitted a fortnight after budding, even though the buds did not take.

Flower paintings provide some of the earliest evidence for the incidence of virus diseases and a particularly fine English example of this genre is the group of broken tulips illustrated in 1798 by Robert John Thornton [?1768–1837], lecturer in medical botany at Guy's Hospital, in his *Temple of Flora* (63). It was a hundred and thirty years later that Miss D. M. Cayley's [d. 1955] investigations at the John Innes Horticultural Institution contributed to the elucidation of this floral abnormality (19).

Potato "curl," that is leaf roll, was well known in Great Britain toward the end of the 18th century (8, 23) and speculation on its cause continued into the 20th. The "degeneration" of Scottish seed potatoes, when grown for several successive seasons in southern regions of Great Britain, also led to much speculation but it was not until about 1914–18 that, as a result of American and Dutch work, the infectious nature of leaf roll, mosaic, etc., of potato was established and their role in potato degeneration through their transmission by aphids was appreciated.

It was in the 1920's that serious virus studies were begun in Britain. Early in the decade, Bewley, at the Cheshunt Research Station, worked on tomato and cucumber mosaics (9) and in 1926 a Potato Virus Research Station (now the Agricultural Research Council's Virus Research Unit) was set up under the direction of R. N. Salaman [1874–1955] with Kenneth M. Smith on the staff. At that time the Empire Marketing Board gave grants for virus work at several centres, the most important being the Rothamsted Experimental Station where, in 1928, a physiologist (J. Caldwell), a cytologist (Frances M. L. Sheffield), and an entomologist [Marion A. Hamilton (Mrs. Watson)] were appointed for virus investigations. Subsequently, F. C. Bawden succeeded J. Henderson Smith [1875–1952] as head of the Rothamsted Plant Pathology Department where he continued his very fruitful collaboration with the biochemist, N. W. Pirie, on the nature of plant viruses by which they established that tobacco mosaic virus protein is a nucleoprotein.

It is somewhat invidious to give examples of discoveries and developments in British plant virology during the inter-war years but two outstanding contributions may be mentioned—both by entomologists: the experimental ecological studies made by Maldwyn W. Davies [?1903–1937] at the university College of North Wales, Bangor (22), and Kenneth Smith's Textbook (61). The former elucidated the part climate plays in aphid movement and, hence, of the spread of viruses in plantings of potatoes. It thus offered a rational explanation for the freedom from viruses of potatoes grown in Scotland and allowed other regions in Great Britain suitable for seed potato production to be defined. Smith's book, even if much abused, was one which every virus worker kept close at hand.

Control

Fungicides.—The few and scattered significant early references in British publications to the successful or attempted control of plant disease by the use of chemicals have been compiled in several modern reviews, most recently in that by McCallan (41). These references include Tull's report (65) of the use of brine as a seed steep to combat wheat bunt, a practice of which he was sceptical; he favoured "a proper change of seed" as a more certain preventive. It was William Forsyth [1737–1804] who, as "Gardener to His Majesty, at Kensington," in 1791 devised a "Composition for curing Diseases, Defects, and Injuries, in all kinds of Fruit and Forest Trees"—a wound dressing prepared from fresh cow dung, lime rubbish from old buildings, wood ashes, and sand (27, 28). Forsythe is famous for his introduction in 1802 of lime sulphur to control mildew on fruit trees (28). In 1834, T. A. Knight reported the control of both leaf curl, caused by *Taphrina deformans* and red spider on peach by sprinkling the trees in early spring with "water holding in solution or suspension a mixture of lime and flowers of sulphur" (38). The same remedy was successfully used by Edward Tucker, the Margate gardener, who first recorded powdery mildew of the grapevine in England in 1845 and who was commemorated by Berkeley in 1847 by the name, *Oidium tuckeri,* given to the pathogen, *Uncinula necator*.

The introduction of copper fungicides into England, and in particular the use of Bordeaux mixture against potato blight during the last quarter of the 19th century, and the development of the now standard schedules for spraying orchards are well covered by Large (39). Most of these introductions were applications of discoveries made elsewhere and their adaptation to local conditions. Among the few minor novelties originating in Great

Britain were Cheshunt Compound (a copper sulphate/ammonium carbonate drench for use against damping off), Shirlan (salicylanilide), which originated at the Shirley Institute at Manchester as a textile preservative, and the antibiotic, griseofulvin, which, in addition to its deployment in medicine, has found experimental use in horticulture. Further details of these materials and much additional information are to be found in H. Martin's *The Scientific Principles of Crop Protection* which, since it first appeared in 1928, has been a standard international text (43).

It should not be forgotten that the somewhat eccentric Professor M. C. Potter [1858–1947] of Armstrong College, Newcastle-on-Tyne, was probably the first to attempt the control of a plant disease by an antibiotic. He found that the culture filtrate on which *Pseudomonas destructans* [*Erwinia carotovora*] (the cause of turnip rot) had been grown would kill the pathogen when invading turnip tissue and that the metabolic products of *Penicillium italicum* would eliminate the same fungus from infected orange rind (53).

One other innovation merits mention. In order to give growers guidance on the choice and reliability of the many fungicides, insecticides and other chemicals on the market for use in plant protection, the Ministry of Agriculture, in 1942, introduced a voluntary Agricultural Chemicals Approval Scheme (later the Crop Protection Products Approval Scheme) whereby manufacturers confidentially disclose the composition of their products and evidence for the claims made for them and in return the products approved by the Ministry are eligible to carry a special label.

Although fungicidal treatments are invaluable in ensuring an adequate crop, alternative, and from a grower's point of view frequently simpler and less costly, approaches to plant disease control must not be forgotten. These include ensuring that the planting material is healthy, the use of resistant varieties, and legislation designed to minimize spread of diseases and the chances of introducing new ones into the country.

Deficiency diseases.—One of the characteristics of modern medicine is the increasing emphasis on maintaining health and the prevention of disease. A similar trend can be noticed in plant pathology. The basic nutritional (manurial) requirements of crops are usually a matter for the agriculturist or horticulturist, but nutritional disorders—"deficiency diseases"—frequently fall into the plant pathologist's province. Pioneering work on the diagnosis of such disorders in Great Britain was done by T. Wallace [d. 1965] at Long Ashton Research Station (66).

Certificated stocks.—Plant pathologists are also consulted on the development and maintenance of disease-free, particularly virus-free, propagating material. In Britain, a good example of the use of healthy planting material is the traditional use of Scotch and other certificated seed potatoes because of their relative freedom from virus infection; the degeneration of such stocks when propagated in southern England has already been touched on. Seed potatoes entered for certification are inspected during growth by trained inspectors and must attain exacting standards of purity and health.

The yield and quality of marketable ware obtained from this seed is consequently improved. Similar certification schemes are in operation for strawberries, loganberries, black currants, tree fruits, and hops.

Seed testing.—The first official attention to the control of the quality of propagating material was seed inspection—for disease as well as for purity and viability. At first based in London, the Ministry's Official Seed Testing Laboratory was moved to Cambridge in the 1920's where it is still housed at the National Institute for Agricultural Botany. Seed testing in Scotland is undertaken at the Department's laboratories at East Craigs, Edinburgh.

Resistant varieties.—Breeding varieties of crops resistant to diverse diseases and with desirable commercial qualities is, in Britain as elsewhere, a major activity. The Plant Breeding Institute at Cambridge, the Welsh Plant Breeding Station, Nr. Aberystwyth, and the Scottish Plant Breeding Station at Pentlandfield, Midlothian, have long been engaged in this field and most of the research stations have at one time and another paid attention to the development of resistant varieties of the crops in which they specialize, as have commercial seed firms. Varietal differences in susceptibility to disease have long been noted. R. Austen, in his *Treatise of Fruit-Trees,* 1657, recorded that "Crab trees . . . are usually free from the canker" (3), and it appears that T. A. Knight in 1806 (37) was among the first to suggest the use of resistant varieties to combat rust of cereals. The major British contribution in this field resulted from the work of Rowland H. Biffen [1874–1949] (15) in the early years of the present century at Cambridge, where he later became professor of agricultural botany and served as the first director of the Plant Breeding Institute. Biffen demonstrated that the inheritance of susceptibility and resistance of wheat varieties to yellow rust (*Puccinia striiformis*) was subject to Mendel's laws of heredity (10, 11) but this finding was at first received with a certain amount of scepticism because the existence of physiologic races of rusts was not appreciated. Two of the best known wheat varieties developed by Biffen were Little Joss and Yeoman.

Legislation.—The first legislative plant protection measure enacted in Britain was the Destructive Insects Act of 1887 designed to keep the Colorado beetle out of the country. In 1907, this Act was replaced by the Destructive Insects and Pests Act which empowered the Board of Agriculture to deal with other insect pests and also with fungal diseases by the Issue of Orders. Within a year, both American gooseberry mildew caused by *Sphaerotheca mors-uvae* (which had been recorded for the first time in Europe by Massee (45) in 1900 from County Antrim, Ireland, from where it had spread widely), and wart disease of the potato caused by *Synchytrium endobioticum* were subject to Orders. The discovery that certain potato varieties were immune to the wart disease led to official tests for resistance, first in the field and later in the laboratory, and to ensure varietal purity of the seed stocks. By this combination of methods wart disease was brought under effective control by the 1930's.

In 1967, the Acts of 1887, 1907, and a later one of 1927 were repealed after consolidation into the Plant Health Act 1967.

Organization

Man's concern for his own health and that of his animals and his plants is a descending series. One result of this is that the private practice of medicine has always been a lucrative profession. In Europe and elsewhere, veterinary medicine is still partly in private hands but in most of the developing and recently developed countries, veterinary medicine is a state service. Growers have always been very reluctant to pay for the diagnosis of disorders of their plants and plant pathological advice and research is characteristically supported by public funds. In Great Britain early studies and advisory writing was undertaken by country clergymen, medical men, and other amateurs with the support of a few members of university staffs.

Although agriculture has always been the largest single industry in Great Britain, it was not until 1889 that agriculture in England and Wales became the concern of a government department on the creation of the Board of Agriculture. This Board introduced a series of advisory leaflets (the forerunners of today's valuable series of bulletins, etc.) those on plant diseases being written by Massee, the Board's adviser at Kew. Space does not allow the development of the Board's involvement in plant pathology or its expansion into the Ministry of Agriculture, Fisheries, and Food to be considered in detail; reference may be made to the reviews by Moore (46) and Marsh (42). It must suffice to note that up to the First World War progress was slow although the fundamental step of providing for both investigational and advisory work on plant disease resulted from the Development and Road Improvement Fund Act of 1909.

A Plant Pathology Laboratory was established at Kew in 1914 and in 1918, A. D. Cotton [1879–1962] (an assistant to Massee and subsequently keeper of the Kew Herbarium) was appointed as the Board's first full-time Mycologist (there had been an Entomologist since 1914). In 1920, the Board's laboratory was moved to an adapted private house in Harpenden where, in 1923, G. H. Pethybridge [1871–1948], well known for his work on potato diseases in Ireland, succeeded Cotton as Mycologist. Finally, in 1960, a move was made to specially built laboratories on the Rothamsted estate on the other side of Harpenden common.

During the inter-war years, the advisory work on plant diseases in England and Wales was divided into thirteen "provinces" in each of which was an "advisory mycologist" based at an appropriate agricultural college or university (29). In 1946 the Ministry took over responsibility for advising growers and established the National Agricultural Advisory Service (NAAS). The country is now divided into eight regions, each provided with a small staff of specialist plant pathologists.

Scotland has an independent advisory service which is administered by the Department of Agriculture and Fisheries for Scotland, and the work, unlike that south of the border, is based on agricultural colleges. For further details for the whole of the British Isles, see (47).

Research stations.—Rothamsted Experimental Station, the senior research station, did not pay particular attention to plant diseases until the research side of the Board of Agriculture activities on pests and diseases was transferred there in 1918, to become eventually the Station's departments of entomology and plant pathology. Earlier plant pathological research had been undertaken at Long Ashton Research Station which dates from 1912 when, as the Department of Agricultural and Horticultural Research of the University of Bristol, it incorporated the National Fruit and Cider Institute founded in 1903 (67), and at Wye College where E. S. Salmon was the plant pathologist. Subsequently, a research station for fruit was initiated at East Malling in 1913, one for glasshouse crops at Cheshunt in 1914 (35), (in 1955 this station moved to Sussex to become the Glasshouse Crops Research Institute), the National Vegetable Research Station, at Wellesbourne, in 1949, and the Scottish Horticultural Research Institute (set up in 1953). Most of these stations originated by private endowment or with the support of growers' associations but today they are all largely financed with public money administered by the Agricultural Research Council which was founded in 1931 to allocate the state aid given to agricultural research. A further and not insignificant contribution to plant pathological research comes from the research stations which large commercial producers of fertilizers, insecticides, and fungicides have set up to test and develop their products.

Disease surveys.—The pressures of war and the need to increase the supply of home grown food so stimulated interest in plant disease that in 1917 a subcommittee of the Technical Committee of the Food Production Department of the Board of Agriculture initiated a survey of the pests and diseases of England and Wales. The results of the first year's survey were published in 1918 as the twopenny "Miscellaneous Publications, No. 21" of the Board, and subsequent reports for yearly or longer intervals have appeared for the period ending 1946 (48); from 1920 onward, pests and diseases being covered in separate publications. The recording of the survey data was, and continues to be, a function of the Ministry's laboratory but the records of plant diseases occurring in England and Wales have never been consolidated as they have for Scotland (25) and many other countries. W. C. Moore's [1960–67] useful compilation (49) does much to fill the gap for fungal diseases.

Education.—Phytopathological teaching has always been neglected in the British Isles, and in Great Britain there is still no full university department of plant pathology. There is, however, a long tradition for plant pathological teaching at two academic centres—the Botany School of Cambridge University and the Imperial College of Science and Technology, London. The first course of lectures and practical work on plant diseases at

Cambridge was that on "Fungoid diseases of plants" given in the autumn term of 1893 by W. G. P. Ellis of St. Catherine's College, university demonstrator in botany, and such teaching was continued and expanded by Marshall Ward and his successor F. T. Brooks. The versatile A. H. R. Buller [1874–1944] also gave a course of twenty weekly lectures, with laboratory work "by arrangement," during the winter of 1903–4 at the University of Birmingham.

At the Imperial College before the First World War, the professor of botany, V. H. Blackman [1872–1967], a fungal cytologist turned physiologist, gave a few generalized lectures on plant pathology which E. S. Salmon, as visiting lecturer, supplemented with accounts of specific diseases. William Brown joined the Imperial College in 1912 and after the war took over from Blackman both investigations on the physiology of fungi and the teaching of plant pathology. In 1938, Brown succeeded Blackman as head of the department of botany. His impact as a teacher and the influence of his school of postgraduate research needs no comment. He wrote the prefatory chapter for volume 3 of the *Annual Review of Phytopathology.* Today, the Imperial College offers a one-year postgraduate diploma course in plant pathology and this example has been followed elsewhere (50).

Scientific societies.—In Great Britain the functions of a phytopathological society, both in organizing meetings and in publishing the results of research, have been undertaken mainly by the British Mycological Society (founded 1896) and the Association of Applied Biologists (founded 1912). Begun as a largely amateur society the BMS membership became more professional after the First War, and from 1919 until 1966 a special committee looked after the Society's phytopathological activities by arranging paper reading meetings and an annual visit to a centre of phytopathological interest. One particularly useful activity of the Plant Pathology Committee was the compilation of a *List of Common Names of British Plant Diseases* first published by the British Mycological Society in 1929 with later editions in 1935, 1944, and 1962 (the last as *List of Common British Plant Diseases*). This publication has done much to standardize nomenclatural usage, particularly of the scientific names of plant pathogenic fungi.

The Association of Applied Biologists had a "Pests and Diseases Committee" for a time (1943–53) but this Association's interests have always been biased toward both pests and diseases of plants, and its journal, the *Annals of Applied Biology,* is still a main outlet for the publication of the results of phytopathological research.

In 1966, these two societies inaugurated the Federation of British Plant Pathologists, an organization able to represent British plant pathologists as a whole and one into which members of either society are eligible to opt.

The Royal Horticultural Society has also made a useful contribution to British plant pathology by publishing phytopathological papers in its journal, and by maintaining a phytopathological laboratory at its gardens at Wisley, Surrey, is able to offer advice on plant disease problems to the Society's large membership.

Overseas Interests

One factor that has greatly influenced plant pathological development in Britain has been the interest that was of necessity shown in the plant diseases of the Colonial Empire. This factor has operated in two directions. Firstly, following the example of Marshall Ward who, in the 1870's, was sent to Ceylon to investigate the outbreak of coffee rust, caused by *Hemileia vastatrix* which was largely responsible for changing Ceylon from a coffee-producing to a tea-producing country, a succession of plant pathologists have gone overseas. During the first half of the present century many such pathologists went abroad on a career basis, e.g., E. J. Butler [1874–1943], a medical man by training, who laid the foundation for plant pathological work in India; T. Petch [1870–1948] who did the same for mycology in Ceylon; and plant disease work in many African countries during the inter-war years is associated with the names of British pathologists including C. G. Hansford [1900–1966] (Uganda), H. H. Storey (Tanganyika), J. C. F. Hopkins (Southern Rhodesia), R. E. Massey (Sudan), F. C. Deighton (Sierra Leone), and H. A. Dade (Gold Coast). The review by Butler (17) may be consulted for further details. This demand for plant pathologists necessitated appropriate training and the Colonial Office offered annually postgraduate scholarships under which, after a year at normally either Cambridge or the Imperial College, the recipients spent a second year at the Imperial College of Tropical Agriculture in Trinidad where supplementary training was given in applying phytopathological knowledge to tropical crops.

As a bonus this pattern provided a supply of experienced workers who, on retirement from their overseas posts, were available for senior appointments in their home country. Today, the position is greatly changed and though plant pathologists still go abroad they usually do so on short-term contracts which rarely allow the continuity of experience so essential for the most productive results.

The second effect was that students came from overseas territories for postgraduate training plant pathology, mainly to Cambridge and the Imperial College, particularly the latter which developed strong ties with both India and Egypt, and phytopathological papers from the Indian subcontinent still frequently show evidence of where the author or the supervisor of the research was trained.

The Commonwealth Mycological Institute.—The more than parochial interests in plant pests led, in 1913, to the establishment of the Imperial Bureau of Entomology, at the British Museum for Natural History in Kensington, to act as a clearing house for information on crop pests throughout the Empire and to summarize the world literature in a journal of abstracts, the *Review of Applied Entomology*. This example stimulated a decision in 1918 by the Imperial War Conference to set up a complementary bureau for plant disease, and in 1920 the Imperial Bureau of Mycology was initiated at 17 Kew Green, the cottage recently vacated by Kew's Plant Pathology Laboratory (16). It appears that the first public suggestion for an Imperial Bureau of Mycology was that made by W. B. Brierley [1889–1963] (then a mycologist at Kew and subsequently the first head of the department of plant pathology at Rothamsted) in 1916 at the Newcastle-on-Tyne meeting of the British Association. In the report of that meeting Brierley adumbrated the functions of such a bureau on a broader basis than has even yet eventuated (13), and characteristically urged "that the two bureaux should work in intimate correlation, and that this would best be achieved were they autonomous subdivisions of an Imperial Bureau of Phytopathology," which some still consider to be an ideal. However, probably every practicing pathologist has at some time or another been grateful to the *Review of Applied Mycology* which, after 48 years, is to be renamed *Review of Plant Pathology*.

I am most grateful to Dr. F. Joan Moore for her interest in this review and for helpful information and suggestions which prevented the perpetuation of certain errors, and to Professor W. Brown and Dr. S. D. Garrett for details of early plant pathological teaching at the Imperial College and Cambridge.

LITERATURE CITED

1. Ainsworth, G. C. The Review of Applied Mycology. *Rept. Commonwealth Mycol. Conf., 6th, 1964,* 19–20 (1964)
2. Ainsworth, G. C., Sampson, K. *The British Smut Fungi (Ustilaginales)* (Commonwealth Mycological Inst., Kew, 137 pp., 1950)
3. Austen, R. *A Treatise of Fruit-Trees, shewing the manner of Grafting, Planting, Pruning, and Ordering of them . . . ,* 54 (Oxford, 1657)
4. Banks, J. *A Short Account of the Cause of the Disease in Corn* (Bulmer, London, 16 pp., 1805); See also Ref. (37)
5. de Bary, A. *Comparative Morphology and Biology of the Fungi, Mycetozoa, and Bacteria* (Trans. by Garnsey, H. E. F., Balfour, I. B., Clarendon Press, Oxford, 525 pp., 1887)
6. Bassi, A. *Del mal del Segna (Phytopathol. Classic,* **10,** 49 pp., 1958) (Trans. by Yarrow, P. J.)
7. Berkeley, M. J. Observations, botanical and physiological, on the potato murrain. *J. Hort. Soc., London,* **1,** 9–34 (1846) (Reprinted in *Phytopathol. Classic,* **8,** 1948)
8. Berkeley, M. J. Vegetable pathology. *Gardeners' Chron., 1854,* 20 et seq. (1854–57) (Reprinted in part in *Phytopathol. Classic,* **8,** 1948)
9. Bewley, W. F. *Diseases of Glasshouse Plants* (Benn, London, 208 pp., 1923)
10. Biffen, R. H. Experiments with wheat and barley hybrids illustrating Mendel's Laws of heredity. *J. Roy. Agr. Soc.,* **65,** 337–45 (1904)
11. Biffen, R. H. Studies in inheritance of disease resistance. *J. Agr. Sci.,* **2,** 109–28 (1907)
12. Bourke, P. M. A. Emergence of potato blight, 1843–46. *Nature,* **203,** 805–8 (1964)
13. Brierley, W. B. The organization of phytopathology. *Rept. Brit. Assoc. Advan. Sci., 1916,* 487 (1917)
14. Brooks, F. T. *Plant Diseases* (Oxford Univ. Press, London, 386 pp., 1928)
15. Brooks, F. T., Professor Sir Rowland Biffen. *Trans. Brit. Mycol. Soc.,* **33,** 166–68 (1950)
16. Butler, E. J. The Imperial Bureau of Mycology. *Trans. Brit. Mycol. Soc.,* **7,** 168–72 (1921)
17. Butler, E. J. Presidential address. The development of economic mycology in the Empire overseas. *Trans. Brit. Mycol. Soc.,* **14,** 1–18 (1929)
18. Cane, H. On the change of colour in grapes and jessamine. *Proc. Roy. Soc.,* **31,** 102 (1720)
19. Cayley, D. M. 'Breaking' in tulips. *Ann. Appl. Biol.,* **15,** 529–39 (1928); **19,** 153–72 (1932)
20. Cooke, M. C. *Fungoid Pests of Cultivated Plants* (Spottiswood, London, 278 pp., 1906)
21. Darwin, E. *The Botanic Garden, Part I, The Economy of Vegetation,* 3rd ed., 202 (footnote). (Johnson, London, 1795)
22. Davies, M. W. Aphis migration and distribution in relation to seed potato production. *Sci. Hort.,* **5,** 47–54 (1937)
23. Dickson, T. Observations on the disease of potato, generally called the curl. *Mem. Caledonian Hort. Soc.,* **1,** 49–59 (1814)
24. Dowson, W. J. *Manual of Bacterial Plant Diseases* (Black, London, 187 pp., 1949)
25. Foister, C. E. Economic plant diseases of Scotland. *Tech. Bull. Dept. Agr. Fish. Scotland,* **1,** 209 pp. (1961)
26. Fontana, F. *Observations on the Rust of Grain* (Trans. by Pirone, P. P., *Phytopathol. Classic,* **2,** 40 pp., 1932)
27. Forsyth, W. *Observations on the Diseases, Defects, and Injuries on all kinds of Fruit and Forest Trees* (Nicol, London, 71 pp., 1791)
28. Forsyth, W. *A Treatise on the Culture and Management of Fruit Trees. . . .* (Longman and Rees, 371 pp., 1802) [Includes a reprint of Ref. (27) together with a supplement]
29. Fryer, J. C. F., Pethybridge, G. H. The phytopathological service of England and Wales. *J. Ministry Agr., London,* **31,** 331–40 (1924)
30. Grove, W. B. *The British Rust Fungi* (Cambridge Univ. Press, 412 pp., 1913)
31. Grove, W. B. *British Stem- and Leaf-Fungi (Coelomycetes)* (Cambridge Univ. Press, **1,** 488 pp., 1935; **2,** 407 pp., 1937)
32. Hales, S. *Vegetable Statics* (Lon-

don, 1727; reprinted, Sci. Book Guild, 216 pp., 1961)

33. Hartig, R. *Text-book of the Diseases of Trees* (Trans. by Somerville, W., Ward, H. M., Ed., Macmillan, London, 331 pp., 1894)
34. Hiley, W. E. *The Fungal Diseases of the Common Larch* (Clarendon Press, Oxford, 204 pp., 1919)
35. *The History and Work of the Experimental & Research Station, Cheshunt* (Cheshunt Press, Cheshunt, 35 pp., 1935)
36. Hooke, R. *Micrographia,* Schem. xii, Fig. 2 (Royal Society, London, 270 pp., 1665; reprinted as a Dover paperback)
37. Knight, T. A. A letter (March 20, 1806) to Sir Joseph Banks on the origin of blight. *Pamphleteer,* **6,** 415–16 (1815) [n.v., after Parris, G. K., *A Chronology of Plant Pathology,* 1968; also Ref. (4), 2nd ed., 1806, fide (55)]
38. Knight, T. A. Upon the causes of the diseases and deformities of the leaves of the peach-tree. *Trans. Hort. Soc. London, Ser. 2,* **2,** 27–29 (1842) (Paper read 15 July, 1834)
39. Large, E. C. *Advance of the Fungi* (Cape, London, 488 pp., 1940; reprinted as a Dover paperback)
40. Laurence, J. *The Clergy-Man's Recreation: showing the pleasure and profit of the art of gardening* (Linott, London, 54–64, 1714 [n.v.]; 3rd ed., 41–42, 1715)
41. McCallan, S. E. A. History of fungicides. In *Fungicides, an advanced treatise,* Chap. 1 (Torgeson, D. C., Ed., Academic Press, New York, 1967)
42. Marsh, R. W. Moulds, mildew, and men. *Nature,* **218,** 1017–19 (1968)
43. Martin, H. *The Scientific Principles of Plant Protection with Special Reference to Chemical Control* (Arnold, London, 1st ed., 316 pp., 1928; 5th ed., 376 pp., 1964)
44. Massee, G. *A Text-book of Plant Diseases caused by Cryptogamic Parasites* (Duckworth. London, 458 pp., 1899; 2nd ed., 1903)
45. Massee, G. *Diseases of Cultivated Plants and Trees* (Duckworth, London, 602 pp., 1910; 2nd ed., 1915)
46. Moore, W. C. Presidential address. Organization for plant pathology in England and Wales—retrospect and prospect. *Trans. Brit. Mycol. Soc.,* **25,** 229–45 (1942)
47. Moore, W. C. Diseases of crop plants 1943–1946. *Bull. Ministry Agr. Fish.,* 139, 90 pp. (1948)
48. Moore, W. C., Frederick Tom Brooks, 1882–1952. *Obit. Notices Fellows Roy. Soc.,* **8,** 341–54 (1953)
49. Moore, W. C. *British parasitic fungi* (Cambridge Univ. Press, 430 pp., 1959)
50. Organization for plant pathology in the British Isles. *Rev. Appl. Mycol.,* **47,** 321–27 (1968)
51. Peace, T. R. *Pathology of Trees and Shrubs, with special reference to Britain* (Clarendon Press, Oxford, 723 pp., 1962)
52. Plowright, C. B. *A Monograph of the British Uredineae and Ustilagineae* (Kegan Paul, London, 347 pp., 1889)
53. Potter, M. C. On a method of checking parasitic diseases in plants. *J. Agr. Sci.,* **3,** 102–7 (1908)
54. Prévost, B. *Memoir on the immediate cause of bunt or smut of wheat* . . . (Trans. by Keitt, G. W., *Phytopathol. Classic,* **6,** 95 pp., 1939)
55. Ramsbottom, J. Some notes on the history of the classification of the Uredinales. *Trans. Brit. Mycol. Soc.,* **4,** 77–105 (1913)
56. Ramsbottom, J., Mordecai Cubitt Cooke (1825-1914). *Trans. Brit. Mycol. Soc.,* **5,** 169–85 (1915)
57. Ramsbottom, J., George Edward Massee (1850–1917). *Trans. Brit. Mycol. Soc.,* **5,** 469–73 (1917)
58. Ré, P. Essay, theoretical and practical, on the diseases of plants. *Gardeners' Chron., 1849,* 228 et seq. (1849–50)
59. Roberts, W. Studies in biogenesis. *Phil. Trans. Roy. Soc., Ser. B,* **164,** 457–94 (1874)
60. Salmon, E. S. A monograph of the Erysiphaceae. *Mem. Torrey Botan. Club,* **9,** 292 pp. (1900)
61. Smith, K. M. *A Textbook of Plant Virus Diseases* (Churchill, London, 615 pp., 1937; 2nd ed., 1957)
62. Smith, W. G. *Diseases of Field and Garden Crops* (Macmillan, London, 353 pp., 1884)
63. Thornton, R. J. *Select Plants. The Temple of Flora* (London, 31 plates, 1799)
64. Tubeuf, K. von. *Diseases of Plants induced by Cryptogamic Parasites* (Trans. by Smith, William G.,

Longmans Green, London, 598 pp., 1897)

65. Tull, J. *Horse-Hoeing Husbandry* [London (fol.), 1733; 3rd ed. (8vo), 432 pp., 1751]
66. Wallace, T. *The Diagnosis of Mineral Deficiencies in Plants by Visual Symptoms. A Colour Atlas and Guide* (H. M. Stationery Office, London, 116 pp.; suppl., 48 pp., 1943, 1944)
67. Wallace, T., Marsh, R. W. *Science and Fruit. Commemorating the Jubilee of the Long Ashton Research Station 1903–1953* (Univ. Bristol, 308 pp., 1953)
68. Ward, H. M. *Diseases of Plants* (Soc. Promotion of Christian Knowledge, London, 196 pp., 1889)
69. Ward. H. M. *Diseases in Plants* (Macmillan, London, 309 pp., 1901)
70. Wilson, M., Henderson, D. M. *British Rust Fungi* (Cambridge Univ. Press, 384 pp., 1966)
71. Woodham-Smith, C. *The Great Hunger* (Hamish Hamilton, London, 510 pp., 1962; reprinted as Four Square paperback)
72. Wormald, H. *Diseases of Fruit and Hops* (Crosby Lockwood, London, 285 pp., 1939; 2nd ed., 1946)
73. Zakon, S. J., Benedek, T. David Gruby and the centenary of medical mycology, 1841–1941. *Bull. Hist. Med.*, **16,** 155–68 (1944) (Includes translations of six of Gruby's papers)

Copyright 1969. All rights reserved

ECONOMIC BASES FOR PROTECTION AGAINST PLANT DISEASES

By George Ordish and David Dufour
St. Albans, England
The Emmanuel School, London, England

INTRODUCTION

The economic bases for the protection of crops against plant diseases are apparently so blindingly simple as scarcely to need any discussion. Reduced almost to a syllogism, they are

1. Disease causes a loss of crop, the loss being valued at so much.
2. A remedy overcomes the disease and prevents all or some of the loss.
3. If the cost of 2 is less than 1, all is well and the procedure should be followed; if 2 is equal to, or more than, 1, the work should not be undertaken.

Unfortunately, the real world is not as simple as this and we have to consider a series of complexities which arise.

THE COMPLEXITIES

Farming.—First, we must note that farming is a most unnatural activity. Man has imposed on the environment a system of survival of what he wants to use over the Darwinian system of the fittest to survive. Consequently, the farmer is engaged in a constant struggle with nature. Rain, drought, frost, and heat are all hazards as are also insect pests, diseases, and weeds. In order to help the farmer, scientists bring various disciplines into play, all very necessary in view of the complexity of the subject and not the least important being economics. This paper is about the economics of diseases, a somewhat neglected theme up to the present. The farmer himself makes very little distinction between pests and diseases. They are all forces of nature which can cause him loss, whether insect, virus, parasitic plant, bacteria, or fungus. For instance, the hop grower in England still uses the word "blight" to describe the hop aphis. Consequently, in studying the economics of the control of their troubles we may have to consider the whole complex in a crop, on a farm, or in a country, rather than any particular disease. As a result, we now tend to use the word "pests" to include all those forms of life and of viruses which can cause crop losses. In the last analysis, man is the worst pest crops have, because, instead of allowing organisms to propagate naturally, he interferes to secure more of that plant for himself. Instead of allowing the wheat to feed rust and smut fungi and to allow such remaining seed to fall to the ground and propagate more wheat plants, he takes steps to destroy these fungi and collect as much as he can of the seed for himself.

Measurement of the loss.—One of the difficulties of an economic analysis of this subject is that of measuring the loss because we have very little idea what the norm is. We are trying to measure something with a yardstick that varies in length from time to time. By means of crop surveys we can get a figure for probable losses. By means of spraying experiments, we can discover the differences between the yields of a sprayed and unsprayed crop. The crop survey can show the returns from crops carrying various amounts of pest running from severe attack to no attack; usually such figures are collected from a wide area and it is not possible to assume a linear relationship. 2x Tons per acre from the unattacked area A and x tons from badly attacked area B does not necessarily mean that area B could have given 2x tons in the absence of this disease because conditions vary; the figure could have been 1.5x or even 3x. Moreover, as Judenko (13) has pointed out, the yield from a sprayed crop is not necessarily the same as the yield from an unattacked crop. Commenting on Judenko's work, Johnson (12) (an entomologist, be it noted) points out that plant pathologists are more interested in the interaction of crops and diseases than are entomologists in the interaction of crops and insects, for the entomologists' interests are more concerned with the insects themselves. Judenko points out that sprays may kill other pests, may affect the physiology of the plant, kill parasites or predators, or restrain or encourage diseases and he is now working on the comparison of yields from individually attacked and pest-free plants, due allowance being made for compensatory effects, in order to find out what losses pests are causing.

The scientist's concern with the difficulties of measurement reached a head in 1967 and led to a Food and Agriculture Organization symposium (4) on the subject. F.A.O. is now preparing a handbook on methods of measuring losses which is most desirable. Nevertheless, we believe that the economics of the subject have been somewhat neglected by this body up to the present, but no doubt these vital economic factors will be considered in due course.

Measurement of the loss depends not only on the individuals making the measurement, but also on who is affected by the loss. The object of measuring the loss can be scientific curiosity or a desire to improve the economic condition of farmers, a country, or the world. The first source of error (i.e., variations between individual investigators) can be overcome by multiple sampling [see, for instance, Sen & Chakrabatry's (19) paper on tea] and similar techniques; the second (i.e., the different aims of different studies) needs an economic approach because it is bound up with supply factors, much influenced by pest attack (using the word "pest" in the full sense defined above).

As early as 1933, Smith et al. (20) pointed out that abundance or scarcity, whether the scarcity was caused by weather or pest, caused different results in different crops and seasons. Using Smith's and his colleagues' figures, we get the following table:

TABLE I

SOME CROP RETURNS IN CALIFORNIA. SMITH ET AL.

Crop	Year	Calamity	Crop 1'000 tons	Return to growers $ million	Result to growers' group
Peaches	1928		414	6.9	—
	1929	Frost	179	12.2	gain
Prunes	1928	—	220	22.0	—
	1929	Frost	103	16.0	loss
Oranges	1927		12.2[a]	50.0	—
	1928	Fruitfly	9.1[a]	52.5	gain
	1929		17.7[a]	47.7	—
	1930	Fruitfly	9.3[a]	54.6	gain
Grapes	1927		355	37.3	
	1928		411	39.1	
	1929		365	42.0	

[a] = Millions of boxes.

In 1929, a smaller crop of peaches increased peach growers' incomes, but in the same year a smaller crop of prunes reduced plum growers' incomes. Smaller crops of oranges increased growers' incomes and a larger crop of grapes first slightly increased returns in 1928, followed in 1929, by a smaller crop than in 1928 which further increased returns. Note that, as Smith pointed out, these were returns to the group; no doubt, in 1929 many peach growers had no crop at all and were ruined, but the group "peach growers" got about twice the income from handling about half the weight of crop.

This is best illustrated by a supply and demand diagram shown in Figure 1. [See reference (16) for a fuller explanation than is given here.] S_1 is the supply curve for two commodities with different demand elasticities. D_1 and D_2 are the demand curves for these two commodities, such as D_1 cereals and D_2 fruit. They both cut the supply curve S_1 at X, consequently, the money received by growers in both cases is the rectangle 6.5×4.5. Obviously, cereal growing is a bigger business than fruit growing, so to secure coincidence of the curves, the scales can be considered to be in different units for the two commodities. This does not affect the argument. If an improvement is made in growing both crops, such as the control of disease complexes, production costs less and the supply curve moves to the right,

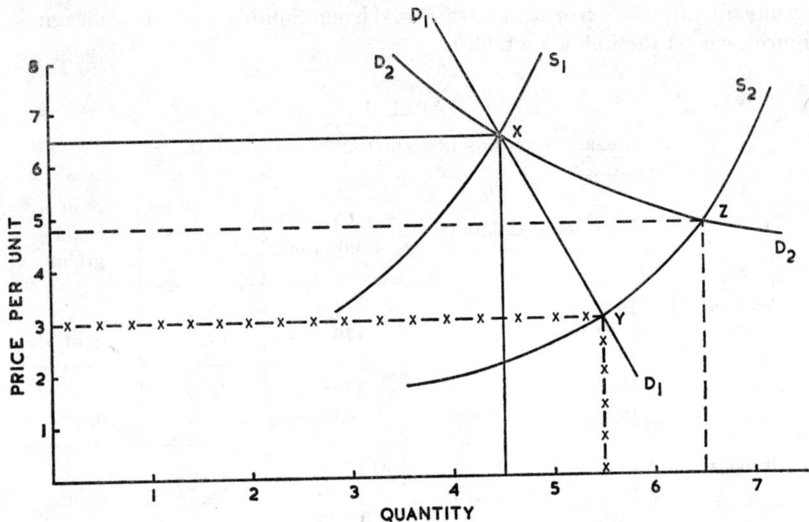

FIG. 1. D_1 is the demand curve for a product with inelastic demand, say wheat.
D_2 is the demand curve for a product with elastic demand, say fruit.
S_1 is the supply curve of wheat and fruit.
X is the point where above curves meet and shows both price and quantity demanded and supplied of both products at the time of consideration.

Both crops are attacked by pests and a new process overcomes them, making production cheaper. The supply curve moves to the right, becoming S_2, cutting D_1 at Y and D_2 at Z.

Money received by producers in the old and new situations is shown by the rectangles on X, Y and Z.

		$	Gain or Loss
OLD SITUATION			
Wheat growers, Fruit growers	$6.5 \times 4.5 =$	29.25	
NEW SITUATION			
Wheat growers	3×5.5	16.5	−43.5%
Fruit growers	4.8×6.5	31.1	+ 6.6%

becoming S_2, now cutting the two demand curves at Y and Z. We now get the position shown in Table II.

Disease control has led to a reduced income for cereal growers and an increased income for fruit growers because of the different demand elasticities of the produce. This is not a purely theoretical concept, but one borne out in practice. See the examples quoted from Smith and the observations of many farmers today; for instance, Tangye (22) points out that cures for lettuce *Botrytis* and bulb fly have made growing lettuce and daffodils so easy in Cornwall that the lowered prices have destroyed a one-time profitable trade for small growers.

Howard (9) records the fact that the cotton growers of Enterprise, Alabama, put up a monument to the boll weevil, as it made cotton growing difficult and kept up the price of the crop.

The different interests.—Following Smith's paper, Ordish (15) postulated the now generally recognised three sets of interests, consumers of agricultural produce, the individual producer, and the group of producers, all of whom may not want the same thing. To this, a fourth group has recently been added (16)—the manufacturers of pesticides. We add a fifth here—the body of research workers engaged in investigating the vast range of pests; we shall have occasion to consider these five interests later in this paper.

Expression of the loss.—Ordish (15) also introduced the concept of the cost/potential benefit ratio, the variation of the ratio according to the value per acre of the crop, and the expression of losses in terms of area lost rather than in money or percentage [see, for instance, Strickland (21)].

The scientists can hope to establish a norm of the yield of an unattacked crop and then indicate the percentage loss caused by a departure from this norm. This can then be translated into financial terms, which are the only terms that mean anything to the majority of farmers. It should be noted that the loss is (*a*) a potential loss which would arise were no steps taken against the disease, and (*b*) the loss still occurring in spite of the steps taken. Moreover, were the remedies completely successful, they all cost money; even resistant varieties of seed have to be created and consequently are a cost of some sort. The philosopher must recognise that the real loss is the area of land used but not really needed. For instance, if x acres are used on the average to produce a country's crop of potatoes, and diseases cause a loss of 15 per cent, then, when the disease is conquered, the area could be $x - 15x/100 = .85x$, to get the same crop, consequently the loss in this case is .15x acres of land and all the seed and effort put into growing the crop. This point did not escape Curtis (3a) in 1860, who complained of

TABLE II

GROWERS' INCOME, GAIN OR LOSS

Crop and position	Rectangle on	Growers' income	Gain or loss
Cereals and fruit,	X		
$S_1/D_1/D_2$	6.5×4.5	29.25	
Cereals	Y		
S_2/D_1	3×5.5	16.5	−43.5%
Fruit	Z		
S_2/D_2	4.8×6.5	31.1	+ 6.6%

the vast increase of the potato acreage after the blight struck. He said it was ridiculous to cultivate land for the purpose of growing rotten potatoes —a remarkably clear-sighted observation.

Quantity, quality, supply, and demand.—Farmers are not in business to feed the world's hungry millions but to make money. Consequently, the quality of a crop can be as important a consideration as the quantity, which has been recognised by the F.A.O. committee studying the measurement of losses. About the only way to compare crops from different plots or areas subject to these two variables—quantity and quality—is to express the results in money terms, that is, in economic terms. Money, moreover, is what impresses politicians, business men, the popular press, and administrators.

Fitting the supply to the demand is more complex in agriculture than in industrial production. It is easier to turn swords into ploughshares at a moment's notice than potatoes into oranges. One of the most important economic effects of disease control is to facilitate the fitting of supply to demand; the gluts and scarcities, so often caused by pests in the past, are now

TABLE III

POTATOES, U.S.A.

Year	Average area ha. '000	Average production met tons '000	Value $ million	Yield met tons per ha
1948–52	662	10,676	502	16.2
1965	568	13,144	660	23.1

largely avoidable. An example of acreage reduction, resulting from increased yields, is seen in Table III, U.S.A. potatoes. The area has been reduced by 14 per cent; production has increased by 23 per cent in weight, 31 per cent in value and 43 per cent in yield per hectare (i.e., 6.9 tons per ha). If we assume, as is most likely, that half the increase in yield (i.e., 3.4 tons per ha) is due to better disease control, then, if the 1948–52 crop had been grown at this same standard, it could have provided 10,676,000 + (662,000 × 3.4) = 13 million tons of potatoes instead of 10.7 million. It could be said that 2.3 million tons of potatoes were lost to disease; this loss was apparently worth $121 million at $52.60 per ton, a lot of money. However, this is not realistic; a 23 per cent increase of potato supply would have depressed prices by at least this figure and probably by more, so that farmers would have received the same or less money for producing and handling 23 per cent more potatoes, a situation similar to that of cereals in Table II. Without disease, the 1948–52 average crop of 10.7 million loss could have been produced from 461,000 ha (instead of from 662,000 ha).

An epidemic may reduce quantity or quality. An example of the latter is the contamination of groundnuts with the fungus, *Aspergillus flavus,* and the poisonous aflatoxin. For instance, in Nigeria the average producer price 1960/61 for groundnuts was $126.50 per metric ton, whereas after the aflatoxin story broke there, the price was $120.00 in 1962 and $111.00 in 1963, a reduction of at first 5 and then 12 per cent (5). In these years, the prices of the uncontaminated United States crop rose by 4 and 2 per cent.

Developed and Developing Areas

In last year's volume of this publication, Paddock (17) pointed to the differences between the developed and developing areas with respect to the economics of crop diseases. Developed areas have considerable industrial resources and can withstand a serious disease loss in crops far more readily than can developing areas. The former overcome the loss by using reserves, buying from other countries, and so on. Developing countries are mostly agricultural and a large percentage of the exports of such countries and thus their foreign exchange, comes from single crops. For instance, 87 per cent of Nigeria's exports are groundnuts; 75 per cent of Colombia's is coffee. It can be seen that any epidemic badly affecting an export crop is a serious economic matter for a developing area. For instance, contamination of Nigeria's groundnut crop with the fungus, *Aspergillus flavus,* reduced that country's returns from exports, the price recovering when the trouble was overcome.

The Economics of Plant Disease Control

The alternatives open to the farmer threatened with attacks of disease in a crop can be summarised as: 1. do nothing and harvest what the disease leaves; 2. substitute another crop; 3. grow resistant or immune varieties; 4. protect the crop by spraying; or 5. protect the crop by mechanical methods.

1. The first is a procedure much used in some agricultures, particularly in less developed areas and where controls have not been found.

2. Substitution is a popular remedy. The classic example is the substitution of tea for coffee in Ceylon in the 1870's because of the rust in the latter, and in Kenya to-day, because of coffee berry disease. Again, the economic effects depend on demand elasticity and the scale of the changeover. Can the market absorb the extra tea?

Grainger (8), who has made many important studies in this field points out that in West Scotland 57 per cent of the land potentially available for turnips is infested with club root fungus, *Plasmodiophora brassicae,* and his recommendation for control is to substitute silage crops, which is just as suitable for winter cattle feed. Grainger estimates that in this small area, the substitution would overcome a loss of some $2 million caused by this disease.

3. Control of disease by resistant or immune varieties naturally appeals to farmers. It is simple for the farmer and appears to be almost costless.

The economic basis is that money, usually raised from taxpayers, must be used to finance unspectacular research. Paper bags, camel-hair brushes, notebooks, and seed trays make a poor showing against computers and other electronic mechanisms demanded by other sciences. It is not, in fact, without cost, because the scientific work required is expensive and the sought-for variety once found may have but a short life before the pathogen develops a strain capable of attacking the hitherto immune plant. Consequently, the economics of this approach turn on the cost of maintaining the laboratories, creating the new varieties, and the chances of success. Up to the present, some successes have been spectacular. Potatoes immune to wart disease are universally grown, but we barely manage to keep one jump head with cereal varieties resistant to the rusts. At Auchincruise, Scotland (8), Pentland Dell potatoes, resistant to blight caused by *Phytophthora infestans* and grown in rich soil, keep their haulm six weeks longer than Kerr's Pink —the former yielding 95 tons per ha against the latter's normal commercial yield of 45 tons per ha. Moreover, today, copyright in newly bred varieties has been created, enabling the creators to draw royalties on the sale of their creations, which encourages the production of new varieties having such advantages as disease resistance. In fact, a variety susceptible to disease has no economic future today.

4. Spraying is a procedure much used against pests in developed agriculture; against diseases it is mostly preventive and sometimes is a routine operation, for instance, in the seed dressing of cereals.

The economics of spraying have been much studied, mostly on a cost-accounting basis, and the concept of the cost/potential benefit ratio has been introduced (15). Often, spraying gives returns of some three times the cost involved, and is thus very popular and worthwhile. However, spraying has a number of much discussed disadvantages which are basically economic: (*a*) Although the cost/benefit ratio may be large, the cost is heavy and regular. (*b*) The advantages are short-term, but long-term disadvantages may arise.

The sales of pesticides in the United Kingdom are about $78 million per annum, to which must be added another $40 million or more for the cost of application. Since all of these costs are borne by farmers, they are passed on to consumers in the form of higher prices. In effect, the United Kingdom population is paying $120 million a year to control pests on farms and not getting 100 per cent results.

Some long-term defects of spraying are well known but most of the examples are in the entomological field. After spraying for some seasons, races of the pest may arise which are resistant to the sprays, or new pests may take the place of the old pests destroyed. Examples of this can be seen in the case of diseases; for instance, Burchell (2), in a recent review of the apple mildew caused by *Podosphaera leucotricha* in England, reports that it became a pest only after the introduction of captan against apple scab caused by *Venturia inaequalis,* and renewal pruning was substituted for treatment with lime-sulphur and the use of spur and tip pruning. Captan controlled the scab and did less damage to foliage but did not control mildew, and the omission of tipping in pruning left an overwintering source of inoculum on the tree. The economic effect is that more elaborate spraying is now needed. For instance, substances such as dodecylguanidine must be added to the sprays and many more treatments given.

Ingram (11) has a number of interesting cost figures, showing that the cost of spray materials has risen from about $25.00 per ha for the simple sulphur spray materials to approximately $50.00 for the more complex programme.

The Politics of Pest Control

The scientist today is supposed to stand aloof from politics and to be concerned solely with elucidating the workings of the natural world, but if he is discussing economics he is bound to impinge on the political world, for an economic event is either caused by a political happening or is going to have a political effect. We contend that the economist must include possible political causes and effects in his thinking if it is to be effective, in addition to striving to bring a detached, objective, scientific judgment to bear on the subject.

In the past, plant diseases have had profound political and social effects. The potato blight in Ireland added a large Irish element to the population of the United States, with considerable political and social consequences. To make a minor social speculation, would there be an annual St. Patrick's day parade in New York if *Phytophthora infestans* had never existed? One doubts it. Britain became an important consumer of tea because rust destroyed the coffee in Ceylon in the 1870's and the plantations had to be replanted with tea. Consequently, we seek to examine social, economic, and political effects of diseases of crops.

In the first place, there is the much discussed question of whether scientists should carry any moral responsibility for the scientific discoveries they make. The case of Dr. Robert Oppenheimer, Sir William (now Lord) Penny and the atom bomb at once springs to mind, but we do not propose to discuss such deep issues here. We do contend, however, that scientists, particularly the plant pathologists, should be fully aware of what are likely to be the effects of their discoveries. That is, they should seriously consider the economic, social, and political consequences whether or not this alters their course of action, for the control of plant diseases can have enormous effects, varying, to a greater or lesser extent, according to the political system under which they operate or the particular laws governing the matter within each political system. Obviously, we can only discuss this matter very briefly.

With respect to political systems, there are roughly three basic ones known in the world today: (*a*) capitalism; (*b*) mixed economy, which we will call socialism; (*c*) communism.

All three claim to secure as much ease and contentment as is possible for their nationals, which economists call "trying to maximise their satisfaction." The systems can be characterised by three brief slogans through which each seeks to increase "utility" (or its "standard of living"). Capitalism relies on "enlightened self interest" (3) and private enterprise; socialism on "the greatest happiness to the greatest number" (1, 10, 18) and considerable public ownership of the means of production. Communism emphasises "from each according to his ability, to each according to his need" (14), often resulting in almost complete direction of the population.

Let us now construct a model and examine the effects of two pests on the respective economies of the three systems. We will take, for example, the spaghetti crop and two very recently discovered pests, the lasagna blight caused by *Trattorispora institutionalis* Ord. and the lesser spotted spaghetti weevil (*Pastivorous econimicarum* Hopf). The blight caused from 5 to 15 per cent lost in a crop potentially of a million tons, and worth, at present prices, potentially $100 million a year, grown on a million hectares; and it is followed by the lesser spotted spaghetti weevil causing an average loss of $5.\dot{5}$ per cent (from 5.25 to 11.5), in spite of measures taken against the pests. That is, on an average, 850,000 tons of crop are grown worth $85 million. Table IV gives us the extremes and average positions of these attacks.

It is noted that the blight attacks first and that three levels of attack are entered, and this is followed by three levels of weevil attack, giving 9 levels of loss in all. There is only one entry for value of crop lost, in this case the average attack level, 10 per cent blight followed by $5.\dot{5}$ per cent weevil. To insert values for the other loss entries, we must know something of the elasticity of demand for the spaghetti crop, the possibility of substitution and the biology of the pest; the latter, for such a newly created species, is sadly unknown. If the demand for spaghetti is high and the pests also attack the substitute crops such as macaroni and tagliatelli (we know from the name that the comparatively lesser grown lasagna is attacked), then the elasticity could be high. For instance, if a spaghetti shortage of 10 per cent increased the price by $15 per ton (from $100 to $115) then the growers would receive for 765,000 tons, not $85 million, but $88 million and would be some $3 million richer. Of course, the blight and the weevil would not attack evenly; some growers would be much advantaged, but others with bad attacks would be worse off—even ruined. On the other hand, if pest control measures (better agronomy, new varieties, better sprays, or acts of God) increase the crop by 10 per cent and (*a*) do not reduce the price or (*b*) reduce the price by $15 per ton, then, in the first case, the growers are better off by $8.5 million and in the second are poorer by $5.5 million for producing and handling a greater weight of crop. What, then, are the effects of less and more pest attack on the farmers and consumers under the three political systems?

Capitalism.—Under *laissez-faire* (classical) capitalism, say, mid-19th

TABLE IV

MODEL OF CROP LOSSES

Full crop is one million tons, worth on average $100 million grown on 1 million hectares. It is attacked first by blight then by weevil and gives on average 850,000 tons worth $85 million, which just about satisfies the market.

Pest	% Loss	Loss		Remainder		Land Lost		Tons left '000
		Tons '000	Value $'000	Tons '000	Value $'000	From individual pest '000 ha	From both pests	
Blight	5	50				50	—	950
followed by								
Weevil	2.2	21				21	71	929
	5.5	52				52	102	989
	11.1	105				105	155	854
Blight	10	100	10,000	900	90,000	100	—	900
followed by	2.2	20				20	120	880
Weevil	5.5	50	5,000	850	85,000	50	150	850
	11.1	100				100	200	800
			15,000 Average					
Blight	15	150				150	—	850
followed by								
Weevil	2.2	19				19	169	831
	5.5	47				47	197	803
	11.1	94				94	244	756

century capitalism in the United Kingdom or France, gluts and scarcities, whether caused by pests, weather or other causes, were mostly allowed to wreak their effect on the market unimpeded. Overproduction led to low prices and forced farmers either into bankruptcy or into other, usually less profitable, crops. Epidemics, such as the potato blight in Ireland or vine phylloxera in France, led to devastating shortages and much distress among farmers, which were dealt with mostly by charities rather than by any interference with the *laissez-faire* system.

Modern democratic capitalist governments need prosperous agricultures for military or for humanitarian reasons, even though farmers tend to be but a small proportion of the electorate. Thus, they seek to soften the blow to farmers caused either by scarcity (pest attack) or occasional glut (the overcoming of pest) by encouraging research on pest control, or even by the introduction of socialist methods, the regulation of crop area sown, which is the very negation of a free enterprise system.

Pest control measures most favoured in the capitalist economy tend to be chemical, for, as Galbraith (6, 7) points out, we have an economy geared to the production of material goods; this therefore gives manufacturers something to sell, whereas biological methods do nothing for the producing industries. On the whole, modern "enlightened self-interest" capitalism tends to overcome pests by the more expensive chemical methods rather than by the cheaper biological ones. This is not necessarily bad for the country's economy as a whole, for what the farmers lose through employing a more expensive method of pest control, manufacturers gain by making and selling pesticides. Great sums are spent on research into pesticides and comparatively small sums on research into biological control. A considerable body of research workers and salesmen gains a livelihood in the pesticide trade.

Socialism.—In a mixed (socialist) economy, greater effort would be made in fitting the supply to the demand by means of legislation restricting acreage where necessary, a device already used in capitalist states to some extent (see above). Research on pest control would be better balanced and directed toward benefiting the farmer and consumer rather than manufacturers and professional pest control workers. The policy is thus to avoid scarcities and gluts by forward planning rather than regulation by the push and pull of the market, but this end will not always be achieved. By allowing more of the benefits of pest control to accrue to farmers, the price of food would tend to be lowered and "satisfaction" increased.

As, on the whole, farmers tend to have a lower income than manufacturers, an extra $1.00 income in the hands of a farmer is likely to be more appreciated, to be more useful, than an extra $1.00 income in the hands of a manufacturer. However, if capital formation is required for some new project (new sprays to replace failing old ones, for instance), then it is more likely to arise from income in the manufacturer's hands than the farmer's. The socialist state raises its capital by taxation.

Communism.—As an example of a communist state we may point to the ancient Inca empire, whose crops were plagued with pests as much as anyone else's, but of which we have few details. The Inca empire was run very strictly on duties and benefits by means of close statistical control. So accurately was food production and potential food production known that the population was fitted to the food available by means of simple birth control techniques, a proposition we are reluctant to accept freely in this day. Pest control would be directed much as in a socialist economy. In the Inca economy, pest attack leading to a shortage would have led to a slowing down of certain activities, such as expansion of the empire or of the religion; abundant crops, resulting from lack of pests (locusts, for instance) or other causes, would lead to expansion of these nonproductive activities. Although the economy was not a monetary one, the statistical knowledge held by the rulers (obtained by the "Quipu" system of notation) was very complete and enabled the economy to be directed efficiently with scarcely a murmur of protest.

However, after this brief consideration of political systems, we must note that pests and diseases of crops are singularly unaware of such systems, though pest incidence can be much influenced by politics. We give three examples; the Colorado beetle (potato beetle) moved quickly across America in the 1860's, partly because the Americans were devoting nearly all their resources to fighting the Civil War rather than the potato beetle. The phylloxera gained such a hold on Europe's vineyards in the 1870's because it was felt that conducting the Franco-Prussian war was more important than controlling the pest. The phylloxera is said to have caused more damage to France in terms of money than the war indemnity demanded by the victors.

The powdery mildew of the vine reached France in the 1840's and a M. Grison, keeper of the king's greenhouses at the palace of Versailles, realised he was going to lose the royal grapes. When they became republican grapes in 1848, they were attacked even more destructively and Grison invented a remedial spray—lime–sulphur. In 1852, the grapes, now being Imperial, were damaged much as ever but the remedy of sulphur was proving effective and has remained so to this day, with two more republics to its name!

The political power latent in pest epidemics is substantial. One but needs to speculate on the effect of a future occurrence of the coffee rust in Colombia and Costa Rica, for instance, most of whose foreign exchange is obtained from coffee sales, to realise that pests are still important politico-economic factors.

The Interests Involved

We may now give more thought to the effects of losses from pests and their control on the five interests mentioned above (page 35). The consumers of agricultural produce want bountiful, pest-free crops. The individual farmer is such a small unit in the market that he wants the same thing; the group of farmers may frequently benefit from a short crop, such as that possibly induced by pests. Manufacturers of pesticides want a moderate level of the kinds of pest attack for which they have remedies that can be profitable to the farmer, and no attack of those troubles for which there are no chemical cures, for instance, cadang-cadang of coconuts, chestnut blight, and pine bark beetle. They wish their customers to remain as prosperous as possible under the circumstances.

The fifth interest, which we add here for the first time, is now very powerful in this pest world; it is the considerable body of men and women engaged in pest control research, whether it be fundamental or applied, or whether it be biological control or chemical control. Such occupation is, on the whole, a varied pleasant intellectual exercise, usually carried out in good laboratories and in interesting farms in different parts of the world among intelligent people, and is much less dull than many routine business or factory jobs. Workers in the field may not earn as much as manual workers in organised industries but, on the whole, they enjoy the stimulus of the job. Consequently, it is to the interest of this group to have the work continue, the older members tending to adhere to the *idées déja reçues* and the younger to be more adventurous, but all wanting the work to go on (the millennium of absence of pests to be delayed). This naturally influences the future economic pattern of pest control for there is a vested professional interest, rather than a financial one, in maintaining the profession on much the same lines as exist at any given moment.

This is merely an example of Galbraith's (7) point, that modern technology is so complex that it can be carried out only by large organisations, so that it is the technologists who now set the pattern of the economy, not the consumers. The economy tends to become one which produces the goods it suits the technologists of large organisations to produce. This appears to be true of pest control.

SURVEYS AND ECONOMICS

So little is known of the economics of pest control, particularly in the developing areas from where most of the world's additional food must come, that we advocate surveys to discover the actual losses which occur because of pests in a small area of the world, and the actual benefits (or additional losses) which accrue (taking costs into account) as a result of measures taken to control them. Whether any such survey is ever conducted will depend upon the resources which interested parties are prepared to devote to this research. The size of the area studied and the complexity of the survey would also depend upon the available finances to do the job. Recently, there has been some activity in the study of crop losses caused by pests (4) as well as the effects of pest control. To some extent, both activities are complementary to one another. However, they are not the same thing and should not be mistaken for each other when they are encountered. To show the essential difference between these two activities, it will perhaps be easiest to provide a brief working explanation of the aims of each of them.

To begin with, it is possible to provide a working definition of "Economics"? This in itself will probably create some controversy among theorists, but it is necessary to have a basis from which to start and the following definition is the one used in this article. The study of Economics is the study of the efficient use of available resources.

Having accepted this definition, it follows that the Economics of Pest Control is the study of the efficient use of available pest controls. It is now useful to assume that the pest controls known to be available are effective in controlling the pests which they were designed to control. If they are not, then the chemist, the agronomist or other experts must return to his drawing board or test tube. One further point is that it is now assumed that the controls being discussed are chemicals which are sprayed on the crops. There are many other types of pest control, but to illustrate the points being made, what is perhaps the commonest form of pest control will be used an an example.

The farmer is offered a number of chemicals at varying prices that will dispose of the different fungi attacking his crops with varying degrees of efficiency. Further, the farmer knows (because it was assumed to be so in the previous paragraph) that if he does what the label of the pesticide instructs him to do, he can expect a particular result in terms of reduced fungal attack. What he probably will not know is the exact result which he can expect. He can discover this by experimenting to find out which pesticide kills most of his particular pest. Obviously, he would not use a herbicide to get rid of powdery mildew, but he might have to decide between a number of competing fungicides. It might also be possible for him to consult the work of scientists who had already assessed the effectiveness of each of these fungicides. An extension officer also might be able to advise him.

The farmer now knows that if he uses chemical A on his powdery mildew, he will get rid of it completely. If he uses chemical B, he will eliminate 90 per cent and if he uses chemical C, he will eliminate 70 per cent, etc. There are two important jobs which the economist can do. First, given this somewhat Utopian situation with regard to information and extension work, he can advise the farmer on the financial benefit of using the different pesticides available.

At first sight, it would seem that in the example given above, chemical A is the most suitable chemical for disposing of powdery mildew; but suppose that A is platinum dust and that C is common salt. Clearly, the difference in the cost of these two "chemicals" is going to affect their suitability. Again, suppose that chemical A is available only in 1 inch cubes, whereas chemical C is a wettable powder and that the farmer already has a machine in which he can use the wettable powder. Also, consider what the position is if the powdery mildew affects 1 per cent of the plants and also what it is if it affects 80 per cent of them.The significance of the technical effectiveness of the different chemicals should now have emerged in true proportion.

The second job which the economist can do is to determine results if (*a*) there are not enough extension officers; (*b*) there are only a few pieces of scientific information available to the farmer who cannot read anyway; and (*c*) if the chemical manufacturers have only limited, inadequate outlets for their products or if the farmers are subject to sales pressure from manufacturers' "slick" salesmen. These situations are frequently closer to reality and introduce a number of new variables for the economist to consider in his assessment of what is and what is not beneficial to the farmer.

The essential work of the economist in this field is basically to compare the cost of the different chemicals as they can practically be obtained and applied with the extra value of any increased quantity or improved quality of crop which results from their use. There are a number of techniques available for accomplishing this. There is the use of cost/benefit ratio, a straightforward input-output comparison, and marginal analysis. All of these require detailed information about the farmer's costs (beyond what he spends on pesticides) and his yields and sales. Marginal analysis is probably impractical because it demands information about virtually everything that the farmer does within these fields of activity. The other two techniques can generally be managed without quite such detailed information. Nevertheless, a fair degree of detail is needed in order to isolate the effect of the input of pesticides on output from the effects of other inputs (e.g., fertilizer, new methods, more labour, more machines, etc.).

So far, some aspects of the economics of pest control have been discussed. Now to consider the study of crop losses. The writers are not phytopathologists, but it seems that the study of crop losses is even more important to the phytopathologist than to the economist. Briefly, the study of crop losses is concerned with the effects of different infestation levels of different pests on different crops. It has already been illustrated that there is significance for the economist in the magnitude and proportion of pest infestations present in a particular crop. The next link in the chain of cause and effect is equally important. This is the depressive effect of these different infestation levels on the yields of different crops. A difference between 2 per cent of plants affected by powdery mildew and 70 per cent of plants affected may make a 1 per cent difference in crop yield or a 90 per cent difference in crop yield, depending on the type of powdery mildew, the crop and the prevailing conditions. This information must be taken into account by the economist, but it is only one, albeit a quite important one, of a number of factors which he must consider. The study of crop losses is an important study in its own right. The results of this work must be used by the economist in his own studies, but it is by no means the only job which the economist should do.

The question remains of how these two pieces of work should be undertaken. For the study of crop losses, Research Station experiments have often been favoured. This is one possible approach to the study of the economics of pest control. An alternative method for the economist, however, is the survey technique which, in general terms, seems to be more suitable.

METHODS OF ASSESSING ECONOMIC FACTORS OF PEST CONTROL

The questions that we mentioned earlier which all groups should rationally ask with regard to the use of pest control measures are: (*a*) How much does it cost? (*b*) How much do I/we benefit? (*c*) Have I/we improved my/our financial position, (i) in the short run, (ii) in the long run? (N.B. Short-term and long-term improvements may not be reconcilable.) Perhaps, at least on a national and world scale, the question should also be asked: (*d*) Have we gained, or will we gain, any moral or humanitarian advantage (i) in the short run, (ii) in the long run? (Whether this last question can be said to relate strictly to economic rationality will be left open for debate by those who are interested.)

What the economist can hope to do is at least to answer the first three questions in positive terms and perhaps venture an opinion on the last.

Some Problems of Providing Answers

The exact cost of any pest control measure will vary with climatic conditions, physical geographical conditions, cultural conditions (i.e., human attitudes), market conditions for both pesticides and farm produce, and agricultural conditions (which last term may, of course, include elements contained in the former terms). The reader might be able to think of additional conditions which impinge upon costs. Each new variable complicates the position.

The exact benefit to be derived from the use of pest control measures will have to be measured according to the particular yardstick of the particular interest (indicated above) that is posing the question (the farmers' meat may be the nation's poison).

Many of the conditions affecting cost assessment may also have a bearing on benefit assessment—particularly market conditions for farm produce.

There is the difficulty, already mentioned, of reconciling long-term and short-term profitability. (Is an increase of $x + y$ in profit tomorrow worthwhile if one faces a decrease of $x + y - a$, or even of $x - y$ next year?)

The methods which at present exist for the study of the economics of producers and producer groups can be broadly divided.

Research Station Experiments

Advantages.—(*a*) Conditions can be controlled to points within fairly precise limits. (*b*) Results are usually fairly easy to obtain from experimenters and can be fairly easily assessed. (*c*) Specific answers can be provided albeit within a strictly limited sphere.

Disadvantages.—In some ways, the very advantages can become disadvantages. (*a*) Results and experiments are usually drops in the ocean of information required. (*b*) Research stations are most numerous in "developed" areas (mainly in temperate zones). Information in the tropics from this source is apt to be localised and scarce. Results may appear insignificant in the pattern of world agriculture or even of tropical agriculture. (*c*) Results and experiments are often of very limited application, referring only to one small place, one specific crop and one specific pest and control. (*d*) Pest control is often only one of the aspects of agriculture being dealt with by the research station, and time and money devoted to this aspect may often be inadequate to provide comprehensive results and statistics. (*e*) Experiments are often planned by pure scientists for pure scientists and the economic data has to be gleaned as a by-product of the main biological results.

Analysis of Farm Accounts

Advantages.—(*a*) The keeping of farm accounts shows what a farmer actually does and consequently gives an indication of what he, with his prejudices, nationality, state of knowledge, considers to be most important. (*b*) Farms are world-wide. Results are of a more comprehensive nature. (*c*) Information can be readily available for long periods if farmers can be induced to keep proper and adequate records.

Disadvantages.—Again, some of the advantages are apt to be double-headed. (*a*) The farmers' point of view might not be the one which is economically rational. While there is value in knowing which attitudes might better be corrected, undesirable attitudes produce unsatisfactory cost analyses. (*b*) Farmers in the "developing" (tropical) areas are less likely to keep adequate records; they are more difficult to convince that they should do so and, in any case, are less capable of doing so than their "developed" counterparts. (*c*) Even organized farm records (by universities, research organizations and commercial undertakings) are apt to rate pest control methods as sufficiently unimportant as to be lumped together in accounts in a manner such as: "Miscellaneous, Fertilizers and Herbicides." (*d*) Even organizers have difficulty in devising cost and statistical methods which will allow for all the variables involved (such as the vagaries of climate, the trends of the market and political attitudes). (*e*) Many farmers are reluctant to divulge what they consider to be nobody's business but their own even if they have the information available.

The methods for studying national and world economies are more comprehensive and sophisticated. They consist mainly of the figures published by world and national authorities (e.g., the Food and Agriculture Organization and Boards of Trade). Here, again, there are certain recurrent problems:

1. The data may be of doubtful accuracy.

2. Organizations which collect and publish information do so for a wide audience and the particular requirements of the study of pest control are often inadequately provided for, on two accounts: (i) The information is not available. (ii) The information is of too narrow interest to justify its inclusion.

The exact elements of cost and benefit of pest control measures vary between the three main types of control measures indicated in the first paragraph of this article. Costs, for example, can include such wide-ranging items as purchase of pesticides, loss of crop due to different time of sowing and harvesting, cost of research into biological control measures (which last may have no corresponding benefit). Benefits have to be assessed (if accuracy is to be obtained) from equally diverse considerations.

What Is Needed

If analyses of the benefits and costs of pest control are to be accurate and dependable, the following factors are important in the obtaining of adequate data:

1. Co-operation by farmers.
2. Comprehension by natural scientists and statisticians.
3. A large expansion both of existing field stations of experiments designed to collect economic data, and also of the number of field stations, particularly in the tropics.
4. The means to ask the relevant questions of the information sources available. This implies co-operative representation of the organization conducting the survey in the field.
5. The means to handle the answers when they have been obtained. This implies an expansion of staff employed on this subject and also a streamlining of methods of work.

THE OBJECTIVES

The objectives of this survey should be fairly straightforward. In order to achieve them it will inevitably be necessary to examine some more refined and complex requirements, but it should be remembered that these are only instrumental in reaching the main goals of the survey.

To turn now to the basic objectives of this survey, it is of primary importance:

1. To discover the costs and benefits of the use of different types of pest control for different types of crop.
2. To analyse this information to discover the economic importance of this for (*a*) each farmer and farm studied; (*b*) the group of producers in the selected area; (*c*) the national economy; and (*d*) the world economy.
3. To analyse the information found under 2(*a*) above, to discover the economic importance for different types of farmer (peasant, subsistence, commercial, etc.).
4. To apply the knowledge gained for (*a*) the immediate and long-term good of farmers in the selected area (of all types); (*b*) the good of the producer group; (*c*) the good of the national economy containing the selected area; (*d*) the world economy; (*e*) (by association) the good of farmers, producer groups and national economies in other areas where comparisons are considered to be valid.
5. To apply the knowledge gained in the hope that it may advance the study of economics, particularly the economics of pest control.

LITERATURE CITED

1. Bentham, J. *Works,* **10,** 142 (Wm. Tait, Edinburgh, 1830)
2. Burchell, R. T. The biology and control of apple mildew. *Federation Brit. Plant Pathol. Meeting, 11th, London, 1968* (Unpublished)
3. Burke, J. *Reflections on the Revolution in France,* **5,** 271 (J. Dodsley, London, 1790)

3a. Curtis, J. *Farm Insects.* Chap. 15 (Blackie & Son, Glasgow, 1860)

4. Food and Agricultural Organization, U.N. Papers presented at *Food Agr. Organ. U.N. Symp. on Crop Losses, Rome 1967*
5. F.A.O. Yearbook of Agriculture, Production. *Food Agr. Organ., U.N.,* **21** (1968)
6. Galbraith, J. K. *The Affluent Society,* 155–74 (Houghton Mifflin. Boston, 1962)
7. Galbraith, J. K. *The New Industrial State,* Chap. 1 (New York and Hamish Hamilton, London, 1967)
8. Grainger, J. Economic aspects of crop losses caused by diseases. *Food Agr. Organ. U.N. Symp. on Crop Losses, Rome, 1967,* 90
9. Howard, L. O. *The Insect Menace,* 321 (Century Co., New York and London, 1931)
10. Hutcheson, F. *Inquiry into the Original of our Ideas of Beauty and Virtue,* Pt. ii, Sect. 3 (J. Darby, London (?), 1720)
11. Ingram, J. Experiments on the control of apple mildew. *Federation Brit. Plant Pathol. Meeting, 11th, London, 1968* (Unpublished)
12. Johnson, C. G., Ordish, G. Supplementary notes to Reference 13, *Pest Articles and News Summaries A,* **11**(3), 366–68 (1965)
13. Judenko, E. The assessment of economic effectiveness of pest control in field experiments. *Pest Articles and News Summaries A,* **11**(3), 359–67 (1965)
14. Marx, C. *Critique of the Gotha Programme* (Lawrence & Wishart, London, 1938)
15. Ordish, G. *Untaken Harvest,* Chap. 3 (Constable, London, 1952)
16. Ordish, G. Some notes on the short-term and long-term economics of pest control. *Pest Articles and News Summaries A,* **14**(3), 343–55 (1968)
17. Paddock, W. C. Phytopathology in a hungry world. *Ann. Rev. Phytopathol.,* **5,** 382 (Table I) (1967)
18. Plato. *The Republic,* Book IV, Chap. 1, Sec. 420B (*Circa* 375 B.C.)
19. Sen, A. R., Chakrabatry, R. P. Estimation of loss of crop from pests and diseases of tea from sample surveys. *Biometrics,* **20,** 492–504 (1964)
20. Smith, H. S., Essig, E. O., Fawcett, H. S., Peterson, G. M., Quayle, H. J., Smith, R. E., Tolley, H. R. The efficiency and economic effects of plant quarantines in California. *Univ. Calif. (Berkeley), Bull. 553* (1933)
21. Strickland, A. H. Persistent organochlorine pesticide usage on crops in England and Wales. *Review of the Persistent Organo-Chlorine Pesticides* (Cook Rept.). (App. E., H.M. Stationery Office, London, 1964)
22. Tangye, D. *The Gull on the Roof,* 98 (Michael Joseph, London, 1961)

Copyright 1969. All rights reserved

EPIDEMIOLOGY AND CONTROL OF BACTERIAL LEAF BLIGHT OF RICE

BY T. MIZUKAMI AND S. WAKIMOTO

National Institute of Agricultural Sciences, Tokyo

INTRODUCTION

The bacterial leaf blight disease of rice was first observed in the latter part of the last century in various localities of southern Japan. Severe damage has been caused annually since then. Scant information is available on the occurrence of this disease in other countries. It has been reported in Korea (105), Taiwan (21), the Philippines (15–17, 19, 77), Indonesia (15, 16), Thailand (19), India (5, 84, 86–88), and China (8, 9, 107). According to investigations carried out in recent years (59, 66, 119), the disease has become of economic importance in many other countries of Asia, namely Ceylon, Cambodia, and Malaysia. It occasionally causes severe damage to rice yields and is recognized as one of the most important diseases of rice growing areas in Asian countries. No reports of the presence of the disease in America, Australia, Africa, and European countries growing rice have been received. Possibly the disease is nonexistent in these countries or is not of any great importance.

Early investigations during 1908 to 1910 indicated that bacterial blight occurred in almost all southwestern prefectures of Japan (73, 79, 104). It gradually spread to northern Japan and in 1962 was observed in Hokkaido. At present, the disease is widespread in all prefectures of Japan, although its severity and damage is limited in the north. The acreage infected by leaf blight disease was 50,000 to 60,000 hectares in 1940, exceeded 100,000 hectares during the Second World War, and is now 300,000 to 500,000 hectares. This corresponds approximately to 10 per cent of the total rice area in Japan.

HISTORICAL

It was believed before 1910 that bacterial blight was caused by a physiological factor, namely acidic soil, as dewdrops appearing on infected leaves were observed to be more acidic than dewdrops on healthy ones (73). Takaishi (104) observed in 1909 that the turbid dewdrops obtained from the diseased leaves consisted of a mass of bacteria. The disease was successfully reproduced by inoculating rice leaves with these dewdrops. From these observations, he concluded that it was caused by a kind of bacterium. Bokura (6) succeeded in isolating a species of the bacterium from diseased leaves which formed yellow colonies on agar medium. He named it *Bacillus oryzae* Hori et Bokura after recording its morphological and physiological properties and confirming its pathogenicity. Further experiments on the bacterium isolated from the diseased leaves were conducted in 1922 by Ishiyama (31) who indicated that the disease was caused by a kind of rod-shaped bacterium which was different from that described by Bokura. He named his isolate as *Pseudomonas oryzae* Uyeda et Ishiyama. This was changed later to *Xanthomonas oryzae* (Uyeda and Ishiyama) Dowson (4).

Since then many experiments have been conducted by workers at different agricultural experiment stations in Japan concerning the resistance of rice varieties to this disease, the relationship between the application of fertilizer and disease occurrence, and measures for its control. Kuwazuka (53–56) particularly conducted extensive investigations on the epidemiology, selection of resistant varieties, and on control measures for bacterial leaf blight.

Considerable information has accumulated since the late fifties in many fields regarding this disease. Mention must be made of the multineedle prick method for inoculation, *Leersia oryzoides* var. *Japonica* as an intermediate host grass of the pathogen, isolation of bacteriophage specific for leaf blight bacteria, utilization of phage for ecological studies of the disease, strain classification of the pathogen, and development of chemicals for its control. These topics will now be considered in detail.

Symptoms

Symptoms on rice plant.—Symptoms develop mainly on leaf blades, leaf sheaths, and sometimes on grains.

At late nursery stage, tiny water-soaked spots appear on the edges of developed lower leaves. These spots enlarge and turn yellow gradually. In the case of early infection, symptoms appear at three or four weeks from transplanting, and gradually spread upwards as the rice plant develops. The lesions are usually initiated at the edge of the upper part of leaves where water pores, through which the bacterium can invade, are distributed more frequently (60). Within two or three days, the lesions enlarge along veins, turning yellow in colour. These enlarged yellow lesions turn white or greyish white later on.

Besides these typical symptoms, the leaves sometimes roll and wither following infection. Symptoms appearance in this case depends on the variety or physiological condition of the rice plant (103), virulence of the pathogen, and climatic conditions. In the diseased plants, either the upper half or the whole leaf dries rapidly and turns pale white before withering eventually. Infected areas of the leaves can be detected before symptom appearance by means of immersing the cut end in a diluted solution of basic fuchsin for one to two days (18). Opaque and turbid drops of bacterial ooze are formed on the surface of lesions. These drops dry up into yellowish small spherical or tendril-like sticky masses which play an important role in overwintering and spread of the disease.

Another symptom in the young growing stage of the rice plant has been reported recently (147). Rice leaves of plants in initial to middle tillering stage, roll completely, droop, turn yellow or greyish brown, and finally wither. Growth of such affected rice plants in paddy fields is retarded, and marked stunting may occur in severe cases. Although such a symptom has been found at present only in Niigata and Nagano prefectures in Japan, it seems common in Indonesia (16), India (66), Malaysia (119), and Ceylon (59). This so-called "Kresek" symptom was first reported by Reitsuma (78) in 1950, and later Goto (15) identified the pathogen as *X. oryzae*. "Kresek" symptoms appear after one month if the roots and the basal part of rice seedlings are dipped into the bacterial suspension just before transplanting (142). From the investigations of Yoshimura & Iwata (141), it appears that infection begins in the basal part of the seedling and spreads into the vascular bundle, destroying the tissues and causing the seedlings to wilt.

Symptoms of pale yellowing of rice leaves which have been observed in the Philippines have been identified as caused by *X. oryzae* (15).

Lesions on leaf blades develop downwards to the basal part, and extend further to the sheath through the midribs. Yellowish or greyish streaks appear according to disease development. On severely affected plants, the whole sheath would be discoloured and killed.

Discoloured spots which are surrounded by water-soaked parts often appear on the glumes of plants in paddy fields. Although these spots are conspicuous while the grains are young and green, they become grey or yellowish white in the middle with an indistinct margin during the ripening stage (135).

Symptoms on weeds.—Three species of wild gramineous plants are known to be infected by *X. oryzae* under natural conditions. They are *Leersia oryzoides* Sw. var. *Japonica* Hack (12, 13), *L. oryzoides* Sw. (138), and *Zizania latifolia* Turez (12, 13, 138). On the *Leersia* plants, water-soaked and discoloured lesions appear first at the edges of leaf tips. They then extend downwards along the vein, increase in size, and often wedge-shaped streaks or large distinct lesions along leaf edges are formed. Margins of the lesions are orange-red in colour, and the central parts of the lesions turn greyish brown. The lesions sometimes enlarge on leaf sheaths as in the case of the rice plant. On *Z. latifolia,* lesions are initiated at the edges of leaf tips. Gradually they enlarge in length and width and are bordered by a yellow wavy margin. Yellow spots along the leaf veins sometimes appear adjacent to main lesions which rapidly enlarge in size.

Effects of Environment on Disease Development

Many factors such as topography, climatic conditions, and cultural practices are related to disease development. Topographic factors or soil conditions are usually constant in contrast to the variable climatic conditions or cultural practices. Naturally, these nonvariable and variable factors are interrelated.

Topographic conditions of an endemic area.—An endemic area usually is a region with an acidic soil and suffers from poor drainage, a relatively high underground water level, and frequent floods (31, 54, 85). Poorly drained areas along rivers or lakes and mountainous basins are conducive to disease development because of the thick morning fog in summer to autumn (31, 54, 56). Soil type appears to be a factor as the disease is severe in sandy loam, clay, or clay loam alluvium soils, and slight in the sandy soil of dune areas. Soil acidity was thought to be an important primary factor (73, 85, 104). However, such an opinion had to be revised in view of further information (6, 36, 54). The disease was found in a severe condition in shale soil areas in Aichi and Gifu prefectures of Japan (33, 44–46, 49).

Climatic conditions.—Climatic conditions greatly influence disease development. The most important factors for disease epidemics are rainfall, humidity, temperature, flood, and typhoon during the rice growing season (6, 11, 54, 56, 69, 85). Kuwazuka (56) concluded that the disease was most prevalent in areas with more than 200 mm rainfall in July and annual mean temperatures higher than 14° C, or July mean temperatures of more than 24° C. A high correlation between disease development and rainfall in the July to August period was observed from 1947 to 1952 according to the climatic data of Saga prefecture, Kyushu. Goto and his co-workers (14) concluded that such climatic conditions as high temperatures in August to early September and a warm autumn are suited for bacterial blight development. Heavy rainfall, a little sunshine and strong winds during this time further increase blight development. Fujikawa and co-workers (11) also reported the correlation between disease development and the temperature, and emphasized the effect of the rainfall and typhoons in the rice growing seasons. Yoshimura (134) considered that the low temperatures in September seem to inhibit the development of the disease in the Hokuriku area.

Judging from these reports, high temperatures during the tillering stage, low temperatures with less sunshine in midsummer, and a warm autumn appear to be conducive for disease development in Japan. Severe winds or typhoons in conjunction with the above conditions accelerate disease occurrence.

Cultural practices.—The cultural practices relating to disease development are preparation methods of the nursery, application of fertilizer, and selection of rice varieties.

The method of preparing the nursery is directly related to primary infection of seedlings. Usually, rice seedlings grown in deeply irrigated or easily flooded nurseries are more contaminated with the bacterium and hence more severe damage occurs in paddy fields than when seedlings are obtained from semi-irrigated or upland nurseries (72, 136).

Application of fertilizers, especially an excess of nitrogen or unsuitable combinations of nitrogen, phosphate, and potassium, increase development of the disease (6, 31, 54, 58, 130). Application of a high dose of silicate, magnesium, and shale also increases the severity of the disease (25, 44, 45). while potassium decreases it (130). The effects of fertilizer application on disease development varies according to the rice varieties used as well as to the soil conditions and many other biological factors.

The depth of irrigation water and drainage also influences disease occurrence to some extent (38, 43, 73, 85). Early transplanting increases disease incidence (6, 28), while deep ploughing decreases it (58).

Damage

Several reports have been published on the damage of rice caused by bacterial leaf blight. Yields of unhulled grains decrease depending upon the severity of the disease. Reductions of 20 to 30 per cent have been observed in yields when infection was moderate, and over 30 per cent when it was severe (31, 79, 85). The weight per 1000 grains of unhulled rice is also reduced by the disease as is the straw weight of affected plants (31, 79, 85). Percentage of husked, sterile, unfilled grains shows an increase in diseased plants (23, 31, 54, 60, 79, 85). The amount of soluble nonnitrogenous substances is less and that of coarse protein more in husked rice from diseased plants, showing incomplete maturity (31). When the disease rapidly develops during harvesting or later, the damage is much less (11, 25).

At present, damage due to bacterial leaf blight has been estimated only for Japan. In tropical countries, however, the severity of the disease has been pronounced in certain areas, and has resulted in almost complete crop failures (66). Loss of yield has never been estimated for the "Kresek" condition which commonly occurs in several Southeast Asian countries.

The Pathogen

Morphological and physiological characteristics.—From electronmicroscope observations, the bacterial cells grown on culture medium appear to be 0.55 to 0.75 × 1.35 to 2.17 μ in size, and those in the host tissues 0.45 to 0.60 × 0.65 to 1.40 μ. Thus the bacteria in the host tissues are somewhat smaller than those cultured on the media (136). The bacterial cell is rod shaped with rounded ends and has a long monotrichous flagellum (31, 136). The bacterial cell is surrounded by a capsule composed of galactose, glucose, xylose, uronic acid, and two other unidentified substances which are probably polysaccharides (60). An early report (31) indicated that the bacterium was aerobic. It did not liquefy gelatin or use nitrates. Slight amounts of hydrogen sulfide were formed, but not ammonia or indole. It did not produce gas or acid from sugar. However, the physiological characteristics of this bacterium have not always followed the original description (68, 84), especially in the case of ammonia, acid production from sugar, gelatin liquefaction, and starch utilization. These discrepancies are probably due to the different basal media and isolates used by these workers.

Media.—The bacteria will grow on PSA (potato semisynthetic agar) medium (111). Owing to the unsuitability of PSA medium for physiological studies, much work has been done for the purpose of developing a suitable synthetic medium for study of the nutritional requirements of this pathogen. Sucrose is a favourable carbon source for *X. oryzae* (7, 67, 126), and dextrose, mannose, and galactose can be utilized also although fructose cannot. Maltose, dextrin, and succinic acid have also been used as carbon sources (126). There is evidence to show that glucose might act as a growth inhibitor in potato broth medium (7). The most favourable nitrogen source appears to be glutamic acid, followed by cystine (126). Only ammonium phosphate has been used among inorganic nitrogenous nutrients. Small quantities of some amino acids and compounds like $MgSO_4$, $FeSO_4$, and $MgCl_2$ promote the growth of the bacteria when added to the basal culture medium (126, 128).

A synthetic medium containing EDTA-Fe (ethylenediamine-tetraacetic acid tetrasodium salt-Fe) was developed recently for the purpose of culturing single cells of the pathogen. About 80 per cent of transferred single cells developed into colonies on this medium according to Suwa (89).

Differentiation of the pathogen.—*X. oryzae* isolates collected from various localities show differences in their characteristics. For the purpose of the classification of the isolates into strains, two criteria, namely virulence and lysotype, were used independently.

Classification by virulence: Differences in virulence against rice varieties have been observed among the isolates. During a severe epidemic of the disease in Kyushu district, the bacterial leaf blight resistant variety "Asakaze" was severely attacked. The pathogen isolated from diseased tissues was virulent to resistant varieties as well as to susceptible ones (48). Subsequently, many workers tried to classify bacterial isolates into strains according to their virulence (20, 47, 52, 80, 118, 124, 127, 138). The criteria for classification, however, are widely different depending upon the individual worker. Some classified the isolates into three groups by observing the virulence of the pathogen against the leaves of matured rice varieties which show differences in resistance (47, 124). Others separated the pathogen into two groups on the basis of tests using seedlings (52, 127). Attempts at classification by using the host grasses (138) and Indica rice varieties (20, 80, 118) have also been carried out. Virulent and less virulent isolates are distributed in every country (20, 118). However, the virulent strains are more frequently distributed in tropical countries (20, 118).

Classification by lysotype: Four kinds of bacteriophage, viz, OP_1 (109, 133), OP_{1h} (22), OP_{1h2} (116), and OP_2 (116, 145), specific to *X. oryzae* were isolated from Japan. The host ranges of these phages are quite different and thus the bacterial isolates can be classified according to their sensitivity to them (47, 106, 116,143). The isolates from Japan were classified into five lysotypes, A B C D, and E (116, 143). Among these lysotypes, A was most widely distributed in Japan followed by B, D, C, and E, in descending order (116). In some cases, certain strains were frequent in certain localities (30, 32, 47, 48). No close correlation was found between the lysotypes and their virulence (47, 116) although the C strain was generally less virulent than others (150).

The bacterial isolates collected from other countries were much different in lysotype. Lysotypes D and C are most frequently distributed in the Philippines (19) and E in India (118).

Various phages have been isolated from Taiwan (50), the Philippines (19), India (65), and other Southeast Asian countries (121). Some of these phages will be employed for the classification of the pathogen into different strains in future.

Ecology of the Pathogen

Host range.—Xanthomonads, generally, are very specific in their pathogenicity under natural conditions. As mentioned earlier, the host of *X. oryzae* had been considered to be the rice plant only, but it has been reported that some other gramineous plants are also attacked (12, 13, 138, 144). By means of multineedle-prick inoculation method, *Leersia oryzoides* var. *Japonica, L. oryzoides, Zizania latifolia,* and *Pharlaris arundinacea* L. could be severely infected. *L. japonica* Mukino, *Phragmites communis* Trinius, and *Isachne globosa* O. Kuntze are slightly infected. Many naturally infected *L. oryzoides* var. *Japonica* have been found in various endemic areas. Sometimes *Z. latifolia* has also shown infection. *L. oryzoides,* which is common in Hokuriku district of Japan, is also attacked by the pathogen (144).

Among these grasses, *L. oryzoides* var. *Japonica* and *L. oryzoides* are extremely important in the ecology of the pathogen which would be discussed later.

Techniques for the ecological study of the pathogen.—Earlier, the only method for diagnosis of the disease or detection of the bacterium was isolation followed by identification. Improved techniques have been devised, however, which play an important role in the study of the ecology of the bacterium. Two of these are the needle-prick inoculation technique (60, 64) and the phage method for detection of the bacteria in a sample (110, 122).

Needle-prick inoculation: Bacteria are washed out of diseased samples with sterilized water, and the suspension is centrifuged at 7500 g for 10 min in order to concentrate the bacteria. Needles dipped in the bacteria are used for inoculation. The presence or absence, and the approximate number of the bacteria can be estimated by the disease development on the inoculated susceptible plants. With high bacterial numbers in the sample, a higher percentage of diseased plants will be obtained (60, 64, 81). Recently, a slight modification of this technique has been devised which consists of an inoculation by a compact group of 100 needles. The observations under the microscope of bacterial exudates from cut specimens collected at 4 to 5 days after inoculation also aid in diagnosis (27).

Bacteriophage method: A bacteriophage specific for *X. oryzae* was first isolated from the paddy field in Fukuoka prefecture, Kyushu. The phage method for studying the ecology of the pathogen is based on the host specificity of the phage. The phage suspension is added to the test sample in PS (potato semisynthetic) or CaVfCh (calcium chloride added vitamin free casein hydrolysate) media at a final concentration of 10^{3-4} phage per ml. A small quantity of this phage-test sample mixture is centrifuged and the supernatant is plated with a bacterial suspension on PSA medium to obtain the "check" value. This check value shows the phage concentration before multiplication. After 5 to 10 hr shaking, the supernatant is again prepared and plated to obtain the "test" value. It can be concluded that pathogenic bacteria were contained in the sample if the result of the number of plaques is greater in the "test" than in the "check" (110). More complicated but accurate methods using antiphage serum have also been employed (122). According to this technique, the approximate number of bacterial cells contained in the sample are estimated through the single-step growth experiment (1, 111) to elucidate the mode of propagation of the phage, such as latent period, rise period, and average burst size.

The phage method was first established using the OP_1 phage. More recently the OP_2 phage has been used for this purpose (136).

Habitats.—*X. oryzae* had earlier been thought to overwinter in the soil of infected areas (31, 54, 55). Since then, seeds and straw have also been considered to be the overwintering habitats and important primary sources of infection, although without any conclusive evidence. The discovery of *L. oryzoides* var. *Japonica* as a host plant of the bacterium stimulated an active study of the life history of the pathogen by means of the above-mentioned techniques. The results obtained up to the present indicate that overwintering of the bacteria in Japan is as follows:

Wild gramineous plants: Several kinds of weeds growing around paddy fields are susceptible to the pathogen. Weeds belonging to the genera *Leersia* grow well on riverbanks, irrigation canals, ditches, and footpaths in the paddy fields. Overwintering of the bacteria on roots and rhizomes of these plants was confirmed by means of the phage techniques (99, 113, 115, 125). Other wild gramineous plants have also been infected during the rice growing season, but the number of bacteria gradually decreases in autumn and few or none exist in winter. None of the hosts other than *Leersia* are important for overwintering of the pathogen (99, 136).

Roots of plants: The immersion of roots of various plants into a concentrated suspension of the bacteria causes the suspension to become clear. This is due to swarming of the bacteria to the surface of the roots (115). The phenomenon seems to be induced by the secretion of sugars, amino acids or oxygen by plant roots and is nonspecific whether the plant is a host or not. Survival of the swarming bacteria around the roots was confirmed by the phage method. Acidic water simulates the swarming of suspended bacteria to the roots. The bacteria are likely to swarm to the root parts having high metabolic activity such as root tips or root hairs. Rice roots have an "activating function" in that they activate and enable the bacteria to bring about infection (61, 63).

Rice straw: The bacteria easily overwinter and survive until spring in dried rice straw kept in farmhouses (24, 99, 110). Usually the bacteria are killed in one or two months when the straw is scattered in the field and ploughed into the soil (24, 99, 120). Straw is often piled up outdoors in Japan, protected from rain. The pathogen in straw which has been protected from rain survives until the next spring (24, 99). It is unwise to spread overwintered straw in spring, because the bacteria will grow and multiply.

Rice stubble: In warm areas of Japan, much of the stubble survives until the next spring after the harvesting of rice if the fields are allowed to lie fallow in winter. The pathogen can survive in this stubble (99). In northern areas of Japan, the stubble withers in winter and therefore the pathogen cannot overwinter on it (134).

Grains: The pathogen readily overwinters on diseased grains stored in farm houses as in the case of rice straw. It normally is found in the rice husk tissues (99, 112), and has not been detected in unhulled rice grains in Japan. A report from China (8) mentions that it has been found in the endosperm in severely infected rice. The importance of seed transmission has been reported and emphasized in China and India (87, 88). However, it is doubtful whether this occurs in Japan. This anomaly may be due to the genetic differences of the rice varieties.

Soil: Bacteria mixed into outdoor soil in autumn have disappeared within one or two months (99, 113, 114). Never has the pathogen been discovered in the soil of an infected field during the spring, except in the rhizosphere of *Leersia* plants. The longevity of the bacteria seems to vary depending upon pH (63), humidity (24, 99, 120), and the antagonistic effects of other microflora. It is probable that the bacteria could overwinter in the field soil under special conditions.

Overwintering form of the bacteria: Two forms of the bacteria should be considered, the dry form and the growth form (114). The pathogen in the dry form is found in the vascular vessels and xylem parenchyma of dried diseased plants, and in ooze which exudes from diseased leaves and rapidly dries in summer. These dry form bacteria gradually die if they are moistened by rain in winter.

Bacterial cells in the growth form are found in stubble and in the root systems of perennial wild plants, especially *Leersia* spp. Pathogens surviving in an inactive state or dry form are activated and turn into the growth form upon receiving moisture under favourable conditions. They are also activated by contact with the roots of rice or other plants (63).

Primary infection of the rice seedlings.—The overwintered pathogen is transferred to the nursery by irrigation water, application of diseased straw, and sowing of diseased rice grains. The bacteria coming in contact with the surface of rice seedlings become activated and multiply. They invade the tissue through wounds which are produced at the basal part of the stem by root development, or through hydathodes on the leaf blade (63, 92). Invasion also occurs through stomata which remain open continuously and are distributed on the coleoptile and sheath of foliage leaf (90, 91).

After invasion, the bacteria multiply in the intercellular spaces of parenchyma without showing any symptoms. They are then exuded, thereby increasing the bacterial population in the nursery (90).

Infection takes place in the late nursery or field stages. The population of the pathogen remains small on seedlings before transplanting, except when an outbreak of the disease has already occurred (93, 98, 99). Severe epidemics result in the paddy fields when seedlings in which the pathogen was detected in the late nursery period are transplanted (98).

Behaviour of the pathogen in paddy fields.—Abundant bacteria were detected in severely infected fields at the tillering stage on upper leaves of rice plants as well as in paddy field water, although symptoms had not yet appeared (60, 98). The bacterial population in the paddy field usually increases on the leaves during May to July, remaining stationary or decreasing in midsummer (August). It shows a rapid increase again at the maximum tillering to young ear formation stages of the rice plant (98, 136). This fluctuation could be due to temperatures, humidity, and the change of resistance of the growing rice plants. The population of the pathogen on the upper leaves generally depends on the severity of the disease. When the population reaches 10^{4-5} per leaf, symptoms begin to appear (98, 136). During the early growth stage of the rice plant, the pathogen is more abundant in lower leaves, but later, it is abundant in the upper leaves also (64, 98). The bacterial population on the leaves gradually decreases as the metabolic activity of the leaves decreases (64, 98).

Resistance of Rice Plant

There are several ways of approaching the subject of resistance of the rice plant to this disease. These are elucidation of varietal resistance, studies on biological, physiological, and physicochemical mechanisms of resistance, and the genetics of the rice plant. Studies on the resistance of the rice plant against leaf blight are of a practical nature and are aimed at finding or breeding resistant varieties. The resistance of the rice plant against leaf blight was studied first from a morphological viewpoint. Rice plants having short narrow leaf blades were considered generally to be resistant because the degree of coverage is low in the later growth stages (40). When the degree of coverage is high, humidity among plants increases, contributing to disease development. In plants with heavy foliage, leaves contact each other, resulting in many wounds which favor the entry of the pathogen.

Resistance estimated by the above criteria is not accurate, and consequently the inoculation method for testing resistance was devised. Several methods for inoculation are now employed (132, 140, 149). These are the multi- or single needle-prick inoculation (132), spray inoculation against seedlings or mature rice plants, and dip inoculation (132, 140) against seedlings. The resistance ratings obtained by these methods are in close agreement with those observed under natural conditions (123, 132, 149). All of these techniques, especially the needle-prick method, are now widely employed not only by plant pathologists but also by rice breeders.

The resistance of many Japanese varieties is now known. Common varieties cultivated in endemic areas of Japan are classified into resistant, moderately resistant, and susceptible groups (37, 80, 123, 124, 132, 140, 149). Genes controlling resistance against the disease have been analysed (80, 124). Washio and his co-workers (124) grouped the rice varieties on an arbitrary scale of I (resistant) to IV (susceptible). They concluded that the resistance of the variety Kogyoku belonging to group III was controlled by two complementary dominant genes, X_1 and X_2. X_1 acted with the gene X_3 found in the variety Shimozuki which belonged to group IV. X_2 does not act with X_3. It was found that the resistance of some varieties against lesion enlargement seemed to be controlled by polygenes. Sakaguchi and others (80) classified the rice varieties including Indica type into three groups; Rantaj-emas group (resistant), Kidama (Kogyoku) group (moderately resistant), and Kinmaze group (susceptible). Rantaj-emas group varieties have both Xa_1 and Xa_2 dominant genes which control resistance. The varieties belonging to the Kidama group have only Xa_1 gene, and Kinmaze group varieties possessed none.

Gene analysis like this is necessary to clarify the genes controlling resistance.

Several kinds of varieties resistant to bacterial blight were bred in Japan. These varieties originated from Kanto 35, Shigasekitori 11, or Syobei (10). The resistant varieties Norin 27, Asakaze, Hayatomo, and Nishikaze were bred from Kanto 35, varieties Hoyoku, Kokumasari, Shiranui, and Ooyodo from Shigasekitori (11), and Nihonbare and Sachikaze from Syobei. Although these resistant varieties are cultivated widely in Japan, it should be emphasized that they are not always immune to attack because very virulent strains of the bacteria occur which can attack every variety under natural conditions.

Disease Forecasting

In general, early occurrence of the disease results in severe epidemics at later growth stages of the rice plant. Exact forecasting is difficult because epidemics are largely dependent on climatic and several other environmental conditions. Several methods used for forecasting the disease include following early disease development, observation of climatic conditions, and periodic investigation of the population of the pathogenic bacteria or phages distributed in the plant or in irrigation water. None of these methods are completely satisfactory at present.

Forecasting by natural infection.—Observation of the disease at the end of the nursery stage is important to forecast any future disease outbreaks. A periodic survey of the disease occurrence is recommended in a paddy field where several representative resistant and susceptible varieties are cultivated, as any disease occurrence in the field could give an indication of its outbreak in the area. The varieties should be selected from those grown in each area (2), and excessive nitrogen should be applied in the forecasting field.

Forecasting by climatic conditions.—Some correlation is found in Japan between rainfall and sunshine in July and August (14, 79), mean temperatures recorded at ten in the morning (36), the number and the severity of typhoons (11, 31), and the disease severity as has been discussed. If typhoons in the middle or later rice growing period were taken into consideration, the degree of the disease in later stages could be forecast more accurately. In the cooler regions of Japan, it seems that low temperatures in summer and high rainfall accompanied by low temperatures in September reduce the disease occurrence (134).

Forecasting by bacteriophage population.—It has been noticed that the bacteriophage specific to *X. oryzae* in the paddy field or irrigation water increases prior to an outbreak of the disease. A method of forecasting by means of phage population in nursery, paddy field, or irrigation water has been recently used (94, 95, 97, 100, 114, 139). Paddy field or irrigation water is collected at selected points, and 0.1 to 1.0 ml of it is plated for plaque counting. The bacterial strain which is susceptible to the phages predominant in the area should be used as indicator for plaque counting. It is suggested that the water sample be collected in the early hours of the morning as the phage population usually decreases during the later hours of the day. This decrease is probably due to sunshine or high temperature. Rainfall increases the population of phages in water. Probably this is caused by the washing of phages from the leaves on which they propagate abundantly.

The relationship between the population of phages in nursery water and the disease severity in the paddy field stage has been investigated (97, 114). Little or no disease was observed in paddy fields where the seedlings were transplanted from a nursery where no phages were detected. The disease in paddy fields was severe when seedlings from the nurseries where phages were detected early had been used. Thus, it is possible to some extent to forecast the disease occurrence in the paddy field by means of investigating phage populations in the nursery water. It should be mentioned, however, that a flood in a paddy field or a typhoon during the later growth stages of the rice plant often disturbs the forecasting accuracy at the nursery stage.

In most of the paddy fields in which phages were not detected until the end of July, the disease was not observed until the middle of October. Generally, the higher the number of phages in the field, the earlier and more severe is the outbreak of the disease (100). The time at which the phage population in the field water should be investigated depends upon the growth stages of the rice plant. This time is not later than the middle of August in the Kyushu district, and is a little earlier in the northern parts of Japan (139). Phage population at the site of outflow of irrigation water is generally higher than that at the inflow, and consequently the sample should be taken at the outflow.

Investigation of the phage population in the irrigation water running along canals or rivers is well suited for forecasting the disease occurrence in a large area (94, 95, 139). In the paddy field stage, the phage population in river water also is closely related to the earliness and severity of the disease in the area. The most effective application of the phage method for forecasting requires that the points for sampling be selected carefully in view of the mutual relationship between irrigation field waters and rivers.

Forecasting of bacterial blight is important for using chemicals for protection.

Chemicals for Control

At first Bordeaux mixture was recommended to be sprayed prior to the occurrence of the disease (6, 21, 54, 79, 85, 104). Copper and mercuric compounds were then tested to control the disease (35, 41, 42, 70, 108, 129). None of them, however, were effective enough for practical use. Since 1955, antibiotics for the control of the disease have also been tested in many experiment stations. Among these, streptomycin and its derivatives were more effective than the other chemicals (57, 82, 146, 148). These antibiotics, however, caused injurious effects on rice yields when sprayed at the heading stage (3, 148). The harmful effects of streptomycin have been considered to be due to inhibition of translocation of accumulated starch (3) and the absorption of manganese (71).

Screening methods for chemicals.—Various chemicals have been tested in the laboratory by the cup method, inhibition zone method using filter paper discs, colorimetric method, and agar dilution methods. The results of laboratory tests, however, do not coincide with applications of chemicals to the diseased plant. Mercuric compounds were found to be more effective than copper compounds in laboratory tests (39), but these were less effective than copper when sprayed on the plant. There are many antibiotics which have been proved effective in laboratory tests, but very few are effective when applied to the plant (117). In addition to these anomalies, the results of plant application tests sometimes differ from those of field tests. For effective screening of chemicals, therefore, the seedling test followed by the field test is commonly used.

Seedling test: Test chemicals are pre- or post-sprayed on inoculated seedlings, and the effects are estimated by the inhibition of disease lesion appearance. Many kinds of inoculation methods have been employed as in the case of the tests for varietal resistance (51, 117, 146, 148). Apart from the method of suppressing lesion enlargement, other methods have also been used for estimating the effectiveness of chemicals. The bacterial exudation method (29) has been reported to be practical for screening. The technique of bacterial swarming to rice roots has been employed to test the physiological activity of chemicals in relation to the rice plant (62). All these techniques can be used easily on a small scale, and many chemicals can thus be tested in a short period.

Field test: Final testing of the effectiveness of the chemicals should be done under natural field conditions. Much useful data may be obtained in this manner concerning the suppression of disease development, phytotoxicity or other harmful effects on yields, optimum concentration, frequency, and periods of spraying (26, 34, 39, 41, 96, 101, 129, 131, 137, 148).

Chemicals for practical use.—In recent years, several kinds of chemicals have been developed from repeated trials in the field. They are Cellocidin, Shirahagen (L-chloramphenicol), Sankel (nickel-dimethyldithiocarbamate), Delan (dithianon), Phenazine, and Celdion (fentiazon). Although these chemicals are available on a commercial scale in Japan, more effective ones are urgently required because of their insufficient effectiveness.

The biochemical mechanisms of some chemicals have been reported. Cellocidin at a concentration of 4 ppm completely inhibits the growth of *X. oryzae* in liquid medium. Exogeneous respiration of *X. oryzae* was almost completely inhibited at 10 ppm. However, inhibition was slight even at 100 ppm when succinate was used as a substrate (75). Use of various metabolic intermediates as substrates indicated cellocidin was an inhibitor of the EMP pathway, especially of the α-ketoglutarate to succinate system. Selective inhibition by cellocidin against NAD dehydrogenase, e.g., α-ketoglutarate dehydrogenase, glutamic dehydrogenase, and malic dehydrogenase has also been reported (76).

Phenazine seems to be incorporated into the respiratory regulating system of the bacterium where it suppresses the rate of respiration (83).

Chloramphenicol is a famous inhibitor of protein synthesis. However, its mechanism of action against *X. oryzae* is unknown. Biochemical mechanisms of other effective chemicals are unknown at present.

Chemical application.—Seed transmission of the disease is not certain in Japan. However, seed disinfection is advisable, as this has been reported from several other countries. Dry seed should be soaked in a mercuric compound or antibiotic solution having a bactericidal effect.

Leaf blight control by chemicals should be done both in the nursery and in the paddy fields. Chemical spraying in the nursery stage is effective in inhibiting bacterial multiplication on seedlings and retards the occurrence of disease in the paddy fields (96, 101, 131). One to two applications of the chemicals just before transplanting is recommended. The method of soaking seedlings in chemical solutions before transplanting has also been recommended (26, 96). This method, though effective to some extent, can not be applied on a practical scale yet.

In paddy fields, the application time and concentration should be determined for each chemical according to its properties and bacterial population in the field (137). Chemical application in general must be done two to three times at one-week intervals during the maximum tillering to booting stage. Some chemicals such as Cellocidin, Sankel, and Phenazine should not be sprayed at the heading stage, because they sometimes show browning of the husks under certain climatic conditions and a decrease in yield.

Control Measures

The use of resistant varieties and improvement of cultural practices such as nursery preparation, management of irrigation water, and application of fertilizer, have already been discussed. Cultural practices required to control the disease are summarized as follows:

Before nursery preparation stage:

(*a*) Use the most resistant variety suitable for each area.

(*b*) Use seed from paddy fields showing no disease.

(*c*) Treat seed with mercury compounds or other seed disinfectants.

(*d*) Nursery should be well drained; no flooding should occur in a heavy downpour; avoid deep irrigation water, or sow in an upland nursery if possible.

(*e*) Remove infection sources such as diseased rice straw from paddy fields, eradicate *Leersia* plants around paddy fields or in irrigation canals. Plough in rice stubble during winter.

In nursery stage:

(*a*) Keep irrigation water level as low as possible to protect the seedlings from flooding.

(*b*) Spray one to two applications of chemicals in late nursery stage as discussed earlier.

In paddy field stage:

(*a*) Avoid excessive nitrogen in fertilizer application; top-dressings should be divided into several small applications.

(*b*) Spray two to three applications of chemicals at the maximum tillering to booting stage. After a typhoon or flood, spraying should be done as soon as possible.

These control measures are recommended generally in Japan (102). They may have to be modified in other countries to some extent to suit local conditions.

LITERATURE CITED

1. Adams, M. H. *Bacteriophages* (International Publishers Inc., New York, 592 pp., 1959)
2. Aoyagi, K., Osaki, M., Kinemuchi, S. Resistance of the predominant rice varieties against bacterial leaf blight in Niigata Prefecture (in Japanese). *Proc. Assoc. Plant Protect., Hokuriku,* **8,** 28–31 (1960)
3. Arata, T., Hori, M., Inoue, Y. Harmful effect of streptomycin against rice plant (abstr. in Japanese). *Ann. Phytopathol. Soc. Japan,* **26,** 78 (1961)
4. *Bergey's Manual of Determinative Bacteriology* (7th ed., Williams & Wilkins, Baltimore, Md., 1094 pp., 1957)
5. Bhapkar, D. G., Kulkarni, N. B., Chavan, V. M. Bacterial blight of paddy. *Poona Agr. Coll. Mag.,* **51,** 36–46 (1960)
6. Bokura, U. Bacterial leaf blight of rice (in Japanese). *Teikoku Nokaiho,* **2**(9), 62–66; **2**(10), 54–57; **2**(12) 58–61 (1911)
7. Fang, C. T., Lin, C. F., Chu, C. L. The inhibition of glucose to the growth of *Xanthomonas oryzae* (in Chinese with English summary). *Acta Phytopathol. Sinica,* **3,** 125–36 (1957)
8. Fang. C. T., Lin, C. F., Chu, C. L. A preliminary study on the disease cycle of the bacterial leaf blight of rice (in Chinese with

English summary). *Acta Phytopathol. Sinica,* **2,** 173–85 (1956)

9. Fang, C. T., Ren, H. C., Chen, T. Y., Chu, Y. K., Faan, H. C., Wu, S. C. A comparison of the rice bacterial leaf blight organism of rice and *Leersia hexandra* Swartz. *Acta Phytopathol. Sinica,* **3,** 99–124 (1957)
10. Fujii, K. Breeding of the resistant rice varieties against bacterial leaf blight (in Japanese). *Plant Protect., Japan,* **22,** 113–15 (1968)
11. Fujikawa, T., Okadome, Z., Utsunomiya, T. Relationship between the climatic conditions and the occurrence of bacterial leaf blight of rice in Oita Prefecture (in Japanese). *Agr. Meteorol., Japan,* **12,** 148–50 (1957)
12. Goto, K., Fukatsu, R., Ohata, K. Natural occurrence of bacterial leaf blight of rice on weeds (abstr. in Japanese). *Ann. Phytopathol. Soc. Japan,* **17,** 154 (1953)
13. Goto, K., Fukatsu, R., Ohata, K. Relationship between the occurrence of bacterial leaf blight disease on rice and that on wild grasses in endemic areas (in Japanese). *Plant Protect., Japan,* **7,** 365–68 (1953)
14. Goto, K., Inoue, Y., Fukatsu, R., Ohata, K. Field observations on the outbreak and fluctuation of severity of bacterial leaf blight of rice plant (in Japanese with English summary). *Bull. Tokai-Kinki Agr. Expt. Sta.,* **2,** 53–68 (1955)
15. Goto, M. "Kreseck" and pale yellow leaf, systemic symptom of bacterial blight of rice caused by *Xanthomonas oryzae* (Uyeda & Ishiyama) Dowson. *Plant Disease Reptr.,* **48,** 858–61 (1964)
16. Goto, M. On the bacterial disease of rice in South-East Asian countries. (Abstr. in Japanese), *Ann. Phytopathol. Soc. Japan,* **29,** 291 (1964)
17. Goto, M. Bacteriophages of *Xanthomonas oryzae,* the pathogen of bacterial leaf blight of rice, collected in the Philippines. *Bull. Fac. Agr., Shizuoka Univ. (Iwata, Japan),* **15,** 31–38 (1965)
18. Goto, M. A technique for detecting the infected area of bacterial leaf blight of rice caused by *Xanthomonas oryzae* before symptom appearance. *Ann. Phytopathol. Soc. Japan,* **30,** 37–41 (1965)
19. Goto, M. Phage-typing of the causal bacteria of bacterial leaf blight (*Xanthomonas oryzae*) and bacterial leaf streak (*X. translucens* f. sp. *oryzae*) of rice. *Ann. Phytopathol. Soc. Japan,* **30,** 253–57 (1965)
20. Goto, M. Resistance of rice varieties and species of wild rice to bacterial leaf blight and bacterial leaf streak diseases. *The Philippine Agriculturalist,* **48,** 329–38 (1965)
21. Hashioka, Y. Bacterial leaf blight of rice and its control (in Japanese). *Agr. Hort. (Tokyo),* **26,** 644–48 (1951)
22. Ikari, H., Wakimoto, S. On the properties of new *Xanthomonas oryzae* phage (OP_{1h}) isolated at the National Agricultural Experiment Station of Kyushu (in Japanese with English summary). *Proc. Assoc. Plant Protect., Kyushu,* **4,** 38–40 (1958)
23. Ikeno, S. Bacterial leaf blight of rice and the damage in Niigata Prefecture (in Japanese). *Proc. Assoc. Plant Protect., Hokuriku,* **6,** 13–14 (1958)
24. Inoue, Y., Goto, K., Ohata, K. Overwintering and mode of infection of leaf blight bacteria on rice plant (in Japanese with English summary). *Bull. Tokai-Kinki Agr. Expt. Sta.,* **4,** 78–82 (1957)
25. Inoue, Y., Tsuda, Y. Assessment of the decrease in yield due to bacterial leaf blight of rice (in Japanese with English summary). *Bull. Tokai-Kinki Agr. Expt. Sta.,* **6,** 154–67 (1959)
26. Inoue, Y., Tsuda, Y. Effective chemicals for control the bacterial leaf blight of rice and their application time (in Japanese). *Proc. Assoc. Plant Protect., Kansai,* **7,** 59 (1965)
27. Isaka, M. Microscopic observation of the bacterial exudation from inoculated rice leaves for detection of *Xanthomonas oryzae* (in Japanese). *Proc. Assoc. Plant Protect., Hokuriku,* **12,** 26–30 (1964)
28. Isaka, M. Relationship between the time of transplanting and the occurrence of bacterial leaf blight of rice (in Japanese). *Proc. Assoc. Plant Protect., Hokuriku,* **14,** 33–36 (1966)
29. Isaka, M. Screening of the chemicals for control of bacterial leaf blight disease of rice (in Japanese).

Plant Protect., Japan, **22,** 24–28 (1968)

30. Isaka, M., Nosaki, S. Lysotypes of *Xanthomonas oryzae* isolates collected in Fukui Prefecture (in Japanese). *Proc. Assoc. Plant Protect., Hokuriku,* **9,** 30–32 (1961)
31. Ishiyama, S. Studies on bacterial leaf blight of rice (in Japanese). *Report, Agr. Expt. Sta.,* **45,** 233–61 (1922)
32. Ito, H., Takahashi, S. Bacteriophages of *Xanthomonas oryzae* in Syonai area, Yamagata Prefecture (in Japanese). *Proc. Assoc. Plant Protect., Kita-Nihon,* **15,** 40–41 (1964)
33. Iwase, S. Environmental conditions related to the occurrence of bacterial leaf blight of rice (in Japanese); *Agr. Hort. (Tokyo)* 35, 505–9 (1960)
34. Iwase, S., Amano, T. Change in sensitivity of rice plant to bacterial leaf blight disease and control effect of chemicals (in Japanese). *Proc. Assoc., Plant Protect., Kansai,* **4,** 52–53 (1962)
35. Kido, Y., Kobayashi, K., Yoshiyama, N. Therapeutical studies on bacterial leaf blight of rice plant (1); (in Japanese). *Kyushu Agr. Res.,* **11,** 114–16 (1953)
36. Kiryu, T. Analysis of environmental conditions related to the occurrence of bacterial leaf blight of rice in endemic areas (abstr. in Japanese). *Ann. Phytopathol. Soc. Japan,* **18,** 168 (1953)
37. Kiryu, T., Kuhara, S. Studies on the varietal resistance trials to the bacterial leaf blight of rice plant (in Japanese with English summary). *Kyushu Agr. Res.,* **13,** 9–14 (1954)
38. Kiryu, T., Kurita, T. Relationship between the time of drainage of paddy field water and the lesion enlargement of bacterial leaf blight of rice (abstr. in Japanese). *Ann. Phytopathol. Soc. Japan,* **20,** 31–32 (1955)
39. Kiryu, T., Mizuta, H. Effects of the various fungicides and chemicals on *Bacterium oryzae* (in Japanese with English summary). *Bull. Kyushu Agr. Expt. Sta.,* **2,** 349–59 (1954)
40. Kiryu, T., Mizuta, H. On the relation between habits of rice plant and varietal resistance against bacterial leaf blight (1); (in Japanese with English summary). *Kyushu Agr. Res.,* **15,** 54–56 (1955)
41. Kiryu, T., Mizuta, H. Control measures for rice diseases after heavy wind or flooding (in Japanese). *Plant Protect., Japan,* **9,** 273–76 (1955)
42. Kiryu, T., Mizuta, H. Results of the experiments on chemical effect against bacterial leaf blight and sheath blight of rice (in Japanese). *Plant Protect., Japan,* **9,** 329–31 (1955)
43. Kojima, K., Iwase, S., Amano, T. Studies on the environmental conditions related to the occurrence of rice blast and bacterial leaf blight (5); (in Japanese). *Bull. Aichi Agr. Expt. Sta.,* **14,** 71–80 (1959)
44. Kojima, K., Iwase, S., Inoki, K. Studies on environmental conditions related to the occurrence of blast and bacterial leaf blight of rice (1); (in Japanese). *Bull. Aichi Agr. Expt. Sta.,* **9,** 157–61 (1954)
45. Kojima, K., Iwase, S., Inoki, K. Relationship between the soil type and the occurrence of bacterial leaf blight disease of rice (abstr. in Japanese). *Ann. Phytopathol. Soc. Japan,* **19,** 163 (1955)
46. Kojima, K., Todo, M., Iwase, S. Studies on environmental conditions related to the occurrence of blast and bacterial leaf blight of rice (2); (in Japanese). *Bull. Aichi Agr. Expt. Sta.,* **10,** 1–13 (1955)
47. Kuhara, S., Kurita, T., Tagami, Y., Fujii, H., Sekiya, N. Studies on the strain of *Xanthomonas oryzae* (Uyeda et Ishiyama) Dowson, the pathogen of the bacterial leaf blight of rice, with special reference to its pathogenicity and phage-sensitivity (in Japanese with English summary). *Bull. Kyushu Agr. Expt. Sta.,* **11,** 263–312 (1965)
48. Kuhara, S., Sekiya, N., Tagami, Y. On the pathogen of bacterial leaf blight of rice isolated from severely affected area where resistant variety was widely cultivated (abstr. in Japanese). *Ann. Phytopathol. Soc. Japan,* **22,** 9 (1957)
49. Kumamoto, Y., Niwa, J. Sensitivities of rice plants grown in the areas having different soil types to bacterial leaf blight (abstr. in Japan-

ese). *Ann. Phytopathol. Soc. Japan,* **24,** 52–53 (1959)

50. Kuo, T. T., Huang, T. C., Wu, R. Y., Yang, C. M. Characterization of three bacteriophages of *Xanthomonas oryzae* (Uyeda et Ishiyama) Dowson. *Botan. Bull. Acad. Sinica,* **8,** 246–54 (1967)
51. Kurita, T., Kuhara, S., Fujii, H., Tagami, Y. The seedling screening method for bactericidal agents against the bacterial leaf blight of rice plant caused by *Xanthomonas oryzae* (in Japanese). *Proc. Assoc. Plant Protect., Kyushu,* **6,** 68–71 (1960)
52. Kusaba, T., Watanabe, M., Tabei, H. Classification of the strains of *Xanthomonas oryzae* (Uyeda et Ishiyama) Dowson on the basis of their virulence against rice plants (in Japanese with English summary). *Nogyo Gijutsu Kenkyusho Hokoku, Ser. C,* **20,** 67–82 (1966)
53. Kuwazuka, K. On the pathogenicity and aggregation reaction of the bacteria causing leaf blight of rice (abstr. in Japanese). *Ann. Phytopathol. Soc. Japan,* **2,** 169 (1928)
54. Kuwazuka, K. Bacterial leaf blight of rice (in Japanese). *Nogyo,* **634,** 14–27 (1933)
55. Kuwazuka, K. On the primary infection of bacterial leaf blight of rice (in Japanese). *Nogei, Aichi Agr. Expt. Sta.,* **175,** 321–25 (1938)
56. Kuwazuka, K. Bacterial leaf blight disease of rice (in Japanese). *Nogyo,* **741,** 64–73 (1942)
57. Maki, Y., Murakami, S., Kono, H. Experiments on the control effect of chemicals against bacterial leaf blight of rice (in Japanese). *Rept. Ehime Agr. Expt. Sta.,* **11,** 20–22 (1960)
58. Matsuda, M. Bacterial leaf blight disease of rice (in Japanese). *Dainihon Noho,* **168,** 33–35 (1928)
59. Matsumoto, S., Tabei, H. (Personal communication, 1968)
60. Mizukami, T. Studies on the bacterial leaf blight of rice plant on the entrance and multiplicating portion of pathogen upon the rice plant leaves (preliminary report); (in Japanese with English summary). *Sci. Bull. Fac. Agr., Saga Univ.,* **4,** 169–75 (1956)
61. Mizukami, T. On the relationship between *Xanthomonas oryzae* and roots of rice seedlings (in Japanese with English summary). *Sci. Bull. Fac. Agr. Saga Univ.,* **6,** 87–94 (1957)
62. Mizukami, T. On the screening of the chemicals for control of bacterial leaf blight disease of rice (in Japanese). *Noyaku,* **7**(6), 12–18 (1960)
63. Mizukami, T. Studies on the ecological properties of *Xanthomonas oryzae* (Uyeda et Ishiyama) Dowson, the causal organism of bacterial leaf blight of rice plant (in Japanese with English summary), *Sci. Bull. Fac. Agr. Saga Univ.,* **13,** 1–85 (1961)
64. Mizukami, T., Seki, M. Studies on the bacterial leaf blight of rice plant. On the distribution of *Bacterium oryzae* (Uyeda et Ishiyama) Nakata upon the rice plants (preliminary report); (in Japanese with English summary). *Kyushu Agr. Res.,* **15,** 57–59 (1955)
65. Mizukami, T., Wakimoto, S., Uematsu, T. On the properties of *Xanthomonas oryzae* phages isolated from India (abstr. in Japanese), *Ann. Phytopathol. Soc. Japan,* **29,** 260 (1964)
66. Mizukami, T., Yoshimura, S. (Personal inspection 1963, 1967)
67. Mizuta, H. Studies on cultivation of *Bacterium oryzae* (Uyeda et Ishiyama) Nakata (preliminary report) (in Japanese). *Ann. Phytopathol. Soc. Japan,* **17,** 73–75 (1953)
68. Muko, H., Isaka, M. Re-examination of some physiological characteristics of *Xanthomonas oryzae* (Uyeda et Ishiyama) Dowson (in Japanese with English summary). *Ann. Phytopathol. Soc. Japan,* **29,** 13–19 (1964)
69. Muko, H., Kusaba, T., Watanabe, M., Tabei, H. Several factors related to the occurrence of bacterial leaf blight disease of rice (in Japanese). *Proc. Assoc. Plant Protect., Kanto-Tosan,* **4,** 7–8 (1957)
70. Nakazawa, M. Studies on the mechanisms of organic mercury compounds as disinfectant (in Japanese). *Bull. Aichi Agr. Expt. Sta.,* **15,** 6–31 (1959)
71. Nasuda, K., Katsumi, F. Studies on the effect of chemicals to disease

resistance of plants (5) (in Japanese). *Proc. Assoc. Plant Protect., Hokuriku,* **8,** 41–43 (1961)

72. Nasuda, K., Takeuchi, Y. Effects of the mode of nursery and time of nutritional deficiency on the occurence of bacterial leaf blight of rice (in Japanese). *Proc. Assoc. Plant Protect., Hokuriku,* **8,** 41–43 (1960)
73. Nishida, T. Bacterial leaf blight of rice (in Japanese). *Noji Zappo,* **127,** 68–75 (1909)
74. Nishimura, J., Udo, T. On the control effect of the chemicals against bacterial leaf blight of rice (in Japanese). *Proc. Assoc. Plant Protect., Kansai,* **9,** 6–14 (1967)
75. Okimoto, Y., Misato, T. Antibiotics as protectant bactericide against bacterial leaf blight of rice plant (2) (in Japanese with English summary). *Ann. Phytopathol. Soc. Japan,* **28,** 209–15 (1963)
76. Okimoto, Y., Misato, T. Antibiotics as protectant bactericide against bacterial leaf blight of rice plant (3) ; (in Japanese with English summary). *Ann. Phytopathol. Soc. Japan,* **28,** 250–57 (1963)
77. Reinking, O. A. Philippine economic-plant diseases. *Philippine J. Sci.,* **13,** 163–274 (1918)
78. Reitsuma, J., Schure, P. S. J. "Kresek", a bacterial disease of rice. *Contr. Gen. Agr. Res. Sta., Bogor,* **117,** 1–17 (1950)
79. *Results of the Experiments on Bacterial Leaf Blight of Rice* (in Japanese) Fukuoka Agr. Expt. Sta., Special Report, 148 pp., 1920)
80. Sakaguchi, S., Murata, N. Studies on the resistance to bacterial leaf blight, *Xanthomonas oryzae* (Uyeda et Ishiyama) Dowson, in the cultivated and wild rice (in Japanese with English summary). *Nogyo Gijutsu Kenkyusho Hokoku, Ser. D.,* **18,** 1–29 (1968)
81. Seki, M., Mizukami, T. On the longevity of *Bacterium* oryzae (Uyeda et Ishiyama) Nakata under some conditions (in Japanese). *Kyushu Agr. Res.,* **16,** 112 (1955)
82. Seki, M., Mizukami, T. Application of antibiotics against bacterial leaf blight of rice plant (in Japanese). *Kyushu Agr. Res.,* **17,** 98 (1956)
83. Sekizawa, Y., Watanabe, T., Oda, M. Effect of phenazine against rice leaf blight bacterium, and its biochemical mechanism. *Ann. Phytopathol. Soc. Japan,* **30,** 145–52 (1965)
84. Shekhawat, G. S., Srivastava, D. N. Variability in Indian isolates of *Xanthomonas oryzae* (Uyeda and Ishiyama) Dowson, the incitant of bacterial leaf blight of rice. *Ann. Phytopathol. Soc. Japan,* **34,** 289–97 (1968)
85. Soga, N. On the bacterial leaf blight disease of rice in Kumamoto Prefecture (in Japanese). *Byochugai Zassi,* **5,** 543–49 (1918)
86. Srivastava, D. N. Epidemiology of bacterial blight of rice and its control in India. *Proc. Symp. Rice Diseases and Their Control by Growing Resistant Varieties and Other Measures, Tokyo, 1967,* 11–18
87. Srivastava, D. N., Rao, Y. P. Epidemic of bacterial blight disease of rice in north India. *Indian Phytopathol.,* **16,** 393–94 (1963)
88. Srivastava, D. N., Rao, Y. P. Seed transmission and epidemiology of bacterial blight of rice in India. *Indian Phytopathol.,* **17,** 77–78 (1964)
89. Suwa, T. Studies on the culture media of *Xanthomonas oryzae* (Uyeda et Ishiyama) Dowson (in Japanese with English summary). *Ann. Phytopathol. Soc. Japan,* **27,** 165–71 (1962)
90. Tabei, H. Anatomical studies of rice plant affected with bacterial leaf blight, with special reference to stomatal infection at the coleoptile and the foliage leaf sheath of rice seedling (in Japanese with English summary). *Ann. Phytopathol. Soc. Japan,* **33,** 12–16 (1967)
91. Tabei, H. Infection and multiplication of the bacteria causing leaf blight disease of rice (in Japanese). *Plant Protect., Japan,* **22,** 9–11 (1968)
92. Tabei, H., Muko, H. Anatomical studies of rice plant leaves affected with bacterial leaf blight, in particular reference to the structure of water exudation system (in Japanese with English summary). *Nogyo Gijutsu Kenkyusho Hokoku, Byori Konchu,* **11,** 37–43 (1960)

93. Tagami, Y. Relationship between the occurrence of bacterial leaf blight of rice and the populations of causal bacteria and its phages in rice growing seasons (in Japanese). *Plant Protect., Japan,* **13,** 389–94 (1959)

94. Tagami, Y., Fujii, H., Kuhara, S., Kurita, T. Relationship between the population of *Xanthomonas oryzae* phage in irrigation water in nursery stage and the occurrence of bacterial leaf blight of rice (abstr. in Japanese). *Ann. Phytopathol. Soc. Japan,* **26,** 56–57 (1961)

95. Tagami, Y., Fujii, H., Kuhara, S., Kurita, T. Relationship between the population of *Xanthomonas oryzae* phage in irrigation water in paddy field stage and the occurrence of bacterial leaf blight of rice (abstr. in Japanese), *Ann. Phytopathol. Soc. Japan,* **26,** 57 (1961)

96. Tagami, Y., Fujii, H., Kuhara, S., Kurita, T. Chemical treatments of rice seedlings for the control of bacterial leaf blight (in Japanese with English summary). *Proc. Assoc. Plant Protect., Kyushu,* **7,** 25–29 (1961)

97. Tagami, Y., Fujii, H., Kuhara, S., Kurita, T. Relationship between the population of *Xanthomonas oryzae* phage in nursery water and the occurrence of bacterial leaf blight of rice in paddy fields (abstr. in Japanese). *Ann. Phytopathol. Soc. Japan,* **24,** 6 (1959)

98. Tagami, Y., Kuhara, S., Kurita, T., Fujii, H., Sekiya, N., Sato, T. Epidemiological studies on the bacterial leaf blight of rice, *Xanthomonas oryzae* (Uyeda et Ishiyama) Dowson (2) (in Japanese with English summary). *Bull. Kyushu Agr. Expt. Sta.,* **10,** 23–50 (1964)

99. Tagami, Y., Kuhara, S., Kurita, T., Fujii, H., Sekiya, N., Yoshimura, S., Sato, T., Watanabe, B. Epidemiological studies on the bacterial leaf blight of rice, *Xanthomonas oryzae* (Uyeda et Ishiyama) Dowson (1) (in Japanese with English summary). *Bull. Kyushu Agr. Expt. Sta.,* **9,** 89–122 (1963)

100. Tagami, Y., Kuhara, S., Kurita, T., Sekiya, N. Relation between the population of *Xanthomonas oryzae* phage in paddy field water and the occurrence of bacterial leaf blight (in Japanese). *Proc. Assoc. Plant Protect., Kyushu,* **4,** 63–64 (1958)

101. Tagami, Y., Kuhara, S., Tabei, H., Kurita, T. On the timing of chemical spraying for the control of bacterial leaf blight of rice (*Xanthomonas oryzae*) ; (in Japanese). *Proc. Assoc. Plant Protect., Kyushu,* **12,** 92–95 (1966)

102. Tagami, Y., Mizukami, T. Historical review of the researches on bacterial leaf blight of rice caused by *Xanthomonas oryzae* (Uyeda et Ishiyama) Dowson. *Special Report of the Plant Diseases and Insect Pests Forecasting Service,* No. 10 (Plant Protection Division, Ministry of Agriculture and Forestry, Japan, 1962)

103. Tagami, Y., Yoshimura, S., Watanabe, B. On the symptoms of bacterial leaf blight of rice (abstr. in Japanese). *Ann. Phytopathol. Soc. Japan,* **22,** 8 (1957)

104. Takaishi, M. First report, studies on bacterial leaf blight of rice (in Japanese). *Dainihon Nokaiho,* **340,** 53–58 (1909)

105. Takeuchi, H. On the occurrence of bacterial leaf blight of rice (in Japanese). *Bull. Chosen Sotokuhu Agr. Expt. Sta.,* **5**(1), 62–64 (1930)

106. Tanaka, Y. Studies on the specialization of the strains in *Xanthomonas oryzae* (Uyeda et Ishiyama) Dowson (in Japanese with English summary). *Proc. Assoc. Plant Protec., Kyushu,* **6,** 66–68 (1960)

107. Tokunaga, Y., Hashioka, Y. List of crop diseases in Hainan island (in Japanese). *Noho, Taiwan Agr. Expt. Sta.,* 2–3 (1949)

108. Tomonaga, Y., Isaka, M. Relationship between the timing of chemical spraying and the control effect against bacterial leaf blight of rice (in Japanese). *Proc. Assoc. Plant Protect., Hokuriku,* **4,** 37 (1956)

109. Wakimoto, S. Biological and physiological properties of *Xanthomonas oryzae* phage (in Japanese with English summary). *Sci. Bull. Fac. Agr. Kyushu Univ.,* **14,** 485–93 (1954)

110. Wakimoto, S. The determination of the presence of *Xanthomonas*

oryzae by the phage technique (in Japanese with English summary). *Sci. Bull. Fac. Agr. Kyushu Univ.*, **14**, 495–98 (1954)

111. Wakimoto, S. Studies on the multiplication of OP_1 phage (*Xanthomonas oryzae* bacteriophage) (1) (in Japanese with English summary. *Sci. Bull. Fac. Agr., Kyushu Univ.*, **15**, 151–60 (1955)
112. Wakimoto, S. Overwintering of *Xanthomonas oryzae* on unhulled grains of rice (in Japanese). *Agr. Hort. (Tokyo)*, **30**, 1501 (1955)
113. Wakimoto, S. Overwintering of *Xanthomonas oryzae* in soil (in Japanese), *Agr. Hort. (Tokyo)*, **31**, 1413–14 (1956)
114. Wakimoto, S. Considerations on the overwintering and the infection mechanisms of *Xanthomonas oryzae* (in Japanese). *Plant Protect., Japan*, **10**, 421–424 (1956)
115. Wakimoto, S. Relationship between the root of some plants and the bacteria causing leaf blight of rice (in Japanese with English summary). *Proc. Assoc. Plant Protect., Kyushu*, **3**, 2–5 (1957)
116. Wakimoto, S. Classification of strains of *Xanthomonas* oryzae on the basis of their susceptibility to bacteriophages (in Japanese with English summary). *Ann. Phytopathol. Soc. Japan*, **25**, 193–98 (1960)
117. Wakimoto, S. Screening of the chemicals effective for control of bacterial leaf blight of rice (in Japanese). *Proc. Assoc. Plan Protect., Kanto-Tozan*, **9**, 3 (1962)
118. Wakimoto, S. Strains of *Xanthomonas oryzae* in Asia and their virulence against rice varieties. *Proc. Symp. Rice Diseases and Their Control by Growing Resistant Varieties and Other Measures, Tokyo, 1967*, 11–18
119. Wakimoto, S. (Personal inspection, 1967)
120. Wakimoto, S., Tamari, K. Studies on the overwintering of *Xanthomonas oryzae* in dried state (preliminary report); (in Japanese with English summary). *Proc. Assoc. Plant Protect., Kyushu*, **2**, 107–9 (1956)
121. Wakimoto, S., Uematsu, T., Mizukami, T. (Unpublished data)
122. Wakimoto, S., Yoshii, H. Quantitative determination of the population of a bacteria by the phage technique (in Japanese with English summary). *Sci. Bull. Fac. Agr. Kyushu Univ.*, **15**, 161–69 (1955)
123. Washio, Y., Kariya, K., Nomura, T., Ishida, T. Applicability of multineedle prick inoculation method for the estimation of varietal resistance against bacterial leaf blight (in Japanese). *Chugoku Agr. Res.*, **2**, 27–30 (1956)
124. Washio, O., Kariya, K., Toriyama, K. Studies on breeding rice varieties for resistance to bacterial leaf blight (in Japanese with English summary), *Bull. Chugoku Agr. Expt. Sta., A*, **13**, 55–86 (1966)
125. Watanabe, B., Kurita, T. Overwintering of *Xanthomonas oryzae* on weeds (abstr. in Japanese). *Ann. Phytopathol. Soc. Japan*, **23**, 60 (1958)
126. Watanabe, M. Studies on the nutritional physiology of *Xanthomonas oryzae* (Uyeda et Ishiyama) Dowson (in Japanese with English summary). *Ann. Phytopathol. Soc. Japan*, **28**, 201–8 (1963)
127. Watanabe, M. Studies on the strains of *Xanthomonas oryzae* (Uyeda et Ishiyama) Dowson, the causal bacterium of the bacterial leaf blight of rice plant (in Japanese with English summary). *Bull. Fac. Agr. Tokyo Univ. Agr. Technol.*, **10**, 1–51 (1966)
128. Watanabe, T., Sekizawa, Y., Oda, M. Biochemical studies of the host-parasite relationship between *Xanthomonas oryzae* and rice plant (1); (in Japanese with English summary). *Ann. Phytopathol. Soc. Japan*, **33**, 32–37 (1967)
129. Yamanaka, I., Hioka, T., Otani, H. Chemical control of bacterial leaf blight of rice (in Japanese). *Bull. Shiga Agr. Expt. Sta.*, **3**, 29–32 (1959)
130. Yamanaka, T., Nakaya, K., Tominaga, T., Uchida, K. Effect of environments on the occurrence of bacterial leaf blight disease of rice (1); (abstr. in Japanese). *Ann. Phytopathol. Soc. Japan*, **16**, 191 (1952)
131. Yokoyama, S., Yoshida, K., Fukano, H. Control effect of chemical treatments in seedling stage against bacterial leaf blight of rice

plant (1); (in Japanese). *Proc. Assoc. Plant Protect., Kyushu,* **9,** 21–22 (1963)

132. Yoshida, K., Muko, H. Multineedle prick method for the estimation of varietal resistance against bacterial leaf blight disease of rice (in Japanese). *Plant Protect., Japan,* **15,** 343–46 (1961)

133. Yoshii, H., Yoshida, T., Matsui, C. Isolation of *Xanthomonas oryzae* phage (abstr. in Japanese). *Ann. Phytopathol. Soc. Japan,* **17,** 177 (1953)

134. Yoshimura, S. Bacterial leaf blight disease of rice in Hokuriku area (in Japanese). *Plant Protect., Japan,* **13,** 395–99 (1959)

135. Yoshimura, S. Considerations on overwintering of *Xanthomonas oryzae* (in Japanese). *Plant Protect., Japan,* **14,** 343–47 (1960)

136. Yoshimura, S. Diagnostic and ecological studies of rice bacterial leaf blight, caused by *Xanthomonas oryzae* (Uyeda et Ishiyama) Dowson (in Japanese with English summary). *Bull. Hokuriku Agr. Expt. Sta.,* **5,** 27–182 (1963)

137. Yoshimura, S. Chemical control of bacterial leaf blight disease of rice (in Japanese). *Plant Protect., Japan,* **22,** 29–32 (1968)

138. Yoshimura, S., Aoyagi, K., Morihashi, T. Comparative studies on *Xanthomonas oryzae* isolates in their pathogenicity against rice varieties and weeds (in Japanese). *Proc. Assoc. Plant Protect., Hokuriku,* **8,** 25–28 (1960)

139. Yoshimura, S., Aoyagi, K., Morihashi, T., Yoshino, M., Nishimura, H., Kinemuchi, S. Relationship between the population of *Xanthomonas oryzae* phage and the occurrence of bacterial leaf blight of rice, with special reference to forecasting disease outbreak for wide area by means of phage population in river water (in Japanese). *Proc. Assoc. Plant Protect., Hokuriku,* **8,** 31–41 (1960)

140. Yoshimura, S., Iwata, K. Studies on the methods for the estimation of varietal resistance of rice against bacterial leaf blight (1); (in Japanese). *Proc. Assoc. Plant Protect., Hokuriku,* **13,** 25–31 (1965)

141. Yoshimura, S., Iwata, K. On the abnormal growth of rice plant caused by bacterial leaf blight (3); (in Japanese). *Proc. Assoc. Plant Protect., Hokuriku,* **13,** 42–47 (1965)

142. Yoshimura, S., Iwata, K., Tahara, K. On the abnormal growth of rice plant caused by bacterial leaf blight (2); (in Japanese). *Proc. Assoc. Plant Protect., Hokuriku,* **13,** 40–42 (1965)

143. Yoshimura, S., Morihashi, T. Classification of *Xanthomonas oryzae* isolates by bacteriophage sensitivity, and their distribution in Hokuriku district (in Japanese). *Proc. Assoc. Plant Protect., Hokuriku,* **7,** 43–52 (1959)

144. Yoshimura, S., Morihashi, T., Suzuki, Y. Distribution of the important grasses for overwintering of *Xanthomonas oryzae,* and their diseased situations in the field (abstr. in Japanese). *Ann. Phytopathol. Soc. Japan,* **24,** 6 (1959)

145. Yoshimura, S., Saito, T., Yoshino, M., Morihashi, T. On the characteristics of OP_2, *Xanthomonas oryzae* phage (in Japanese). *Proc. Assoc. Plant Protect., Hokuriku,* **8,** 15–20 (1960)

146. Yoshimura, S., Sekiya, N. Inhibition effect of some antibiotics on the growth of rice bacterial leaf blight pathogen in the pot experiment (in Japanese). *Proc. Assoc. Plant Protect., Kyushu,* **3,** 5–7 (1957)

147. Yoshimura, S., Tahara, K. On the abnormal growth of rice plant caused by bacterial leaf blight (1); (in Japanese). *Proc. Assoc. Plant Protect., Hokuriku,* **10,** 29–32 (1962)

148. Yoshimura, S., Tabara, K., Aoyagi, K. On the control of bacterial leaf blight disease of rice by antibiotics (in Japanese). *Proc. Assoc. Plant Protect., Hokuriku,* **9,** 24–26 (1961)

149. Yoshimura, S., Yamamoto, T. Studies on the methods for the estimation of varietal resistance of rice against bacterial leaf blight (2); (in Japanese). *Proc. Assoc. Plant Protect., Hokuriku,* **14,** 23–25 (1966)

150. Yoshimura, S., Yoshino, M., Morihashi, T. Lysotypes of *Xanthomonas oryzae* and their pathogenicity (in Japanese). *Proc. Assoc. Plant Protect., Hokuriku,* **8,** 21–24 (1960)

Copyright 1969. All rights reserved

INSECT TISSUE CULTURES AS TOOLS IN PLANT VIRUS RESEARCH

By L. M. Black

Department of Botany, University of Illinois
Urbana, Illinois

Introduction

This review is not so much a survey of a well explored field as an introduction to an area in which research activity has just begun to expand. The progress and success of that work may depend to a considerable extent upon whether or not a discipline requiring facilities for both plant virus and insect virus investigations can find a niche in departments already established for work in only one field or the other.

During recent years, a number of reviews have appeared concerning the culture of insect tissues (18, 25, 26, 33, 41, 67, 69). Some recent reviews (26, 69) included sections on the subject of plant viruses in insect tissue cultures, and all contain much information relevant to workers in this area. This review concerns itself specifically with the subject of plant viruses in cultures of tissues and cells from vectors that serve as hosts to the viruses they transmit to plants. In addition to the conventional sections it contains four short original joint articles by either Schlegel, Whitcomb, Reddy, or Gamez, and myself, that present relevant experimental material accumulated by collaborative work over many years and not previously reported or only briefly reported before.

A number of early contributions have been of notable importance in helping to achieve the study of some plant viruses in cultured cells of their vectors. Coons' (17) fluorescent antibody staining technique has proved to be very important for the study of plant virus infections of insect cell cultures. It is also difficult to exaggerate the importance of Grace's demonstration (24) that insect cells could be grown continuously in tissue culture. More recently and more specifically the work of Hirumi & Maramorosch (29, 30) was most valuable in demonstrating the superiority of leafhopper embryos, in a specific stage of development in the egg, as a source of explants for tissue cultures. The medium developed by Schneider (56) for the maintenance of *Drosophila* cells proved, on slight modification, to be very suitable for the continuous culture of monolayers of cells from some agallian leafhoppers.

Much of this review will deal with wound-tumor virus (WTV) and potato yellow dwarf virus (PYDV). The techniques now available for work with these viruses are the product of a number of forward steps and findings over a long period, for example: density-gradient zonal centrifugation

(9), multiplication of virus in the vector (6), loss of transmissibility (3), production of antisera for the above leafhopper-borne viruses (7, 73), development of the ring-time test (71), and the application of immunofluorescence (49). Twelve weeks were once required to complete relatively crude infectivity assays of WTV by insect injection, whereas very precise assays are now possible on monolayers of host vector cells within 2 days. In the case of WTV and PYDV, most of the advances depended very much and very obviously upon prior advances. However, in future work with other such systems, many of these steps can probably be bypassed. This may be very important because most of the viruses with propagative transmission have not yet been transmitted by injecting vectors or by ingestion of virus from cell-free extracts. Consequently, most of these viruses have not yet been studied *in vitro*.

In this review, the loss of transmissibility of propagative plant viruses will be thoroughly discussed since this has direct and important bearings upon the use of host vector cells for work with these viruses.

Definitions.—It is desirable to define at the outset certain terms that will be used repeatedly:

(*a*) Host vector: a vector in which the virus multiplies before being transmitted.

(*b*) Propagative transmission of virus to plants: that in which the ingested virus multiplies in the vector.

(*c*) Vectorial (VI) virus: one which is fully transmissible by its vector. (Such a virus is also referred to as the field or wild strain of the virus on the assumption that most collections of virus from the vector in the field or from plants recently infected by vectors in the field will have maximal transmissibility by the vector.)

(*d*) Exvectorial (EV) virus: a virus strain which has completely lost its ability to be transmitted by the vector of the virus from which it has been derived. Exvectorial is suggested as preferable to vectorless, a term used in earlier reports from our laboratory.

(*e*) Subvectorial (SV) virus: one with any partial transmissibility somewhere between that of the vectorial and that of the exvectorial virus. For practical reasons it is suggested that wherever there is reason to doubt that a virus is vectorial or exvectorial, it should be termed subvectorial.

Theoretical considerations.—The researchers on bacteriophages showed that it was possible to inoculate large numbers of bacterial cells synchronously by using appropriate concentrations of bacteriophage particles. This made possible the study of single cycles of development of the bacteriophage, the determination of one-step growth curves, and investigations of development on various parts of a single growth cycle. Later, similar techniques were developed for animal cells. Subsequently, such techniques have been used extensively for investigations of the molecular biology of bacterial and animal viruses (36).

Bacteriophages attach themselves to host bacteria and establish infection in the bacteria by injecting their nucleic acid into the host cell. The animal viruses are absorbed to the surface of animal cells and may be taken into animal cells by the pinocytic activity of the cell membrane. Plant viruses, on the other hand, enter plants through wounds almost without exception. The evidence of Schneider & Worley (57, 58) to the contrary indicates that entrance without wounds is a very rare event. Even in such exceptions, one cannot eliminate the possibility that the viruses gain entrance to the plant cells through accidental wounds or wounds which occurred in them during processes of growth. Roberts & Price (54) confirmed the findings on the entrance of certain viruses into uninjured leaf parenchyma cells from the xylem. However, they noticed that the lesions which developed in such experiments did so only on leaves that were rapidly expanding during the period between the introduction of virus into xylem and the appearance of the lesions. In this period wounds caused by growth might have occurred. In their experiments on dipping uninjured leaves into solutions containing virus, they point out that one cannot be certain of the absence of wounds in spite of careful controls.

Plant virologists have lacked any technique for synchronous inoculation of cells comparable to that employed by phage and animal virologists. Inoculation of leaf surfaces by abrasion initially infects only a very small proportion of the cells in the leaf, and the viral components developing in these cells constitute such a small fraction of the mass of healthy cells that, during a single cycle of virus replication, such components are difficult or impossible to separate from the mass of host components. By the time they occur in sufficient concentration to be separable, so many replicative cycles have occurred and overlapped one another that there is no longer any synchrony to the cycles. Inoculation of plants by any of the various vectors of plant viruses initially infects an even smaller proportion of cells.

The possibility that plant tissue cultures might provide a satisfactory system for the study of plant viruses has been rather extensively investigated. However, until recently (48) tissue cultures of plant cells had shown little promise of providing a superior or even an acceptable alternative to inoculations by leaf abrasions. The assessment of Kassanis (34, p. 560) that plant "tissue cultures can be infected by mechanical inoculation but only in a manner that cannot be used for quantitative assay" was, I think, an accurate evaluation of the work up to that time.

The cells of a number of plant species have been grown in tissue culture. With such species it is not difficult to isolate plant cells infected with virus from certain plants already infected. However, minimal success has been obtained in inoculating healthy cultured plant cells with virus. Cultured plant tissues can even be inoculated with pathogens axenically by means of "aseptic" vector insects (45), but such inoculations do not induce synchronous and total inoculation of large populations of cells. Plant cells, when growing in tissue culture, tend to adhere together, to form masses, or to differentiate into organelles or plantlets. Moreover, the cells are normally surrounded by cell walls.

Murakishi (48) dissociated plant callus cells by a vibration treatment and within 10 min inoculated the separated components with tobacco mosaic virus RNA. The fact that no virus infectivity was detected in the inoculated cells between the first day and the fourth day but that the infectivity increased eightyfold between the fifth and fourteenth day may indicate that initially a small proportion of cells was inoculated. However, the success in inoculating tomato callus tissue in culture is most encouraging and may be developed and improved until effective synchronous inoculations of enough cells in a population may permit studies of single cycles of virus development in plant cells.

Efforts have also been made to achieve systems similar to those of animal cells and their viruses by removal of the plant cell wall before inoculation. Although some success has been achieved in obtaining living protoplasts from some higher plant cells (55), the inoculation of such protoplasts with plant virus (16) has not achieved sufficient success to be employed as a method for studying plant viruses. In the light of Benda's results below, the finding of virus particles in what were considered pinocytic vesicles (16) may not mean that the protoplasts containing them were infected. Mundry (47, p. 182) has discussed the passage of plant virus through the plasmalemma, the exterior membrane of the plant protoplast, and through the tonoplast or vacuolar membrane. Perhaps most significant are the failures of virus to pass through these membranes into the cytoplasm and the failures to produce infection. Benda (2) obtained only 70 lesions at the base of 694 leaf hairs in which one cell had been inoculated by micropuncture through a 10^{-8} ml drop containing about 10^{6} tobacco mosaic virus particles. Benda included in these data only inoculations that showed volume changes in the drop of inoculum at time of puncture that indicated a small volume of the cell contents was first extruded and then retracted. He also included only leaf hairs that remained alive for at least a day after inoculation. van Hoof (68) showed that abrasive inoculations of plant cells were successful before plasmolysis, unsuccessful while the cells were plasmolyzed, and successful once again when the cells had recovered from plasmolysis. Mundry (47) cites Zech as recording failures when virus was introduced into the liquid separating plasmolyzed protoplasts from the cell wall. Mundry (47, p. 184) concluded that pinocytosis in plant cells could be deduced from a number of researches. However, the plant plasmalemma and tonoplast may differ in important respects from those of animal cells and if pinocytosis occurs in the plant protoplast, it may differ in kind or degree from that in animal cells. The cell wall appears as an obvious obstacle to the introduction of virus. It is penetrated, however, by plasmodesmata and ectodesmata, and the cell wall may be less an obstacle than the composition and behavior of the plant cell membrane. As a result of his experiments on plasmolysis, van Hoof (68) considered that successful abrasive inoculation of plant cells might require that the plasmalemma be appressed to the cell wall and that the plasmalemma itself be wounded (68, p. 63 to 65, p. 92). Mundry (47, p. 183) also postulated that successful inoculation might require wounding of the plasmalemma. Injury to the plasmalemma is implicit in the estimate by Nault & Gyrisco (51) that an aphid making an inoculative 15 sec stylet probe 5 μ deep between the anticlinal epidermal cells of a leaf was likely to rupture at least 3 plasmodesmata.

To return to the topic of this review: Plant viruses which are propagative in their vectors probably behave towards susceptible vector cells as do animal viruses. Thus there is a prospect that large populations of vector cells can be inoculated simultaneously, that synchronous cycles of development may be studied, and that a valuable means of studying the molecular biology of such viruses may thus be provided. This might make the technique worthwhile, even if a minority of the same viruses may also be transmissible by abrasive leaf inoculations. The prospect is especially attractive for viruses propagative in their vectors, because only a few of them can be transmitted from plant to plant mechanically. The PYDV's are among the exceptions. Transmissions by pinprick inoculation, as in WTV, are very rare and have had virtually no application. Studies depending on infectivity measurements of propagative viruses in cell-free preparations usually had to be made by injection of the solutions into virus-free insects. In these respects, these viruses have been among the most difficult plant viruses to study, and the errors of infectivity assays have been large. Host vector cell culture, however, may convert these viruses from those most difficult to work with to those most quickly and precisely investigated by infectivity measurements.

The use of host vector tissue cultures may also be expected to have important practical applications by providing better methods of detection and identification of propagative plant viruses. Detection of virus more rapidly or at lower concentration has practical advantages. The possibilities of detecting relationships and of measuring degrees of relationship by neutralization tests with antisera and virus inocula on monolayers of host vector cells have not yet been explored. Because of the probable range of concentrations of inocula and antisera that can be used and the prospect for quantification, such neutralization tests are likely to have fruitful applications in identification of viruses. The fact that monolayers of vector and nonvector cells may exhibit measurable differences of susceptibility to a virus may be useful in indicating in what taxa vectors not yet discovered may occur. Plant quarantine regulations often forbid bringing viruses or vectors into uninfested areas. Monolayers offer means for studies that are without danger of escape of the vector, and that make escape of the virus easy to prevent.

Theoretically, there are reasons for supposing that only plant viruses which multiply in their vectors will be capable of multiplication in insect tissue cultures. On the basis of the knowledge available in 1958 (4, p. 183) it was pointed out that there was some evidence that the viruses which multiply alternately in plants and insects may be larger than other plant viruses. It was also suggested that the ability to multiply in such widely different hosts might have reduced the number of physiological functions that were dispensable in the evolution of such viruses, and that this in turn might have limited an evolutionary reduction in size. If one reexamines this concept today, one finds that among the agents now known to be capable of such alternate cycles of multiplication, we can include in order from the largest to the smallest: agents of the aster yellows type which may very well be mycoplasma (19, 32), the potato yellow dwarf virus group (about 200 to 300 mμ long by 50 to 100 mμ wide) (23) which multiply in aphids and leafhoppers (14, 63, 66), and the wound-tumor virus group (70 mμ in diameter). It is likely that tomato spotted wilt virus (80 mμ in diameter) multiplies in thrips and plants. In contrast, the evidence for the multiplication of small plant viruses, like potato leaf roll virus and pea enation mosaic virus (25 to 30 mμ in diameter), in their vectors is still conflicting or uncertain of interpretation (27, 28, 59, 64, 65). To date then the evidence for multiplication in plant and insect vector is clear only for the entities as large as wound-tumor virus (WTV) or larger and it may be that these unusual abilities cannot be incorporated into the smaller viruses. This may mean that the plant viruses that can be studied in host vector cell monolayers are those of the size of WTV and larger.

We have no evidence that any of the plant viruses that do not multiply in their vectors are capable of multiplication in vector cell cultures. However, it is hoped that this review will make the advantages of insect tissue culture for certain plant viruses sufficiently obvious so that readers will not be content to accept the above theoretical limitations without subjecting them to a test.

Plant Viruses and Tissue Cultures from Host Vectors

In 1956 it was demonstrated (39) that the agent of aster yellows increased over a period of 10 days in explants from leafhopper vectors that had recently ingested the pathogen. This experiment showed that fragments of the vectors could be maintained alive for this period in a suitable medium and indicated that multiplication of the pathogen had occurred. However, it did not provide any information on inoculation of tissue cultures. Operationally, therefore, it could not provide techniques for the kinds of study outlined above and has not been so employed.

In attempts to grow host vector cells in our laboratory with a variety of media, various tissue explants had remained alive up to 3 weeks as revealed by the maintenance of peristaltic movements. However, little progress was made until after the work of Hirumi & Maramorosch (29, 30). Extensive sheets of cells, such as they describe, did not grow from explants made by us prior to the use of their discovery that the embryonated leafhopper egg was the best source of explants from which cells would grow. Moreover, the easily seen movement of the eyespot from the posterior to the anterior end of the egg is very convenient in identifying the critical stage of development which gives the best growth of cells. Hirumi & Maramorosch described several different types of cells growing from their explants. They considered extended pseudopodia and plasma membranes as evidence of a healthy state of the cells and managed to keep some of them alive in primary culture for as long as 40 days. They did not achieve successful subcultures free from the explants.

Maramorosch et al. (40) reported that they had maintained primary cultures from explants from embryos, nymphs, and adults of *Macrosteles fascifrons* (Stal) and *Agallia constricta* (Van Duzee) for over 5 months during which the explants exhibited "contractive" movements. In these experiments, they used their axenic nymphs and adults as sources of tissue. Following inoculations with crude extracts containing the aster yellows pathogen, epithelial cells stopped growing, deteriorated, and granular inclusions appeared within 12 days. The fact that the pathogen could not be recovered from such primary cultures indicates that the deteriorative changes may have been caused not by the pathogen but by another factor, such as toxic inoculum.

Mitsuhashi & Maramorosch (44) achieved extensive growth of cells in primary cultures from three species of leafhoppers and described symbionts as common inhabitants of such primary cultures. Mitsuhashi (42) inoculated primary cultures of the vector *Nephotettix cincticeps* (Uhl.) with rice dwarf virus and observed shortly thereafter that the sheets of cells in the primary explants showed granulation and other deteriorative changes. However, he did not regard these as specific cytopathic effects of the virus; instead, he related them to deteriorative changes commonly observed in insect cell cultures that are degenerating. Ultra-thin sections examined in the electron microscope revealed numerous small particles, 30 mμ in diameter, which had earlier been associated with cells infected with the 70 mμ rice dwarf virus by Nasu (50). "Not many" particles of the size of the complete virus were found.

Sinha (60) showed that the organs of *A. constricta* were infected sequentially with WTV. One could therefore expect that many different cells from the vector could be infected in culture. Moreover, the detection of the primary focus in the filter chamber on the fourth day after ingestion clearly indicated that infections in monolayers might well be detected by fluorescent antibody staining within a similar interval. In some other studies of this time, much longer incubation periods were allowed, periods corresponding more to the incubation period of the virus in the vector itself.

Chiu, Reddy & Black (15) inoculated primary cultures of the vector *A. constricta* with WTV. The host vector cells consisted of monolayer sheets extending from the explants. The initial inocula were taken from adult leafhoppers containing a peak concentration of the virus. It had been shown (53) that subsequent to ingestion of the virus by the vector this peak is reached after about 28 days at 27°. Viruliferous insects were homogenized in medium, and the suspension clarified by centrifugation. After Millipore filtration, adsorption of the inoculum was allowed for 2 hr at 24°. The inoculum was replaced with growth medium, and incubation allowed to proceed at 24°. Treatment with fluorescent anti-WTV serum specifically stained the cytoplasm on the third day after inoculation (15); the intensity of staining and the number of stained cells increased during later days. Control cultures failed to stain.

Relative infective virus concentrations in the inoculated cells were determined by the injection of extracts into virus-free insects, followed by the determination of the concentration of soluble antigen of the virus (WTSA) in the injected insects after 21 days at 27° according to the technique of Reddy & Black (53). This virus concentration in the cells was shown to increase at least one hundredfold during 8 days of incubation (15).

Similar tests showed that infective virus added to medium decreased to about 0.1 per cent during a 24 hr period at 24°. In spite of this, it was demonstrated that in the medium over inoculated monolayers the concentration of infective virus was maintained at a value of about $10^{6.0}$ (on the relative scale) in spite of the inactivation that was occurring and in spite of the replacements of the medium every 2 or 3 days. These measurements indicated that at a minimum there was about $10^{2.4}$ times as much virus present in the medium at the end of 8 days as one would expect when the losses by inactivation and replacement of medium were estimated very conservatively (15). In another experiment, the virus was passed in series from culture to culture through six passages with a dilution of about $10^{-1.5}$ at each passage. The relative infectivity titer at the end of the series was $10^{5.6}$, whereas it should have fallen to at least $10^{-2.5}$ by dilution if there had been no multiplication. The multiplication of virus by $10^{8.1}$ was again a conservative estimate (15).

There was no evidence of any cytopathic effect in the inoculated cultures when they were compared with the uninoculated controls under the phase contrast microscope. Incidentally, in these experiments it was also demonstrated that plant tumors, viruliferous insects, and infected primary explants could all provide infective inoculum (15).

Mitsuhashi & Nasu (46) reported futher electron microscope study of primary cultures from embryo fragments of the leafhopper vector, *N. cincticeps,* inoculated with rice dwarf virus. Complete rice dwarf virus particles, 70 mμ in diameter, were found dispersed in viral matrices. Near the periphery of such matrices, virions occurred in linear arrangements within sheath-like substructures. They also occurred in crystalline aggregates in the inoculated cells. The 30 mμ particles again were observed, often in the viral matrix. Contrary to observations in an earlier report (42), no granulation of the cells occurred shortly after inoculation, and the authors considered that this confirmed Mitsuhashi's earlier interpretation that the granulation was not a cytopathic effect of the virus infection.

Chiu & Black (12) reported continuous culture in monolayers of cells isolated from embryonic explants of *A. constricta.* In early attempts to achieve continuous culture the medium developed by Mitsuhashi & Maramorosch (44) was employed, but for continuous culture a modification of the medium of Schneider (56) was better. For convenience these cell cultures were dubbed AC to indicate the source. The cell population was heterogeneous, but most of the cells were of epithelial type. The cells were subcultured through at least 55 passages and similar monolayers were obtained from three other species, *Aceratagallia sanguinolenta* (Provancher), *Agalliopsis novella* (Say), and *Agallia quadripunctata* (Provancher).

The AC monolayer cells in continuous culture were susceptible to inoculation with WTV as revealed by specific staining with fluorescent antibody. Infection could be detected as early as 12 hr after inoculation when incubation was at 30°. Cells counts showed that there was no significant difference in the rate of multiplication of the healthy cells and those infected with WTV, although there was some evidence that the infected cells did not attach quite as well as did the healthy. There was nothing particularly noteworthy about the medium, except that the inclusion of 20 per cent fetal bovine serum made hemolymph of insects or lobsters unnecessary for satisfactory growth. Extensive cell growth in the primary explants could be obtained in 3 or 4 weeks and established cell lines within another month. The fact that this medium has been used successfully to obtain subcultures of monolayers from three genera of agallian leafhoppers indicates that it may have a wider utility in culturing leafhopper cells.

The monolayer cultures provide a method for testing components of media for toxic contaminants that may be harmful to other host vector cells, and they may provide a means of conditioning media that might be advantageous in the culture of other insect cells.

Gamez & Black (20, 21) reported direct virion counts on WTV in *A. constricta* between the eighteenth and forty-second day after acquisition of the virus from plants and incubation at 27°. These counts revealed the same peak concentration reported by Reddy & Black (53). The two peaks were equated and the log of relative infective virus concentration (53) converted to a log scale of the absolute number of virions/ml by multiplying by $10^{2.5}$ (21). At the peak concentration, there was an average of about $10^{9.26}$ virions per infected leafhopper. As the leafhoppers weigh about 1 mg, the concentration per gram was about $10^{12.26}$.

Tenfold dilutions of WTV inoculum from tumors were tested on coverslip monolayer cultures (22). The minimal concentration giving positive results was between $10^{6.1}$ and $10^{5.4}$ virions/ml, and the minimum number of virions in the 0.02 ml applied per coverslip was $10^{4.44}$. The fact that this value is 10^2 times greater than the estimated minimum infective dose for insect injection or the cell-infecting unit (CIU) value (see next paragraph) probably means that the two coverslip monolayers used for each dilution in these tests were insufficient for an accurate determination.

The fluorescent cell counting technique as a means of assaying WTV was studied in detail (13). Monolayers in circles of about 30 mm² were prepared on the center of coverslips 15 mm in diameter. Virus adsorption was nearly maximal after 2 hr at 30° and the optimal time for counting the cells was after about 27 hr at 30°. The distribution of infected cells in the monolayer was found to follow a Poisson distribution, and the number of fluorescent cells bore a linear relationship to the relative virus concentration of the inoculum, indicating that a single virus particle produced each infection. Under the conditions employed for inoculation, the CIU was determined as 405 virions. The minimum infective dose for inoculating leafhopper vectors according to Figure 6 of Gamez & Black (21) was 0.1 μl of a solution containing $10^{6.6}$ virions/ml or $10^{2.6}$ virions (i.e., 398 virions).

Reddy & Black (53) and Chiu, Reddy & Black (15) estimated that the limit of sensitivity in assay by injection of vectors was a virus concentration which measured 10^4 on the relative scale. This is equivalent to $10^{6.5}$ virions/ml. In the paper by Gamez & Chiu (22), the minimum concentration giving infection of monolayers varied from $10^{6.14}$ to $10^{6.37}$ in three tests.

There is a rather remarkable consistency in certain of these parameters established by different workers working independently of one another. Nearly exact agreement is no doubt in part fortuitous, but the general agreement seems significant.

When cells of the nonvector *A. sanguinolenta* (AS cells) were inoculated with WTV, the number of infected cells was 1 or 2 log units lower than with the vector AC cells (13).

New York potato yellow dwarf virus (N.Y.-PYDV) is transmitted by *A. sanguinolenta* but not by the related *A. constricta;* the vector relationships are reversed with the New Jersey virus (N.J.-PYDV). When the viruses are inoculated on cell monolayers in all combinations the infectivity of each virus on nonvector cells was about 10 per cent of that on the related vector cells (35). Heterologous antisera stained only 0.2 to 4 per cent as many infected cells as did homologous antisera.

Earlier, Mitsuhashi (43) infected primary cultures of cells from the leafhopper *N. cinticeps* by inoculation with the iridescent virus which occurs naturally in the rice stem borer *Chilo suppressalis.* Symptoms of infection appeared in the cells within 24 hr and later the cell sheets showed the iridescence typical of infections by this group of viruses. Virus was demonstrated in the cells by electron microscopy of sectioned material. The iridescent viruses are known to have a wide host range among insects but had not been previously known to infect cells of homopterous insects.

Earlier experience with animal viruses in tissue culture had shown that cell cultures are sometimes susceptible to viruses which will not cause disease in the vertebrates from which the cells were derived (11). More specifically, results of this kind were indicated by earlier work in which it was shown that the aster yellows pathogen multiplied in nonvector leafhoppers, but at a lower level than in the vector (38) or in only some of the organs that apparently must be infected for transmission to occur (62). Even in its vector *A. constricta,* WTV may cause abortive infections in insects acquiring virus as adults or in insects exposed to high temperatures (61) so that the sequence of organ infections that is presumed to be necessary for transmission is not completed.

Chiu et al. (14) used N.Y.-PYDV in studies with AS monolayer cultures from the vector *A. sanguinolenta.* It was much more difficult to obtain inoculation of host vector monolayers with PYDV than with WTV. The first successful inoculations of monolayers of vector cells with PYDV were detected by immunofluorescent staining. Only a small proportion of cells (less than 1 per cent) were stained. It would have been difficult to detect such minimal infection by electron microscopy, and therefore impractical to assay modifications of the techniques of inoculation by this means. The availability of specific immunofluorescent staining at this stage was a great advantage. PYDV (measuring about 75 mμ by 380 mμ) is much larger than WTV, an icosahedron about 60 mμ across. At first Millipore filtration to remove possible microbial contaminants from virus inoculum also removed most of the infective virus and gave unsatisfactory results. Such filtration was often circumvented by various means such as extracting inoculum from small portions of infected stems of *Nicotiana rustica* L. which had been surface sterilized before they were triturated. Diethylaminoethyl-dextran at 25 μg/ml greatly increased the efficiency of inoculation. Again, a linear relationship was established between the virus concentration in the inoculum and the incidence of infected cells in the inoculated monolayers. The slopes of the regression lines for N.Y.- and N.J.-PYDV were virtually identical although the former was tested on AS cells and the latter on AC cells. On the other hand, the slopes for WTV- and N.J.-PYDV both of which were determined on AC cells were definitely different.

The first observations of immunofluorescent staining of cells infected with PYDV showed that there was a marked contrast with the corresponding staining of cells infected with WTV. This difference is illustrated in Figures 1 and 2. In both figures, the monolayers are comprised of AC20 cells cultured from *A. constricta,* which is a vector of both viruses. Figure 1 shows cells of AC20 inoculated with N.J.-PYDV. The monolayer was stained with fluorescent antiserum to N.J.-PYDV. In this infection, staining is at first confined to the nucleus and occurs in spots near its periphery. Later it begins to appear in the cytoplasm. Figure 2 shows cells of AC20 after inoculation with WTV and staining with fluorescein-conjugated anti-WTV serum. The cytoplasm is intensely stained whereas the nuclei are unstained.

Examination of heavily infected cultures in the electron microscope revealed complete PYDV particles and virions in various stages of formation. Many particles were located between the perinuclear membranes as is characteristic of the virus in plants (37). Although the long incubation period and the persistence of PYDV in the vector had indicated multiplication therein, these observations have provided the first critical evidence that PYDV multiplies in its vector.

Cells of AC20 and AS1 can be stored at −80° and can be revived at will (31). This minimizes the danger of losing these cell lines and means that they can be shipped any distance over which they can be kept at this low temperature.

It is fortunate that we have had the opportunity to work with two very different viruses, WTV and PYDV, which represent two groups of plant viruses that are known to multiply in their vectors.

It should be noted that we have not as yet found plaques produced by either WTV or PYDV in inoculated monolayers. The finding of plaques would have important advantages for certain kinds of work such as the isolation of virus mutants. Cell monolayers from nonvectors have not as yet given plaques when infected with viruses transmissible by insects related to the nonvectors.

Environmental and cinematographic studies of cells by D. Schlegel & L. M. Black[1].—The effect of temperature on growth of AC2 cells in the medium of Chiu & Black (12) was tested in two experiments. Cells were sown in a number of disposable plastic flasks, each with 25 cm^2 of growing surface. The cells were allowed to attach and to grow for about 3 days at room temperature before two flasks were placed at each temperature to be tested. At this time the cell population per flask was 3.9×10^6 and 1.2×10^6 in experiments 1 and 2, respectively. After a period of 72 hr to 21 days during which a measurable increase in growth had occurred at the different temperatures, the flasks were removed and the cells in each counted. The calculated hours required for doubling of the population at the different temperatures were as follows:

Temperatures:	12°C	15	18	21	24	27	30	33
Hr (expt. 1):	23,600	475	265	156	101	70	72	77
Hr (expt. 2):	—	—	—	—	89	78	90	97

[1] Department of Plant Pathology, University of California, Berkeley, California.

FIGS. 1 AND 2: Contrast in the staining of potato yellow dwarf virus infections (Fig. 1 *above*) and of wound tumor virus infections (Fig. 2 *below*) in *Agallia constricta* cell monolayers treated with specific fluorescent antisera and photographed under a fluorescence microscope. [Reproduced with permission of the authors (14).]

above: AC20 cells inoculated with New Jersey PYDV. The staining of spots is apparent in some nuclei; in others because of the large number of spots or because the nuclei are somewhat out of focus the staining appears solid. In the lower left quadrant, and less so elsewhere, many uninfected cells appear as ghosts with black circles, the nuclei, surrounded by faintly autofluorescing cytoplasm. (×400)

below: AC20 cells inoculated with WTV. The black circles are the unstained nuclei surrounded by stained cytoplasm. Sometimes stain in the cytoplasm above or below the nucleus gives the appearance of some stain in the nucleus. All the cells are infected. (×400)

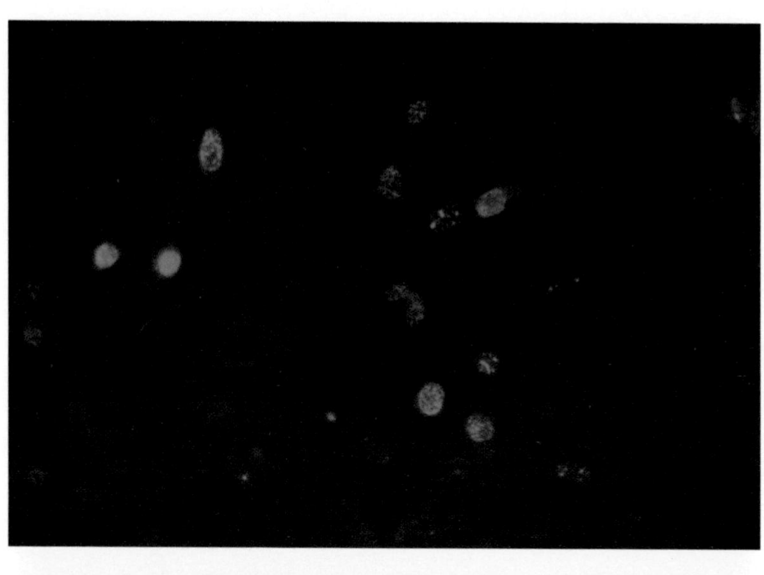

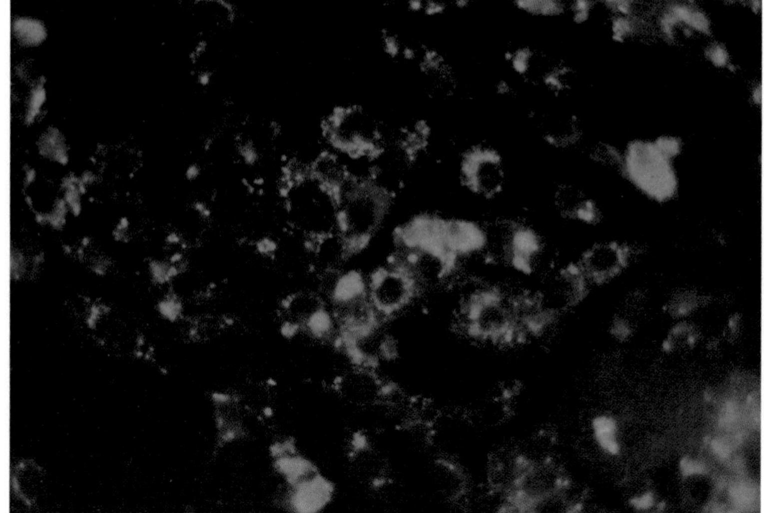

No growth occurred in flasks placed at 9° for 30 days, but upon returning the flasks to an optimum temperature, growth was vigorously resumed. In flasks held in a refrigerator at 3 to 5°, the cells began to detach in 2 days and very few were attached at the end of a week. Only a few of the detached cells reattached and grew when the flasks were subsequently placed at optimum temperature.

The counts were made in these experiments by trypsinizing the monolayers, suspending the cells in a known volume, and counting them on a hemocytometer slide. Four counts were made on each flask, but the variability in the counts was high—in part due to clumping of cells. In spite of this variability, it was clear that the optimum temperature was about 27° and that cultures of AC2 cells could be held for long periods at 9°. (Since these experiments were carried out, Mr. Ho-yuan Liu has kept AC cultures as long as 375 days at 9° with only one or two changes of medium.)

In the above temperature tests, the medium had a pH of 7.2 at the start, and the maximal changes were to pH 6.9 after 96 hr at 21° and to pH 7.0 after 30 days at 9°. Growth of AC2 cells at 28° was nearly optimal at starting pH's between 7.0 and 7.6 and measurably less at 5.8 and 7.9. It was very difficult to stabilize the medium at pH's between 6.0 and 7.0.

Cinematography of the monolayers of AC cells revealed cell movement, mitosis, active pseudopodia, and what appeared to be pinocytosis by the plasmalemma. Figure 3 represents a sequence in which every tenth frame was printed from a movie in which the filming rate was four frames/min, that is, each frame shown was taken 2.5 min after the previous one. Examination of different points, some of which are indicated by arrows, shows the motility of the surface membrane even though such examination is a poor substitute for viewing the motion picture. It seems probable that the activity of the membrane is related to its high susceptibility to inoculation.

Loss of Transmissibility

One of the important considerations involved in plant virus work with cultured vector cells is the degree to which any particular inoculum may have lost its infectivity for the vector. It is probably safe to assume that virus being carried in the field by the vector, or virus recently transmitted to plants in the field by the vector, usually has a nearly maximal infectivity for the vector and its cells. The phrase "recently transmitted" is used advisedly, because many infections in the field, particularly in clonally propagated plants, or in long-lived perennials, may be produced by virus that was transmitted to them by the vector many months or years earlier. Many plants in agriculture are clonally propagated. Many cultivated plants, including fruit trees, bush fruits, small fruits, many ornamentals, crops like potatoes and sugar cane, etc. are so propagated. Many wild plants are perennials, e.g., trees. Moreover, one should be on guard against the possibility that plants which are normally annuals or biennials may have assumed the perennial habit under certain circumstances. Nasturtium, *Tropaeolum majus* L., is an annual susceptible to tomato spotted wilt virus, and it might be convenient to collect the virus from the field in this plant. However, in some areas, e.g., the San Francisco Bay area in California, it grows as a perennial, and in such cases an infected nasturtium plant provides no assurance that the virus was received from the vector in the current year. Tomato plants, *Lycopersicum esculentum* Mill., under ordinary cultivation in the same area do furnish such assurance.

The importance of these discriminations should become evident from the following account of our experience with the loss of transmissibility in propagative viruses. A rather thorough discussion of this phenomenon in WTV will be given even though it was first reported for PYDV (3). Following the treatment of WTV, a brief account of the same condition in PYDV will be presented, but only where this provides an opportunity to describe additional aspects of the subject.

Demonstration of exvectorial wound-tumor virus by R. F. Whitcomb and L. M. Black.—An exvectorial (EV) strain of WTV was found in *Melilotus officinalis* (L.) Lam., clone C10, which had been infected with WTV for many years (8). This virus isolate was probably introduced into clone C10 in 1949, hence the designation EV49. During routine vegetative propagation of the infected clone by rooting stem-cuttings, opportunities for multiplication of the virus in the vector were incidentally eliminated. When the virus isolate was later tested by injecting tumor extracts into the vector *A. constricta,* it was found to have lost its transmissibility.

Between October 10, 1955 and May 10, 1957, 12 experiments were carried out in which 807 *A. constricta,* injected with vectorial (VI) WTV, infected 79 of 274 plants on which they were tested; a corresponding 705 vectors injected with WTV isolate EV49 infected none of 180 plants; 457 uninoculated control vectors were tested on 97 plants, one of which developed wound tumor. This one plant was the last to develop symptoms and this and other circumstances indicated that the plant had been accidentally contaminated by a viruliferous insect not employed in the experiment. Of the 79 plants infected with vectorial virus, 15 plants and 4 plants had been infected with inoculum at a dilution of 1/10 and 1/100, respectively. In most of the experiments the concentration of vectorial virus injected into the leafhoppers was equal to or less than that of the exvectorial virus. This was arranged by adjusting concentrations on the basis of many serological ring tests on rate or quasi-equilibrium zonal fractions which showed that the concentration of vectorial virus in crude plant extracts was usually four times greater than that of EV49 virus. On the other hand, the concentration of soluble antigen (WTSA) of EV49 was two to four times the WTSA concentration of the VI virus. These experiments showed conclusively that as early as 1957 EV49 had lost its ability to be transmitted by its original vector. The distinctive symptoms, the retention of serological relationships, and the laboratory history of EV49 left no doubt as to the origin of the virus. The control injections of the propagative strain resulted in the development of WTSA in the injected vectors to a titer of 1/176 24 days later. After comparable injections with EV49, no WTSA could be detected in extracts of the vectors at a dilution of 1/11. This result suggested that the EV virus cannot multiply in the vector or undergoes only limited multiplication and therefore cannot be transmitted. During this period (1955–57), EV49 could be isolated from the tumors it incited and could be measured serologically. Later (Reddy & Black, below), it was not possible to isolate the virus from these tumors by methods which yielded purified VI virus.

TABLE I

MEASUREMENT OF WTSA AND INFECTIVE WTV IN TUMORS CAUSED BY VARIOUS WTV ISOLATES

Experiment	Isolate[a]	WTSA titer[b] in tumors	WTSA titer[b] in insects injected with tumor extract dilutions of			Infectivity titer in terms of: WTV concentration in tumor extract at $10^{-1.5}$ dilution Relative scale[c]	Infectivity titer in terms of: Apparent extracted infective virions from 1 g tumors Absolute scale[d]	WTV RNA per 100 g tumors OD_{260}	Per unit WTV RNA[e]
			10^{-1}	10^{-2}	10^{-3}				
	EV49	235	0	0		0	0	—[f]	—
	SV55	287	687	401		$10^{6.8}$	$10^{10.8}$	15.5	1/163
	SV57	213	490	362		$10^{6.5}$	$10^{10.5}$	6.1	1/80
1	SV58		—	—		—	—	—	—
	SV60a	128	0	0		0	0	1.2	0
	SV60b	124	0	0		0	0	1.8	0
	SV62	105	543	429		$10^{6.7}$	$10^{10.7}$	3.3	1/28
	VI64	110	—	931	710	$10^{8.4}$	$10^{12.4}$	6.0	1
	EV49	475	0	0		0	0	—	—
	SV55	444	219	88		$10^{4.8}$	$10^{8.8}$	43.2	1/32,400
	SV57	339	128	41		$10^{3.8}$	$10^{7.8}$	19.0	1/142,000
2	SV58	287	0	0		0	0	9.0	0
	SV60a	317	0	0		0	0	7.1	0
	SV60b	268	112	31		$10^{3.5}$	$10^{7.5}$	8.5	1/127,000
	SV62	287	429	219		$10^{6.0}$	$10^{10.0}$	18.8	1/889
	VI64	320	—	1825	1258	$10^{9.3}$	$10^{13.3}$	42.2	1

[a] The numbers indicate the year when the wild virus was introduced into sweet clover clone C10. The actual dates for the beginning of the inoculation period by the vectors are given in parentheses: EV49 (7–12), SV55 (11–4), SV57 (4–11), SV58 (12–5), SV60a (1–7), SV60b (10–20), SV62 (10–26), VI64 (8–5). For isolates similarly inoculated into clone C10 in later years the dates were 1965 (12–21), 1966 (2–1), 1967 (12–1), 1968 (5–21).

[b] The WTSA titers were determined by precipitin ring-time test as described in Reddy & Black (1966) and are given as reciprocals of the computed end-point titers.

[c] Relative scale of Reddy & Black (1966).

[d] Absolute scale of Gamez & Black (1968).

[e] In each experiment VI64 was taken as a standard with a value of 1.0. Each of the other values was calculated in the same manner as the following example for SV62, experiment 1: the infectivity of SV62 extracts was $10^{-1.7}$ times that of the VI64 extracts, and the WTV-RNA isolated from the SV62 extracts was 3.3/6.0 that from the VI64 extracts. ($10^{-1.7} \times 6.0/3.3 = 1/28$.)

[f] A dash indicates that no measurement was made.

Comparative infectivity of WTV isolates by D. V. R. Reddy and L. M. Black.—Once the exvectorial nature of EV49 was evident (52), the vectorial virus was inoculated by the vector into sweet clover, clone C10, at various times and subsequently maintained in this clone by vegetative propagation. In each case, no passage through the vector was allowed subsequent to the introduction of the virus into the sweet clover. Consequently, the inocula were separated one from another and are hereafter referred to as WTV isolates. Each isolate was given a number indicating the year that it was inoculated into sweet clover. Actually, as shown above (Table I), there is a question about the transmissibilities of SV58 and SV60a because they failed to produce WTSA in injected vectors in the experiments reported. However, there also is doubt about their complete inability to be transmitted. Conceivably, an isolate of the virus might be able to stimulate WTSA production in the vector without being able to be transmitted by the vector, or conversely, the virus might be transmissible without producing detectable WTSA in the vector, although this latter possibility seems less likely. Such distinctions have not been demonstrated in this work. Failure to produce WTSA in the vector probably means that the virus does not multiply in the vector, and SV58 and SV60a may well be unable to do so at the present time.

In 1965, the infectivities of various WTV isolates for the vector were compared by growing all isolates under the same environmental conditions and harvesting the root-tumors at the same time. Two separate experiments were carried out. Tumors were thoroughly ground with three times their weight of a solution containing 0.05 *M* K_2HPO_4 and 0.01 *M* Na_2SO_3, and the extract clarified by 5 min centrifugation at 7200 rpm. A solution of 0.1 *M* glycine and 0.01 *M* $MgCl_2$ was added to make various dilutions of the virus in terms of the starting weight of tumors. Two different appropriate dilutions of each isolate were injected into virus-free *A. constricta*. The injected insects were held at 27° for 21 days on crimson clover plants and were transferred to fresh crimson clover plants at the end of the first 10 days. Under these conditions, they could not inoculate virus into and acquire virus from any of these crimson clover plants. The WTSA titers of the tumors and of the insects 21 days after injection were determined by the techniques described earlier (53), and from the latter titers the relative quantities of infective virus that had been injected were calculated.

All WTSA titers are presented in Table I, columns 3, 4, 5, and 6, and the treatment of the data and derivative values are given in columns 7, 8, 9, and 10 of the same table. Column 7 contains the relative infective virus concentration calculated for tumor extracts at a dilution of $10^{-1.5}$; column 8 gives the corresponding estimates of extracted infective virions from 1 g of tumors according to an absolute scale (21). The absolute values are derived by multiplying the relative values by $10^{1.5}$ to give dilution 10^{0} and by $10^{2.5}$ to convert from the relative to the absolute scale. The term "apparent" is given to the absolute values because the interaction between noninfective and infective virus has not been determined, and it seems probable that noninfective virus interferes with infections by infective virus.

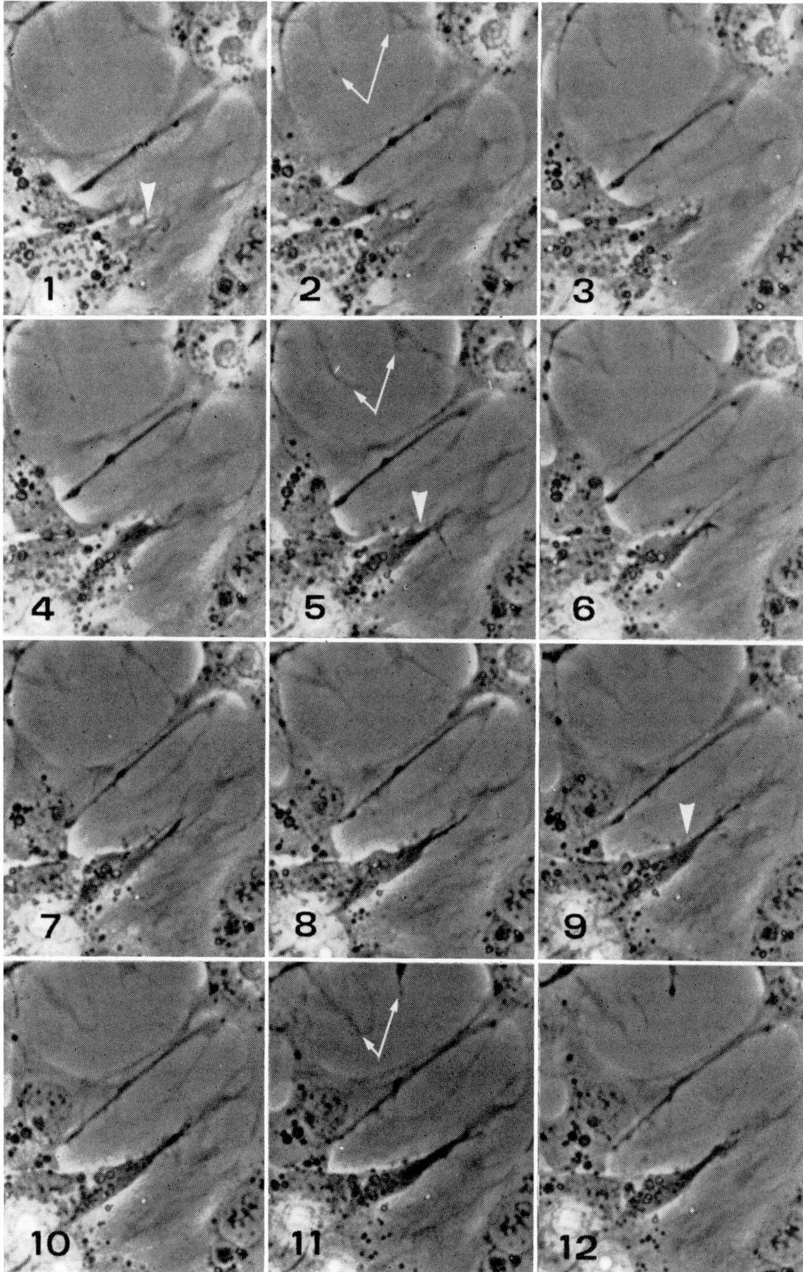

FIG. 3. A sequence of pictures of AC2 cells taken from a movie so that an interval of 2.5 min separates each frame. The cytoplasmin strand extending diagonally across the center frame 1 contains granules whose movements can be followed readily through succeeding frames. Some other changes are indicated by arrows. (Phase contrast microscopy, ×400).

An attempt was made to ascertain if there were measurable differences in the specific infectivities of the different isolates by extracting and measuring the relative amounts of WTV-RNA in the samples of tumors from each isolate. The virus in each extract was purified through the first equilibrium density gradient zonal centrifugation (5, 10). The virus zone was removed by puncturing the bottom of the tube with a hypodermic needle and collecting the zone into a Spinco rotor 40 tube which was then filled with a protective solution of 0.1 *M* glycine and 0.01 *M* $MgCl_2$ at pH 7.0. After mixing the contents, the virus was pelleted at 36,000 rpm in a Spinco Model L centrifuge for 2 hr. After the supernatant was drained completely, the outside of the tube next to the pellet was thoroughly washed and wiped with cleansing tissue paper. The virus pellet and the part of the tube on which it rested were carefully cut out and transferred to a clean glass tube; the virus was suspended in a small quantity of distilled water and the suspension then transferred by means of a Pasteur pipette to a cleaned and weighed clinical centrifuge tube with a conical bottom. Residual virus suspension was transferred to the tube by successive washings with additional small volumes of distilled water until the virus extract transferred to the tube weighed 0.5 g. Slowly, an equal weight of 2 *N* HC1 was then added to the centrifuge tube. The tubes were then capped with parafilm and shaken gently to disperse the WTV in the I *N* HC1. After standing overnight at room temperature, the virus protein, which appeared in the form of a white precipitate, was centrifuged out and the spectrum of the RNA hydrolysate in the supernatant was measured in a Cary Model 14 spectrophotometer. The RNA hydrolysates gave maximum absorption at 260 mμ and minimum at 232 mμ; the ratio between the maximum and minimum absorption varied between 3.5 and 4.5. Most tumor samples weighed between 15 and 25 g, but the optical density value obtained for each tumor sample was computed to a standard 100 g weight of tumors. These calculated OD_{260} maxima are given in Table I, column 9. Infectivity values, calculated from the data in columns 7 and 9 and presented in column 10, are an approximation of the infectivity per unit WTV-RNA of the different WTV isolates. In each experiment, the isolate VI 64 was assumed to have a standard infectivity of 1.0 and all other values in column 10 are expressions of the infectivity of the other isolates in terms of the infectivity of this strain. The values show a rough agreement in the two experiments for isolates VI64, SV60b, SV60a, and EV49. The infectivity of SV60b in experiment 1 was not detectable, whereas in experiment 2 it was barely detectable. There are major differences in these values for SV62, SV57 and SV55 in the two experiments, although even here the ranking of the infectivities is the same except for the relative positions of SV55 and SV57 with regard to each other. The minimal infectivities of SV55 and SV57 in experiment 2 are markedly different from their infectivities in experiment 1.

The results in column 3, showing positive WTSA titers in the tumor-extracts from every isolate where enough tumors were produced to permit a test, clearly demonstrate that the inciting agent in all cases was some form of WTV, even in those cases where no infective virus was demonstrated.

The results also show that when WTV is maintained in vegetatively propagated sweet clover for long periods of time without passage through the vector, it loses its ability to be transmitted by the vector. This change occurs at irregular intervals and to varying degrees, but it seems to be an inevitable consequence of perpetuation in sweet clover without passage through the vector. Although these changes are accompanied by lower concentrations of virus in the vector, the differences in concentration are minor compared with the differences in infectivity. There must be great differences in the specific infectivity of the different virus isolates.

It is obvious that for work with cultured cells of the vector the wild strain of WTV would be the preferred strain unless the studies contemplated were directed towards some aspect of the loss of transmissibility.

Counts of virions in extracts from tumors of WTV isolates by R. Gamez and L. M. Black.—Virions were counted in extracts of tumors induced by the different WTV isolates in sweet clover clone C10. The techniques of Williams, Backus & Watson (1, 70, 72) were used by Gamez & Black (21, 22). In these comparisons, the plants infected with the different isolates were grown under the same greenhouse conditions, and all root-tumors were harvested from 60 to 64 days after setting the infected cuttings. The root-tumors from infected plants were first washed and all rootlets removed from the large tumors at the base of the stem. These basal tumors were triturated with mortar and pestle in three times their weight of a solution containing 0.05 M K_2HPO_4 and 0.01 M Na_2SO_3, until the pulp attained the consistency of a finely ground meal. The crude extract, at approximately pH 7.0, was squeezed through one layer of cheesecloth and then clarified in a Servall SS-1 centrifuge for 5 min at 7200 rpm. A mixture was prepared containing 0.1 ml of the supernatant, 0.1 ml of a polystyrene latex suspension of known concentration, 0.2 ml of 2 per cent potassium phosphotungstate at pH 7.0, and 0.05 ml of a 1 per cent solution of bovine serum albumin. This mixture was sprayed onto specimen grids for electron microscopy. The WTV particles were readily distinguished under the electron microscope from the latex particles, both by size and morphology.

The results of two independent experiments are presented in Table II. A tendency for virus concentration to decline with the length of time in sweet clover is obvious. SV57 is a notable exception, in that its concentration remained higher than that of three or possibly four isolates which were maintained in sweet clover for as much as 5 years less. Although the virus EV49 could not be isolated by our standard purification technique in 1965, these virion counts showed that EV49 still had a concentration of about 10^{11} extractable virions/g of tumor or about 4 to 11 per cent of that of VI66.

In Table III, the WTV-RNA concentrations of Reddy & Black (Table I, above) and the virion concentrations of each isolate in two experiments are expressed as fractions of those of the corresponding wild strain (VI64 for

TABLE II

WTV CONCENTRATION IN TUMORS OF DIFFERENT WTV ISOLATES[a]

WTV Isolate	Number of virions $\times 10^{11}$/g of tumors	
	Experiment 1	Experiment 2
EV49	1.60±0.08	0.39±0.06
SV55	5.80±0.55	2.56±0.35
SV57	9.45±1.66	5.58±0.99
SV58	4.45±0.55	3.78±0.79
SV60a	6.79±0.96	2.56±0.28
SV60b	7.69±1.09	5.94±1.22
SV62	6.07±0.54	5.04±0.90
SV64	10.80±1.65	7.22±1.15
SV65	13.77±1.14	8.60±1.11
VI66	14.22±1.37	11.16±1.45

[a] The periods in 1966 during which the different plant tumors were grown were as follows: Exp. 1: VI66, 6–1 to 8–3; SV65, 6–23 to 8–22; others, 6–22 to 8–22; Exp. 2: VI66, 8–15 to 10–17; EV49, 8–26 to 10–24; others, 8–22 to 10–21.

TABLE III

COMPARATIVE YIELD OF WTV-RNA AND OF COUNTED WTV PARTICLES FROM DIFFERENT WTV ISOLATES[a]

WTV Isolate	WTV-RNA Experiment		WTV Isolate	Virion Count Experiment	
	1	2		1	2
EV49	—	—	EV49	0.11	0.04
SV55	2.6	1.0	SV55	0.41	0.23
SV57	1.0	0.45	SV57	0.67	0.50
SV58	—	0.21	SV58	0.31	0.34
SV60a	0.20	0.17	SV60a	0.48	0.23
SV60b	0.30	0.20	SV60b	0.54	0.53
SV62	0.55	0.45	SV62	0.43	0.45
VI64	1.0	1.0	SV64	0.76	0.65
			SV65	0.97	0.77
			VI66	1.0	1.0

[a] In the WTV-RNA experiments, isolate VI64 was the wild strain of WTV. It was given a standard value of 1.0, and other results were calculated as proportionate values. In the virion count experiments, all results were similarly expressed proportionate to VI66, which at that time was the wild strain used as a standard; it is likely that at that time the 1964 and 1965 isolates had become subvectorial, but that was not actually determined.

the WTV-RNA determinations and VI66 for the virion determinations) which have been given a standard value of 1.0. Considering that in each case the two experiments were independent of each other, and that the two kinds of experiments were carried out independently 2 years apart by different investigators, the results are consistent in showing that with the passage of time in sweet clover the virus concentration decreases consistently but irregularly in the different isolates. The only important exceptions are the WTV-RNA values for SV55 in both experiments and those for SV57 in experiment 1. These three values may be in error. The data also show that the variation in virus concentration is relatively insignificant in relation to the tremendous variation in the infectivity of tumor extracts incited by the different isolates and that although those infectivities are so divergent in the two experiments (Table I), the important differences are not attributable to virus concentration but must be caused by differences in the specific infectivities. However, other influences not properly understood at this time probably affected the variation in infectivity determinations.

Loss of transmissibility in potato yellow dwarf virus.—Loss of transmissibility in WTV occurred under conditions where the virus after introduction into sweet clover multiplied for years—without intervention of the vector—in successive vegetative generations derived from the originally infected plants by the propagation of cuttings. In nature, the persistence of virus in perennials, such as trees, or in clonally propagated plants, such as potatoes, probably provides the same conditions for loss of transmissibility. Loss of transmissibility of PYDV occurred in the laboratory during years of multiplication of the virus in the annual, *N. rustica* (3, 8, 74). The virus was maintained by rubbing extracts from infected leaves upon healthy leaves of young *N. rustica* plants. Such conditions often occur in the maintenance of plant viruses in the laboratory.

With WTV, the transmissibility of the virus was tested by injecting it into vectors. With PYDV, it was tested by allowing vectors opportunities to ingest and to transmit virus. When PYDV not demonstrably transmissible by such tests was injected into vectors, only two cases of transmission were obtained. Both of these virus isolates were then tested for transmission following ingestion, and it was found that each isolate was transmitted about 16 per cent as frequently as the wild strain. This low transmissibility persisted unchanged through four cycles of transmission from insect to plant in series.

Field isolates of PYDV produce numerous yellow primary lesions on *N. rustica* leaves. The EV strain characteristically produces a milder lesion than does the field virus. Nevertheless, a necrotic strain of virus was readily isolated from the EV virus by selecting and subculturing virus from a single necrotic lesion. The strain so isolated proved to be vectorless. Conversely, a strain producing mild lesions could be similarly isolated from the field strain. It proved to be tranmissible.

The desirability of using vectorial strains, rather than subvectorial or exvectorial ones, for studies on host vector monolayers is obvious. The importance of the phenomenon of loss of transmissibility in the field is not so obvious, but the extent to which clonally propagated plants are used in agriculture suggests that it is important.

Conclusion

The monolayer culture of host vector cells provides an important means for studying the molecular biology of propagative plant viruses and a sensitive technique for more rapid and precise investigation of many practical problems concerning these viruses.

Addenda

Peters & Black[2] prepared primary cultures of aphid cells on cover slips. Ovarian tissues and embryos from apterous viviparous sowthistle aphids, *Hypermyzus lactucae* (L.), were removed, cut into fragments and treated with pronase. Fragments and cells were washed free of the enzyme in a suitable culture medium and sown on cover slips. Cells attached to the cover slip and maintained good attachment and good pseudopodial extensions for up to 10 days. Cells in such primary cultures were inoculated with samples of sowthistle yellow vein virus (SYVV) purified from infected plants. Inoculations were carried out by replacing medium with inoculum and allowing adsorption for 2 to 3 hr at room temperature. Subsequently, infected cells were detected by staining with fluorescent SYVV antiserum. Staining was first evident about 37 hr after inoculation and later became more intense and also occurred in more cells. As is the case with the PYDV infections, staining was prominent in the cell nuclei. PYDV and SYVV belong in the same virus taxon. No staining occurred in uninoculated control cells of *H. lactucae* or in inoculated cell cultures from the aphid *Acyrthosiphon pisum* (Harris) or from the leafhopper *A. constricta* Van Duzee. As many as 1400 infected cells were counted on a single cover slip bearing cells of *H. lactucae* so that even with primary cultures the results indicate that these new techniques provide the best assay available for this virus.

(Peters and Black acknowledge the assistance of Dr. James E. Duffus in obtaining the virus and vector for the above work.)

Recently a paper by Hirumi & Maramorosch[3] came to my attention. The authors inoculated primary cultures of *M. fascifrons* with WTV. Although this insect is not a vector of WTV they found that sections of tissue taken seven days after inoculation showed WTV viroplasms and WTV cores without capsids when examined in the electron microscope. Such structures were

[2] Peters, D., Black, L. M. Infection of primary cultures of cells from sowthistle aphids with inoculum of sowthistle yellow vein virus. (to be submitted to *Virology*)

[3] Hirumi, H., Maramorosch, K. Electron microscopy of wound tumor virus in cultured embryonic cells of the leafhopper *Macrosteles fascifrons. Intern. Colloq. Invertebrate Tissue Culture, 1967, 2nd, Inst. Lombardo,* 203-17 (1968)

not found in control tissues. Hirumi & Maramorosch also reported that degenerative effects, including the development of holes in the cell sheets, appeared in the inoculated cultures. However, degenerative changes occurred in control cultures treated with extracts from virus-free vectors of WTV. The changes in these control cultures were only sporadic and not as pronounced as in those cultures that were inoculated with extracts from vectors carrying WTV. Elucidation of the nature of the degenerative changes seems desirable.

Acknowledgment

The author is indebted to Dr. D. Peters and Mrs. Isobel Windsor for suggestions and discussions which led to the adoption of the new terms used in this review. During these discussions we considered the term "avectorial" which was not adopted for any of the conditions described here. It was thought to be more appropriate for viruses that perhaps have never had a vector, e.g. tobacco mosaic virus.

I am also much indebted to Mr. Peter E. Bloom and Mr. Charles Bussman whose careful propagation of infected and healthy sweet clover cuttings over 20 years without mishap merits confidence in the record of the wound-tumor isolates. The record is supported by the consistency of the results over the long term.

LITERATURE CITED

1. Backus, R. C., Williams, R. C. The use of spraying methods and of volatile suspending media in the preparation of specimens for electron microscopy. *J. Appl. Phys.*, **21,** 11–15 (1950)
2. Benda, G. T. A. Infection of *Nicotiana glutinosa* L. following injection of two strains of tobacco mosaic virus into a single cell. *Virology,* **2,** 820–27 (1956)
3. Black, L. M. Loss of vector transmissibility by viruses normally insect transmitted. *Abstr. Phytopathology,* **43,** 466 (1953)
4. Black, L. M. Biological cycles of plant viruses in insect vectors. In *The Viruses,* **2,** 157–85 (Burnet, F. M., Stanley, W. M., Eds., Academic Press, Inc., New York, 408 pp., 1959)
5. Black, L. M. Physiology of virus induced tumors in plants. In *Encyclopedia of Plant Physiology,* **15/2,** 236–66 (Ruhland, W., Lang, A., Eds., Springer-Verlag, New York, 1965)
6. Black, L. M., Brakke, M. K. Multiplication of wound-tumor virus in an insect vector. *Phytopathology,* **42,** 269–73 (1952)
7. Black, L. M., Brakke, M. K. Serological reactions of a plant virus transmitted by leafhoppers. *Abstr. Phytopathology,* **44,** 482 (1954)
8. Black, L. M., Wolcyrz, S., Whitcomb, R. F. A vectorless strain of wound-tumor virus. *Intern. Congr. Microbiol., 7th, Stockholm,* 255 (1958)
9. Brakke, M. K. Density gradient centrifugation. A new separation technique. *J. Am. Chem. Soc.,* **73,** 1847 (1951)
10. Brakke, M. K., Vatter, A. E., Black, L. M. Size and shape of wound-tumor virus. *U. S. At. Energy Comm., Symp. Biol., No. 6, Brookhaven National Lab., Upton, N.Y.,* 137–56 (1954)
11. Chaproniere, D. M., Andrews, C. H. Cultivation of rabbit myxoma and fibroma viruses in tissues of nonsusceptible hosts. *Virology,* **4,** 351–65 (1957)
12. Chiu, R. J., Black, L. M. Monolayer cultures of insect cell lines and their inoculation with a plant virus. *Nature,* **215,** 1076–78 (1967)
13. Chiu, R. J., Black, L. M. Assay of wound tumor virus by the fluorescent cell counting technique. *Virology,* **37,** 667–77 (1969)
14. Chiu, R. J., Liu, H. Y., MacLeod, R., Black, L. M. Potato yellow dwarf virus in leafhopper cell culture (To be submitted to *Virology*)
15. Chiu, R. J., Reddy, D. V. R., Black, L. M. Inoculation and infection of leafhopper tissue cultures with a plant virus. *Virology,* **30,** 562–66 (1966)
16. Cocking, E. C. An electron microscopic study of the initial stages of infection of isolated tomato fruit protoplasts by tobacco mosaic virus. *Planta,* **68,** 206–14 (1966)
17. Coons, A. H., Creech, H. J., Jones, R. N., Berliner, E. The demonstration of pneumococcal antigen in tissues by the use of fluorescent antibody. *J. Immunol.,* **45,** 159–70 (1942)
18. Day, M. F., Grace, T. D. C. Culture of insect tissues. *Ann. Rev. Entomol.,* **4,** 17–38 (1959)
19. Doi, Y., Teranaka, M., Yora, K., Asuyama, H. Mycoplasma- or PLT group-like microorganisms found in the phloem elements of plants infected with mulberry dwarf, potato witches' broom, aster yellows, or *Paulownia* witches' broom. *Ann. Phytopathol. Soc. Japan,* **33,** 259–66 (1967)
20. Gamez, R., Black, L. M. Application of particle-counting to a leafhopper-borne virus. *Nature,* **215,** 173–74 (1967)
21. Gamez, R., Black, L. M. Particle counts of wound-tumor virus during its peak concentration in leafhoppers. *Virology,* **34,** 444–51 (1968)
22. Gamez, R., Chiu, R. J. The minimum concentration of a plant virus needed for infection of monolayers of vector cells. *Virology,* **34,** 356–57 (1968)
23. Gibbs, A. Plant virus classification. *Advan. Virus Res.,* **14,** 263–328 (1969)
24. Grace, T. D. C. Establishment of four

strains of cells from insect tissues grown *in vitro*. *Nature,* **195,** 788–89 (1962)

25. Grace, T. D. C. Insect cell culture and virus research. In *Differentiation and defense in lower organisms,* 104–17 (Sigel, M. M., Ed., The Tissue Culture Assoc., Inc., Williams & Wilkins Co., Baltimore, Md., 1968)
26. Grace, T. D. C. Insect tissue culture and its use in virus research. *Advan. Virus Res.,* **14,** 201–20 (1969)
27. Harrison, B. D. Studies on the behavior of potato leaf roll and other viruses in the body of their aphid vector *Myzus persicae* (Sulz.). *Virology,* **6,** 265–77 (1958a)
28. Harrison, B. D. Ability of single aphids to transmit both avirulent and virulent strains of potato leaf roll virus. *Virology,* **6,** 278–86 (1958b)
29. Hirumi, H., Maramorosch, K. Insect tissue culture: Use of blastokinetic stage of leafhopper embryo. *Science,* **144,** 1465–67 (1964a)
30. Hirumi, H., Maramorosch, K. Insect tissue culture: Further studies on the cultivation of embryonic leafhopper tissues *in vitro*. *Contrib. Boyce Thompson Inst.,* **22,** 343–52 (1964b)
31. Hsu, H. T., Black, L. M. Preservation of viable leafhopper vector cell cultures under liquid nitrogen. *Abstr. Phytopathology* (In press)
32. Ishiie, T., Doi, Y., Yora, K., Asuyama, H. Suppressive effects of antibiotics of tetracycline group on symptom development of mulberry dwarf disease. *Ann. Phytopathol. Soc. Japan,* **33,** 267–75 (1967)
33. Jones, B. M. The cultivation of insect cells and tissues. *Biol. Rev.,* **37,** 512–36 (1962)
34. Kassanis, B. Plant tissue culture. In *Methods in Virology,* **1,** 537–66 (Maramorosch, K., Koprowski, H., Eds., Academic Press, Inc., New York and London, 640 pp., 1967)
35. Liu, H. Y., Black, L. M. Infectivity of varieties of potato yellow dwarf virus on vector and related nonvector cell monolayers. *Abstr. Phytopathology* (In press)
36. Luria, S. E., Darnell, J. E., Jr. Role of tissue culture in the study of animal viruses. In *General Virology,* 2nd ed., 291–93 (Wiley & Sons, Inc., N.Y., 512 pp., 1967)
37. MacLeod, R., Black, L. M., Moyer, F. H. The fine structure and intracellular localization of potato yellow dwarf virus. *Virology,* **29,** 540–52 (1966)
38. Maramorosch, K. Studies on the nature of the specific transmission of aster-yellows and corn-stunt viruses. *Phytopathology,* **42,** 663–68 (1952)
39. Maramorosch, K. Multiplication of aster yellows virus in *in vitro* preparation of insect tissues. *Virology,* **2,** 369–76 (1956)
40. Maramorosch, K., Mitsuhashi, J., Streissle, G., Hirumi, H. Animal and plant viruses in insect tissues *in vitro*. *Abstr. Bacteriol. Proc.,* **120** (1965)
41. Martignoni, M. E. Problems of insect tissue culture. *Experientia,* **16,** 125–28 (1960)
42. Mitsuhashi, J. Preliminary report on the plant virus multiplication in the leafhopper vector cells grown *in vitro*. *Japan. J. Appl. Entomol. Zool.,* **9,** 137–41 (1965)
43. Mitsuhashi, J. Infection of leafhopper and its tissues cultivated *in vitro* with *Chilo* iridescent virus. *J. Invert. Pathol.,* **9,** 432–34 (1967)
44. Mitsuhashi, J., Maramorosch, K. Leafhopper tissue culture: embryonic, nymphal, and imaginal tissues from aseptic insects. *Contrib. Boyce Thompson Inst.,* **22,** 435–60 (1964)
45. Mitsuhashi, J., Maramorosch, K. Inoculation of plant tissue cultures with aster yellows virus. *Virology,* **23,** 277–79 (1964)
46. Mitsuhashi, J., Nasu, S. An evidence for the multiplication of rice dwarf virus in the vector cell cultures inoculated *in vitro*. *Japan. J. Appl. Entomol. Zool.,* **2,** 113–14 (1967)
47. Mundry, K. W. Plant virus-host cell relations. *Ann. Rev. Phytopathol.,* **1,** 173–96 (1963)
48. Murakishi, H. H. Infection of tomato

callus cells in suspension with TMV-RNA. *Phytopathology,* **58,** 993–96 (1968)

49. Nagaraj, A. N., Sinha, R. C., Black, L. M. A smear technique for detecting virus antigen in individual vectors by the use of fluorescent antibodies. *Virology,* **15,** 205–8 (1961)
50. Nasu, S. Electron microscopic studies on transovarial passage of rice dwarf virus. *Japan. J. Appl. Entomol. Zool.,* **9,** 225–37 (1965)
51. Nault, L. R., Gyrisco, G. G. Relation of the feeding process of the pea aphid to the inoculation of pea enation mosaic virus. *Ann. Entomol. Soc. Am.,* **59,** 1185–97 (1966)
52. Reddy, D. V. R., Black, L. M. Specific infectivity of different isolates of wound-tumor virus. *Abstr. Phytopathology,* **55,** 1072 (1965)
53. Reddy, D. V. R., Black, L. M. Production of wound-tumor virus and wound-tumor soluble antigen in the insect vector. *Virology,* **30,** 551–61 (1966)
54. Roberts, D. A. Price, W. C. Infection of apparently uninjured leaves of bean by the viruses of tobacco necrosis and southern bean mosaic. *Virology,* **33,** 542–45 (1967)
55. Ruesink, A. W., Thimann, K. V. Protoplasts: Preparation from higher plants. *Science,* **154,** 280–81 (1966)
56. Schneider, I. Differentiation of larval *Drosophila* eye-antennal discs *in vitro. J. Exptl. Zool.,* **156,** 91–104 (1964)
57. Schneider, I. R., Worley, J. F. Upward and downward transport of infectious particles of southern bean mosaic virus through steamed portions of bean stems. *Virology,* **8,** 230–42 (1959a)
58. Schneider, I. R., Worley, J. F. Rapid entry of infectious particles of southern bean mosaic virus into living cells following transport of the particles in the water stream. *Virology,* **8,** 243–49 (1959b)
59. Shikata, E., Maramorosch, K., Granados, R. R. Electron microscopy of pea enation mosaic virus in plants and aphid vectors. *Virology,* **29,** 426–36 (1966)
60. Sinha, R. C. Sequential infection and distribution of wound-tumor virus in the internal organs of a vector after ingestion of virus. *Virology,* **26,** 673–86 (1965)
61. Sinha, R. C. Response of wound-tumor virus infection in insects to vector age and temperature. *Virology,* **31,** 746–48 (1967)
62. Sinha, R. C., Chiykowski, L. N. Multiplication of aster yellows in a nonvector leafhopper. *Virology,* **31,** 461–66 (1967a)
63. Sinha, R. C., Chiykowski, L. N. Multiplication of wheat striate mosaic virus in its leafhopper vector *Endria inimica. Virology,* **32,** 402–5 (1967b)
64. Stegwee, D., Ponsen, M. B. Multiplication of potato leafroll virus in the aphid *Myzus persicae* (Sulz.). *Entomol. Exptl. Appl.,* **1,** 291–300 (1958)
65. Sylvester, E. S., Richardson, J. "Recharging" pea aphids with pea enation mosaic virus. *Virology,* **30,** 592–97 (1966)
66. Sylvester, E. S., Richardson, J. Additional evidence of multiplication of the sowthistle yellow vein virus in an aphid vector—serial passage. *Virology,* **37,** 26–31 (1969)
67. Vago, C. Invertebrate tissue culture. In *Methods in Virology,* **1,** 567–602 (Maramorosch, K., Koprowski, H., Eds., Academic Press, N.Y. & London, 640 pp., 1967)
68. van Hoof, H. A. *An Investigation of the Biological Transmission of a Non-Persistent Virus* (Doctoral thesis, Wageningen, Netherlands, 1958, van Putten & Oortmeijer, Alkmaar, 110 pp., 1958)
69. Vaughn, J. L. A review of the use of insect tissue culture for the study of insect-associated viruses. *Current Topics Microbiol. Immunol.,* **42,** 108–28 (1968)
70. Watson, D. H. Electron-micrographic particle counts of phosphotungstate-sprayed virus. *Biochim. Biophys. Acta,* **61,** 321–31 (1962)
71. Whitcomb, R. F., Black, L. M. A precipitin ring time test for estimation of relative soluble-antigen concentrations. *Virology,* **15,** 507–8 (1961)
72. Williams, R. C., Backus, R. C.

Macromolecular weights determined by direct particle counting. I. The weight of the bushy stunt virus particle. *J. Am. Chem. Soc.*, **71/3,** 4052–57 (1949)

73. Wolcyrz, S., Black, L. M. Serology of potato yellow-dwarf virus. *Abstr. Phytopathology,* **46,** 32 (1956)

74. Wolcyrz, S., Black, L. M. Origin of vectorless strains of potato yellow-dwarf virus. *Abstr. Phytopathology,* **47,** 38 (1957)

Copyright 1969. All rights reserved

CELLULAR RESPONSES OF PLANTS TO NEMATODE INFECTIONS

By Victor H. Dropkin

United States Department of Agriculture, Agricultural Research Service, Crops Research Division, Nematology Investigations, Plant Industry Station Beltsville, Maryland[1]

Introduction

The purpose of this review is to present a view of the whole range of plant cell responses to nematode infection. Literature citations therefore represent selections to illustrate types of responses and not a complete list of publications. The review is organized into seven sections:

1. Nematode feeding behavior and apparatus.
2. Destructive cell changes, arranged in order from least to most destructive.
3. Adaptive cell changes; cell responses to *Tylenchulus, Nacobbus, Heterodera,* and *Meloidogyne.*
4. Mechanisms of nematode action—a summary of current thinking.
5. Resistant cell changes.
6. Alteration of host resistance to other pathogens—a short statement.
7. Summary and conclusions.

Recent reviewers of phytonematology have focused on various aspects of the subject, including some of the topics listed above (13, 17, 31, 41, 47, 51, 63).

Phytonematodes comprise three distinct taxonomic groups: the Tylenchidae, the Aphelenchidae, and several genera of the Dorylaimida. All possess a stylet used to penetrate cell walls, a set of glands emptying into the esophageal region of the alimentary canal, and a pump between the stylet and intestine. Most phytonematodes damage plant cells to various degrees by removing cell contents, or by inducing cell lysis. A few parasites, however, induce host cells to provide specific feeding sites containing specialized nutritive cells or syncytia upon which the nematode feeds. In addition to changes in those cells penetrated by the nematode, there frequently are effects in cells at some distance from the invader, as in galled tissue.

Feeding: Behavior and Apparatus

Nematodes have a well developed sensory and behavior system that enables them to seek out specific parts of plants. *Ditylenchus dipsaci* pene-

[1] Present Address: Department of Plant Pathology, University of Missouri, Columbia, Mo.

trates the stem-root junction of oats (7). *Meloidogyne* larvae enter the growing root of tomato seedlings, often close to the root cap, and they soon locate the area of differentiating xylem (43). *Pratylenchus,* however, more commonly attacks well behind the root cap (49). *Aphelenchoides ritzemabosi* ascends the stem to invade leaves. *Trichodorus,* an ectoparasite, feeds on cells of the zone of elongation in roots of apple and cherry trees; as the root grows, a population of nematodes accumulates at this zone and follows its progression through the soil until the root is damaged and the nematodes disperse (64).

Phytonematodes vigorously attack the cell wall barrier of the plant's surface by repeated thrusts of the stylet. The opposing force may be provided by lip suction (15), or by a resistant surface against which the nematode's body is braced (29). Stylet penetration into certain fungus hyphal cells is rapid (1, 81), but root-knot larvae make repeated attempts to penetrate plant cells on the root surface (43). Nematodes may also enter tissues through natural openings such as stomata, or at breaks in the surface such as the points of emergence of secondary roots.

In some associations, the stylet acts simply as the penetrating organ and suction tube. *Paraphelenchus acontiodes* (81) penetrates cells of *Pyrenochaeta terrestris* and withdraws their contents within 2 to 3 sec (Diagram II). In other associations, however, the nematode remains at a single cell for long periods during which fluid probably enters the host cell through the stylet. The complete feeding cycle of *D. myceliophagus* feeding on *Botrytis cinerea* includes an injection period of about 40 min during which fluid from the dorsal esophageal gland empties from a reservoir just behind the stylet and passes through the stylet into a hyphal cell. Following this, the posterior esophagus pumps gently for about 40 min. The termination of feeding is marked by very rapid action of the conspicuous muscular median bulb (Diagram III) for a brief period (18).

Typically three large gland cells are associated with the esophagus in Tylenchoid nematodes, one dorsal and two subventral. The duct from the dorsal gland empties anteriorly, either directly behind the stylet, or in the region of the median bulb. The other two glands empty posteriorly, in the region of the median bulb. Some forms have a large dorsal esophageal gland that opens to the alimentary tract a few microns behind the stylet. This duct commonly has an enlargement close to its junction with the esophagus. The system seems designed to deliver secretions to the plant cell via the stylet. Those few forms that have been observed to feed by removing cell contents rapidly do not have an enlargement of the dorsal gland duct. The duct, in these instances, opens farther back in the region of the median bulb. We are tempted to conclude that the former arrangement implies extra-oral digestion while the latter implies absence of this. But in *Seinura,* a predator on other nematodes, the dorsal gland opens to the esophagus in the anterior region of the median bulb, and something that paralyzes the prey is

EFFECTS OF NEMATODES ON PLANT CELLS

I. No visible destructive effect

Cyclosis continues during prolonged feeding; dome of granules forms at stylet tip.

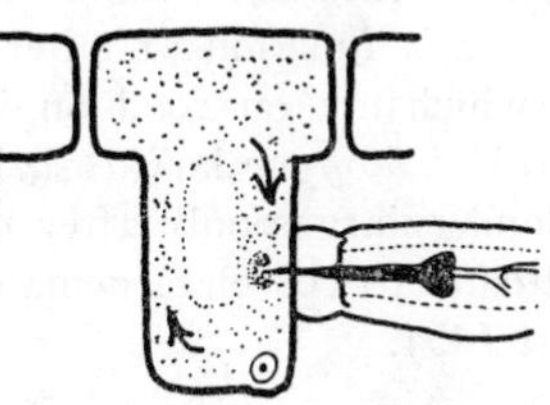

II. Immediate removal of cell contents

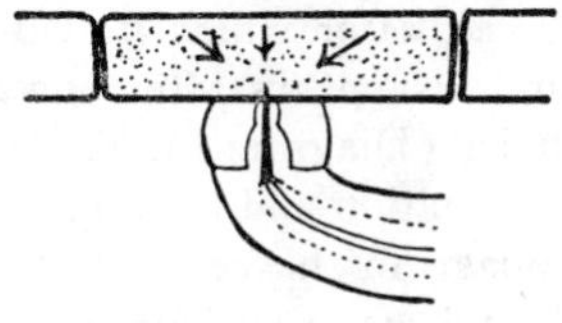

III. Delayed removal of cell contents

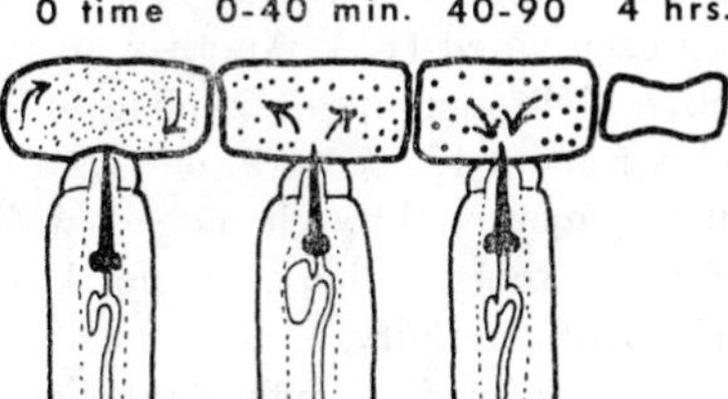

Penetration - cyclosis stops

Injection - cytoplasm becomes granular

Ingestion - granulation increases

Withdrawal - cell shrinks

IV. Progressive penetration

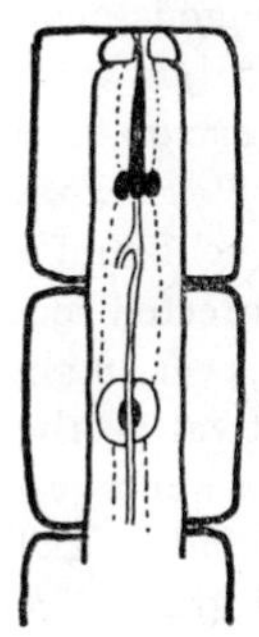

injected through the stylet. Presumably this is a toxic chemical secreted by the dorsal gland (34).

Few observations have been published on the action of glands of Dorylaimoids. *Xiphinema* has a gland opening into the esophagus close to the base of the spear as well as glands opening farther back (69).

These observations on the dorsal esophageal gland leave unanswered the question of the function of other esophageal glands. They may contribute enzymes which act on food on its way to the intestine. The subventral glands of *Meloidogyne* larvae shrink in size after penetration, whereas the dorsal gland enlarges only after the nematode has entered its host (6). A recent attempt to correlate nematode morphology with feeding behavior is of interest (45).

Destructive Cellular Changes

The feeding of some nematodes produces only slight trauma in host cells. When *Tylenchorhynchus dubius* feeds upon root hairs of *Lolium perenne,* a spherical mass forms at the stylet tip within the cells. The nematode then remains quiet for 30 sec, after which its bulb pulsates: the spherical mass diminishes in size and disappears after one min. No visible changes are evident (37) in root hairs (Diagram I). *T. claytoni* feeds on epidermal cells in the region of cell elongation and between root hairs of alfalfa without causing observable damage to the cells (38). *Paratylenchus projectus* may remain attached to one tobacco root hair cell for days. A characteristic dome of granular material develops around the stylet tip, but normal cytoplasmic streaming continues. The cells survive this prolonged feeding (68).

In contrast to the gentle action of these nematodes, most plant-parasitic forms affect host cells more drastically. Direct removal of cell contents by a mycophagous nematode has already been noted (81). Another mycophagous species, *Aphelenchus avenae,* removes cell contents of hyphae of *Thanatephorus cucumeris* in a feeding cycle of about 15 sec duration (29). One species of *Trichodorus* makes a shallow puncture through the cell wall of epidermal cells and removes part of the cytoplasm, but does not kill the cell (70). Infected roots lose their meristematic activity.

However, many nematodes require a period of extra-oral digestion before activating the median bulb pump to remove cell contents. *D. destructor* takes an average of 10 to 15 min per feeding period on the fungus *Chaetomium indicum.* The injection phase lasts 3 to 4 min during which the cytoplasm gradually becomes more translucent. The cell collapses after spear withdrawal (1).

The effect on hyphal cells varies from one fungus to the next. Cells in an unidentified genus of *Eurotiales* do not shrink after feeding and there are no visible changes other than arrest of cyclosis (1). In hyphae with small cells, those adjacent to the target cell also show evidence of damage, but such damage is not apparent in hyphae with large cells.

Cells of higher plants also suffer destruction by direct feeding. *Rotylen-*

chus uniformis was observed to feed upon root hairs and cortical parenchyma cells of *Lolium perenne* (37). The stylet penetrated through the cell wall into the vacuole, and the nematode remained with its stylet in the same cell for long periods. The affected cells, but not neighboring ones, became yellow, then brown after the nematode withdrew. The same author reported that *Pratylenchus crenatus* pierced epidermal cells of *Poa annua* rootlets. After feeding on a given cell, the nematode penetrated an adjacent cell, fed again, and moved to another cell (Diagram IV). Rootlets displayed lightly colored necrotic spots.

In another study, *Amaryllis* roots infected with *P. scribneri* had extensive areas of necrosis in the cortex with "nests" of eggs and larvae. Tracheal elements of the stele showed no histological changes. In the same host, the ectoparasite *Scutellonema brachyurum* confined itself to more superficial cells of *Amaryllis* roots. It did not seem to move as frequently as *Pratylenchus* (54). *Belonolaimus longicaudatus* produced elliptical lesions in the cortex of grapefruit roots. Each lesion consisted of a cavity with a narrow neck to the exterior, and a region of damaged cells bordering the cavity. The damage extended longitudinally in both directions from the cavity (79). In similar lesions in bean roots, the necrotic cells adjacent to the cavity extended as much as 1 mm longitudinally in each direction (78). *Helicotylenchus multicinctus* fed in the superficial cortex of banana, forming small local areas of necrotic cells, with no evidence that the nematode migrated within the cortex. In the same host, however, *Radopholus similis* moved into cavities formed by the collapse of cortical cells upon which it had fed. These cavities enlarged as the nematodes continued to destroy cells of the cortex. *Radopholus* appeared to be a restless feeder; it made extensive tunnels throughout the cortex, and there was little evidence of the death of cells at a distance (9). In citrus, however, *R. similis,* probably of another biotype, invaded the stele as well as the cortex. It caused extensive tunnels by destroying cells during feeding. Starch disappeared in cells adjoining the lesion, and nematodes which penetrated beyond the endodermis caused hypertrophy and division of the pericycle cells. Gnotobiotic cultures of grapefruit seedlings infected with *Radopholus* had no generalized cell necrosis. They did, however, show strong evidence of lytic action, presumably from enzymes secreted by the nematode (25).

Nematodes may also alter cell permeability. Roots of carrots infected with *Pratylenchus* have lesions of necrotic cells in the cortex as described above in *Amaryllis.* Healthy carrot roots suspended in a dilute solution of rhodamin B take up the dye only in the central cylinder. But in roots infected with *Pratylenchus,* this dye leaves the central cylinder and moves to tissues around the nematodes. Infection of the carrot with *Paratylenchus,* which remains attached to the root epidermis, does not alter dye penetration. We must therefore conclude that *Pratylenchus* affects permeability of tissues at some distance to itself while *Paratylenchus* does not (90).

Many phytoparasitic nematodes, e.g. *Rotylenchulus,* affect only those cells upon which they feed, or a limited number of cells in the immediate vicinity of the feeding site (3). *D. dipsaci,* however, causes changes in cells at great distances from itself. This nematode is an important pathogen of the bulb industry, both ornamental and vegetable, and of forage crops such as alfalfa. Infected plants are dwarfed and misshapen, and underground storage structures are subject to rot. Cells of infected tissues separate, undergo marked hypertrophy, and lose chloroplasts (Diagram V). In aseptic infections with one or two nematodes, leaf palisade parenchyma of alfalfa leaflets contained masses of cells with dense, granular cytoplasm (40). The conspicuous galling of young alfalfa stems, caused by cell hypertrophy, was visible within 24 hr after inoculation. As early as 12 hr after inoculation, cavities developed in the cotyledonary cortex, the epidermal cells had enlarged, and cell reaction to dyes had changed. Vascular damage was not conspicuous, but did occur in old infections. Necrosis was not common in the young alfalfa tissues. The outstanding feature of these infections is that only a few nematodes affect many cells.

Adaptive Cellular Changes

The nematodes mentioned thus far feed on cells with either mild or destructive effects. However, there is another type of host-parasite adjustment in which the parasite stimulates changes in cells resulting not in cell destruction, but in alterations of metabolism vital to the parasite's growth and development. The larvae of *Tylenchulus semipenetrans* feed on superficial cells of citrus roots. The young adult female, however, penetrates more deeply into the cortex where its head becomes surrounded by a "feeding site" consisting of six to ten altered cortex cells. These have dense cytoplasm without a vacuole, and a much enlarged nucleus and nucleolus (64 to 125 × normal volume). Older "nurse cells" develop abnormally thick walls. The altered cells are normal in size, and the immediately adjacent cells show no departures from their normal morphology (Diagram VI). Neither hypertrophy nor hyperplasia are present (83). In addition to these effects, *T. semipenetrans* leaves tracks of necrotic cells as it penetrates the cortex. Eventually the "nurse cells" disintegrate into a mass of necrotic tissue. There is some evidence of starch depletion at the feeding site (11).

Nematodes of the genera *Anguina* and *Nothanguina* stimulate gall formation in leaves and flower parts of grasses and other plants. These galls develop by hypertrophy and hyperplasia of parenchyma tissues and usually have a central cavity containing the nematodes. There are no syncytia, and with one exception, no specialized cell types. The exception is *N. cecidoplastes* which induces a well-differentiated structure resembling a cynipid gall with distinctive cell types (33). These galls seem especially suitable for study of nematode-induced pathology. The tissues are readily accessible and the reaction to only a few parasites is spectacular.

Among the phytoparasitic nematodes, members of the family Heterode-

V. Cell lysis

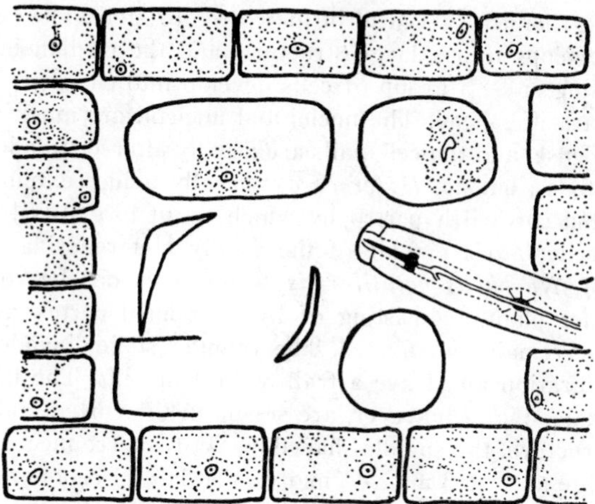

Cells near nematode enlarge and separate; cytoplasm withdraws, walls collapse; cavities form in tissues.

VI. Nurse cells of Tylenchulus

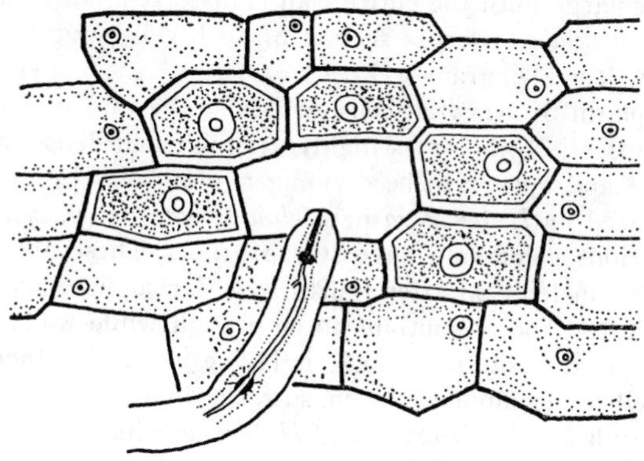

Nuclei and nucleoli enlarge, cytoplasm becomes dense, walls thicken, vacuole disappears.

ridae show the greatest morphological adaptations to parasitism. All members of the family have strong sexual dimorphism: the male is a typical elongate nematode, while the female becomes a flask-shaped sessile adult, embedded in a root with its head in the cortex or stele. The dorsal esophageal gland develops into a large structure, and the median bulb is prominent. At feeding sites, a group of cells develop into characteristic syncytia around the parasite's head. The nuclei and nucleoli are much enlarged, the cytoplasm is dense, and the cell walls are usually altered. In addition, *Meloidogyne* nematodes, but not *Heterodera,* typically induce extensive pericycle hyperplasia and cortical hypertrophy which result in galls (44).

One other nematode outside of the family Heteroderidae also induces syncytia (75). *Nacobbus batatiformis* forms galls on the roots of sugar beets and other plants, consisting of hypertrophied cortex and epidermal cells; the stele remains unaffected. The younger stages invade roots by intracellular migration and leave a trail of broken cells. The last larval and the adult female stages, however, are sessile in the cortex and feed upon a specialized structure, the spindle-shaped mass of interconnected cells, which may be as large as 2×3 mm. As many as 50 lateral roots arise from galls containing adult females. The syncytial mass forms by the alteration of cells immediately adjacent to the nematode's head, and extends longitudinally along the root. Many cells retain their individuality, but cell walls are broken down to interconnect them into a unit. Dissolution of walls is greatest close to the nematode's head, and the syncytial cells at the periphery of the system intergrade gradually with normal cells so that there is no sharp boundary between cell types. A channel leads from the nematode's head to the interior of the mass. Some cell walls break down in a series of perforations which enlarge until the entire wall is dissolved. Other walls are thickened so that in places a cross section of such cells superficially resembles scalariform xylem (Diagram VII).

The cytoplasm of syncytial cells, as is also the case in *Meloidogyne* and *Heterodera* infections, becomes highly granular and dense. Nuclei and nucleoli enlarge and appear to be dividing amitotically. In sharp contrast to syncytia induced by *Heterodera* or *Meloidogyne,* the syncytial cells of *Nacobbus* infections commonly have starch grains. These appear first in the vicinity of the nuclei soon after fourth stage larvae have initiated syncytial formation. Starch bodies continue to be present while walls are dissolving, but may disappear while the female is laying eggs. No other nematode induces host cells to accumulate starch, so far as known.

To recapitulate, *Meloidogyne* and *Heterodera* induce syncytia primarily in the stele, *Meloidodera* (72) and *Nacobbus* in the cortex. The syncytia of *Meloidogyne* infections usually consist of discrete units, the thick-walled "giant cells" containing hundreds of nuclei; *Heterodera*-induced syncytia tend to retain remnants of cell walls, but also have large numbers of nuclei; those of *Nacobbus* are groups of interconnected cells retaining much of their individuality.

VII. Syncytium of Nacobbus

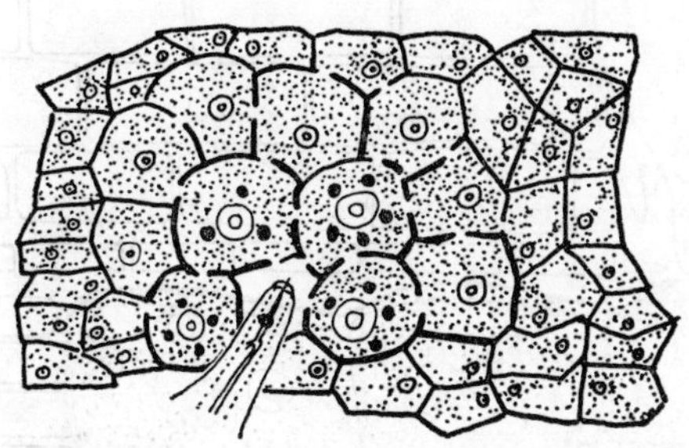

Cells hypertrophy; nucleus and nucleolus enlarge, divide amitotically; cytoplasm becomes dense; starch accumulates, walls dissolve in part but also thicken, cells retain identity.

Tylenchulus, Anguina, Nacobbus, and members of the Heteroderidae all apparently stimulate cells to increase synthesis of products on which the nematodes feed. Direct evidence of this exists only for *Meloidogyne*. Increased protein content of syncytia has been demonstrated histochemically, and increased DNA synthesis has been shown by autoradiography (71). In addition, increased dehydrogenase and diaphorase activity has been localized histochemically around the head of *M. incognita acrita* in soybean roots (27). The nuclei of syncytia are polyploid. Mitochondrial changes have also been observed in infections by *Heterodera* (61). Details of cellular changes in infections by the other genera of nematodes are not available.

In addition to the nematodes already mentioned, many others induce gall formation. Representative species are: *D. radicicola* (32), *Dolichodorus heterocephalus* (56), *Hemicycliophora arenaria* (84), *Longidorus maximus* (80), and *Xiphinema diversicaudatum* (74) associated with root galls; *Ditylenchus dipsaci* characteristically forms stem and leaf galls; and *Anguina* and *Nothanguina* induce leaf and flower galls.

Mechanisms of Nematode Action

The evidence for nematode-induced changes in plant growth regulatory substances is fairly strong. The multiple side roots arising from galls of *Nacobbus* and *Meloidogyne,* the greening of galls under *in vitro* cultivation (22), and the occasional appearance of stem buds in *Nacobbus* and *Meloidogyne* infections in roots, all point to plant growth regulators. Analysis of affected plant tissues or of nematode parasites has not yet revealed definite patterns to explain the observed growth changes. *Ditylenchus dipsaci* (par-

VIII. Syncytium of Heterodera

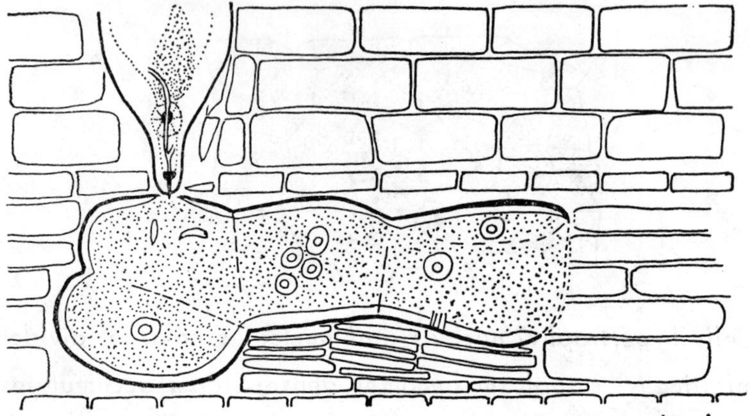

Nuclei enlarge; cell walls dissolve, but not completely; cytoplasm merges into one unit; nuclei disintegrate; cytoplasm becomes dense and granular; wall thickens.

IX. Giant cells of Meloidogyne

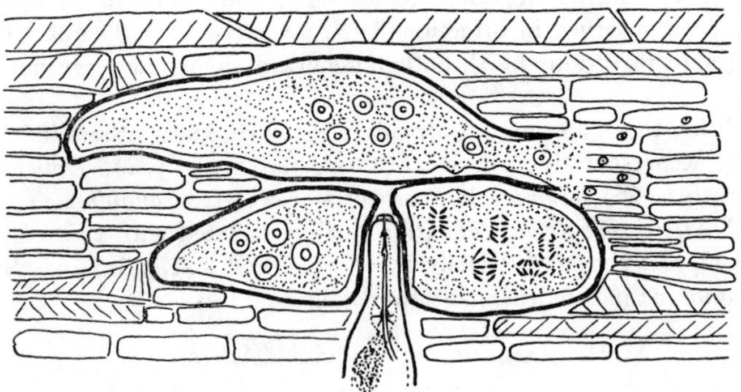

Nuclei enlarge, become polyploid, undergo synchronous mitoses; cytoplasm becomes granular; new cells incorporated by cell-wall dissolution; walls of syncytium thicken; cells of pericycle divide repeatedly.

asitic on plants) and *D. triformis* (myceliophagous) contain a growth-promoting compound tentatively identified as indole-3-acetic methyl ester. No growth-promoting activity could be found in extracts from *Turbatrix aceti* (microbivorous). The nematodes probably synthesize the biologically active compounds since host tissues do not contain them (12). There are also reports of auxin activity in extracts of *Meloidogyne* (77, 92) and of auxin inactivators in *Ditylenchus* and *Meloidogyne* (86); however no auxin inactivators were found in exudates of *Meloidogyne* by another investigator (5). There are several reports of changes in levels of endogenous growth regulators in the presence of nematode infection: enhanced auxins of several types, according to R_F values in paper chromatography of extracts of tomato galls induced by *Meloidogyne* (4, 87, 92), depressed auxins in alfalfa-shoot tips infected with *D. dipsaci* (86), and depressed polar transport of auxins in tobacco stem segments infected with *Meloidogyne* (73). It has not been possible to distinguish between the introduction of a growth regulator by the nematode, and the induction of changes in levels or gradients of endogenous growth regulators of the host, or inhibition of enzymes that degrade growth regulators. *Aphelenchoides* provides an interesting experimental subject. These nematodes invade buds of plants and feed upon the protected embryonic tissues. Infected plants are dwarfed and deformed with damaged leaves and flowers (10). There are no detailed descriptions of the histopathology of such plants. It would be most interesting to imitate the damage to buds by destroying tissues mechanically, and thus to determine whether the observed growth abnormalities result from bud destruction per se, or from an additional disturbance by the nematode of the plant's own growth regulators.

It is tempting to explain the extensive cell lysis observed around *D. dipsaci,* or the cell wall dissolution in syncytia of *Meloidogyne* or *Heterodera* infections, by suggesting that the nematodes emit hydrolytic enzymes into plant tissues. Plant parasitic nematodes have relatively high concentrations of enzymes that degrade modified cellulose. This enzyme is absent or present only in low concentrations in microbivorous forms (19, 39, 52). Some parasites also have a pectinase, but related parasitic forms may lack the enzyme. Homogenates of *Ditylenchus dipsaci* contain polygalacturonase and pectin transeliminase, but homogenates of microbivorous species and of species of nematodes that are strictly mycophagous lack pectinases. Separation of plant cells, which is characteristic of tissues harboring *D. dipsaci,* cannot result solely from the action of pectinases, since these enzymes also occur in phytonematodes which do not cause cell separation in their hosts (42, 52). The data at hand are not sufficient to account for the observed differences among nematode species in the details of damage to plant tissues.

Another difficulty is that in the syncytia of *Meloidogyne* or *Heterodera* infections, the site of active cell wall dissolution is usually the end of the syncytium distal to the nematode. Cell walls in the immediate region of the nematode's head are intact. At the same time that cells are being incorporated into a syncytium by cell wall dissolution, thickened walls are being deposited in other parts of the same syncytium. The whole process of plant cell wall formation is surely altered in the presence of *Meloidogyne*. In certain soybean plants resistant to *Meloidogyne incognita,* anomalous cellulose-containing spiral structures were observed throughout the cells, as if the orderly deposition of cellulose in cell walls had given way to aberrant deposition in the cytoplasm (24).

Recent work on decapitated pea epicotyls indicates that IAA plays a central role in the control of cellulase synthesis (28). The authors postulate that IAA derepresses that part of the genome which synthesizes RNA for cellulase, and therefore acts on transcription. The cellulase presumably moves to the cell walls where it is rapidly inactivated. The enzyme may change wall plasticity and may influence cellulose deposition by providing sites for attachment of new cellulose. Many of the effects of nematodes are parallel to those reported in the work on IAA. Hyperplasia, hypertrophy, side root induction, and the breakdown of cell walls all developed in decapitated pea epicotyls treated with exogenous IAA. The pattern of cell wall lysis in *D. dipsaci* infections of stems and in *Meloidogyne, Heterodera* or *Nacobbus* root infections does not seem to result from passive response to enzymes emanating from the nematode, but rather suggests an active host participation in response to some controlling force from the parasite. The observations on disturbed wall synthesis in resistant soybean plants also suggests some control by the nematode of the host's cellulose formation.

The nematode, therefore, may introduce or stimulate the formation of certain plant growth regulators in its host, which in turn regulate some of the observed changes.

The effect of nematodes on host nuclear activity seems to be the key action by the genera *Anguina, Tylenchulus, Nacobbus,* and by members of the Heteroderidae. There is strong evidence of nuclear destruction by *H. rostochiensis* in cells close to the larva (60). However, nuclear destruction by *Meloidogyne* is not evident, at least during the early development of the parasite in the host, but host nuclei divide, become polyploid, and increase in volume in hypertrophied cells which do not subdivide. Nuclei in cells immediately adjacent to syncytia are normal in size although once incorporated into a syncytium the nuclei swell to many times the original volume (55). The nematode must therefore exert some control over nuclear activity, probably indirectly. The strong protein-degrading activity demonstrated in homogenates of *Heterodera* larvae may be part of the control mechanism (16). The syncytia of *Meloidogyne* and *Heterodera* infections deteriorate at the end of the life cycle when the adult female dies (26). They also deteriorate if the developing nematode is killed by puncture (4) or is removed (16). The nematode's secretions must continually regulate the altered metabolism of its host's cells.

Meloidogyne has two distinct effects in hosts: (*a*) syncytia develop from cells directly penetrated by the nematode's stylet and from immediately adjacent cells incorporated by cell wall dissolution; (*b*) cells, especially of the pericycle but also of the cortex, enlarge and divide. The stimulus for syncytial development does not cross cell boundaries, except as cell walls are lysed and new cells incorporated into the developing unit. However, the stimulus to hyperplasia and hypertrophy reaches several hundred cells in the general vicinity of the parasite. Galls are visible within 24 hr after root penetration by infective larvae (21).

The size and character of the galls are influenced by both host and parasite. *Meloidogyne incognita* induces comparatively large galls in okra, for example, but small or no galls in certain soybean varieties and grasses. Moreover, *M. hapla* infections characteristically develop small galls with many side roots, whereas other species induce larger galls with fewer side roots.

Resistant Cell Changes

The cells of many plants (both host and non-host) react to the feeding of nematodes by shrinking and becoming brown. This has been noted in infections with *Pratylenchus, Radopholus, Belonolaimus, Rotylenchus, Aphelenchoides* and others. Exactly what kills these cells is not understood. *Pratylenchus* releases enzymes that hydrolyze glycosides to phytotoxic compounds (50). Quantities of phenols in tissues are correlated with the degree of browning observed in *Pratylenchus* infections. Large amounts of phenolic substances appear only in dermal and endodermal cells of apple roots according to histochemical tests. Peach roots, however, react strongly positive for phenols in cortical parenchyma as well as in dermal and endodermal layers. *Pratylenchus* causes rapid browning of all tissues of peach roots but only of dermal and endodermal tissues of apple roots (66). Similarly, *P. penetrans* reproduces without inducing brown lesions in roots of turfgrass which have low concentrations of phenolic compounds, but induces brown lesions in leaves of ryegrass with high concentrations of phenolic substances (82).

Certain, but not all types of genetic resistance also depend upon a necrotic response to nematodes. The most common resistance to *Meloidogyne* and *Heterodera* is of this type. Differences in phenolic substrates in hosts or in beta-glucosidases in parasite have been invoked to account for resistance reactions (91). The developing tracheid of a susceptible tomato infected with *Meloidogyne* loses its spiral wall thickenings and enlarges into a multinucleate cell within a day or two after receiving the stimulus. Nuclei become polyploid and divide synchronously. Meanwhile, cell walls fail to develop so that a typical multinucleate syncytium is clearly discernible 48 hr after invasion. Several days later, the syncytium incorporates additional cells through cell wall lysis. The cells of a resistant tomato, however, do not survive the nematode's intrusion. The larva's head is quickly surrounded by a sleeve of necrotic, dark brown cells, and the nematode remains quiescent in the root until it presumably starves to death. This reaction is observable as early as 11 hr after seedling roots are exposed to larvae. The total amount of browning is very slight in the susceptible root, although the larva may damage tissue along its path of penetration and migration (20).

The resistant reaction depends upon the genetic constitution of the affected cells. Cells from resistant and susceptible tomato roots maintain their reaction types in cultures (59). The majority of reports on grafting experiments indicate no influence of the scion upon the stock. Several reports of such an influence are open to the objection that lack of vigor of root growth in the grafts confused the results (85).

The necrotic response is blocked and syncytia develop when resistant tomato seedlings are grown in the presence of exogenous cytokinins. Young seedlings grown on agar with several levels of indoleacetic acid, naphthaleneacetic acid, and gibberellic acid exhibited the necrotic reaction. In contrast, when kinetin, 6-benzyl adenine, zeatin, or 6-(γ,γ dimethylallylamino) purine were added to the medium, a large proportion of *M. incognita acrita* larvae began to grow, and syncytia developed as they do in a susceptible host. Purines and pyrimidines without cytokinin activity failed to suppress or alter the expression of the resistance (23).

Temperature also changes the resistance cell reactions in some, but not in all host-parasite combinations (35, 85). In the temperature-labile combinations, resistance drops as temperature is elevated.

In a series of inoculated seedlings incubated at temperatures rising in one-degree increments from 28° to 33°C, the resistance of Nematex tomatoes fell progressively as the temperature rose (20).

The genera that induce adaptive cell changes in their hosts probably deliver some type of signal to the receptor cell which responds by increasing protein synthesis. In a resistant plant of the type described, the nematode kills the cell which subsequently develops its brown color, possibly by polyphenol oxidation and polymerization. Cytokinins are involved in some way (at least in the case studied) so that the genetically resistant cell in the presence of added cytokinins responds to the signal by proceeding on the path of syncytial development. This correlates with the suspected role of cytokinins in nucleic acid metabolism, although this role is not yet fully established. Similarly, elevated temperature permits the host cell to react to the parasite's stimulus with increased synthesis.

The study of hypersensitivity to fungus and bacterial infections has not yet revealed the mechanism by which host cells die. The hypothesis that some type of antigen-antibody reaction occurs is attractive. The possibility that virulent plant pathogens and their hosts share antigens, whereas avirulent parasites do not have these antigens suggests that hypersensitivity may result from an antigen-antibody response at the cellular level (14). No work of this type has been done with nematodes of plants. Suitable combinations for test of the hypothesis can be found in resistant and susceptible varieties of host plants such as soybeans or tobacco and biotypes of *Meloidogyne*.

Another hypothesis is that *Meloidogyne* and *Heterodera* induce the formation of syncytia by secreting proteolytic enzymes into the plant. The plant, in turn, synthesizes inhibitors more or less specific to the particular nematode species. An equilibrium is established in this way. In resistant plants, the inhibitors predominate. Proof of this hypothesis will depend on the isolation of the postulated inhibitors (48).

The larvae of *Meloidogyne* in the roots of a resistant variety of tomato mentioned above are not killed for at least 6 days by the products of host cell destruction. They may be removed by dissection and subsequently will invade the roots of susceptible plants to establish normal infections. But in the hypersensitive root those larvae do not move from the site of the reaction. *Pratylenchus* nematodes, on the other hand, typically induce strong necrotic reactions in a group of cells immediately surrounding the feeding site. They move to undamaged areas of the root where the cycle is repeated. Perhaps *Meloidogyne* larvae move in response to a gradient of stimuli emanating from differentiating tissues within the root. The larvae remain quiescent when the gradient is interrupted by cell necrosis.

Cells of resistant plants also respond in other ways to nematode attack. The early stages of syncytial development in soybeans resistant to *Heterodera glycines,* parallel those of the susceptible varieties; within a few days, however, development of both host syncytia and of the parasite come to a halt, and parenchyma tissue replaces the damaged cells (26). A similar situation differing only in minor details exists in soybean varieties resistant to *Meloidogyne* spp. (24). Potato tubers resistant to *D. dipsaci* produce an ovoid structure of altered parenchyma cells filled with a gummy substance around the invading nematodes. A true cork cambium develops so that the tuber effectively seals off the parasites (76). The early reaction of resistant Alaska 14A garden pea seedlings to two populations of *D. dipsaci* resembles infections in susceptible alfalfa seedlings. Cells enlarge and separate in both, galls are formed, and large intercellular spaces appear. The most noteworthy difference, however, is that in later infections, the nuclei enlarge in susceptible alfalfa but not in resistant peas (36). In Wando garden peas, however, a susceptible reaction is induced by one of these populations and a hypersensitive-resistant reaction by the other (2). *D. dipsaci,* however, induces much less tissue destruction in resistant oats than it does in susceptible varieties (8). The cellular response, therefore, reflects both host and parasite genetics.

Alteration of Host Resistance to Other Plant Pathogens

Nematodes not only affect plant cells directly, but they also influence the reaction of plants to other pathogens. Although very little information about the mechanism of such action is in hand, many papers report that pathogenic nematodes increase the susceptibility of plants to bacterial and fungal pathogens. The extensive early literature on nematode-fungus interactions was reviewed by Dittman (17). Other reviews are listed (47, 62, 63,

67). Fungus hyphae have been seen to invade galls and syncytia more heavily than normal tissue, perhaps in response to higher levels of amino acids or other compounds in these cells (46). Nematodes probably also affect symptom expression of virus diseases, but there are no reports. In at least one disease, the "cauliflower disease" of strawberries, nematodes, and bacteria together produce symptoms not produced by either organism alone (65).

Nematodes also serve as vectors of other plant pathogens, principally viruses (13). We may note that those nematodes proven to be virus vectors have relatively mild effects on host cells. No cells are destroyed when *Trichodorus christiei,* an ectoparasite, inserts its stylet into epidermal cells of the root tip. The root however, develops a "stubby root" symptom. Similarly, *Xiphinema index* does not induce cell necrosis (30). Both of these species transmit viruses. Russian nematologists consider microbivorous nematodes to be important vectors of bacteria that degrade plant tissues (57).

Summary and Conclusions

The effects of nematodes upon plant cells are varied. Some nematodes withdraw cell contents at such a slow rate and with such finesse that they do not permanently damage the cells. At the opposite extreme, some nematodes remove cell contents with great force, emptying the cell by vigorous suction. Many nematodes, although feeding upon cells with destructive results, do not remove cell contents completely and leave enough residue to impart a brown color to the tissue. Several groups of nematodes belonging to different taxonomic categories have developed a special relationship in which they stimulate the host cells to increased synthesis. Highly specialized structures appear on which the parasites feed.

The mechanism of nematode action for all of the effects except the last seem to offer no theoretical problem. The parasite possesses a suitable organ to penetrate cell walls and to channel fluid from host cell to its alimentary canal. The muscular pumps and associated glands of the esophagus have obvious functions. Some nematodes introduce fluids into cells and wait for a time to withdraw cell contents. Presumably these are periods of extra-oral digestion.

The induction of specialized structures in those cells directly penetrated, and the extensive changes in cells at a distance from the parasite, are more complex phenomena. If the nematode does indeed direct protein synthesis by the host cell, then it assumes a role analogous to that of a virus, albeit with a much lengthened life cycle.

Changes in host growth regulatory substances are probably part of many nematode infections. The ability of certain growth regulators and of temperature to alter the course of infection in genetically resistant plants suggests that resistance is the result of subtle metabolic differences between varieties. The extensive top growth changes observed in some nematode infections are probably responses to changes in growth regulators only indirectly related to nematode activities. Growth regulators also stimulate nematode growth and reproduction in callus cultures (89).

The observed effects of nematodes are usually considered to result from substances discharged from the stylet and emanating from associated glands. However, the excretory activities of nematodes must not be overlooked. Both the anus and the excretory pore discharge products of nematode metabolism into the surrounding plant tissues. Particularly during molting, probably fluids also escape through the cuticle. These products certainly contain nitrogenous and other compounds which influence host cells (27, 53).

The gross pathology of nematode infections reflects the cellular effects resulting from the individual parasite's activities. Many nematodes simply remove cell contents of a limited number of cells. Others appear to be destructive because of their tendency to move about in tissues. Perhaps this indicates lack of adequate nutrition. Wallace has suggested this in the case of *Aphelenchoides ritzemabosi* in chrysanthemum leaves (88). Part of the pathology may also result from the release of toxic products by the action of nematode enzymes on host substrates (50). In addition, excretions from nematodes must affect the cells with which they come into contact.

The series described under adaptive cell changes suggests to me that nematodes have developed systems for directing the host's metabolism. *Tylenchulus semipentrans* apparently affects only the cells into which it thrusts its stylet; it probably lacks the factors that enable *Meloidogyne* to direct a relatively large effort by the plant on behalf of the parasite. Perhaps *Meloidogyne* acts by influencing the plant's growth regulators, but *Tylenchulus* does not. The gall induced by *Meloidogyne* is the external sign of altered metabolism and especially of changes leading to increased protein synthesis such as the increased content of nucleic acids, amino acids, and the increased dehydrogenase and diaphorase enzyme activity (27, 55). How nematodes produce such effects is completely unknown. The plant cell nucleus seems to be an immediate target of stimuli emanating from the parasite.

It will be difficult to determine the nature of such stimuli, but the attempt should be made. We must focus on the first few hours after penetration. We must also learn to recover nematodes and to separate them from host tissues after penetration. Since the dorsal esophageal glands become active (at least in *Meloidogyne*) after penetration, it seems useless to look in freshly emerged larvae for stimuli which will induce the polyploidy, increased protein synthesis, cell wall dissolution, and other effects (yet to be determined). An immediate problem is to produce accurately timed infections to provide sufficient quantities of tissue for study by modern techniques of biochemistry and molecular biology. Very young seedlings exposed to large numbers of infective nematodes will be suitable for *Ditylenchus, Meloidogyne, Heterodera, Pratylenchus,* and other root- and stem-inhabiting forms. Leaves and buds of plants inoculated with *Aphelenchoides* should serve equally well for the collection of useful amounts of tissue. On the cellular level, the techniques of cytochemistry seem especially adapted for study of the large syncytia produced by *Meloidogyne* and *Heterodera*. We must delineate the biochemical changes in host metabolism that result in starch accumulation in *Nacobbus* infections. Since giant cells of *Meloidogyne* and syncytia of *Heterodera* infections are discrete units that undoubtedly differ in density from the surrounding cells, we may learn to separate them *en masse* from unaltered cells by enzyme maceration and differential sedimentation. Cell culture techniques offer an avenue for the study of the effects of ectoparasites such as *Criconema* and *Belonolaimus*.

This review will possibly be the last one to be written on the cellular effects of nematodes in plants that is based almost exclusively on observations made with the light microscope and classical histological procedures. A few investigators are beginning to use electron microscopy, histochemistry, and autoradiography to study the dynamic aspects of the subject (58, 71). Cinematography has already begun to show us how phytonematodes spend their time in and around plants (18).

To state the situation simply, we can no longer rely on the classical methods of light microscopy. The nematologist must now develop systems for managing host and parasite so that he can apply modern techniques to unravel the intricacies of the relationships between host and parasite—and this implies the organization of teams of qualified people focused on the problems posed in this review.

LITERATURE CITED

1. Andersen, R. V. Feeding of *Ditylenchus destructor*. *Phytopathology,* **54,** 1121–26 (1964)
2. Barker, K. R., Sasser, J. M. Biology and control of the stem nematode, *Ditylenchus dipsaci*. *Phytopathology,* **49,** 664–70 (1959)
3. Birchfield, W. Host-parasite relations of *Rotylenchulus reniformis* on *Gossypium hirsutum*. *Phytopathology,* **52,** 862–65 (1962)
4. Bird, A. F. The inducement of giant cells by *Meloidogyne javanica*. *Nematologica,* **8,** 1–10 (1962)
5. Bird, A. F. Some observations on exudates from *Meloidogyne* larvae. *Nematologica,* **12,** 471–82 (1966)
6. Bird, A. F. Changes associated with parasitism in nematodes. I. Morphology and physiology of preparasitic and parasitic larvae of *Meloidogyne javanica*. *J. Parasitol.,* **53,** 768–76 (1967)
7. Blake, C. D. Some observations on the orientation of *Ditylenchus dipsaci* and invasion of oat seedlings. *Nematologica,* **8,** 177–92 (1962)
8. Blake, C. D. The etiology of tulip-root disease in susceptible and in resistant varieties of oats infested by the stem nematode, *Ditylenchus dipsaci* (Kühn) Filipjev II. Histopathology of tulip-root and development of the nematode. *Ann. Appl. Biol.,* **50,** 713–22 (1962)
9. Blake, C. D. The histological changes in banana roots caused by *Radopholus similis* and *Helicotylenchus multicinctus*. *Nematologica,* **12,** 129–37 (1966)
10. Christie, J. R. *Plant nematodes, their bionomics and control* (Agr. Expt. Sta., Univ. Florida, Gainesville, 256 pp., 1959)
11. Cohn, E. On the feeding and histopathology of the citrus nematode. *Nematologica,* **11,** 47–54 (1965)
12. Cutler, H. G., Krusberg, L. R. Plant growth regulators in *Ditylenchus dipsaci, Ditylenchus triformis* and host tissues. *Plant Cell Physiol. (Tokyo),* **9,** 479–97 (1968)
13. Dalmasso, A. Connaissance actuelles sur les Nematodes phytophages et leurs relations avec les maladies a virus. *Ann. Épiphyties,* **18,** 249–72 (1967)
14. DeVay, J. E., Schnathorst, W. C., Foda, M. S. Common antigens and host-parasite interactions. In *The Dynamic Role of Molecular Constituents in Plant-Parasite Interaction,* 313–28 (Mirocha, C. J., Uritani, I., Eds., Am. Phytopathol. Soc., St. Paul, Minnesota, 372 pp., 1967)
15. Dickinson, S. The behavior of larvae of *Heterodera schachtii* on nitrocellulose membranes. *Nematologica,* **4,** 60–66 (1959)
16. Dieter, A. Vergleichende experimentelle Untersuchungen an zoophagen und phytophagen Nematoden. *Wiss. Z. Martin-Luther Univ., Halle-Wittenberg, Math-Nat. Reihe,* **5,** 157–86 (1955–56)
17. Dittmann, A. L. Ätiologische Zusammenhänge zwischen Nematoden und Pilzkrankheiten. *Zentr. Bakteriol. Parasitenk. Abt. II.,* **116,** 716–49 (1963)
18. Doncaster, C. C. Nematode feeding mechanisms. 2. Observations on *Ditylenchus destructor* and *D. myceliophagus* feeding on *Botrytis cinerea*. *Nematologica,* **12,** 417–27 (1966)
19. Dropkin, V. H. Cellulase in phytoparasitic nematodes. *Nematologica,* **9,** 444–54 (1963)
20. Dropkin, V. H. (Unpublished data)
21. Dropkin, V. H., Boone, W. R. Analysis of host-parasite relationships of root-knot nematodes by single-larva inoculations of excised tomato roots. *Nematologica,* **12,** 225–36 (1966)
22. Dropkin, V. H., Davis, D. W., Webb, R. E. Resistance of tomato to *Meloidogyne incognita acrita* and to *M. hapla* (root-knot nematodes) as determined by a new technique. *Proc. Am. Soc. Hort. Sci.,* **90,** 316–23 (1967)
23. Dropkin, V. H., Helgeson, J. P., Upper, C. D. The hypersensitivity reaction of tomatoes resistant to *Meloidogyne incognita:* reversal by cytokinins. *J. Nematol.,* **1,** 55–61 (1969)
24. Dropkin, V. H., Nelson, P. E. The histopathology of root-knot nematode infections in soybeans. *Phytopathology,* **50,** 442–47 (1960)

25. DuCharme, E. P. Morphogenesis and histopathology of lesions induced on citrus roots by *Radopholus similis*. *Phytopathology,* **49,** 388–95 (1959)

26. Endo, B. Y. Histological responses of resistant and susceptible soybean varieties and backcross progeny to entry and development of *Heterodera glycines*. *Phytopathology,* **55,** 375–81 (1965)

27. Endo, B. Y., Veech, J. A. The histochemical localization of oxidoreductive enzymes of soybeans infected with the root-knot nematode *Meloidogyne incognita acrita*. *Phytopathology,* **59** (in press)

28. Fan, Der-Fong, Machlachlan, G. A. Massive synthesis of ribonucleic acid and cellulase in the pea epicotyl in response to indoleacetic acid, with and without concurrent cell division. *Plant Physiol.,* **42,** 1114–22 (1967)

29. Fisher, J. M., Evans, A. A. Penetration and feeding by *Aphelenchus avenae*. *Nematologica,* **13,** 425–28 (1967)

30. Fisher, J. M., Raski, D. J. Feeding of *Xiphinema index* and *X. diversicaudatum*. *Proc. Helminthol. Soc. Wash. D.C.,* **34,** 68–72 (1967)

31. Flegg, J. J. M. The plant pathogen relationship (Blackman Essay). *Rept. East Malling Res. Sta.,* **1964,** 62–70 (1965)

32. Goodey, T. Some observations on the biology of the root-gall nematode, *Anguillulina radicicola* (Greef, 1872). *J. Helminthol.,* **10,** 33–44 (1932)

33. Goodey, T. *Anguillulina cecidoplastes,* n. sp., a nematode causing galls on the grass, *Andropogon pertusus* Willd. *J. Helminthol.,* **12,** 225–36 (1934)

34. Hechler, H. C. Description, developmental biology, and feeding habits of *Seinura tenuicaudata* (De Man) J. B. Goodey, 1960 (Nematoda: *Aphelenchoididae*), a nematode predator. *Proc. Helminthol. Soc. Wash. D.C.,* **30,** 182–95 (1963)

35. Holtzmann, O. Effect of soil temperature on resistance of tomato to root-knot nematode (*Meloidogyne incognita*). *Phytopathology,* **55,** 990–92 (1965)

36. Hussey, R. S., Krusberg, L. R. Histopathology of resistant reactions in Alaska pea seedlings to two populations of *Ditylenchus dipsaci*. *Phytopathology,* **58,** 1305–10 (1968)

37. Klinkenberg, C. H. Observations on the feeding habits of *Rotylenchus uniformis, Pratylenchus crenatus, P. penetrans, Tylenchorhynchus dubius* and *Hemicycliophora similis*. *Nematologica,* **9,** 502–6 (1963)

38. Krusberg, L. R. Investigations of the life cycle, reproduction, feeding habits and host range of *Tylenchorhynchus claytoni* Steiner. *Nematologica,* **4,** 187–97 (1959)

39. Krusberg, L. R. Hydrolytic and respiratory enzymes of species of *Ditylenchus* and *Pratylenchus*. *Phytopathology,* **50,** 9–22 (1960)

40. Krusberg, L. R. Studies on the culturing and parasitism of plant parasitic nematodes in particular *Ditylenchus dipsaci* and *Aphelenchoides ritzemabosi* on alfalfa tissues. *Nematologica,* **6,** 181–200 (1961)

41. Krusberg, L. R. Host response to nematode infection. *Ann. Rev. Phytopathol.,* **1,** 219–40 (1963)

42. Krusberg, L. R. Pectinases in *Ditylenchus dipsaci*. *Nematologica,* **13,** 443–51 (1967)

43. Linford, M. B. The transient feeding of root-knot nematode larvae. *Phytopathology,* **32,** 580–89 (1942)

44. Mankau, R., Linford, M. B. Host-parasite relationships of the clover cyst nematode, *Heterodera trifolii* Goffart. *Illinois Agr. Expt. Sta. Bull. 667,* 1–50 (1960)

45. McElroy, F. D., Van Gundy, S. D. Observations on the feeding processes of *Hemicycliophora arenaria*. *Phytopathology,* **58,** 1558–65 (1968)

46. Melendez, P. L., Powell, N. T. Histological aspects of the Fusarium wilt-root-knot complex in flue-cured tobacco. *Phytopathology,* **57,** 286–92 (1967)

47. Miller, H. N. Interactions of nematodes and other plant pathogens. *Soil Crop Sci. Soc. Florida Proc.,* **24,** 310–25 (1965)

48. Mjuge, S. G. On the relation of the phytohelminths of the Heteroderidae family to the host-plant. In *Problems of Phytohelminthology,* 146–60 (Skriabin, K. L., Turlygina, E. S., Eds., Acad. Sci., U.S.S.R., Moscow, 1961, in Russian)

49. Mountain, W. B. Studies of nematodes in relation to brown root rot of tobacco in Ontario. *Can. J. Botan.*, **32,** 737–59 (1954)

50. Mountain, W. B. Studies on the pathogenicity of *Pratylenchus. Recent Advances in Botany, Intern. Botan. Congr., 9th, Montreal, 1959,* 414–17 (1961)

51. Mountain, W. B. Pathogenesis by soil nematodes. In *Ecology of Soil-borne Plant Pathogens,* 285–301 (Baker, K. F., Snyder, W. C., Eds., Univ. California Press, Berkeley, Calif., 571 pp., 1965)

52. Myers, R. F. Amylase, cellulase, invertase, and pectinase in several free-living, mycophagus, and plant-parasitic nematodes. *Nematologica,* **11,** 441–48 (1965)

53. Myers, R. F., Krusberg, L. R. Organic substances discharged by plant-parasitic nematodes. *Phytopathology,* **55,** 429–37 (1965)

54. Nong, L., Weber, G. F. Pathological effects of *Pratylenchus scribneri* and *Scutellonema brachyurum* on amaryllis. *Phytopathology,* **55,** 228–30 (1965)

55. Owens, R. G., Specht, H. N. Biochemical alterations induced in host tissues by root-knot nematodes. *Contrib. Boyce Thompson Inst.,* **23,** 181–98 (1966)

56. Paracer, S. M., Waseem, M., Zuckerman, B. M. The biology and pathogenicity of the awl nematode, *Dolichodorus heterocephalus. Nematologica,* **13,** 517–24 (1967)

57. Paramonov, A. A. *Plant-parasitic nematodes,* Vol. 1. *Origin of nematodes, ecological and morphological characteristics of plant nematodes, principles of taxonomy* (Skriabin, K. L., Ed., Trans. from Russian, Israel Prog. Sci. Transl., Jerusalem, 390 pp., 1968) Available from U. S. Dept. of Commerce, TT-67-51219, Clearinghouse for Federal Scientific and Technical Information, Springfield, Va. 22151

58. Paulson, R. E., Webster, J. H. Ultrastructural response of plant cells to gall-forming nematodes (Abstr.) *J. Nematol.,* **1,** 22–23 (1969)

59. Peacock, F. C. The development of a technique for studying the host-parasite relationship of the root-knot nematode *Meloidogyne incognita* under controlled conditions. *Nematologica,* **4,** 43–55 (1959)

60. Piegat, M., Wilski, A. Changes observed in cell nuclei in roots of susceptible and resistant potato after their invasion by potato root eelworm (*Heterodera rostochiensis* Woll.) larvae. *Nematologica,* **9,** 576–80 (1963)

61. Piegat, M., Wilski, A. Cytological differences in root cells of susceptible and resistant potato varieties invaded by potato root eelworm (*Heterodera rostochiensis* Woll.) larvae. *Nematologica,* **11,** 109–15 (1965)

62. Pitcher, R. S. Role of plant-parasitic nematodes in bacterial diseases. *Phytopathology,* **53,** 35–39 (1963)

63. Pitcher, R. S. Interrelationships of nematodes and other pathogens of plants. *Helminthol. Abstr.* **34,** 1–17 (1965)

64. Pitcher, R. S. The host-parasite relations and ecology of *Trichodorus viruliferus* on apple roots, as observed from an underground laboratory. *Nematologica,* **13,** 547–57 (1967)

65. Pitcher, R. S., Crosse, J. E. Studies in the relationship of eelworms to certain plant diseases. II. Further analysis of the strawberry cauliflower disease complex. *Nematologica,* **3,** 244–56 (1958)

66. Pitcher, R. S., Patrick, Z. A., Mountain, W. B. Studies on the host-parasite relations of *Pratylenchus penetrans* (Cobb) to apple seedlings. *Nematologica,* **5,** 309–14 (1960)

67. Powell, N. T. The role of plant-parasitic nematodes in fungus diseases. *Phytopathology,* **53,** 28–35 (1963)

68. Rhoades, H. L., Linford, M. B. Biological studies on some members of the genus *Paratylenchus. Proc. Helminthol. Soc. Wash. D.C.,* **28,** 51–59 (1961)

69. Roggen, D. R., Raski, D. J., Jones, N. O. Further electron microscopic observations of *Xiphinema index. Nematologica,* **13,** 1–16 (1967)

70. Rohde, R. A., Jenkins, W. R. Host range of a species of Trichodorus and its host-parasite relationships on tomato. *Phytopathology,* **47,** 295–98 (1957)

71. Rubinstein, J. H., Owens, R. G. Thymidine and uridine incorporation in relation to the ontogeny of

root-knot syncytia. *Contrib. Boyce Thompson Inst.,* **22,** 491–502 (1964)

72. Ruehle, J. L. Histopathological studies of pine roots infected with lance and pine cystoid nematodes. *Phytopathology,* **52,** 68–71 (1962)
73. Sandstedt, R., Schuster, M. L. The role of auxins in root-knot nematode-induced growth on excised tobacco stem segments. *Physiol. Plantarum,* **19,** 960–67 (1966)
74. Schindler, A. F. Parasitism and pathogenicity of *Xiphinema diversicaudatum,* an ectoparasitic nematode. *Nematologica,* **2,** 25–31 (1957)
75. Schuster, M. L., Sandstedt, R., Estes, L. W. Host-parasite relations of *Nacobbus batatiformis* and the sugar beet and other hosts. *J. Am. Soc. Sugar Beet Technologists,* **13,** 523–37 (1965)
76. Seinhorst, J. W., Dunlop, M. J. De Antasting van enige Solanum soorten en enige Kruisingen tussen *Solanum demissum* en *S. tuberosum* door het Stengelaaltje *Ditylenchus dipsaci* (Kühn) Filipjev. *Tijdschr. Plantenziekten,* **51,** 73–81 (1945)
77. Setty, K. G. H., Wheeler, A. W. Growth substances in roots of tomato (*Lycopersicon esculentum* Mill.) infected with root-knot nematodes (*Meloidogyne* spp.). *Ann. Appl. Biol.,* **61,** 495–501 (1968)
78. Standifer, M. S. The pathologic histology of bean roots injured by sting nematodes. *Plant Disease Reptr.,* **43,** 983–86 (1959)
79. Standifer, M. S., Perry, V. G. Some effects of sting and stubby root nematodes on grapefruit roots. *Phytopathology,* **50,** 152–56 (1960)
80. Sturhan, D. Der pflanzenparasitische Nematode *Longidorus maximus,* seine Biologie und Ökologie, mit Untersuchungen an *L. elongatus* und *Xiphinema diversicaudatum. Z. Angew. Zool.,* **50,** 129–93 (1963)
81. Taylor, D. P., Pillai, J. K. *Paraphelenchus acontiodes* n. sp. (Nematoda: Paraphelenchidae), a mycophagus nematode from Illinois, with observations on its feeding habits and a key to the species of *Paraphelenchus. Proc. Helminthol. Soc. Wash. D.C.,* **34,** 51–54 (1967)
82. Troll, J., Rohde, R. A. Pathogenicity of *Pratylenchus penetrans* and *Tylenchorhynchus claytoni* on turfgrasses. *Phytopathology,* **56,** 995–98 (1966)
83. Van Gundy, S. D., Kirkpatrick, J. D. Nature of resistance in certain citrus rootstocks to citrus nematode. *Phytopathology,* **54,** 419–27 (1964)
84. Van Gundy, S. D., Rackham, R. L. Studies on the biology and pathogenicity of *Hemicycliophora arenaria. Phytopathology,* **51,** 393–97 (1961)
85. Viglierchio, D. R. Resistance in Beta species to the sugar beet nematode, *Heterodera schachtii. Exptl. Parasitol.,* **10,** 389–95 (1960)
86. Viglierchio, D. R., Yu, P. K. Plant parasitic nematodes: a new mechanism for injury of hosts. *Science,* **147,** 1301–3 (1965)
87. Viglierchio, D. R., Yu, P. K. Plant growth substances and plant parasitic nematodes. II. Host influence on auxin content. *Exptl. Parasitol.,* **23,** 88–95 (1968)
88. Wallace, H. R. The nature of resistance in chrysanthemum varieties to *Aphelenchoides ritzemabosi. Nematologica,* **6,** 49–58 (1961)
89. Webster, J. M. The influence of plant-growth substances and their inhibitors on the host-parasite relation of *Aphelenchoides ritzemabosi* in culture. *Nematologica,* **13,** 256–62 (1967)
90. Weischer, B. Neuere Gesichtspunkte zur Frage der Biologie und Ökologie wer wandernden Wurzelnematoden. *Nematologica,* **2** (suppl.), 2, 406–12 (1957)
91. Wilski, A., Giebel, J. Beta-glucosidase in *Heterodera rostochiensis* and its significance in resistance of potato to this nematode. *Nematologica,* **12,** 219–24 (1966)
92. Yu, P. K., Viglierchio, D. R. Plant growth substances and parasitic nematodes. I. Root knot nematodes and tomato. *Exptl. Parasitol.,* **15,** 242–48 (1964)

Copyright 1969. All rights reserved

CULTURING OF RUST FUNGI

By K. J. Scott[1] and D. J. Maclean[1]
Department of Biochemistry, University of Sydney
Sydney, Australia

Introduction

Members of the Uredinales have usually been considered the classical examples of obligate parasites amongst fungal plant pathogens. Recently Williams et al. (97, 98) reported the culture from uredospores of an Australian isolate of wheat stem rust in an artificial medium. This accomplishment, together with the previous reports of Hotson & Cutter (55) and Cutter (29, 30–33) on the saprophytic culture from infected callus tissue of three other rust species, necessitates that a new appraisal of the culture of these obligate parasites be attempted.

The field of obligate parasitism has been comprehensively reviewed by Brian (14) and Yarwood (101); briefer treatments have been included in reviews by Allen (2, 4). This review describes work that led to the axenic culture of some rust fungi, and considers the implications of this work in relation to the concept of obligate parasitism. Axenic culture is defined as the growth of organisms of a single species in the absence of living organisms or living cells of any other species (41), that is, growth on nonliving substrata.

Why Culture the Rusts?

Some of the more important reasons for culturing the rusts are considered below.

(*a*) The axenic culture of an obligate parasite is an interesting biological problem in itself.

(*b*) Present knowledge of the nutrition, metabolism, and physiology of the rusts has been restricted by the materials available for study. Much work has been performed on the host-parasite complex (2, 4, 35, 52, 53, 67, 84, 102) and significant changes have been observed in the metabolism of infected plants. However, it has not been possible to distinguish unequivocally between the relative contribution of host and pathogen to these changes.

Resting or germinating spores, usually uredospores, have provided the only material for metabolic studies on the free-living fungus (5, 85, 88). As sporelings have failed to initiate saprophytic growth, most of these studies

[1] Present address: Botany Department, University of Queensland, St. Lucia, Queensland, Australia.

have been restricted to investigating biochemical and physiological events associated with spore germination and germ tube elongation. However, if failure of the rusts to grow axenically were associated with a metabolic block in an essential pathway, it was hoped that these studies would reveal it.

Some metabolic studies have been performed on mycelium isolated from infected host tissues. Turel & Ledingham (91) developed procedures for obtaining aerial mycelium of *Melampsora lini* from infected leaf cultures. Although mycelium was excised and used for radioactive feeding experiments (68, 92, 99), they did not know how excision affected the metabolism of this material (99). Dekhuijzen et al. (38, 39) isolated hyphal segments from rust-infected bean leaves by grinding, sieving, and settling in sucrose gradients. This method suffers from the disadvantages mentioned above for excised mycelium, and may lead to loss of soluble constituents as well (39). To obtain definitive data on the nutrition and metabolism of rusts it seems necessary to grow these organisms axenically.

(*c*) A 1951 editorial in *Nature* (7) stressed the importance of cultivating pathogens free of their hosts and other organisms. It stated that "when we know how the parasite breathes, feeds, excretes and uses, within its host and at its host's expense, whatever biochemical processes link it to that host, we may learn to control it."

An exchange of nutrients, toxins, or both between the rust and host may determine whether the host is resistant or susceptible. Saprophytic cultures should prove useful in investigating these interactions, and so aid our understanding of mechanisms determining host resistance or susceptibility.

Approaches to Axenic Culture

Attempts to culture the rusts saprophytically have been based on two methods: firstly, the initiation of saprophytic growth from germinating uredospores and secondly, saprophytic growth from mycelium formed in infected host callus tissue.

Spores are the obvious starting material in attempts to obtain axenic cultures, since they are produced by the fungus for its natural dissemination. Characteristic spores are produced by the rusts at different stages in their life cycle (1). By far the greatest amount of work on the reaction of rust spores to nutrient media has been carried out with uredospores. Amongst species of major interest, they are the most prolific and uniformly produced spores and may be readily obtained by reinfection of plants of the same host species. While relatively little work has been done with the other spore forms, a résumé of some of this work is given, since these forms may be possible starting materials for future culture experiments.

Teliospores and basidiospores.—Work on the response of teliospores to nutrients, chemicals, moisture, temperature, etc., has been mainly concerned with overcoming dormancy (9, 19, 60, 82, 90). If growth were to be obtained from teliospores presumably the promycelium would form basidiospores which would germinate and grow, or else the promycelium itself would initiate saprophytic growth directly. It has been reported (78) that germinating teliospores of *Gymnosporangium juniperi-virginianae* formed promycelia which occasionally produced hyphae without basidiospore formation. Bailey (10) mentioned a frequent "indefinite growth" and branching of promycelia of *Puccinia helianthi.* Other workers (24, 87) described the formation of a narrow twisted hypha without basidiospore production when teliospores of *Cronartium ribicola* were germinated under water. In the light of these observations, promycelia may be useful material for studies on the saprophytic growth of some rusts.

It is possible that the first axenic culture of a rust from spores was accomplished by Cutter (33, 34). Using callus tissue of *Althaea* infected with the microcyclic rust, *P. malvacearum,* telial pustules bearing mature teliospores were observed in three-month-old callus tissue. In one instance a minute fungal colony was noted growing on the agar surface about 8 mm from the base of the callus stem. Cutter surmised that the colony had arisen from the germination of a basidiospore or directly from the promycelium of a teliospore which had fallen to the agar surface since no visible connection to the stem was noted. Later attempts to obtain axenic growth by the deliberate seeding of basidiospores onto nutrient agar were unsuccessful (34).

Aeciospores.—Work on these spores has been limited to studies of germination on water or host decoctions (10, 19, 59, 78, 87). These spores are often difficult to obtain in abundance.

Spermatia.—Although many workers failed in attempts to germinate spermatia (37, 78), Plowright (73) observed limited yeast-like budding when spermatia of several rusts were placed on nutrient solutions containing sucrose or honey. Other workers (13, 26) have observed similar phenomena. Carleton (19) obtained germ tubes rather than budding, and their formation was stimulated by sucrose or honey. Spaulding (87) cited unpublished work of York & Overholts, who investigated the germination of *C. ribicola* spermatia on various nutrient solutions. It was noted that preincubation in the cold for up to 18 days was a necessary prerequisite for germination.

The possibility of yeast contamination (9) and the male role of spermatia (28) have been mentioned (101) as reasons for discounting the usefulness of such observations, although it appears that at least some of these reports were carefully described, particularly that of Plowright (73).

Cultures from Uredospores

The morphology and physiology of resting and germinating uredospores has been reviewed by Mains (65), Arthur (9), and more recently by Allen (5), Cochrane (22, 23), Shaw (85), and Staples & Wynn (88). We shall here attempt to relate some of the more significant observations to our present perspective.

In an exhaustive search of the 19th century literature, Mains (65) found few published accounts of attempts to grow the rust fungi saprophytically. The work of the period was summarized when de Bary (37) stated that uredospores germinated best when sufficiently supplied with water in a humid atmosphere, noting that nutrient solutions often inhibited germination.

Ezekiel (45) and Stock (89) observed that when uredospores were germinated on distilled water, germ tubes 150 to 800μ in length were obtained depending on the batch of uredospores and species of rust studied, but the process generally ceased within 24 hr of incubation (45, 89). Higher percentage germination and greater elongation were obtained when uredospores of *P. coronata* were floated on a liquid rather than suspended within it (66). The viability of these spores was influenced by the age and vigor of the host (66). Ezekiel (45) effected a several-fold increase in germ tube length by increasing the uredospore inoculum density in distilled water. Host decoctions usually had to be diluted before germination was possible, and greatly diluted before a statistical increase in germ tube elongation was observed (45).

In general, the optimum temperature for germination of uredospores of different species is between 12 to 23° C, (9, 89). The optimum pH always included the range 4.5 to 6.5 for any species (9, 89). Uredospore germination is affected by self-inhibitors and self-stimulators (5, 85, 88). Blue light may delay or inhibit germ tube growth (9, 94) and in at least some species caused a negative directional response (47).

The above observations helped define the environmental conditions most suitable for early development of sporelings. If further growth was to occur, the major variables to be contended with were selection of nutrients, period of incubation, and degree of asepsis necessary. It would seem in retrospect that any particular group of workers usually neglected at least one of these variables. For example, Brefeld (12), who was so successful in early work on the axenic culture of other fungi, encountered trouble with contamination. Stock (89) incubated uredospores for only 14 hr and did not use aseptic conditions, hence his nutritional results are largely inconclusive. Many authors contented themselves with descriptions of germ tube branching or assorted vesicle and fusion-body formation by these germ tubes (45, 79, 81). Ezekiel's (45) interpretation of apical swellings was that the germ tube entered a reproductive stage almost immediately after germination which inhibited further vegetative growth.

An example of the better work was that of Mains (65) who attempted to solve the problem of obligate parasitism by parallel studies of uredospore infection of the host and uredospore development *in vitro*. Cultures of *P. sorghi* were maintained upon sterile seedlings of corn in the light. Whilst detached portions of these seedlings could be infected if maintained in the light, the addition of a carbohydrate source such as starch, sucrose, dextrose, maltose, or dextrin was necessary for infection in the dark. Germ tubes of *P. sorghi* ceased elongating and were dead in 3 to 4 days when supplied directly with sucrose, maltose, dextrose, asparagine, leucine, or peptone; each with or without Knop's mineral solution. Decoctions of the host were no better in producing growth. Growth could not be obtained on pieces of autoclaved corn leaf floated on various solutions, whereas infection and rust development occurred on nonautoclaved controls set up at the same time. He concluded that the obligate parasitism of rusts could be explained by their requirement for some "transitory or nascent organic products related to the carbohydrates which they obtain in the living host."

Contrary to this, Ray (76, 77) reported the axenic growth of rusts on agar-solidified media containing decoctions of host tissues or carrots. Lack of experimental evidence makes it difficult to assess his work and it has not been repeated. Gretschushnikoff (51) reportedly maintained saprophytic cultures of species of *Puccinia* on a medium containing compounds which absorbed ammonia and urea released by the growing mycelium. This report has not been confirmed.

Repeated failures to obtain axenic cultures of the rusts accounted for such comments as made by Duggar (42), who stated in 1909 that "it would be useless to try to cultivate these fungi in non-living substrata." In 1928, Arthur (8) related that "repeated attempts to grow the rust on artificial media or even to accelerate the germination of the spores by using nutritive solutions have met with uniform failure." Similar statements of implied and overt hopelessness have been made by other authorities. For example, Chester (21) in 1946 wrote of many investigators seeking in vain "the philosophers' stone underlying the problem; a means of culturing rusts on nutrient media other than the living plant."

It is clear from Yarwood's (101) review that the "hit and miss" method of finding the correct nutrients for saprophytic cultures to develop from spores had not yet been successful. Because of this failure, the idea developed that infection structures, as formed on host plants, may be necessary prerequisites for saprophytic growth. For example, Brown (15) suggested that much of the nutritional specificity of obligate parasites might disappear if haustorial formation could be induced under artificial conditions.

Infection structures and axenic growth.—The formation of infection structures and the process of infection were first described by Pole Evans (74) and Allen (6). Using cereal rusts with susceptible hosts, these workers observed that the uredospore germ tube grew towards a stoma, and that the protoplasm migrated to the tip to form an appressorium commonly containing four nuclei. Entrance of the fungus into the substomatal cavity was effected by the infection peg, a narrow branch from the appressorium, which passed through the stomatal slit and formed a substomatal vesicle. The nuclei in the appressorium migrated into the substomatal vesicle and divided to give eight daughter nuclei. An infection hypha was formed from the substomatal vesicle and produced a haustorium mother cell when its growth was checked by contact with a mesophyll cell. After forming the haustorium mother cell, which in turn gives rise to a uninucleate haustorium, the infection hypha usually contained six nuclei.

Dickinson (40) and Chakravarti (20) showed that infection structures are necessary for successful infection and that the epidermis is apparently required for their formation on the host plant. In Dickinson's studies, uredospores were placed either on intact leaves or exposed mesophyll cells, and in both instances germination was high. Formation of infection structures and infection of intact leaves occurred as previously described. However, in instances where the uredospore germ tubes were formed in direct contact with the mesophyll cells, no infection structures were formed and penetration of the mesophyll cells was not observed. Chakravarti (20) placed uredospores directly on exposed mesophyll cells, replaced the stripped epidermis, and observed two types of germ tube growth. In one type, formation of appressoria was observed on the under surface of the epidermis and on the walls of mesophyll cells, and successful infection occurred. In the other type, germ tubes failed to form appressoria and subsequently did not grow or infect.

The formation of infection structures in artificial medium was first reported by Hurd-Karrer & Rodenhiser (56). Primary and secondary vesicles corresponding to appressoria and substomatal vesicles were formed from uredospores seeded on a glucose-agar medium. Hyphae, corresponding to infection hyphae with occasional septa, and up to 300μ in length, developed from secondary vesicles. The frequency of infection-structure formation was only one per 1,000 germ tubes. Dickinson (40) showed that germ tube contact with collodion membranes increased the frequency of infection structure formation *in vitro,* particularly when the membrane was impregnated with cell wall debris or certain types of paraffin wax. He observed nuclear division in structures which corresponded to appressoria and substomatal vesicles. A ball-like head was often observed at the tip of the infection hypha which was analogous to a haustorium. Using similar membranes impregnated with mineral oil or hydrocarbons isolated from the surface wax of leaves, Maheshwari et al. (63) reported up to 100 per cent infection-structure formation with some rusts. Cellulose membranes have also been shown to increase the frequency of infection-structure formation (72). Emge (44) has shown that light and temperature play an important role in their formation.

A number of workers have studied the effectiveness of different chemicals in inducing formation of infection structures from germinating uredospores under artificial conditions (3, 27, 46, 63, 83). Unidentified volatile components isolated from wheat stem rust uredospores induced the formation of infection structures in germinating uredospores (3, 63). French & Weintraub (46) reported that pelargonaldehyde stimulated formation of infection structures, but other workers (63) failed to obtain this result. Of a wide range of heavy metals tested in gelatin media, only zinc stimulated vesicle formation in germ tubes of *P. coronata* (83). The effect of zinc was related to the sample and concentration of gelatin used, and evidence was presented that the cation composition and pH of the medium modified the influence of zinc (27).

Maheshwari et al. (64) have made a detailed study of nuclear behavior during development of infection structures in *Uromyces phaseoli* on nitrocellulose membranes impregnated with mineral oil. The two uredospore nuclei divided in the germ tube before the formation of the appressorium. A second division of the four daughter nuclei occurred in the appressorium, and occasionally the eight nuclei divided again in the vesicle and infection hypha. Haustorium mother cells formed from infection hyphae contained from two to five nuclei. In contrast, nuclear division was not observed in germ tubes which did not differentiate.

It has recently been shown (75) that RNA synthesis is six to ten times greater in differentiated germ tubes of bean rust uredospores than in undifferentiated germ tubes. It appears that some of this synthesis is related to messenger RNA.

It is clear from the foregoing that formation of infection structures under artificial conditions is accompanied by synthesis of DNA and RNA. Such synthetic reactions would be necessary prerequisites for saprophytic growth. However, in the experiments described above, little emphasis was placed on asepsis or on the addition of nutrients which may have enabled saprophytic growth to occur. It rather appears that this work on infection structures was concerned only with gaining a more complete understanding of germ tube differentiation under artificial conditions.

Fuchs & Gaertner (48) studied the effect of different amino acids on the formation of infection structures and the growth of infection hyphae. These workers incubated contaminant-free uredospores of wheat stem rust for up to ten days on a basal medium containing glucose, egg yolk, Knop's mineral solution, pyridoxine, and iron saccharate solidified in a silica gel matrix. By adding cysteine, leucine, or glutathione to this medium, formation of infection structures was increased from 10 to 30 to 40 per cent and growth of infection hyphae, including branches, was stimulated. The maximum observed length of an infection hypha was 3.3 mm formed on the basal medium supplemented with cysteine. The average length of infection hyphae on basal medium alone was 0.77 mm. Although growth had always ceased by 10 to 13 days, their observations were perhaps the most promising of all attempts to obtain axenic growth from uredospores until the work of Williams et al. (97) in 1966.

The axenic culture of an Australian isolate of wheat stem rust.—The first successful saprophytic culture of a rust from uredospores was reported by Williams et al. (97, 98). Working on the assumption that rusts, like many other organisms previously considered to be "obligate parasites," would grow if given the correct nutritional environment, these workers incubated uredospores on various natural products under sterile conditions. The necessity for using aseptic techniques was emphasized by comparing rates of elongation of germ tubes from uredospores (approximately 75μ/hr) with growth rates of intercellular hyphae in susceptible hosts (approximately 5 to 10μ/hr). They anticipated that growth rates on artificial media would be slow and unlikely to be faster than that of intercellular hyphae. Extended periods of incubation would thus probably be necessary for significant saprophytic growth to occur. Uncontaminated uredospores were obtained by a method of aseptic leaf culture (97) modified from the methods of Turel & Ledingham (91).

Whereas many previous workers had used germ tube length after short periods of incubation as an index of growth, Williams et al. (97) looked for septa formation and continued branch growth over extended periods of incubation. It was observed that septate vegetative mycelium was formed from uredospores of *P. graminis* var. *tritici* race 126-ANZ-6,7 incubated on an agar medium containing Czapek's minerals, 0.1 per cent yeast extract, and 3 per cent sucrose. In tracing the development of this mycelium it was observed that germination and germ tube elongation had occurred after 2 days. Initial branching of germ tubes was observed 4 days after seeding (Fig. 1A), and this branching continued for 3 to 4 weeks. In subsequent work (98), it was observed that the addition of 0.1 per cent Evan's peptone to the medium enhanced vegetative growth, and some cultures formed stromata (Fig. 1B) bearing uredospores (Fig. 1C) and teliospores (Fig. 1D) 2 to 4 weeks after inoculation.

Bushnell (17) has subsequently repeated this work, using peptone medium containing glucose and the same Australian isolate of wheat stem rust. He noted that yeast extract medium without the addition of peptone did not support growth. It has been observed that the replacement of sucrose by glucose and the addition of citrate to the yeast extract-peptone medium has resulted in more consistent growth (62).

FIG. 1. A. A branched germ tube grown from a uredospore incubated on Czapek's medium containing yeast extract (×160). B. Section through a uredial stroma showing the original inoculum (a) and the uredospore-forming layer (b). The section was cut at right angles to the agar surface, fixed in ethanol-acetic acid, embedded in paraffin, and stained with crystal violet. C. Water mount of uredospores formed *in vitro.* D. Glycerine-jelly mount of teliospores formed *in vitro.* E. 51-Day-old colonies growing on cellophane on agar-solidified medium containing Czapek's minerals, yeast extract, peptone, and glucose. Colonies are seventh serial subculture of original axenic isolate, isolated 15 months previously (×1.25). Reinfection of a leaf segment resulting in sori production. The lower epidermis of the leaf had been removed and the mesophyll exposed to a mycelial colony (similar to E) for 12 days. A "green island" was formed on the leaf above the fungus (×8.5). A is reproduced from Williams, Scott & Kuhl (97). B, C, and D are reproduced from Williams et al. (98).

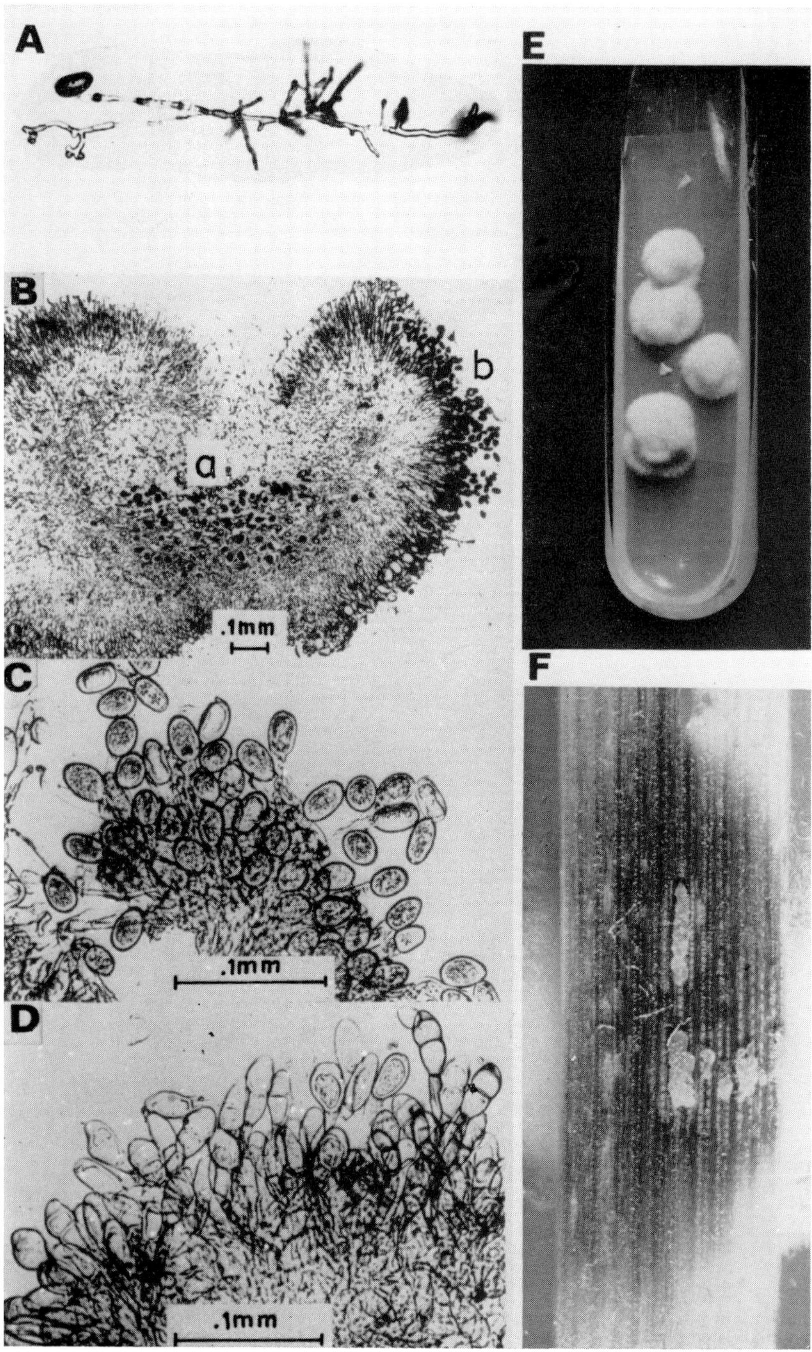
A
B
b
a
.1mm
C
.1mm
D
.1mm
E
F

Three types of growth response have been obtained when rather massive inocula of approximately 10^6 uredospores of *P. graminis tritici* were floated on 25 to 35 ml of liquid medium containing glucose, yeast extract, and peptone (62). (*a*) Germination occurred but saprophytic growth did not initiate even after extended periods of incubation. (*b*) Vigorous saprophytic growth occurred, leading to formation of a dense white hyphal mat on the liquid surface 2 to 3 weeks after seeding, followed by sporulation or staling. Staling was considered to have occurred when mycelium assumed a dull brown necrotic appearance with the concomitant excretion of a brown pigment into the medium. Once begun, the process of staling could not be reversed by subsituting fresh medium and it would appear that autolysis of the fungus ensued. (*c*) Germinated uredospores exhibited little or no saprophytic growth within the first 2 to 3 weeks. However, on further incubation, up to 30 mycelial tufts per flask became macroscopically visible 30 to 60 days after seeding. These tufts were transferred to agar slants of the same medium and have now been maintained for over 18 months by serial subculture at 1- to 2-month intervals (Fig. 1E).

Culture of other rusts from uredospores.—Axenic cultures of other cereal rusts have since been attempted using the above procedures. Bushnell & Stewart (18) have tested 16 North American races of *P. graminis tritici* and have obtained vegetative growth from 12 of these races. Most of these cultures survived initial subculturing, but development from subsequent subcultures was erratic. Large quantities of teliospores have been produced by most of these axenic isolates, and one isolate produced uredospores abundantly. Macko et al. (61) obtained rather limited saprophytic growth of *P. graminis tritici* race 56 but Bushnell & Stewart (18) have not been successful in obtaining any growth from this race. Williams (96) in this laboratory has obtained sporulation of an Australian isolate of *P. graminis avenae.* Macroscopic colonies, but no sporulation, of three other Australian races of *P. graminis tritici,* and limited post-sporeling growth of *P. graminis secalis* and *P. recondita tritici* have been obtained (96). Singleton & Young (86) have reported vegetative growth of an American isolate of *P. recondita,* using the yeast extract-peptone medium. They claimed that the less virulent races produced the best growth response.

Cultures from Infected Callus Tissue

The first successful axenic culture of a rust was not obtained directly from spores but from the mycelium of infected host callus tissue (55).

The original idea was to form callus tissue of a susceptible host and to infect it with rust spores. Cultural conditions of the undifferentiated host-parasite complex would then be varied with the ultimate aim of rendering the parasite independent of its host. Morel (69) applied the principles of tissue culture as developed by Gautheret (50) and White (95) to establish callus tissue of *Vitus vinifera* which he infected with *Plasmopara viticola.* Subsequent attempts to culture several rusts on undifferentiated host tissues were unsuccessful. Morel (70) attributed his failure partly to high contamination rates and partly to an inability of uredospore germ tubes to penetrate and infect callus tissues. He suggested that both these difficulties might be overcome by starting cultures from systemically infected plant tissues which could be surface sterilized.

Hotson (54), Hotson & Cutter (55), and Cutter (31) successfully adopted this suggestion using telial galls formed by *G. juniperi-virginianae* on *Juniperus*. These workers considered a culture successful if the fungus continued to grow within its host upon serial subculture of the callus. Their methodology was tedious. For example, in one series of experiments with this cedar-apple rust, Cutter (31) obtained 358 infected primary calluses from 13,504 cultures started from 8,840 galls collected in the eastern U.S.A. The medium used was Gautheret's nutrient No. 4 solution modified by the addition of 3 per cent dextrose and 500 ppm ascorbic acid. After about 4 to 9 months the fungus in one of the original cultures and six of the many subcultures produced a necrotic reaction in the surface host tissue and subsequently grew out onto the agar medium (31). Cutter originally suggested that this rust first behaved as an obligate parasite, then became destructively pathogenic, and finally saprophytic (29). Later he thought that a prior change in the host callus could select for variant nuclei with saprophytic tendencies (31).

Altogether, a total of 41 rust species infecting 60 vascular plants were investigated (34). Of some 23,151 attempted two-membered cultures, 13 strains representing three species of rust were finally obtained in axenic culture (34). Apart from the seven isolates of *G. juniperi-virginianae,* Cutter obtained five isolates of *U. ari-triphylli* (30, 32) and one isolate of *P. malvacearum* in axenic culture (33, 34). As described for *G. juniperi-virginianae,* he observed that a necrotic reaction in the host tissue preceded the emergence of all saprophytic strains of *U. ari-triphylli.* The circumstances associated with the isolation of *P. malvacearum* have already been discussed (cf. p. 124).

Unfortunately, other workers have not been able to obtain axenic cultures of *G. juniperi-virginianae* (25, 80, 91) or of any of the following rusts from two-membered callus systems: *P. antirrhini* (80), *M. lini* (91), *P. helianthi* (71), *P. minutissima* (80), or *P. tatarica* (11, 100).

Because of the failure of others to repeat Cutter's experiments his work has not received the recognition it deserves.

Properties of the Fungus in Axenic Culture

Factors affecting the initiation of saprophytic growth.—The extremely low incidence of axenic fungal isolation obtained by Hotson & Cutter (55) from infected host callus has already been noted. A major problem we have encountered in obtaining saprophytic growth from uredospores has been the erratic behavior of these spores fron one experiment to the next under apparently identical conditions. This variability has been due partly to spore germination, which may range from zero to 80 per cent in different experiments, and partly to the growth potential of germ tubes from different batches of uredospores. Some of this erratic behavior may be due to the conditions under which contaminant-free uredospores are produced. Such factors as age of leaf at time of inoculation and detachment; techniques of surface sterilization; the effect of the medium on leaf metabolism during culture; the position of the sori on leaf segments; the age of the uredospores in the sori, self-inhibitors; and self-stimulators may all affect the subsequent behavior of the uredospores on nutrient medium. These factors require further study.

On media cited above, colony formation is best in regions of high inoculum density (17, 97). Bushnell (17) obtained saprophytic growth only in the presence of high spore densities (100 to 200 spores/mm^2). However, we have observed that a few sporelings did initiate saprophytic growth in sparsely populated areas on yeast extract-peptone medium. The frequency of colony formation in regions of low inoculum density can be increased by modifying the medium (62), for example, by the addition of cysteine, or by using peptone which had been acid hydrolysed to its constituent amino acids. This frequency was also increased when sporelings in regions of low inoculum-density were adjacent to densely populated areas (62).

The simplest interpretation of these observations is that essential metabolites, perhaps amino acids, are lost from the germ tubes in amounts exceeding their biosynthetic capacity. In this connection the work of Eagle & Piez (43) is of interest. They showed that, at low inoculum densities, cultured mammalian cells require serine, asparagine, cysteine, glutamine, homocysteine, inositol, and pyruvate for survival. This requirement was population-dependent and disappeared at cell densities sufficiently large to bring the concentration of these compounds, in the medium and in the cellular pool, to metabolically effective levels.

"Leakiness" of uredospores and germ tubes has been reported by a number of workers (36, 57, 58, 88). Staples & Wynn (88) reported that when bean rust uredospores were incubated with P^{32}-orthophosphate for 30 min and then with unlabelled phosphate for 60 min, all the ethanol-soluble P^{32}-labelled compounds had been leached from the spores. Jones & Snow (58) showed that when S^{35}-labelled uredospores were germinated in water, 22 free amino acids, including four that were radioactive, were found in the germinating solution. Jones (57) subsequently showed that when primary leaves of oat plants were inoculated with S^{35}-labelled uredospores of the crown-rust fungus, the S^{35} was released by the fungus into the host tissue and translocated into the primary and secondary leaves and roots. Daly et al. (36) reported on amounts of carbon lost from uredospores to the germinating medium. Using C^{14}-labelled uredospores, these workers showed that approximately 7 per cent of the total spore carbon was lost to the medium during early stages of germination. Most of the carbon in the medium was soluble in 80 per cent ethanol, and included sugar alcohols and acidic and basic compounds. At the end of the third hour of germination, amounts of carbohydrates external to the sporelings were greater than amounts of carbohydrates within the sporelings. Since the external amounts of carbohydrates subsequently decreased, these workers suggested that they were taken up again for subsequent metabolism. The pools of acidic and basic compounds in the medium did not change appreciably with germination, and were approximately equal to the pools in the spores at all stages of germination.

These results are consistent with the view that essential metabolites are lost to the external medium during germination, and this may account in part for the subsequent failure of many germ tubes to initiate saprophytic growth, especially in regions of low inoculum density.

Nutrient requirements.—Cutter carried out extensive nutritional studies on his isolated strains of *G. juniperi-virginianae* and *U. ari-triphylli* (29 31, 32, 34). Both rusts have been cultured on a number of nonspecific media including potato agar, potato-dextrose agar, oat-digest agar, yeast-beef agar, cornmeal agar, peptone agar, soya-digest agar, carrot-dextrose agar and the more defined Czapek's-Dox agar (34). In liquid shake cultures of any of these media, *G. juniperi-virginianae* formed minute pink or brown colonies thickly coated with mucilaginous materials (34).

Axenic cultures of *U. ari-triphylli* and *G. juniperi-virginianae* were capable of growth and subculture on a defined basal medium consisting of potassium phosphate, magnesium sulphate, and Burkholder & Nickell's (16) trace elements, solidified by purified agar, and supplemented with dextrose and ammonium nitrate (34). One strain of *G. juniperi-virginianae* was further tested for other carbon and nitrogen requirements, using as inoculum a mycelial suspension prepared by fragmentation and washing of 30-day-old cultures grown on the above medium (34). In the presence of ammonium nitrate or dextrose, respectively, a wide range of carbon[2] and nitrogen[3] sources supported growth. In no case was growth significantly better than on the original dextrose and ammonium nitrate medium, nor was growth stimulated by the presence of any of the common vitamins singly or in combination, or the bases adenine, guanine, uracil, or xanthine.

Saprophytic growth has been initiated from uredospores of *P. graminis tritici* on agar medium containing yeast extract and peptone with sucrose (98), glucose, fructose, mannose, or mannitol (62) as added carbohydrate

[2] Dextrose, fructose, mannose, galactose, sorbose, sucrose, maltose, lactose, melezitose, L-arabinose, L-rhamnose, amylose, inulin, sorbitol, mannitol, dextrin, glycerin. D-arabinose and D-xylose supported less growth and growth was very poor on dulcitol. Acetate failed to support any growth.

[3] α-Amino-*n*-butyric acid, arginine, asparagine, aspartate, hydroxyproline, phenylalanine, proline, serine, alanine, ammonium salts, nitrates. Creatinine, cystine, glutamate, glycine, leucine, lysine, methionine, ornithine, threonine, and haemoglobin supported less growth. Growth with cysteine, urea, and tryptophan was poor whilst no growth occurred with nitrite.

source. On a more defined chemical basis, uredospores have initiated vigorous growth on liquid media containing Czapek's minerals, glucose, aspartate, citrate, and either cysteine or glutathione (62). Staling reactions terminated growth on this medium after 3 to 4 weeks.

Subculture, staling, and sporulation.—

(*a*) Colonies derived directly from uredospores. We would like to distinguish between three types of axenic cultures of *P. graminis tritici* as judged by their morphology and ability to persist as vegetative mycelium when exposed to similar nutritional environments. In the first type (98), cultures grew vegetatively for 2 to 4 weeks after seeding, then either staled or sporulated, and growth stopped. Subcultures made when this mycelium was still actively growing likewise did not persist through more than two to three transfers. In those cultures which sporulated, stromatic mycelium usually covered the external surface of the colony with abundant production of uredospores and teliospores (Fig. 1B, C, D).

The second type (62) of culture developed much more slowly, becoming macroscopically visible only after spores had been incubated for 4 to 8 weeks. Vegetative colonies formed after this time can apparently be maintained indefinitely by serial subculture. Although sporulation occurred in these cultures, the whole colony did not enter a reproductive phase, and uredospores and teliospores were produced only in localized stromatic pigmented regions within the colony. The remainder of the colony continued to grow vegetatively.

The third type, described by Bushnell (17), appears to be intermediate between the first two types in both morphology and time required for colony formation. Heavily pigmented sporophore-like hyphae have been observed at the periphery of these cultures 11 weeks after seeding. Only a few spores resembling uredospores or aeciospores have been noted. Vegetative subcultures of this isolate have been maintained for at least 6 months by effecting transfers at 30- to 60-day intervals (17).

(*b*) Colonies derived from infected callus tissue. Cutter (31, 34) maintained his axenic strains of *G. juniperi-virginianae* and *U. ari-triphylli* for extended periods by subculture and did not report any difficulty with staling reactions. One of the seven isolates of *G. juniperi-virginianae* grew much more slowly than the others, however, and died after 18 months in axenic culture. It had been serially subcultured six times during this period (31).

U. ari-triphylli did not produce spores in axenic culture (32, 34, 40). Isolated abortive spores were formed in axenic cultures of *G. juniperi-virginianae,* but fruiting bodies were not formed (31, 34). Cutter noted that, although adenine inhibited vegetative growth, it greatly enhanced sporulation (29); he observed a somewhat similar effect when salicin was used as a carbon source (34).

Sectoring and nuclear content.—Extensive sectoring was observed in axenic cultures of *G. juniperi-virginianae* (34). Colonies were pale pink to orange during the first 3 weeks of growth, after which red, purple, brown, grey, and pure white sectors frequently appeared. None of the other rusts yet cultured exhibit this pigmented sectoring, although the vegetative hyphae are often colored (34, 62, 80).

Telial galls, the material from which Cutter isolated his cultures of *G. juniperi-virginianae,* characteristically contain binucleate fungal mycelium. Four of the seven axenic strains were predominantly binucleate within 30 days of isolation, the other three being completely uninucleate (31, 34, 40). Two of the previously binucleate strains became uninucleate after serial subculture. Another strain, although mostly binucleate in young cultures, exhibited patches of predominantly uninucleate cells in sectors of older cultures. These patches were not consistently associated with any type of sector.

Observations on *U. ari-triphylli* under phase contrast, indicated that a majority of the cells were binucleate. All of the isolates were remarkably stable in culture (32, 34, 40). The unstable strain of *P. malvacearum* was uninucleate, with occasional bi- or multinucleate cells in the interior of the colonies (34).

Examination of cultures of *P. graminis tritici* has shown binucleate cells at an early stage after seeding uredospores on nutrient media (97). Although no further examination has been made, the formation of uredospores or teliospores after maintenance for 15 months in axenic culture would indicate that the binucleate condition is not completely lost (62).

Growth rates.—The rate of colony expansion of axenic cultures of *P. graminis tritici* has been shown to vary between 30 to 200μ/day (17). Although Cutter did not present specific data on the growth rates of strains of *G. juniperi-virginianae* and *U. ari-triphylli,* his results indicate that they were several-fold faster than those of *P. graminis tritici* (34). He observed that *U. ari-triphylli* grew faster than *G. juniperi-virginianae* on all media tested (34).

Reinfection and Reisolation

A critical question is whether these axenic isolates retain their pathogenicity and, if so, whether they can be reisolated as axenic cultures from infected hosts.

Reinfection by axenic cultures derived directly from uredospores.—Reinfection has been achieved using both mycelium and uredospores formed in axenic culture (17, 62, 98). Mycelium has been placed under the epidermis of the primary leaf of wheat seedlings in some experiments (17, 79, 98). In other instances, segments of wheat leaves with part of the lower epidermis removed have been placed with exposed mesophyll in contact with mycelium (62, 98) (Fig. 1F). In both instances, sori ruptured the epidermis on the side of the leaf opposite the inoculum after 8 to 12 days. These sori may contain uredospores (17, 62, 98), and teliospores and aerial hyphae (62). Bushnell (17) noted that exposed mesophyll of leaf segments induced formation of teliospores in aging axenic cultures which did not reinfect. In no case has reinfection been observed on unstripped leaves. Studies on fungal morphology in reinfected hosts have not been reported.

"Green island" formation, a common disease symptom under natural conditions, has been observed in wheat leaves after exposure of mesophyll cells to subcultures of vegetative mycelium of *P. graminis tritici* (62) (Fig. 1F). It was noted that "green islands" occurred only where the fungus produced fruiting bodies on the host. In further experiments it was shown that a hydrolysate of RNA obtained from this axenic strain produced similar "green islands" on uninfected detached leaves (93).

Successful reinfection by spores formed in axenic culture was obtained only when uredospores were placed under the epidermis in contact with mesophyll cells (98). Although spore germination was high on intact leaves, germ tube elongation was limited. Occasional appressoria formed over stomata did not develop further. It was suggested (98) that the inability of these uredospores to infect the host by stomatal invasion was due to a low potential for independent growth rather than a lack of pathogenicity.

Reinfection by axenic cultures derived from infected host callus tissue.—For reinfection studies Hotson & Cutter (55) and Cutter (31, 32) used fragmented mycelial suspensions and applied these to host callus or to intact plants. In later experiments, Cutter (31) reinfected host plants by inoculating with a paste of mycelial suspension and carbowax, the carbowax apparently being biologically inert. The ability of any particular axenic isolate to reinfect and form characteristic fructifications was related to two factors: firstly, to the nuclear condition of the fungal mycelium used for reinfection and secondly, to the host species tested (31). Cutter used formation of haustoria or fructifications as criteria for reinfection of callus tissue. Successful reinfection of intact plants or leaf segments was judged by the production of characteristic fruiting bodies.

In the case of *G. juniperi-virginianae,* the primary host, *Juniperus,* was reinfected only by binucleate mycelium (31). Cutter tested the ability of his cultures to infect *Pyrus,* the expected alternate host (31, 34). Uninucleate mycelium infected this host and produced pycnia but not aecia. In other instances reinfection of *Pyrus* resulted only in the formation of aecia, a result consistent with a completely binucleate mycelial inoculum. The concurrent production of aecia and pycnia was correlated with an inoculum containing both uni- and binucleate elements. Cross-spermatisation was not carried out in the above experiments.

One of the isolated strains of *G. juniperi-virginianae* could also infect *Crataegus* (hawthorn) as alternate host (31, 55), a phenomenon which created taxonomic difficulties. The aecial fructifications on some species of *Crataegus* resembled the hawthorn rust *G. globosum,* and on other species of *Crataegus* they resembled the quince rust *G. clavipes.* However, infection of *Pyrus* and *Juniperus* was characteristic of the cedar-apple rust *G. juniperi-virginianae.* It was not known if the natural host range of this strain of *G. juniperi-virginianae* was less strict than previously assumed, or whether host specificity was broadened by axenic culture. An interesting question arising from these results was whether the characteristic aecial structures were determined by the host rather than by the fungus (31).

The five axenic strains of *U. ari-triphylli* obtained by Cutter (32) displayed a low but definite incidence of reinfection of *Arisaema.*

During the early part of its independent life *P. malvacearum* successfully reinfected *Althaea,* producing teliospores, but attempts at reinfection during the latter period that this rust was in culture were unsuccessful (33, 34).

Except in one instance, reinfection of callus tissue by *G. juniperi-virginianae* and *U. ari-triphylli* did not result in the necrotic host reaction observed in the original isolation of the fungus (31, 32). The exception was when one strain of *U. ari-triphylli* was used for reinfection soon after its isolation, an occurrence which could not be repeated after maintenance of this strain in axenic culture (32).

Reisolation of the fungus.—It is important to know whether properties of the axenically cultured fungus are retained after repassage through host tissue. When uredospores of *P. graminis tritici,* formed as a result of reinfection by vegetative mycelium, were placed on nutrient agar, saprophytic growth was initiated and led to the formation of uredial stromata (98). It would be of interest to determine whether these spores produce saprophytic colonies in a more predictable manner than did uredospores used in obtaining the original axenic cultures.

Cutter (31, 32) said that studies were in progress to determine the percentage of axenic fungus isolation that could be obtained from calluses set up from reinfected plants. He did not publish any reports of success before his death.

Discussion

As it is now possible to culture some rusts axenically, their classification as obligate parasites must be reconsidered. In the original definition, de Bary (37) defined obligate parasites as species which require a parasitic existence for the attainment of their full development. He recognised two groups of obligate parasites: firstly, "facultative" saprophytes, which go through the whole course of their development as parasites but can live as saprophytes for at least part of their life cycle, and secondly, "strict obligate" parasites, which as far as is known live only as parasites. It is interesting to note that de Bary (37) thought fungi classified as strict obligate parasites may, in some cases, be cultured under artificial conditions as nonparasites, but he was more concerned with what happened under natural conditions.

Mains (65) postulated that the rust fungus obtains transitory compounds from the living host necessary for its growth. This hypothesis provided an attractive explanation of the obligate parasitism of the rusts and was consistent with the almost unique relationship existing between host and rust fungus. Unlike many other fungal parasites of plants, the rusts, certainly in the early stages of infection, stimulate synthetic activities within the leaf rather than cause degenerative changes.

With repeated failure to grow the rusts axenically, it is not surprising that later definitions of obligate parasitism came to be based on the reaction of the fungus to non-living substances under artificial conditions rather than being founded on its relationship to its host and natural surroundings. Yarwood (101) thus thought of obligate parasites as organisms which could not be cultured axenically.

The whole Order, Uredinales, was cited by Brian (14) as an example of "physiologically obligate" parasites. He used this term to refer to those organisms which apparently could not live saprophytically even in axenic culture in the most complex media. The corollary to this statement is that such organisms can overcome their biochemical deficiencies only by association with living host cells. These were distinguished from other strict obligate parasites which he termed "ecologically obligate." This group includes organisms which are incapable of saprophytic growth under natural conditions because they lack what Garrett (49) has called "competitive saprophytic ability." In pure culture however these organisms can be grown on artificial media.

Those members of the Uredinales which have been cultured axenically can no longer be called physiologically obligate parasites. They may well be ecologically obligate parasites in the sense that Brian (14) uses this term, but to establish what happens under natural conditions would require a special study in itself. We have observed in this connection that, when axenic cultures become contaminated, the rust fungus is always overrun by the contaminants.

Now that some rusts have been cultured axenically and their properties studied, it is informative to compare the rust as a parasite with it as a saprophyte. One obvious difference is that as a parasite it behaves in a highly predictable manner whereas as a saprophyte, growth patterns are often erratic. Thus, on susceptible hosts, uredospores germinate, differentiate to form infection structures, and establish a vegetative parasitic mode of existence leading to formation of fruiting bodies containing uredospores and teliospores. However, as a saprophyte, following uredospore germination there is a lag phase before growth initiates. Growth proceeds directly from germ tubes without the formation of infection structures and subsequent growth patterns may be highly variable as previously described.

It appears that development of the fungus as a parasite is under strict control, and that part of this control is lost when the fungus grows as a saprophyte. Vegetative rust mycelium is closely associated with host cells under normal parasitic conditions. This rust mycelium may have part of its genome repressed, and so avoid wasteful duplication of some metabolic processes already operative in the plant. Spore germination, germ tube elongation, and differentiation during the infection process probably require some metabolic syntheses not needed once infection is established. One could thus expect infection to be associated with successive repression of those parts of the sporeling genome not required for parasitic growth. A concomitant derepression of other parts of the genome may also be necessary.

We suggest that a different expression of the fungal genome is necessary for saprophytic growth to occur. The ability of any rust isolate to effect this change would be determined by the nutritional environment to which it is exposed, together with the composition of its cytoplasm, for example, carry-over nutrients. Different species of rust could be expected to vary in their ability to adapt to a particular nutritional environment. Genetic variation within a given rust species may also be important. It has already been noted that some races of *P. graminis tritici* are more easily cultured than others under conditions so far studied. If some rusts are incapable of such adjustments, they would indeed be physiologically obligate parasites.

The following observations are consistent with the above suggestions. Studies on saprophytic growth from uredospore germ tubes has revealed that initiation of growth is often erratic, but always requires an extended period of incubation of at least a few days after germination. This lag phase may be explained in part by the leakage hypothesis used earlier to explain inoculum-density effects. It is possible, however, that this lag phase prior to initiation of saprophytic growth is due to metabolic changes in the fungus in response to the range of nutrients provided by the artificial medium. Although the formation of infection structures is a necessary prerequisite for parasitic growth, their formation is not required for saprophytic growth. This indicates that changes in control mechanisms necessary for saprophytic growth can occur in the undifferentiated germ tubes. A detailed study of the response of differentiated germ tubes to the same nutrient media would be of interest.

The many varied forms of growth observed in axenic cultures (e.g., sectoring, nuclear content, aberrantly produced spores, the extent of vegetative growth, and the occurrence of staling) can all be explained by loss of some of the metabolic control associated with parasitic growth.

Some axenic colonies of *P. graminis tritici* may become completely stromatic and bear uredospores and teliospores. These reproductive structures are similar to those found on the host. It is interesting to note that once colonies enter this reproductive phase they are no longer capable of growth. The formation of these reproductive bodies both in the host and in axenic culture must be under strict control. Factors associated with control mechanisms leading to their formation are not understood.

A most puzzling feature of Cutter's work is the extremely low incidence of saprophytic growth he obtained from infected callus tissue. He suggested that this was due to selection for genetically variant nuclei with saprophytic tendencies (29, 31, 32). If this were the case then it may well be expected that reisolation of these axenic strains after repassage through the host would be much higher than that observed in the original isolations. Initial attempts to reisolate the fungus apparently were unsuccessful.

When saprophytic cultures of *G. juniperi-virginianae* which had exhibited aberrant spore production were used to reinfect host tissue, fruiting bodies characteristic of the parasitic state of the fungus were produced (31). We suggest that upon reinfection, expression of the fungal genome reverted to that characteristic of the parasitic state. Reisolation would thus rely on the same unknown mechanisms affecting changes in metabolic control which resulted in the original isolation.

One explanation for the isolation of saprophytic cultures consistent with Cutter's observations (31, 32) is that a prior degenerative change in the host cells allowed the fungus to adjust to a saprophytic mode of existence. In this regard, isolation of the fungus was always preceded by changes in host cells which led to eventual necrosis. This necrosis was not observed, however, upon reinfection of host callus. The low incidence of saprophytic growth obtained by Cutter may well reflect a low tendency for change in expression of the hyphal genome.

The suggestion that saprophytic growth of rust fungi involves repression and derepression of the genome can be tested experimentally by competitive hybridisation studies using rust DNA and messenger RNA isolated from sporelings, axenic cultures, and infected plants.

Acknowledgements

We are indebted to Dr. Lois Cutter for making available an unpublished manuscript of the late Victor M. Cutter, Jr. We wish to thank Professors K. F. Baker and N. H. White for criticisms of the manuscript and Miss P. M. Whyte for aid in its preparation.

LITERATURE CITED

1. Alexopoulos, C. J., *Introductory Mycology,* 2nd ed. (John Wiley & Sons, Inc., New York, 613 pp., 1962)
2. Allen, P. J. Physiological aspects of fungus diseases of plants. *Ann. Rev. Plant Physiol.,* **5,** 225–48 (1954)
3. Allen, P. J. Properties of a volatile fraction from uredospores of *P. graminis* var. *tritici* affecting their germination and development. I. Biological activity. *Plant Physiol.,* **32,** 385–89 (1957)
4. Allen, P. J. Metabolic considerations of obligate parasitism. *Plant Pathol. Probl. Progr. 1908–1958,* 119–29 (Holton, C. S. and others, Eds., Univ. of Wisconsin Press, 588 pp., 1959)
5. Allen, P. J. Metabolic aspects of spore germination in fungi. *Ann. Rev. Phytopathol.,* **3,** 313–42 (1965)
6. Allen, R. F. A cytological study of *Puccinia triticina* physiologic form 11 on Little Club wheat. *J. Agr. Res.,* **33,** 201–22 (1926)
7. Anonymous. Future of parasitology. *Nature,* **168,** 527–29 (1951)
8. Arthur, J. C. Progress of rust studies. *Phytopathology,* **18,** 659–74 (1928)
9. Arthur, J. C., *The Plant Rusts (Uredinales)* (John Wiley & Sons, Inc., New York, 446 pp., 1929)
10. Bailey, D. L. Sunflower rust. *Minn. Univ. Agr. Expt. Sta. Tech. Bull.,* **16,** 1–31 (1923)
11. Bauch, R., Simon, V. Kulturversuche mit Rostpilzen. *Ber Deut. Botan. Ges.,* **70,** 145–56 (1957)
12. Brefeld, O. Untersuchungen aus dem Gesammtgebiete Heft 5. (Die

Brandpilze, A. Felip, Leipzig, 28 pp., 1883)
13. Brefeld, O. Untersuchungen aus dem Gesammtgebiete der Mykologie. Heft 9, *Die Hemiasci und die Ascomyceten* (Munster, 159 pp., 1891)
14. Brian, P. W. Obligate parasitism in fungi. *Proc. Roy. Soc. (London), Ser. B,* **168,** 101–18 (1967)
15. Brown, W. The physiology of host-parasite relations. *Botan. Rev.,* **2,** 236–81 (1936)
16. Burkholder, P. R., Nickell, L. G. Atypical growth of plants. I. Cultivation of virus tumours of *Rumex* on nutrient agar. *Botan. Gaz.,* **110,** 426–37 (1949)
17. Bushnell, W. R. *In vitro* development of an Australian isolate of *Puccinia graminis* f. sp. *tritici. Phytopathology,* **58,** 526–27 (1968)
18. Bushnell, W. R., Stewart, D. M. (Unpublished data)
19. Carleton, M. A. Culture methods with Uredineae. *J. Appl. Microscopy Lab. Meth.,* **6,** 2109–14 (1903)
20. Chakravati, B. P. Attempts to alter infection processes and aggressiveness of *Puccinia graminis* var. *tritici. Phytopathology,* **56,** 223–29 (1966)
21. Chester, K. S., *The Nature and Prevention of the Cereal Rusts as Exemplified in the Leaf Rust of Wheat* (Chronica Botanica Co., Waltham, Mass., 269 pp., 1946)
22. Cochrane, V. W., *Physiology of Fungi* (John Wiley & Sons, Inc., New York, 524 pp., 1958)
23. Cochrane, V. W., Spore germination. *Plant Pathology, An Advanced Treatise* **II,** 169–202 (Horsfall, J. G., Dimond, A. E. Eds., Academic Press, New York, 715 pp., 1960)
24. Colley, R. H. Parasitism, morphology, and cytology of *Cronartium ribicola. J. Agr. Res.,* **15,** 619–59 (1918)
25. Constabel, F. Ernährungsphysiologische und manometrische Untersuchungen zur Gewebekultur der *Gymnosporangium*-Gallen von *Juniperus*-Arten. *Biol. Zbl.,* **76,** 385–413 (1957)
26. Cornu, M. Ou doit-on chercher les organes fécondateurs chez les urédinées et ustilaginées? *Bull. Soc. Botan. France,* **23,** 120–21 (1876)
27. Couey, H. M., Smith, F. G. Effect of cations on germination and germ tube development of *Puccinia coronata* uredospores. *Plant Physiol.,* **36,** 14–19 (1961)
28. Craigie, J. H. Discovery of the function of the pycnia of the rust fungi. *Nature,* **120,** 765–67 (1927)
29. Cutter, V. M. Jr. The isolation of plant rusts upon artificial media and some speculations on the metabolism of obligate plant parasites. *Trans. N.Y. Acad. Sci.,* **14,** 103–8 (1951)
30. Cutter, V. M. Jr. Observations on the growth of *Uromyces caladii* in tissue cultures of *Arisaema triphyllum. Phytopathology,* **42,** 479 (1952)
31. Cutter, V. M. Jr. Studies on the isolation and growth of plant rusts in host tissue cultures and upon synthetic media. I. *Gymnosporangium. Mycologia,* **51,** 248–95 (1959)
32. Cutter, V. M. Jr. Studies on the isolation and growth of plant rusts in host tissue cultures and upon synthetic media. II. *Uromyces ari-tryphylli. Mycologia,* **52,** 726–42 (1960)
33. Cutter, V. M. Jr. An axenic culture of *Puccinia malvacearum. Bull. Assoc. Southern Biologists.,* **7,** 26 (Abstr.) (1960)
34. Cutter, V. M. Jr. (Unpublished manscript provided posthumously by Dr. Lois Cutter)
35. Daly, J. M. Some metabolic consequences of infection by obligate parasites. In *The dynamic role of molecular constituents in plant-parasite interaction,* 144–61 (Mirocha, C. J., Uritani, I., Eds., Am. Phytopathol. Soc., St. Paul, Minnesota, 372 pp., 1967)
36. Daly, J. M., Knoche, H. W., Wiese, M. V. Carbohydrate and lipid metabolism during germination of uredospores of *Puccinia graminis tritici. Plant Physiol.,* **42,** 1633–41 (1967)
37. De Bary, A., *Comparative Morphology and Biology of the Fungi, Mycetozoa, and Bacteria* (Eng. trans. by Garnsey, H. E. F.; rev. Balfour, I. B., Clarendon Press, Oxford, 525 pp., 1887)
38. Dekhuijzen, H. M., Singh, H., Staples, R. C. Some properties of hyphae isolated from bean leaves

infected with the bean rust fungus. *Contrib. Boyce Thompson Inst.,* **23,** 367–72 (1967)

39. Dekhuijzen, H. M., Staples, R. C. Mobilization factors in uredospores and bean leaves infected with bean rust fungus. *Contrib. Boyce Thompson Inst.,* **24,** 39–52 (1968)
40. Dickinson, S. Studies in the physiology of obligate parasitism. II. The behaviour of the germ tubes of certain rusts in contact with various membranes. *Ann. Botany (London),* NS **13,** 219–36 (1949)
41. Dougherty, E. C. Problems of nomenclature for the growth of organisms of one species with and without associated organisms of other species. *Parasitology,* **42,** 259–61 (1953)
42. Duggar, B. M., *Fungous Diseases of Plants* (Ginn & Company, Boston, 508 pp., 1909)
43. Eagle, H., Piez, K. The population-dependent requirement by cultured mammalian cells for metabolites which they can synthesize. *J. Exptl. Med.,* **116,** 29–43 (1962)
44. Emge, R. G. The influence of light and temperature on the formation of infection-type structures of *Puccinia graminis* var. *tritici* on artificial substrates. *Phytopathology,* **48,** 649–52 (1958)
45. Ezekiel, W. N. Studies on the nature of physiologic resistance to *Puccinia graminis tritici.* Minn. Univ. *Agr. Expt. Sta. Tech. Bull.,* **67,** 1–62 (1930)
46. French, R. C., Weintraub, R. L. Pelargonaldehyde as an endogenous germination stimulator of wheat rust spores. *Arch. Biochem. Biophys.,* **72,** 235–37 (1957)
47. Fromme, F. D. Negative heliotropism of urediniospore germ tubes. *Am. J. Botany,* **2,** 82–85 (1915)
48. Fuchs, W. H., Gaertner, A. Untersuchungen zur Keimungsphysiologie des Schwarzrostes *Puccinia graminis tritici* (Pers) Erikss. u. Henn. *Arch. Mikrobiol.,* **28,** 303–9 (1958)
49. Garrett, S. D., *Biology of Root-infecting Fungi* (Cambridge Univ. Press, London, 293 pp., 1956)
50. Gautheret, R. J., *Manual Technique de la Culture des Tissus Végétaux* (Masson & Cie, Paris, 172 pp., 1942)
51. Gretschushnikoff, A. I. Toxins of rust *(Puccinia). Compt. Rend. Acad. Sci. U.R.S.S. (II),* **8,** 335–40 (1936) (*Abstr. Rev. Appl. Mycol.,* **15,** 710 (1936)
52. Hare, R. C. Physiology of resistance to fungal diseases in plants. *Botan. Rev.,* **32,** 95–137 (1966)
53. Heitefuss, R. Nucleic acid metabolism in obligate parasitism. *Ann. Rev. Phytopathol.,* **4,** 221–44 (1966)
54. Hotson, H. H. The growth of rust in tissue culture. *Phytopathology,* **43,** 360–63 (1953)
55. Hotson, H. H., Cutter, V. M. Jr. The isolation and culture of *Gymnosporangium juniperi-virginianae* Schw. *Proc. Natl. Acad. Sci. U.S.,* **37,** 400–3 (1951)
56. Hurd-Karrer, A., Rodenhiser, H. A. Structures corresponding to appressoria and substomatal vesicles produced on nutrient solution agar by cereal rusts. *Am. J. Botany,* **34,** 377–84 (1947)
57. Jones, J. P. Absorption and translocation of S^{35} in oat plants inoculated with labelled crown rust uredospores. *Phytopathology,* **56,** 272–75 (1966)
58. Jones, J. P., Snow, J. P. Amino acids released during germination of S^{35} labelled crown rust spores. *Phytopathology,* **55,** 499 (Abstr.) (1965)
59. Klebahn, H., *Die Wirtwechselnden Rostpilzen* **XXXVII** (Bornträger, Berlin, 447 pp., 1904)
60. Kucharek, T. A., Young, H. C. Jr. Teliospore germination of *Puccinia recondita* f. sp. *tritici. Phytopathology,* **58,** 1056 (Abstr.) (1968)
61. Macko, V., Woodbury, W., Stahmann, M. A. The effect of peroxidase on the germination and growth of mycelium of *Puccinia graminis* f. sp. *tritici. Phytopathology,* **58,** 1250–54 (1968)
62. Maclean, D. J., Scott, K. J. (Unpublished data)
63. Maheshwari, R., Allen, P. J., Hildebrandt, A. C. Physical and chemical factors controlling the development of infection structures from uredospore germ tubes of rust fungi. *Phytopathology,* **57,** 855–62 (1967)
64. Maheshwari, R., Hildebrandt, A. C., Allen, P. J. The cytology of infection structure development in uredospore germ tubes of *Uromyces phaseoli* var. *typica* (Pers)

Wint. *Can. J. Botany,* **45,** 447–50 (1967)

65. Mains, E. B. The relation of some rusts to the physiology of their hosts. *Am. J. Botany,* **4,** 179–220 (1917)
66. Melhus, I. E., Durrell, L. W. Studies on the crown rust of oats. *Iowa Agr. Expt. Sta. Res. Bull.,* **49,** 113–44 (1919)
67. Millerd, A., Scott, K. J. Respiration of the diseased plant. *Ann. Rev. Plant Physiol.,* **13,** 559–74 (1962)
68. Mitchell, D., Shaw, M. Metabolism of glucose-C-14, pyruvate-C-14, and mannitol-C-14 by *Melampsora lini.* II. Conversion to soluble products. *Can. J. Botany,* **46,** 435–60 (1968)
69. Morel, G. Le développement du mildiou sur des tissus de vigne cultivés *in vitro. Compt. Rend. Acad. Sci. Paris,* **218,** 50 (1944)
70. Morel, G. Recherches sur la culture associée de parasites obligatoires et de tissus végétaux. *Ann. Épiphyties (II),* **14,** 1–112 (1948)
71. Nozzolillo, C., Craigie, J. H. Growth of the rust fungus *Puccinia helianthi* on tissue cultures of its host. *Can. J. Botany,* **38,** 227–33 (1960)
72. Pavgi, M. S., Dickson, J. G. Influence of environmental factors on development of infection structures of *Puccinia sorghi. Phytopathology,* **51,** 224–26 (1961)
73. Plowright, C. B., *A Monograph of the British Uredineae and Ustilagineae* (Kegan Paul, Trench & Co., London, 347 pp., 1889)
74. Pole Evans, I. B. The cereal rusts. I. The development of their uredo mycelia. *Ann. Botany (London),* **21,** 441–46 (1907)
75. Ramakrishnan, L., Staples, R. C. Template activity and RNA synthesis from germinating bean rust uredospores. *Phytopathology,* **58,** 1064 (Abst.) (1968)
76. Ray, J. Cultures et formes aténuées des maladies cryptogamiques des végétaux. *Compt. Rend. Acad. Sci. Paris,* **133,** 307–9 (1901)
77. Ray, J. Étude biologique sur le parasitisme: *Ustilago maydis. Compt. Rend. Acad. Sci. Paris,* **136,** 567–70 (1903)
78. Reed, H. S., Crabill, C. H. The cedar rust disease of apples caused by *Gymnosporangium juniperi-virginianae* Schw. *Virginia Agr. Expt. Sta. Tech. Bull.,* **9,** 1–106 (1915)
79. Rodenhiser, H. A., Hurd-Karrer, A. Evidence of fusion bodies from uredospore germ tubes of cereal rusts on nutrient solution agar. *Phytopathology,* **37,** 744–56 (1947)
80. Rossetti, V., Morel, G. Le développement du *Puccinia antirrhini* sur tissu de Muflier cultivés *in vitro. Compt. Rend. Acad. Sci. Paris,* **247,** 1893–95 (1958)
81. Sappin-Trouffy, P. Recherches histologiques sur la famille des Urédinées. *Le Botaniste,* **5,** 59–244 (1896)
82. Schneider, W. Zur Biologie einiger liliaceenbewohnender Uredineen. *Zentr. Bakteriol. Parasitenk.,* **72,** 246–65 (1927)
83. Sharp, E. L., Smith, F. G. The influence of pH and zinc on vesicle formation in *Puccinia coronata avenae corda. Phytopathology,* **42,** 581–82 (1952)
84. Shaw, M. The physiology and host-parasite relations of the rusts. *Ann. Rev. Phytopathol.,* **1,** 259–94 (1963)
85. Shaw, M. The physiology of rust uredospores. *Phytopathol. Z.,* **50,** 159–80 (1964)
86. Singleton, L. L., Young, H. C. The *in vitro* culture of *Puccinia recondita* f. sp. *tritici. Phytopathology,* **58,** 1068 (Abstr.) (1968)
87. Spaulding, P. Investigations of the white-pine blister rust. *U.S. Dept. Agr., Agr. Infor. Bull.,* **957,** 1–100 (1922)
88. Staples, R. C., Wynn, W. K. The physiology of uredospores of the rust fungi. *Botan. Rev.,* **31,** 537–64 (1965)
89. Stock, F. Untersuchungen über Keimung und Keimschlauchwachstum der Uredosporen einiger Getreideroste. *Phytopathol. Z.,* **3,** 231–80 (1931)
90. Thiel, A. F., Weiss, F. The effect of citric acid on the germination of the teliospores of *P. graminis tritici. Phytopathology,* **10,** 448–52 (1920)
91. Turel, F. L. M., Ledingham, G. A. Production of aerial mycelium and uredospores by *Melampsora lini* (Pers) Lév. on flax leaves in tissue culture. *Can. J. Microbiol.,* **3,** 813–19 (1957)
92. Turel, F. L. M., Ledingham, G. A. Utilisation of labelled substrates

by the mycelium and uredospores of flax rust. *Can. J. Microbiol.*, **5,** 537–45 (1959)

93. Waddy, C. T., Scott, K. J. (Unpublished data)
94. Ward, H. M. On the relations between host and parasite in the bromes and their brown rust, *Puccinia dispersa* (Erikss). *Ann. Botany (London),* **16,** 233–315 (1902)
95. White, P. R., *A Handbook of Plant Tissue Culture* (Jacques Cattell Press, Lancaster, Pa., 277 pp., 1943)
96. Williams, P. G. (Unpublished data)
97. Williams, P. G., Scott, K. J., Kuhl, J. L. Vegetative growth of *Puccinia graminis* f. sp. *tritici in vitro. Phytopathology,* **56,** 1418–19 (1966)
98. Williams, P. G., Scott, K. J., Kuhl, J. L., Maclean, D. J. Sporulation and pathogenicity of *Puccinia graminis* f. sp. *tritici* grown on an artificial medium. *Phytopathology,* **57,** 326–27 (1967)
99. Williams, P. G., Shaw, M. Metabolism of glucose-C-14, pyruvate-C-14, and mannitol-C-14 by *Melampsora lini.* I. Uptake. *Can. J. Botany,* **46,** 435–40 (1968)
100. Witkowski, R., Grümmer, G. Beobachtungen an *Puccinia tatarica* Trazsch. auf Gewebekulturen von *Lactuca tatarica* (L.) C. A. Meyer. *Z. Allgem. Mikrobiol.,* **1,** 79–82 (1960)
101. Yarwood, C. E. Obligate parasitism. *Ann. Rev. Plant Physiol.,* **7,** 115–42 (1956)
102. Yarwood, C. E. Response to parasites. *Ann. Rev. Plant Physiol.,* **18,** 419–38 (1967)

Copyright 1969. All rights reserved

PARASEXUALITY IN PLANT PATHOGENIC FUNGI

BY R. D. TINLINE
Canada Department of Agriculture, Research Station
Saskatoon, Saskatchewan

AND

B. H. MACNEILL
Department of Botany, University of Guelph
Guelph, Ontario

The capacity of plant pathogens to survive, reproduce, and extend their host range is of fundamental importance in plant pathology. In this regard, the versatility of the asexual filamentous fungi has been attributed largely to mutations and their subsequent clonal propagation. More recently, heterokaryosis has been invoked as another mechanism of variation and adaptation. Today, mutation, heterokaryosis, and somatic recombination are advocated as nuclear mechanisms conferring versatility in the *Fungi imperfecti*, and, together with sexual reproduction, in fungi of other classes.

Genetic recombination in which there is no fine coordination between recombination, segregation, and reduction as there is in meiosis, has been termed parasexual (109, 110). Parasexuality has been the subject, wholly or in part, of a number of recent reviews (11, 18, 30, 42, 43, 108, 116, 122). The present paper, therefore, is not an exhaustive treatment but, rather, it concentrates on some aspects that we believe are pertinent to the plant pathologist. We will, however, make frequent reference to non-plant pathogens; indeed parasexuality, as it is known in *Aspergillus nidulans*, serves as a model to which somatic recombination in other fungi is compared. In *A. nidulans* three steps appear essential: (*a*) heterokaryosis; (*b*) formation of heterozygous diploids; and (*c*) segregation and recombination at mitosis. Each of these steps is briefly treated at the outset to facilitate a condensed and uniform coverage of the parasexual phenomenon in plant pathogens.

HETEROKARYOSIS

The condition wherein genetically unlike nuclei occupy a common cytoplasm has been termed heterokaryosis. Several excellent reviews on this subject have recently appeared (28, 102) and consequently only certain aspects need be considered here. The prior condition of heterokaryosis followed by fusion of two nuclei to produce a heterozygous diploid is implicit in the parasexual cycle. The underlying problem concerns the origin of the dissimilar nuclei that make up the heterokaryotic condition. Unquestionably, heterokaryosis may have its genesis in mutation. Ishitani et al. (76) reported that heterokaryons occurred in *Aspergillus* following treatment of homokaryons with ultraviolet. Tinline (134) was able to obtain both prototrophic and auxotrophic clones of *Cochliobolus sativus* from random hyphal tips of homokaryotic propagules which had been irradiated with ultraviolet. But, while it is undoubtedly true that mutations can occur within a clone, the detection of such genetic changes either in the heterokaryotic state or as recombinant genotypes is seldom realized except as a laboratory procedure involving highly selective techniques.

Sexual spores may be heterokaryotic from the time they are formed. Nonsister nuclei cooperate to delimit ascospores in some fungi such as *Neurospora tetrasperma* (36) and *Gelasinospora tetrasperma* (37), and unlike nuclei occur in the basidiospores of some Basidiomycetes (117).

Heterokaryosis may also be instituted by nuclear exchange following anastomosis both between sex cells and between vegetative cells. Outbreeding which leads to the formation of heterokaryons in sexual reproduction is an obvious means of maintaining a high level of genetic diversity. Unfortunately we have very little information concerning the forces which control anastomosis and heterokaryon formation in the reproductive cycle of asexual fungi. Anastomosis occurs often between strains of a species (9, 56, 76, 95, 103), but some workers have indicated that this occurs less frequently than anastomosis within a strain (135). Anastomosis has also been reported between different fungal species (59, 76, 95). Considerable support for the view that the environment influences the incidence of anastomosis is now available (67, 76, 127, 133). Flentje & Stretton (47) and Flentje (46) conclude that anastomosis in *Thanatephorus cucumeris* is a complex phenomenon. When pairs of isolates were opposed they exhibited a variety of reactions: (*a*) they repelled one another; (*b*) they came in contact but did not fuse; (*c*) the cell walls fused but no cytoplasmic connection was established; (*d*) they anastomosed but soon the anastomosed cells collapsed and died; and (*e*) they anastomosed without any detrimental effects. It is not known how widespread such hyphal reactions may be in the fungi; obviously such studies merit increased attention since they are fundamental to an understanding of heterokaryosis and nuclear exchange.

Heterokaryosis has been reported in many fungi including numerous plant pathogens (22, 102, 133). It has been demonstrated or inferred between strains (129), isolates (39, 97, 133), form species (17, 54), and species (17, 34, 79, 95).

Strain specificity in the formation of heterokaryons, a concept of vegetative incompatibility (42), has been investigated in a few fungi in which a sexual state permits genetic analysis (28, 42, 43). Incompatibility was found to have a genetic basis in the heterothallic species, *Neurospora crassa* (55, 70, 151); in *T. cucumeris* (56, 139, 147), a species in which some isolates at least are homothallic (47); and the homothallic species, *A. nidulans* (57, 58, 78). Jinks et al. (77) suggested that identical alleles may be necessary at a minimum of five loci for heterokaryon formation between an offspring of an outcross and one of its parents in *A. nidulans.* Pittenger & Brawner (106) described a pair of alleles that influence nuclear ratios in heterokaryons. Caten & Jinks (22), discussing the significance of heterokaryosis, concluded that where incompatibility occurs in homothallic and imperfect fungi it is restrictive to gene exchange. However, sexual incompatibility does not always preclude the formation of heterokaryons. In *Ustilago maydis,* incompatibility of alleles at a single locus prevents formation of the dikaryon (69) but in *Coprinus lagopus* and *Schizophyllum commune,* fungi with tetrapolar sexuality, heterokaryons develop in both compatible and incompatible matings (117). In *S. commune,* Leary & Ellingboe (85) reported that anastomosis and nuclear exchange appeared to be independent of mating type but that subsequent events were determined by mating-type genes. Incompatibility has provoked the frequent utilization of mutants from a single isolate in studies of heterokaryosis and parasexuality within species. Although this orthodoxy of approach may facilitate detection of the phenomena, it probably obscures the real potential of heterokaryosis and parasexuality as they may occur in nature.

Heterozygous Diploids

Evidence for the occurrence of heterozygous diploids in vegetative cells of filamentous fungi has been almost exclusively genetic. Cytological investigations of somatic cells have been inconclusive (4) because of the difficulties which attend the resolution of mitotic nuclear divisions. These problems are discussed later.

The detection of a relatively rare event, such as a heterozygous diploid, is most readily accomplished on the basis of phenotype characteristics. Diploids synthesized from two mutant strains differing in spore color may well exhibit the wild-type spore color. If in addition, a number of genetic markers are included in the heterokaryon, especially such factors as govern nutrition of the strains and tolerance to chemicals, the heterozygous or recombinant diploids may be detected either by direct inspection of the cultures or by plating out the presumptive diploids on selective media. This essentially is the procedure pioneered by Roper (121) in his studies with *A. nidulans* and it has since been widely adopted by many other workers. While the initial identification of the diploids is commonly based on phenotype, the recovery of somatic recombinants constitutes the primary evidence for the diploid event. In most instances, the diploid persists as a clone and is subject to analyses over a period of time; but obviously a rare diploid event followed by immediate haploidization could produce much of the genetic diversity implicit in parasexuality, with no direct evidence of the presence of a fusion nucleus.

Detection of diploids sometimes can be based on the size of the conidia, diploid spores being larger than haploid spores (5, 6, 15, 19, 29, 44, 45, 61, 73, 88, 113, 124, 132). But MacDonald et al. (91) have pointed out the need to assess spore sizes anew in each combination of strains in order to apply the size criterion to *Penicillium crysogenum.* Size differences have not been noted between haploid and diploid spores in *Aspergillus oryzae* and *A. sojae* (75), *C. sativus* (135), and *U. maydis* (69). Diploid spores may contain more deoxyribonucleic acid (DNA) than haploid spores (64, 69), but in determinations of DNA content the number of nuclei per spore is also an important consideration (75). Nuclear size as a criterion in estimating ploidy may also be useful; however, Clutterbuck & Roper (25) found that while diploid and haploid nuclei differed in mean area, an overlap in size distribution precluded differentiating them in heterokaryons. Nuclear size differences were also noted in *C. sativus* (137).

The frequency of occurrence of heterozygous diploids as determined by spore platings from heterokaryons is variable within investigations of a single species as well as between species. For the most part, the spontaneous occurrence of diploids is relatively low and several agents such as camphor (5, 113, 114) and ultraviolet irradiation (29, 72, 74) have been employed to increase the frequency of this event. A high incidence of spontaneous diploids has been found in three Basidiomycetes—*C. lagopus* (19), *U. maydis* (69), and *U. violaceae* (29). It is intriguing to speculate that the frequency of diploids may be related to nuclear ratios; since Basidiomycetes are generally dikaryotic, the opportunity for diploidy to occur may be more highly favored than in other fungi in which the majority of cells of the thallus are homokaryotic. In the absence of nuclear or cytoplasmic suppressors of nuclear fusion, dikaryons may be expected to generate more diploids than other heterokaryons.

Mitotic Recombination and Segregation

The relative stability of heterozygous diploid nuclei through many generations of somatic cells is almost a general characteristic of the parasexual cycle in fungi studied to date. Perhaps, as suggested previously, a transient diploid stage may occur during this cycle, but it is obvious that the opportunity for mitotic crossing-over and segregation is enhanced when the diploid phase is prolonged. Mitotic crossing-over was invoked in 1952 to explain the occurrence from heterozygous diploids in *A. nidulans* of segregants that were homozygous for some markers and which remained heterozygous for others (112). Subsequent studies have shown that the recombination event represents a reciprocal exchange in a chromosome arm of homologous chromosomes at the four-strand stage (80, 123). Depending upon the pattern of mitotic segregation, daughter nuclei will be either homozygous for all markers distal to the point of exchange, or heterozygous for all markers as in the parent. Since more than one exchange in the whole chromosome complement is rare and markers distal to an exchange segregate together, the order of genes along chromosome arms can be determined.

Haploid (114) and diploid (111) segregants and unstable segregants that appeared to be aneuploids (111) in which all markers on individual chromosomes remained linked were found in *A. nidulans.* Käfer (80) postulated a single process i.e., nondisjunction, as the basis for all these recombinant types. Since all markers in a single linkage group move together in these chromosomal segregants and since haploids can often be distinguished from diploids, haploidization provides a convenient means of determining linkage relationships. Criteria often used to indicate haploidy are: no further segregation of markers; small size of haploid spores; the expression of recessive markers in each linkage group; and mating-type reaction. Various chemical agents and ultraviolet irradiation have been employed to increase the frequency of recombination. *p*-Fluorophenylalanine has been especially useful in inducing haploidization (87).

Cytological Considerations

Roper (122) pointed out that the discovery (107) of the parasexual cycle which leads to genetic recombination was not supported by prior microscopic evidence suggesting that such recombination might be found. One might argue, even today, that the genetic evidence for the parasexual cycle in fungi is so well substantiated that there is little need to seek microscopic proof of fusion nuclei in vegetative hyphae, or cytological data which might elucidate chromosome mechanics during nuclear division. Cytological evidence of this kind is hard to find primarily because fungal nuclei are small and somatic chromosomes have a low affinity for stains and are considerably smaller than they are during the meiotic cycle (120). Thus, among workers engaged in cytological studies of fungi, a great deal of controversy has developed around the details of nuclear division and chromosome distribution. A review by Weisberg (145) encompasses most of the major theories of somatic nuclear division as they have been applied to fungi in general.

At this juncture, the work of Weijer et al. (143) appears pertinent for the reason that it has introduced a new and controversial element into fungal cytology. These workers described a method of division (karyokinesis) in *Neurospora* which is based upon the longitudinal division of a nucleus consisting of a Feulgen-positive triangular centriole and seven chromosomes attached end-to-end in linear order by a Feulgen-positive thread. Recently they reported that in young, fast-growing hyphae of *A. nidulans,* the nucleus assumes a filamentous nature and is particulate, containing nine Feulgen-positive bodies (144, 146). Weijer et al. assume that these chromatin bodies represent the eight chromosomes of *A. nidulans,* together with a DNA-containing centriole. Prior to nuclear division, the centriole divides. Since no spindle is detected, the forces generated by cytoplasmic streaming presumably act to separate the daughter nuclei. In older hyphal cells the nuclei become very large and filamentous, approximately 100 μ in length. These nuclei divide longitudinally without a spindle, and a nuclear fiber attaches the centriole and eight chromosomes in an end-to-end fashion.

Other results that support the interpretations of Weijer and co-workers can presently be found. Laane (84) observed that dividing nuclei of *Peni-*

cillium expansum contain a number of bodies connected by a filament. He called these nuclear fragments. The fact that similar nuclear bodies and the configuration into which they assemble were found in living spores and mycelium was taken by Laane as strong indication that they are real and not artifacts. Heale et al. (65) reported that nuclear division in conidia and hyphae of *Verticillium albo-atrum* is characterized by a double-stranded filament composed of two parallel rows of chromosomes connected by fine threads. In addition to offering support for the idea that in certain fungi nuclear division seems to have little in common with classical mitosis, they attempted to solidify the nomenclature suggested by Weijer et al. (142) to describe the multiplicity of karyokinetic configurations observed.

A good deal of implied and expressed disagreement with the Weijer interpretations, however, obtains. Somers et al. (128) and Ward & Ciurysek (140) described anaphase and metaphase figures in *N. crassa* and concluded that mitosis is classical. Knox-Davies (81) reported the presence of centrioles and spindles in dividing nuclei in spores of *Macrophomina phaseoli,* but noted certain variations from what might be considered as the normal pattern of chromosome behavior, e.g., lagging chromosomes at anaphase, chromosome bridges, and aneuploidy. In hyphal cells, Knox-Davies (82) found no centrioles or spindles; again, there were lagging chromosomes and chromosome bridges between separating daughter chromosome groups. Although many elongated nuclei were seen, they were interpreted either as migrating nuclei or as artifacts resulting from squashing. The author indicates that there was no complicated nuclear cycle *sensu* Weijer, nor was there "tear-drop" (118) or "taffy-pull" (98) separation of daughter nuclei. He concluded that true mitosis occurs in the vegetative hyphae of *M. phaseoli,* and that with improved techniques it should be possible to demonstrate a mitotic apparatus in the somatic cells of a variety of fungi. Recently Aist & Wilson (2) who formerly had reported elongated, double-stranded somatic nuclei as characteristic of nuclei in division in 11 fungi (12), published a reinterpretation based primarily on new findings. Studies with living and stained preparations of *Fusarium oxysporum* allowed them to follow nuclear division within the nuclear envelope. Artifacts which might otherwise have been interpreted as amitotic figures were avoided by improved cytological techniques. They emphasized that misleading interpretations easily arise out of improperly fixed and stained material, and implied that many disputed aspects of nuclear division could be resolved by observation of the living cell with phase-contrast optics.

Perhaps several mechanisms are operative in the division of somatic nuclei, but the cytological evidence is not fully convincing. Certainly, many of the modern proposals seem inadequate to explain chromatin exchange and genetic recombination any more satisfactorily than the classical scheme of mitosis. The work of Wilson and co-workers, however, is impressive on such topics as motility of fungal nuclei, chromosome division within the nuclear envelope, synchronous division of nuclei in hyphal cells, and fusion of protuberances from nuclear envelopes which might allow the direct exchange of bits of chromatin (1, 149, 150). In addition, a wide variety of fungi now are used as models for cytologic and genetic studies, e.g., *Coprinus* (26, 90), *Marasmius* (38), *Schizophyllum* (99), *Phymatotrichum* (71), *Verticillium* (73, 92), *Ceratocystis* and *Fusarium* (2), and *Fomes* (149). In 1967, Bracker (10) suggested that the incomplete nature of the information on nuclear division precluded a general hypothesis and apparently this situation still obtains. Consequently, the fundamental mechanics of parasexuality remain unresolved.

Plant Pathogens

It is regrettable that we can only speculate on the importance of parasexuality in the survival and evolution of plant pathogens in nature. Still, the deciphering of basic processes in the laboratory and an assessment of their potentials remain a fundamental approach, and additional work appears warranted to provide models which might lead to a better understanding of variability, genetic controls, and the molecular bases of pathogenicity and virulence. Certainly, variation in plant pathogens is a dynamic and complex process; it is to be expected, then, that very valuable results will continue to arise in the laboratory environment where strict controls of known variables can be exercised.

Phycomycetes.—Unequivocal examples of parasexuality are wanting in this class so far as the authors can determine, thereby making it unique among the classes of Eumycetes. Still, parasexuality has been considered as a possible mechanism of variation in some species such as *Phytophthora* (13). Sansome (125) suggested that the Oomycetes in general may be predominantly diploid. Savage et al. (126) speculated that homothallic species may be diploid, heterothallic ones haploid; they [and Gallegly in his review of the genetics of *P. infestans* (51)] pointed out the need for additional studies to resolve the question of ploidy. If some or all species are in fact habitually diploid, then recombinants arising by nondisjunctive processes should occur. However, as Caten & Jinks (23) stated, we do not know what to expect in the way of recombination frequencies in such fungi since rates have been estimated only from transient diploids.

An indication of parasexuality derives from the work of Park et al. (101) with a nonpathogenic Zygomycete, *Phycomyces blakesleeanus*. Heterokaryons synthesized from two auxotrophs showed an altered morphology; some sporangiospores from the heterokaryon appeared to have recombinant phenotypes, two were diauxotrophic and formed colonies with normal morphology.

Ascomycetes: Cochliobolus sativus (imperfect state *Bipolaris sorokiniana* syn. *Helminthosporium sativum*).—This species causes common root rot, spot blotch, and head blight in various cereals and grasses. It is heterothallic with one pair of alleles controlling incompatibility. Although compatible isolates appear randomly distributed in nature, the sexual stage is known only from laboratory cultures. Using spore color and anisomycin-resistant and auxotrophic markers, Tinline (135) obtained heterokaryons, heterozygous diploids, and somatic recombinants. Vegetative cytoplasmic units of the fungus are multinucleate. First order segregants selected from the diploids gave rise to second order recombinants. These results indicate that parasexuality occurs in this species.

Glomerella cingulata (imperfect state, *Colletotrichum gloeosporioides*).—This fungus is considered to be homothallic but it is sexually complicated. It causes anthracnose in a wide range of herbaceous and woody plants. Stephan (129) demonstrated heterokaryosis by combining strains carrying morphological markers, and subsequently (130) by using auxotrophic markers. Most conidia are uninucleate. A heterokaryon comprised of two auxotrophic homokaryons produced some prototrophic conidia which maintained a stable phenotype through several generations. A few recombinants were found but, since they differed from the parents in only one marker, it is not known if they arose by mutation or parasexual recombination.

Leptosphaeria maculans (imperfect state, *Plenodomus lingam*).—This fungus causes a stem and corrosive root disease called blackleg in a wide range of cruciferous plants including rape and mustard. It is not known if the species is homo- or heterothallic. In pairings of diauxotrophs, some of which carry color markers, Petrie (105) obtained slow-growing colonies on minimal medium which occasionally produced rapidly growing sectors. Uninucleate conidia from these sectors were larger than the parental spores and were prototrophic. Single auxotrophs were recovered as first-order segregants from several of the presumptive diploids, and from some of the segregants second-order recombinants were obtained. Apparently, parasexuality occurs in this species.

Basidiomycetes: Rusts.—New strains arising on host plants from uredospore mixtures of two races have been reported in *Puccinia graminis* f. sp. *tritici, P. recondita* f. sp. *tritici, Melampsora lini,* and between *P. graminis* f. sp. *tritici* and f. sp. *secalis*. The new phenotypes were identified by their virulence to differential varieties and sometimes, in *Puccinia,* by spore color. Christensen (24), Flor (49), and Watson & Luig (141) have cited most of the work and it will not be detailed here. Several mechanisms including parasexuality have been proposed in explanation for the origin of the new strains. Anastomosis was observed between races and formae of *P. graminis,* and nuclear exchange between strains appears a likely mechanism. In several investigations with *Puccinia* however, the multiplicity of new strains exceeded the number that could be accounted for on the basis of nuclear exchange, and somatic or parasexual recombination was invoked. Flor (49) paired F_1 cultures of known genotype and attributed all authenic variants that he observed in *M. lini* to nuclear exchange and single mutations in loci for which the parental culture was heterozygous. He did not find variants differing in pathogenicity at several loci; such variants might be expected if a parasexual process were involved. Further, as Ellingboe

(40) pointed out, some rust races are known to be heterozygous for several virulence genes and hence recombination might occur in single uredospore cultures, but such recombinants have not been detected. Watson & Luig (141) who have favored a mechanism analogous to parasexuality also suggested that cytoplasmic effects may be involved in changes of virulence. Recently, Bugbee et al. (14) obtained somatic recombinants from mixtures of uredospores of *P. graminis* f. sp. *tritici;* their results do not conform to the hypothesis of a simple nuclear exchange. Some authors (41, 49) have indicated the need for analysis of genotypes of uredospore cultures through selfing and crossing studies before mechanisms of variation can be clarified. Certainly, however, parasexuality remains a likely explanation but further work is necessary to resolve the question. Possibly *in vitro* studies of variation in laboratory cultures of rusts (148) will expedite resolution.

Basidiomycetes: Smuts.—Some smuts of the genus *Ustilago* have exhibited parasexual recombination.

U. maydis causes a disease in corn and teosinte that is apparent as galls in embryonic tissues. Holliday (69) infected corn seedlings with auxotrophic strains differing in mating type, and isolated from gall tissue stable prototrophic, solopathogenic strains that were demonstrated to be heterozygous diploids. Homokaryons and diploids grew in culture; dikaryons were restricted to host tissue. He obtained segregants from the diploids rarely as spontaneous events but frequently following ultraviolet irradiation. Except for one segregant that appeared to be aneuploid and which later gave rise to a stable diploid, recombination was associated with mitotic crossing-over. Reciprocal products of an exchange were often detected. Diploid segregants facilitated an analysis of the function of the mating-type loci and Holliday found that the *a* locus determines the fusion compatibility of sporidia and the *b* locus controls pathogenicity.

U. violacea causes anther smut in plants of the Caryophyllaceae; unlike *U. maydis,* it is systemic in its host. The fungus has bipolar incompatibility. Day & Jones (29) obtained prototrophic colonies from fusion of sporidia on water agar from auxotrophic strains differing in mating type. The prototrophs did not mate with either a_1 or a_2 haploid lines and were shown to be heterozygous, solopathogenic diploids. As in *U. maydis,* few segregants from the diploids arose spontaneously but many occurred after treatment with ultraviolet rays; all segregants appeared to be either aneuploids or diploids. Haploid segregants were obtained, however, by treating diploid sporidia with fluorophenylalanine. Also as in *U. maydis* the diploids and haploids, but not the dikaryons, could grow saprophytically.

U. hordei, causing covered smut of barley, is a well-known fungus which infects young seedlings and ultimately aborts the kernels. Dinoor & Person (35) obtained growth of the dikaryotic phase in culture by pairing auxotrophic sporidia of opposite mating type on minimal medium; the fungus has bipolar incompatibility. The authors indicate that haploid recombinant sporidia have been obtained from dikaryons. Thus, there is some indication in this species of somatic recombination, and in view of known parasexuality in the two previous species, we presume that the recombination is parasexual.

Hymenomycetes.—Intensive studies on somatic recombination in some species of this sub-class have not fully resolved the mechanisms involved. *Schizophyllum commune,* a wound fungus of various hardwoods, has complex tetrapolar incompatibility comprised of factors A and B, each a complex of two or more loci with multiple alleles. Matings of compatible homokaryons form dikaryons and matings of incompatible homokaryons produce heterokaryons. Further, dikaryon-homokaryon matings can occur and, out of these, new combinations may arise. Ellingboe (41), in a review, indicated that at least two mechanisms of somatic recombination have been demonstrated. Recombination between linked genes at a frequency comparable to that in meiosis has been reported, and the phenomenon termed precocious meiosis (117). In some dikaryon-monokaryon matings, strains were obtained which differed from the original components only in incompatibility factors despite linkage between the factors and other markers. This phenomenon has been called specific factor transfer. Some results indicate, however, that standard parasexuality also may occur in *S. commune.* Middleton (96) recovered aneuploids and recombinant haploids from common AB heterokaryons. Further, stable diploids apparently occur. Parag (100) synthesized a common B heterokaryon and found it was a stable diploid from which he obtained a somatic recombinant; when he crossed the diploid with a haploid the progeny segregated as expected from a diploid-haploid cross. Koltin & Raper (83) have reported an exciting discovery concerning diploids in *S. commune.* They found that in the absence of a dominant gene *dik*$^{+}$ in both partners of a dikaryon, nuclear fusion occurred and a stable diploid resulted. The dominant allele present in one or both partners inhibited the formation of a stable diploid. Such genetic controls of nuclear fusion in fungi have important implications.

Apparently, parasexuality has been demonstrated in the saprophytic, tetrapolar *C. lagopus.* Casselton (19) obtained stable heterozygous diploids from a common A heterokaryon and in turn selected aneuploid and haploid recombinants. Cowan & Lewis (26) obtained evidence that recombination may occur through gradual haploidization. Casselton (19) and Casselton & Lewis (20) reported that diploids were unstable in a diploid-haploid dikaryon but stable when alone.

Fungi imperfecti: Ascochyta imperfecta.—A fungus in the order, Sphaeropsidales, causes black stem in a number of leguminous hosts including alfalfa and red clover. Sanderson & Srb (124) isolated heterokaryons via hyphal tips from paired strains differing in acriflavine resistance, nutritional requirements, and morphological features, or a combination of these markers. Heterozygous diploids obtained as prototrophs from uninucleate conidia of heterokaryons gave rise to recombinant strains, one of which was unstable and segregated further. All recombinants had diploid spore size; some were presumed to be nondisjunctional diploids arising from aneuploids. Mitotic crossing-over may also have been involved.

Aspergillus.—Parasexuality is known to occur in several species of this genus in addition to *A. nidulans,* namely, *A. amstelodami* (86) which, with some other members of the *A. glaucus* group, may cause mouldiness and deterioration of stored and damp grain, *A. fumigatus* (8, 131) which also has been associated with mouldy grain, *A. rugulosus* (27) a member of the *A. nidulans* group, *A. oryzae* (72) and *A. sojae* (75) two industrially important fungi, and *A. niger* (87, 88, 113) which causes various plant diseases such as crown rot of ground-nuts, a boll rot of cotton, a seedling blight of sorghum, and black mould of onion and shallot. In each of the above species, strains having different spore color and auxotrophic markers were used to synthesize heterokaryons or presumptive heterokaryons and from these, heterozygous diploids were isolated. Recombinants were obtained from the diploids, and the majority of those that appeared spontaneously in *A. sojae* and *A. niger* appeared to be still diploid; presumably they arose by mitotic crossing-over. Haploidization also is known to occur in most, if not all, of these species. Lhoas (88) found mitotic crossing-over and haploidization after chemical treatment to be frequent events in *A. niger.*

Fusarium oxysporum.—Parasexuality has been reported in three *formae speciales,* namely, *pisi* (15, 45, 138), *cubense* (16), and *callistephi* (68), that cause a wilt disease in pea, banana, and China aster, respectively. Auxotrophic and morphologic markers were employed in all the investigations; in addition to these, Buxton (15, 16) utilized differences in actinomycin tolerance and virulence to pea varieties in *pisi* and differences in haemagglutinating properties and virulence to seedlings of *Musa balbisiana* in *cubense.* Prototrophic, heterozygous diploids were obtained from heterokaryons, and recombinant segregants were selected from the diploids in *pisi* and *cubense.* Hoffmann (68) demonstrated heterokaryosis in *callistephi;* viable heterokaryons occurred chiefly in older cultures of paired auxotrophic strains. Although a stable, heterozygous diploid state was not apparent, he obtained recombinants by hyphal-tipping sectors that appeared in some cultures. Results of these various investigations indicate that parasexuality is an operative mechanism of variation in *F. oxysporum.*

Penicillium.—Parasexual recombination has been described in three species, *P. chrysogenum* (91, 115), *P. expansum* (5, 6, 44, 53) and *P. italicum* (132). The latter two species cause storage rots of pome and citrus fruits respectively. Diploid and haploid segregants from prototrophic heterozygous diploids were identified in *P. expansum.* The recombination processes in *Penicillium* sp. appear analogous to those elucidated in *A. nidulans.*

Phymatotrichum omnivorum.—This pathogen has a broad host range comprising some 2000 plant species, but it is best known for causing root rot in cotton. We are unaware of any genetic evidence for parasexuality in the organism; however, Hosford & Gries (71) from their study of nuclear phenomena hypothesized that it may occur. They observed regular mitosis, anastomosis of hyphae, and nuclear migration. The haploid number of chromosomes was four but some nuclei, presumed to be diploid, contained eight while other nuclei, aneuploids, contained more or less than eight chromosomes. Further studies aimed at relating cytological behavior with genetical evidence for parasexuality should be rewarding in *P. omnivorum.*

Pyricularia oryzae.—This fungus causes an important stem and foliage disease in rice called blast. Comprehensive studies on variation have been reported recently by Suzuki (133) and Yamasaki & Niizeki (153). Hyphal anastomosis and nuclear migration were observed. Suzuki (133) found that conidiophore initials and young conidia were multinucleate and that heterokaryosis of a persistent nature occurred. Yamasaki & Niizeki (153) reported that the conidiophore possessed a single generative nucleus and most cells were uninucleate in the strains they used. A high incidence of nuclear fusion was suspected (153) and aneuploid and polyploid nuclei were observed (133). Yamasaki & Niizeki (153) mixed pairs of strains differing in color, copper sulphate tolerance, nutritional requirements, and hydrogen sulphide production on minimal medium. Prototrophic growth occurred and conidia presumed to be heterozygous diploids were isolated. Recombinants were obtained by plating spores from the prototrophic diploid; consequently parasexuality would appear to occur in this fungus.

Verticillium.—Parasexuality has been described in two species, *V. albo-atrum* (60) and *V. dahliae* var. *longisporum* (73) and between the species *V. albo-atrum* and *V. dahliae* (50). Fordyce & Green (50) suggested that since anastomosis, heterokaryosis, and parasexualism occur between the species, they should be combined into one, *V. albo-atrum. Verticillium* is a vascular invader and produces wilt in a large number of dicotyledenous plants. Hastie (61, 62, 63) with *V. albo-atrum* found that heterokaryons and heterozygous diploids were very unstable and the diploids very frequently formed diploid segregants homozygous for some markers, apparently by mitotic crossing-over. Haploid recombinants likely originating through nondisjunction occurred at a high incidence in older cultures; three-week old cultures of diploids yielded about 95 per cent haploid conidia (61). Recently, Hastie (62, 63) has conducted analyses of recombination in phialide families. In this work, all the spores produced on some single phialides of heterozygous diploids were isolated and those families assessed in which diploidy was still extant. Since a phialide and the spores are uninucleate and the spores thus represent products of the phialide nuclear divisions, the timing and frequency of mitotic crossing-over could be determined. Reciprocal products of an exchange were predictably recovered in segregating families, and because there was no delay between mitotic crossing-over and marker segregation, his results support the hypothesis of crossing over as an exchange between whole chromatids. Further, he found no association between crossing-over and haploidization. His estimate of mitotic recombination was about 0.2 per nuclear division, a value far in excess of that reported by Käfer (80) for *A. nidulans*. In *V. dahliae* var. *longisporum*, Ingram (73) obtained prototrophic heterozygous diploids readily from mixtures of auxotrophic strains. These diploids, unlike those of *V. albo-atrum* were very stable.

Conformity Between Species

In many fungi in which parasexual recombination has been reported, the detailed processes of recombination were not resolved, and some, in which haploidization and mitotic crossing-over or both were invoked, differed in degree from those in *A. nidulans*. In our view, parasexuality analogous to that in *A. nidulans* has been acceptably demonstrated wherever there is a lack of coordination between the occurrence of a heterozygous diploid and recombination, where recombination processes occur independently, and where first-order segregants give rise to second-order recombinants. We are incognizant of reasons for differences in the stability of diploids and in the modes of recombination. However, as Pontecorvo (110) pointed out, the expression but not the structure of the genetic material is subject to regulation both qualitatively and quantitatively at all levels of organization. Since cells with identical genotypes may show different properties as a result of regulation of gene expression, it is not surprising that species with diverse genotypes exhibit wide differences. In *U. maydis,* mitotic recombination appeared predominant, whereas in *C. lagopus,* haploidization appeared to be the major process. The high frequency of parasexual recombination in *V. albo-atrum* prompts speculation that precocious meiosis in certain other fungi may also be of a parasexual nature. The occurrence of stable diploids in *S. commune* may facilitate clarification of its recombination processes.

Possibly too, some anomalies in the frequency of recombination events in various fungi that preclude cohesive genetic analysis may be due to nonoptimum selective techniques. The growth of some recombinants may be inimical to the growth and detection of others. Still, we should expect differences between species where independent processes of recombination maintain. These generalities, however, fail to provide an adequate explanation for the difficulty encountered by Strømnaes & Garber (131) in determining linkage in *A. fumigatus,* where 23 of 26 markers could not be assigned.

In plant pathogens generally, linkage relationships of markers have not progressed rapidly and more work along this line is warranted. We suspect that an elucidation of both the genetic nature and the biochemical basis of virulence will necessitate well-marked arms in all chromosomes.

Although currently recognized processes of parasexuality largely conform to those in *A. nidulans,* other viable processes of recombination are not precluded.

HETEROKARYONS AND DIPLOIDS

The assumption is sometimes made that balanced heterokaryons are analogous to diploids because of identical genotype, but this surely is erroneous. Although heterokaryons may better fit the ideal of flexibility for adaptation to new and immediate situations than diploids, they do not offer the long-term evolutionary benefits which diploids can confer. In many of the fungi discussed in this review, the diploid is phenotypically distinguishable from the heterokaryon and often appears in a heterokaryotic culture as a sector of more rapid growth. Several investigations to detect other differences between heterokaryons and diploids have been conducted, and although differences were observed their nature is not completely resolved. In *A. nidulans* Roberts (119) reported that recessive sorbitol mutants were complementary in heterozygous diploids but not in balanced heterokaryons. Similar results in complementation were obtained by Apirion (3) with fluoracetate mutants located at three distinct loci. Both he and Roberts (119) indicated that failure of complementation in heterokaryons was not due to unfavorable nuclear ratios. Casselton & Lewis (21) compared complementation in the heterokaryons, dikaryons, and diploids of *C. lagopus,* using linked auxotrophic mutants in both *cis* and *trans* arrangements. Complementation was incomplete in the heterokaryons. They further found that diploids and dikaryons heterozygous for a recessive suppressor of an auxotrophic marker were unable to grow on minimal medium whereas the heterokaryon did grow, indicating a difference in dominance effects between the culture types. They attributed the differences in complementation and dominance to an irregular distribution of nuclei in a heterokaryon and a dilution of the gene products in the cytoplasm. Day (32), using adenineless mutants, observed complementation in some diploids but not in dikaryons. In *U. maydis* (69) and *U. violacea* (29), differences between dikaryons and diploids were pronounced; the dikaryons appeared obligately parasitic whereas diploids were solopathogenic and could be cultured.

If we assume that virulence in plant-pathogenic fungi is comprised of one to many, simple or compound loci, and is a function of gene products via biosynthetic pathways, then we might expect differences in virulence between homokaryons, heterokaryons, and diploids, and within heterokaryons.

VIRULENCE

Attempts to correlate nutritional deficiencies with changes in virulence have been largely negative. In several fungi, nutritional mutants (especially some amino acid auxotrophs), appeared nonpathogenic, but when the mutants were combined in heterokaryons and heterozygous diploids, pathogenicity was restored (6, 7, 136). Prototrophic recombinants from the diploids also were pathogenic. Tinline (137) found an association between leakiness of auxotrophic mutants of *C. sativus* and their invasiveness in wheat seedlings. Yamasaki et al. (152) observed that mutants of *P. oryzae* and *Xanthomonas oryzae,* having the same nutritional requirements, differed in ability to attack a common host. Beraha & Garber (6, 7) found that avirulent auxotrophs of *P. italicum* and *P. digitatum* grew at the site of inoculation and concluded that nutritional requirements and virulence were not related. MacNeill & Barron (94) found no consistent difference between auxotrophy and avirulence in *P. expansum,* and MacNeill (93) reported an independent segregation of avirulence and nutritional markers in the recombinants from heterozygous diploids. Loprieno (89) found all auxotrophic mutants of *Colletotrichum coccodes* to be pathogenic. Thus, while auxotrophy affects parasitism, the latter is quite distinct from virulence and pathogenicity.

Changes in virulence and host range, however, have been reported, and some have been attributed to mutation, heterokaryosis, and parasexuality. Several reviews pertinent to this subject have appeared (31, 104). Yang & Hagedorn (154) found two induced variants of *F. solani* f. sp. *phaseoli* that were pathogenic on radish, a host resistant to the parental isolates. Holliday (69), by parasexual analysis in *U. maydis,* found that pathogenicity was controlled by multiple alleles at the *b* locus. Recently Day & Puhalla (33) have sought to determine the structure of the *b* locus and the nature of the control it exerts on pathogenicity. They irradiated homozygous *b* diploids which were nonpathogenic and obtained five mutants that were pathogenic to corn seedlings, and from this they inferred that either a mutant form of the *b* allele or a dominant suppressor had occurred. In *T. cucumeris,* Flentje and co-workers (46, 48) obtained mutants in which the infection process in radish was blocked at various stages: attachment of hyphae to the stem, growth along the stem, production of an infection cushion, penetration, and inter- and intra-cellular development of hyphae in the host. Heterokaryons consisting of pairs of the mutants were pathogenic. Infection appeared controlled by a number of genes. Garza-Chapa & Anderson (56) found that virulence of heterokaryons exceeded that of either of the parents. Similarly, Vest & Anderson (139) reported that when single basidiospore lines with low virulence were paired, the majority of the ensuing heterokaryons had an enhanced virulence. Heterokaryons comprised of strains of low and high virulence exhibited a virulence at least equal to that of the most virulent component. Assuming that alleles at several loci control virulence as has been suggested (56), then complementation of gene products may result in an increased virulence. Ming et al. (97) isolated from one wild-type culture of *F. fujikuroi* three components which differed in production of gibberellin and hence in virulence. One component was almost avirulent, while another was more virulent than the wild type. Possibly a gene dilution effect is evident here. These results are of especial interest in demonstrating the possible means of survival of a weakly virulent strain.

Still another aspect of virulence is being explored by Garber (52) and

Henry & Garber (66). Apparently with a view to elucidating distinctions between pathogenic races, they are investigating intra- and extra-cellular enzymes from virulent and avirulent strains.

The work of Buxton (15) remains the outstanding example of the role of parasexuality in modifying the host range of a pathogen. Heterozygous diploids of *F. oxysporum* f. sp. *pisi* obtained from heterokaryons comprised of race 1, virulent to the pea cultivar Onward, and race 2, virulent to Onward and Alaska, produced recombinants which were virulent to both cultivars as well as to a third, Delwiche Commando, which neither of the parental races could attack. Many more explorations, using parasexual recombination into the modification of host range, inter-genic effects on virulence, and blocks in pathogenicity, appear warranted.

Possible Occurrence and Role of Parasexuality in Nature

Parasexuality seems to be of fairly common occurrence in laboratory cultures of fungal pathogens. Presumably, where it has been sought and highly selective techniques used for its detection, it has usually been found. It is not confined to any particular taxonomic unit nor is it restricted to any special group of pathogens, occurring as it does in fungi that cause diseases in a wide range of hosts. Further, parasexuality seems not to be hindered by the nuclear number in somatic cells although possibly the frequency of its occurrence may be greatly influenced by this. Where details of recombination processes have been resolved they generally conform to the basic pattern established in *A. nidulans.* In the standard *A. nidulans* model of parasexuality, a heterokaryon appears requisite for subsequent recombinational events. While diploids may arise from an ephemeral heterokaryon, such as might occur between vegetatively incompatible strains following anastomosis, it appears logical to relate parasexuality with a relatively stable heterokaryotic state. The occurrence of heterokaryosis in nature therefore is meaningful in projections of the occurrence of parasexuality. Caten & Jinks (22) have concluded that, while heterokaryosis occurs in wild populations, its frequency is probably overestimated. However, although reliable estimates are wanting, one might speculate conversely that its frequency is underestimated. Extensive work with *T. cucumeris* (47, 48) has shown that there is a wide variety of different strains which exist in nature as heterokaryons. Ming et al. (97) have presented convincing evidence for the occurrence of heterokaryosis in wild types of *F. fujikuroi.* Suzuki (133) found that almost all single spore isolates of *P. oryzae* from host plants were heterokaryotic, and heterokaryosis was of a persistent nature. Further, in Ascomycetes and Basidiomycetes dikaryosis is a step in the sexual process. If genic controls operate to prevent nuclear fusion in somatic cells in these special heterokaryons, as the *dik*$^+$ gene is reported to do in *S. commune* (83), mutations in these genes may impair their function and nuclear fusions would ensue to form diploids.

Evidence for parasexuality in natural populations, however, has been wanting. The results of Ingram (73) indicate that the phenomenon may occur. Large spore size is a stable characteristic of *V. dahliae* var. *longisporum*. When such spores were treated with *p*-fluorophenylalanine, a haploidizing agent (87), some small spores were recovered. Auxotrophs were readily obtained from these small spores but not from the large spores, following their irradiation with ultraviolet. These results were explicable if the large spores were diploid. Further, from mixtures of small-spored diauxotrophs, she isolated prototrophic heterozygous diploids which had a spore size similar to that of the original parent. The diploid was stable in culture.

A major difficulty in assessing the occurrence of parasexuality in nature is due to the lack of criteria for its detection. Since various mechanisms of variation may be operational and since variation is dynamic, the identification of any one mechanism when all are interrelated, is problematic. Still, further analyses such as Ingram's may be useful in ascertaining the general importance of parasexuality. Cytological studies also appear pertinent. Where diploidy appears likely among isolates, a comparison of interphase nuclear volume or DNA content per nucleus may be useful.

Currently, we know little about the significance of parasexuality in plant pathogens. It may be the major evolutionary mechanism at least in the imperfect fungi. However, a more complete realization of its potential requires research not only in the cytology and genetics of the recombination events per se, but also with regard to problems of anastomosis, incompatibility, heterokaryosis, and extranuclear inheritance. Further search for standard parasexuality also may uncover processes other than chromatid exchange and non-disjunction. Can viruses mediate transduction in a fungal host? What is the nature and prevalence of the specific factor transfer? Despite many unknowns, we believe that standard parasexuality also may be a useful tool for the elucidation of genetic controls and the biosynthetic processes of virulence.

LITERATURE CITED

1. Aist, J. R. Mitotic apparatus in *Ceratocystis fagacearum* and *Fusarium oxysporum*. *J. Cell Biol.,* **40,** 120–35 (1969)
2. Aist, J. R., Wilson, C. L. Interpretation of nuclear division figures in vegetative hyphae of fungi. *Phytopathology,* **58,** 876–77 (1968)
3. Apirion, D. Recessive mutants at unlinked loci which complement in diploids but not in heterokaryons of *Aspergillus nidulans*. *Genetics,* **53,** 935–41 (1966)
4. Bakerspigel, A. Cytological investigations of the parasexual cycle in fungi. I. Nuclear fusion. *Mycopathol. Mycol. Appl.,* **26,** 233–40 (1965)
5. Barron, G. L. The parasexual cycle and linkage relationships in the storage rot fungus *Penicillium expansum*. *Can. J. Botany,* **40,** 1603–13 (1962)
6. Beraha, L., Garber, E. D. Genetics of phytopathogenic fungi. XI. A genetic study of avirulence due to auxotrophy in *Penicillium expansum* by means of the parasexual cycle. *Am. J. Botany,* **52,** 117–19 (1965)
7. Beraha, L., Garber, E. D., Strømnaes, Ø. Genetics of phytopathogenic fungi. Virulence of color and nutritionally deficient mutants of *Penicillium italicum* and *Penicillium digitatum*. *Can. J. Botany,* **42,** 429–36 (1964)
8. Berg, C. M., Garber, E. D. A genetic analysis of color mutants of *Aspergillus fumigatus*. *Genetics,* **47,** 1139–46 (1962)
9. Boyer, M. G. Variability and hyphal anastomosis in host-specific forms of *Marssonina populi* (Lib.) Magn. *Can. J. Botany,* **39,** 1409–27 (1961)
10. Bracker, C. E. Ultrastructure of fungi. *Ann. Rev. Phytopathol.,* **5,** 343–74 (1967)
11. Bradley, S. G. Parasexual phenomena in microorganisms. *Ann. Rev. Microbiol.,* **16,** 35–52 (1962)
12. Brushaber, J. A., Wilson, C. L., Aist, J. R. Asexual nuclear behavior of some plant-pathogenic fungi. *Phytopathology,* **57,** 43–46 (1967)
13. Buddenhagen, I. W. Induced mutations and variability in *Phytophthora cactorum*. *Am. J. Botany,* **45,** 355–65 (1958)
14. Bugbee, W. M., Line, R. F., Kernkamp, M. F. Pathogenicity of progenies from selfing race 15B and somatic and sexual crosses of races 15B and 56 of *Puccinia graminis* f. sp. *tritici*. *Phytopathology,* **58,** 1291–93 (1968)
15. Buxton, E. W. Heterokaryosis and parasexual recombination in pathogenic strains of *Fusarium oxysporum*. *J. Gen. Microbiol.,* **15,** 133–39 (1956)
16. Buxton, E. W. Parasexual recombination in the banana-wilt *Fusarium*. *Trans. Brit. Mycol. Soc.,* **45,** 274–79 (1962)
17. Buxton, E. W., Ward, V. Genetic relationships between pathogenic strains of *Fusarium oxysporum, Fusarium solani* and an isolate of *Nectria haematococca*. *Trans. Brit. Mycol. Soc.,* **45,** 261–73 (1962)
18. Casselton, L. A. Somatic recombination in fungi. *Sci. Progr. (Oxford),* **53,** 107–15 (1965)
19. Casselton, L. A. The production and behavior of diploid strains of *Coprinus lagopus*. *Genet. Res.,* **6,** 190–208 (1965)
20. Casselton, L. A., Lewis, D. Compatibility and stability of diploids in *Coprinus lagopus*. *Genet. Res.,* **8,** 61–72 (1966)
21. Casselton, L. A., Lewis, D. Dilution of gene products in the cytoplasm of heterokaryons in *Coprinus lagopus*. *Genet. Res.,* **9,** 63–71 (1967)
22. Caten, C. E., Jinks, J. L. Heterokaryosis: Its significance in wild homothallic ascomycetes and fungi imperfecti. *Trans. Brit. Mycol. Soc.,* **49,** 81–93 (1966)
23. Caten, C. E., Jinks, J. L. Spontaneous variability of single isolates of *Phytophthora infestans*. I. Cultural variation. *Can. J. Botany,* **46,** 330–48 (1968)
24. Christensen, J. J. Somatic variation in the rust fungi. In *Recent Advances in Botany,* **1,** 571–74 (Univ. Toronto Press, Toronto,

Ontario, 947 pp., 1961)

25. Clutterbuck, A. J., Roper, J. A. A direct determination of nuclear distribution in heterokaryons of *Aspergillus nidulans*. *Genet. Res.*, **7,** 185–94 (1966)
26. Cowan, J. W., Lewis, D. Somatic recombination in the dikaryon of *Coprinus lagopus*. *Genet. Res.*, **7,** 235–44 (1966)
27. Coy, D. O., Tuveson, R. W. The effects of supplementation and plating densities on apparently aberrant meiotic and mitotic segregation in *Aspergillus rugulosus*. *Genetics,* **50,** 847–53 (1964)
28. Davis, R. H. Mechanisms of inheritance. 2. Heterokaryosis. In *The Fungi,* **II,** 567–88 (Ainsworth, G. C., Sussman, A. S., Eds., Academic Press, New York, 805 pp., 1966)
29. Day, A. W., Jones, J. K. The production and characteristics of diploids in *Ustilago violacea*. *Genet. Res.*, **11,** 63–81 (1968)
30. Day, P. R. Variation in phytopathogenic fungi. *Ann. Rev. Microbiol.*, **14,** 1–16 (1960)
31. Day, P. R. Recent developments in the genetics of the host-parasite system. *Ann. Rev. Phytopathol.*, **4,** 245–68 (1966)
32. Day, P. R. Complementation in dikaryons and diploids. (Abstr.) *Phytopathology,* **57,** 808 (1967)
33. Day, P. R., Puhalla, J. E. (Personal communication)
34. Dhillon, T. S., Garber, E. D., Wyttenbach, G. Genetics of phytopathogenic fungi. VI. Heterokaryons involving *Gibberella fujikuroi* and formae of *Fusarium oxysporum*. *Can. J. Botany,* **39,** 785–92 (1961)
35. Dinoor, A., Person, C. Genetic complementation in *Ustilago hordei*. *Can. J. Botany,* **47,** 9–14 (1969)
36. Dodge, B. O. Nuclear phenomena associated with heterothallism and homothallism in the ascomycete *Neurospora*. *J. Agr. Res.*, **35,** 289–305 (1927)
37. Dowding, E. S. *Gelasinospora,* a new genus of Pyrenomycetes with pitted spores. *Can. J. Res.*, **9,** 294–305 (1933)
38. Duncan, E. J., MacDonald, J. A. Nuclear phenomenon in *Marasmius androsaceus* (L. ex Fr.) Fr. and *Marasmius rotula* (Scop. ex Fr.) Fr. *Trans. Roy. Soc. Edinburgh,* **66,** 129–41 (1965)
39. Dutta, S. K., Garber, E. D. Genetics of phytopathogenic fungi. III. An attempt to demonstrate the parasexual cycle in *Colletotrichum lagenarium*. *Botan. Gaz.*, **122,** 118–21 (1960)
40. Ellingboe, A. H. Somatic recombination in *Puccinia graminis* var. *tritici*. *Phytopathology,* **51,** 13–15 (1961)
41. Ellingboe, A. H. Somatic recombination in Basidiomycetes. In *Incompatibility in Fungi,* 36–48 (Esser, K., Raper, J. R., Eds., Springer-Verlag, Berlin, 124 pp., 1965)
42. Esser, K., Kuenen, R. *Genetics of Fungi* (Trans. Steiner, E., Springer-Verlag, New York, 500 pp., 1967)
43. Fincham, J. R. S., Day, P. R. *Fungal Genetics* (Blackwell Scientific Publications, Oxford, 326 pp., 1965)
44. Fjeld, A., Strømnaes, Ø. The parasexual cycle and linkage groups in *Penicillium espansum*. *Hereditas,* **54,** 389–403 (1966)
45. Fleischmann, G. Studies on the wilt of peas caused by *Fusarium oxysporum* Schl. f. *pisi* (Linf.) S. & H., race 1. *Can. J. Botany,* **41,** 1569–84 (1963)
46. Flentje, N. T. (Personal communication)
47. Flentje, N. T., Stretton, H. M. Mechanisms of variation in *Thanatephorus cucumeris* and *T. praticolus*. *Australian J. Biol. Sci.*, **17,** 686–704 (1964)
48. Flentje, N. T., Stretton, H. M., McKenzie, A. R. Mutation in *Thanatephorus cucumeris*. *Australian J. Biol. Sci.*, **20,** 1173–80 (1967)
49. Flor, H. H. Genetics of somatic variation for pathogenicity in *Melampsora lini*. *Phytopathology,* **54,** 823–26 (1964)
50. Fordyce, C., Jr., Green, R. J., Jr. Mechanisms of variation in *Verticillium albo-atrum*. *Phytopathology,* **54,** 795–98 (1964)
51. Gallegly, M. E. Genetics of pathogenicity of *Phytophthora infestans*. *Ann. Rev. Phytopathol.*, **6,** 375–96 (1968)

52. Garber, E. D. Genetics of phytopathogenic fungi XVII. An electrophoretic study of extracellular and intracellular endopolygalacturonases from virulent and avirulent strains of *Penicillium italicum*. *Phytopathol. Z.*, **59,** 147–52 (1967)

53. Garber, E. D., Beraha, L. Genetics of phytopathogenic fungi. XVI. The parasexual cycle in *Penicillum expansum*. *Genetics*, **52,** 487–92 (1965)

54. Garber, E. D., Wyttenbach, E. G., Dhillon, T. S. Genetics of phytopathogenic fungi. V. Heterokaryons involving formae of *Fusarium oxysporum*. *Am. J. Botany*, **48,** 325–29 (1961)

55. Garnjobst, L., Wilson, J. F. Heterokaryosis and protoplasmic incompatibility in *Neurospora crassa*. *Proc. Natl. Acad. Sci. U.S.*, **42,** 613–18 (1956)

56. Garza-Chapa, R., Anderson, N. A. Behavior of single-basidiospore isolates and heterokaryons of *Rhizoctonia solani* from flax. *Phytopathology*, **56,** 1260–68 (1966)

57. Grindle, M. Heterokaryon compatibility of unrelated strains in the *Aspergillus nidulans* group. *Heredity*, **18,** 191–204 (1963)

58. Grindle, M. Heterokaryon compatibility of closely related wild isolates of *Aspergillus nidulans*. *Heredity*, **18,** 397–405 (1963)

59. Hansen, H. N., Smith, R. E. Interspecific anastomosis and the origin of new types in imperfect fungi. (Abstr.) *Phytopathology*, **24,** 1144 (1934)

60. Hastie, A. C. Genetic recombination in the hop-wilt fungus, *Verticillium albo-atrum*. *J. Gen. Microbiol.*, **27,** 373–82 (1962)

61. Hastie, A. C. The parasexual cycle in *Verticillium albo-atrum*. *Genet. Res.*, **5,** 305–15 (1964)

62. Hastie, A. C. Mitotic recombination in conidiophores of *Verticillium albo-atrum*. *Nature*, **214,** 249–52 (1967)

63. Hastie, A. C. Phialide analysis of mitotic recombination in *Verticillium*. *Molec. Gen. Genet.*, **102,** 232–40 (1968)

64. Heagy, F. C., Roper, J. A., Deoxyribonucleic acid content of haploid and diploid *Aspergillus* conidia. *Nature*, **170,** 713 (1952)

65. Heale, J. B., Gafoor, A., Rajasingham, K. C. Nuclear division in conida and hyphae of *Verticillium albo-atrum*. *Can. J. Genet. Cytol.*, **10,** 321–40 (1968)

66. Henry, C. E., Garber, E. D. Genetics of phytopathogenic fungi. XVIII. Detection of esterases of *Colletotrichum lagenarium* in culture filtrates and diseased tissue-extracts by starch-gel zone electrophoresis. *Acta Phytopathol.*, **2,** 89–94 (1967)

67. Hoffmann, G. M. Untersuchungen über die Heterokaryosebildung und den Parasexualcyclus bei *Fusarium oxysporum*. I. Anastomosenbildung im Mycel und Kernverhältnisse bei der Conidienentwicklung. *Arch. Mikrobiol.*, **53,** 336–47 (1966)

68. Hoffmann, G. M. Untersuchungen über die Heterokaryosebildung und den Parasexualcyclus bei *Fusarium oxysporum*. III. Paarungsversuche mit auxotrophen Mutanten von *Fusarium oxysporum* f. *callistephi*. *Arch. Mikrobiol.*, **56,** 40–59 (1967)

69. Holliday, R. Induced mitotic crossing-over in *Ustilago maydis*. *Genet. Res.*, **2,** 231–48 (1961)

70. Holloway, B. W. Genetic control of heterokaryosis in *Neurospora crassa*. *Genetics*, **40,** 117–29 (1955)

71. Hosford, R. M., Jr., Gries, G. A. The nuclei and parasexuality in *Phymatotrichum omnivorum*. *Am. J. Botany*, **53,** 570–79 (1966)

72. Ikeda, Y., Ishitani, C., Nakamura, K. A high frequency of heterozygous diploids and somatic recombination induced in imperfect fungi by ultra-violet light. *J. Gen. Appl. Microbiol. (Tokyo)*, **3,** 1–11 (1957)

73. Ingram, R. *Verticillium dahliae* var. *longisporum*, a stable diploid. *Trans. Brit. Mycol. Soc.*, **51,** 339–41 (1968)

74. Ishitani, C. A high frequency of heterozygous diploids and somatic recombination produced by ultraviolet light in imperfect fungi. *Nature*, **178,** 706 (1956)

75. Ishitani, C., Ikeda, Y., Sakaguchi, K., Hereditary variation and genetic recombination in Koji-molds

(*Aspergillus oryzae* and *Asp. sojae*). VI. Genetic recombination in heterozygous diploids. *J. Gen. Appl. Microbiol. (Tokyo)*, **2**, 401–30 (1956)

76. Ishitani, C., Sakaguchi, K. Hereditary variation and genetic recombination in Koji-molds (*Aspergillus oryzae* and *Asp. sojae*). V. Heterocaryosis. *J. Gen. Appl. Microbiol. (Tokyo)*, **2**, 345–400 (1956)
77. Jinks, J. L., Caten, C. E., Simchen, G., Croft, J. H. Heterokaryon incompatibility and variation in wild populations of *Aspergillus nidulans. Heredity*, **21**, 227–39 (1966)
78. Jinks, J. L., Grindle, M. The genetical basis of heterokaryon incompatibility in *Aspergillus nidulans. Heredity*, **18**, 407–11 (1963)
79. Jones, D. A. Heterokaryon compatibility in the *Aspergillus glaucus* Link group. *Heredity*, **20**, 49–56 (1965)
80. Käfer, E. The processes of spontaneous recombination in vegetative nuclei of *Aspergillus nidulans. Genetics*, **46**, 1581–1609 (1961)
81. Knox-Davies, P. S. Nuclear division in the developing pycnospores of *Macrophomina phaseoli. Am. J. Botany*, **53**, 220–24 (1966)
82. Knox-Davies, P. S. Mitosis and aneuploidy in the vegetative hyphae of *Macrophomina phaseoli. Am. J. Botany*, **54**, 1290–95 (1967)
83. Koltin, Y., Raper, J. R. Dikaryosis: Genetic determination in *Schizophyllum. Science*, **160**, 85–86 (1968)
84. Laane, M. M. The nuclear division in *Penicillium expansum. Can. J. Genet. Cytol.*, **9**, 342–51 (1967)
85. Leary, J. V., Ellingboe, A. H. Nuclear exchange and complementation in matings of *Schizophyllum commune.* (Abstr.) *Genetics*, **60**, 196 (1968)
86. Lewis, L. A., Barron, G. L. The pattern of the parasexual cycle in *Aspergillus amstelodami. Genet. Res.*, **5**, 162–63 (1964)
87. Lhoas, P. Mitotic haploidization by treatment of *Aspergillus niger* diploids with *para*-fluorophenylalanine. *Nature*, **190**, 744 (1961)
88. Lhoas, P. Genetic analysis by means of the parasexual cycle in *Aspergillus niger. Genet. Res.*, **10**, 45–61 (1967)
89. Loprieno, N. I. Mutanti nutrizionali nello studio dei rapporti ospite-patogeno nelle fitopatie da microorganismi. *Agr. Ital.*, Marzo-Aprile, 3–15 (1964)
90. Lu, B. C. (Personal communication)
91. MacDonald, K. D., Hutchinson, J. M., Gillett, W. A. Formation and segregation of heterozygous diploids between a wild-type strain and derivatives of high penicillin yield in *Penicillium chrysogenum. J. Gen. Microbiol.*, **33**, 385–94 (1963)
92. MacGarvie, Q., Isaac, I. Structure and behaviour of the nuclei of *Verticillium* spp. *Trans. Brit. Mycol. Soc.*, **49**, 687–93 (1966)
93. MacNeill, B. H. Genetic control of virulence in *Penicillium expansum. Proc. Can. Phytopathol. Soc.*, **34**, 22 (1967)
94. MacNeill, B. H., Barron, G. L. Avirulence in prototrophs of *Penicillium expansum. Can. J. Botany*, **44**, 355–58 (1966)
95. Menzinger, W. Zur Variabilität und Taxonomie von Arten und Formen der Gattung *Botrytis* Mich. II. Untersuchungen zur Variabilität des Kulturtyps unter konstanten Kulturbedingungen. *Zentr. Bakteriol. Parasitenk. Abt. II.*, **120**, 179–96 (1966)
96. Middleton, R. B. Sexual and somatic recombination in common-AB heterokaryons of *Schizophyllum commune. Genetics*, **50**, 701–10 (1964)
97. Ming, Y. N., Lin, P. C., Yu, T. F. Heterokaryosis in *Fusarium fujikuroi* (Saw.) Wr., *Sci. Sinica (Peking)*, **15**, 371–78 (1966)
98. Moore, R. T. Fine structure of mycota. 12. Karyochorisis—somatic nuclear division—in *Cordyceps militaris. Z. Zellforsch.*, **63**, 921–37 (1964)
99. Parag, Y. Phase-microscopic observations of fusions of nuclei in somatic cells of a heterokaryon of *Schizophyllum commune. Am. J. Botany*, **55**, 984–88 (1968)
100. Parag, Y., Nachman, B. Diploidy in the tetrapolar heterothallic basidiomycete *Schizophyllum commune. Heredity*. **21**, 151–54 (1966)
101. Park, S., Kenehan, P., Goodgal, S. Heterocaryons and recombination

in *Phycomyces blakesleeanus.* (Abstr.) *Genetics,* **60,** 209–10 (1968)

102. Parmeter, J. R., Jr., Snyder, W. C., Reichle, R. E. Heterokaryosis and variability in plant-pathogenic fungi. *Ann. Rev. Phytopathol.,* **1,** 51–76 (1963)
103. Parmeter, J. R., Jr., Whitney, H. S., Platt, W. D. Affinities of some *Rhizoctonia* species that resemble mycelium of *Thanatephorus cucumeris. Phytopathology,* **57,** 218–23 (1967)
104. Person, C. Genetical adjustment of fungi to their environment. In *The Fungi,* **III,** 395–415, (Ainsworth, G. C., Sussman, A. S., Eds., Academic Press, New York, 738 pp., 1968)
105. Petrie, G. A. Variability in *Leptosphaeria maculans* (Desm.) Ces. & De Not., the cause of blackleg of rape (Doctoral thesis, Univ. Saskatchewan, Saskatoon, Saskatchewan, 1969)
106. Pittenger, T. H., Brawner, T. G. Genetic control of nuclear selection in *Neurospora* heterokaryons. *Genetics,* **46,** 1645–63 (1961)
107. Pontecorvo, G. Mitotic recombination in the genetic systems of filamentous fungi. *Carylogia (Suppl.),* **6,** 192–200 (1954)
108. Pontecorvo, G. The parasexual cycle in fungi. *Ann. Rev. Microbiol.,* **10,** 393–400 (1956)
109. Pontecorvo, G. *Trends in Genetic Analysis* (Columbia Univ. Press, New York, 145 pp., 1958)
110. Pontecorvo, G. Microbial genetics: retrospect and prospect. *Proc. Roy. Soc. (London), Ser. B,* **158,** 1–23 (1963)
111. Pontecorvo, G., Käfer, E. Genetic analysis by means of mitotic recombination. *Advan. Genet.,* **9,** 71–104 (1958)
112. Pontecorvo, G., Roper, J. A. Genetic analysis without sexual reproduction by means of polyploidy in *Aspergillus nidulans. J. Gen. Microbiol.,* **6,** vii (1952)
113. Pontecorvo, G., Roper, J. A., Forbes, E. Genetic recombination without sexual reproduction in *Aspergillus niger. J. Gen. Microbiol.,* **8,** 198–210 (1953)
114. Pontecorvo, G., Roper, J. A., Hemmons, L. M., MacDonald, K. D., Bufton, A. W. J. The genetics of *Aspergillus nidulans. Advan. Genet.,* **5,** 141–238 (1953)
115. Pontecorvo, G., Sermonti, G. Recombination without sexual reproduction in *Penicillium chrysogenum. Nature,* **172,** 126–27 (1953)
116. Raper, J. R. Parasexual phenomena in Basidiomycetes. In *Recent Advances in Botany,* **1,** 379–83 (Univ. Toronto Press, Toronto, Ontario, 947 pp., 1961)
117. Raper, J. R. *Genetics of Sexuality in Higher Fungi* (Ronald Press Co., New York, 283 pp., 1966)
118. Raper, J. R., Esser, K. The fungi. In *The Cell,* **6,** 139–244 (Brachet, J., Mirsky, A. E., Eds., Academic Press, New York, 1964)
119. Roberts, C. F. Complementation in balanced heterokaryons and heterozygous diploids of *Aspergillus nidulans. Genet. Res.,* **5,** 211–29 (1964)
120. Robinow, C. F., Bakerspigel, A. Somatic nuclei and forms of mitosis in fungi. In *The Fungi,* **I,** 119–42 (Ainsworth, G. C., Sussman, A. S., Eds., Academic Press, New York, 748 pp., 1965)
121. Roper, J. A. Production of heterozygous diploids in filamentous fungi. *Experientia,* **8,** 14–15 (1952)
122. Roper, J. A. Mechanisms of inheritance. 3. The parasexual cycle. In *The Fungi,* **II,** 589–617 (Ainsworth, G. C., Sussman, A. S., Eds., Academic Press, New York, 805 pp., 1966)
123. Roper, J. A., Pritchard, R. H. The recovery of complementary products of mitotic crossing-over. *Nature,* **175,** 639 (1955)
124. Sanderson, K. E., Srb, A. M. Heterokaryosis and parasexuality in the fungus *Ascochyta imperfecta. Am. J. Botany,* **52,** 72–81 (1965)
125. Sansome, E. R. Meiosis in diploid and polyploid sex organs of *Phytophthora* and *Achlya. Cytologia,* **30,** 103–17 (1965)
126. Savage, E. J., Clayton, C. W., Hunter, J. H., Brenneman, J. A., Laviola, C., Gallegly, M. E. Homothallism, heterothallism, and interspecific hybridization in the genus *Phytophthora. Phytopathology,* **58,** 1004–21 (1968)
127. Schreiber, L. R., Green, R. J., Jr.

Anastomosis in *Verticillium albo-atrum* in soil. *Phytopathology,* **56,** 1110–11 (1966)

128. Somers, C. E., Wagner, R. P., Hsu, T. E. Mitosis in vegetative nuclei of *Neurospora crassa. Genetics,* **45,** 801–10 (1960)
129. Stephan, B. R. Untersuchungen über die Variabilität bei *Colletotrichum gloeosporioides* Penzig in Verbindung mit Heterokaryose. III. Versuche zum Nachweis der Heterokaryose. *Zentr. Bakteriol. Parasitenk. Abt. II,* **121,** 73–83 (1967)
130. Stephan, B. R. Untersuchungen zum Nachweis der Heterokaryose bei *Colletotrichum gloeosporioides* Penzig unter Verwendung auxotropher Mutanten. *Zentr. Bakteriol. Parasitenk. Abt. II,* **122,** 420–35 (1968)
131. Strømnaes, Ø., Garber, E. D. Heterocaryosis and the parasexual cycle in *Aspergillus fumigatus. Genetics,* **48,** 653–62 (1963)
132. Strømnaes, Ø., Garber, E. D., Beraha, L. Genetics of phytopathogenic fungi. IX. Heterocaryosis and the parasexual cycle in *Penicillium italicum* and *Penicillium digitatum. Can. J. Botany,* **42,** 423–27 (1964)
133. Suzuki, H. Studies on biologic specialization in *Pyricularia oryzae* Cav. *Inst. Plant Pathol. Tokyo,* 235 pp. (1967)
134. Tinline, R. D. *Cochliobolus sativus.* IV. Drug-resistant, color, and nutritionally exacting mutants. *Can. J. Botany,* **39,** 1695–704 (1961)
135. Tinline, R. D. *Cochliobolus sativus.* V. Heterokaryosis and parasexuality. *Can. J. Botany,* **40,** 425–37 (1962)
136. Tinline, R. D. *Cochliobolus sativus.* VII. Nutritional control of the pathogenicity of some auxotrophs to wheat seedlings. *Can. J. Botany,* **41,** 489–97 (1963)
137. Tinline, R. D. (Unpublished data)
138. Tuveson, R. W., Garber, E. D. Genetics of phytopathogenic fungi. I. Virulence of biochemical mutants of *Fusarium oxysporum* f. *pisi. Botan. Gaz.,* **121,** 69–74 (1959)
139. Vest, G., Anderson, N. A. Studies on heterokaryosis and virulence of *Rhizoctonia solani* isolates from flax. *Phytopathology,* **58,** 802–7 (1968)
140. Ward, E. W. B., Ciurysek, K. W. Somatic mitosis in *Neurospora crassa. Am. J. Botany,* **49,** 393–99 (1962)
141. Watson, I. A., Luig, N. H. Asexual intercrosses between somatic recombinants of *Puccinia graminis. Proc. Linnean Soc., N.S. Wales,* **87**(2), 99–104 (1962)
142. Weijer, J., Koopmans, A., Weijer, D. L. Karyokinesis *in vivo* of the migrating somatic nucleus of *Neurospora* and *Gelasinospora* species *Trans. N. Y. Acad. Sci.,* **25,** 846–54 (1963)
143. Weijer, J., Koopmans, A., Weijer, D. L. Karyokinesis of somatic nuclei of *Neurospora crassa. Can. J. Genet. Cytol.,* **7,** 140–63 (1965)
144. Weijer, J., Weisberg, S. H. Karyokinesis of the somatic nucleus of *Aspergillus nidulans.* I. The juvenile chromosome cycle (Feulgen staining). *Can. J. Genet. Cytol.,* **8,** 361–74 (1966)
145. Weisberg, S. H. The somatic nuclear events during hyphal differentiation in *Aspergillus nidulans* (Eidam) Wint. (Ph.D. thesis, Univ. Alberta, Edmonton, Alberta, 1968)
146. Weisberg, S. H., Weijer, J. Karyokinesis of the somatic nucleus of *Aspergillus nidulans.* II. Nuclear events during hyphal differentiation. *Can. J. Genet. Cytol.,* **10,** 699–722 (1968)
147. Whitney, H. S., Parmeter, J. R., Jr. Synthesis of heterokaryons in *Rhizoctonia solani* Kühn. *Can. J. Botany,* **41,** 879–86 (1963)
148. Williams, P. G., Scott, K. J., Kuhl, J. L., Maclean, D. J. Sporulation and pathogenicity of *Puccinia graminis* f. sp. *tritici* grown on an artificial medium. *Phytopathology,* **57,** 326–27 (1967)
149. Wilson, C. L. (Personal communication regarding his film *Organelles in living cells of* Fomes annosus)
150. Wilson, C. L., Aist, J. R. Motility of fungal nuclei. *Phytopathology,* **57,** 769–71 (1967)
151. Wilson, J. F., Garnjobst, L. A new incompatibility locus in *Neurospora crassa. Genetics,* **53,** 621–31 (1966)
152. Yamasaki, Y., Murata, N., Suwa, T. The effect of mutations for nutritional requirements on the pathogenicity of two pathogens of rice.

Proc. Japan Acad., **40,** 226–31 (1964)

153. Yamasaki, Y., Niizeki, H. Studies on variation of the rice blast fungus, *Piricularia oryzae* Cav. I. Karyological and genetical studies on variation. *Bull. Natl. Inst. Agr. Sci. (Japan),* **13,** 231–73 (1965)

154. Yang, S. M., Hagedorn, D. J. Cultural and pathogenicity studies of induced variants of bean and pea root rot *Fusarium* species. *Phytopathology,* **58,** 639–43 (1968)

Copyright 1969. All rights reserved

BIOCHEMISTRY OF THE CELL WALL IN RELATION TO INFECTIVE PROCESSES[1]

BY PETER ALBERSHEIM, THOMAS M. JONES, AND PATRICIA D. ENGLISH[2]
Department of Chemistry, University of Colorado
Boulder, Colorado

INVOLVEMENT OF CELL WALLS IN THE DISEASE PROCESS

THE HYPOTHESIS

Pathogens find themselves most commonly in the presence of plants other than their hosts. Under such circumstances, a pathogen fails in its efforts to initiate infection. What is it that renders a plant's environment inhospitable to a pathogen except in that rare instance when the plant happens to be a susceptible host? It is this question which we propose to answer by our hypothesis that, in many instances of pathogenesis by bacteria or fungi, it is an interaction between the pathogen and the carbohydrates of the host which determines the pathogen's ability to produce enzymes capable of degrading the host's cell walls. The production of these enzymes, then, determines whether or not a successful infection will be initiated. It should be stressed that only the initiation of the infective process is considered in this hypothesis. Once the plant actively responds to the presence of a pathogen, the ensuing metabolic responses are known to be both complex and varied.

EVIDENCE FROM SYSTEMS NOT DIRECTLY RELATED TO PHYTOPATHOLOGY

At the most fundamental level, a disease syndrome represents a summation of the interactions between molecules which are characteristic of two or more types of cells. This interpretation of the disease process is perhaps best described by illustration. Several systems will be considered in which interactions of molecular structures on the cell surface determine the establishment of a morbid state.

A classic example of the involvement of surface polysaccharides in pathogenesis is the requirement for *Pneumococcus* (*Diplococcus pneumoniae*) to possess capsular polysaccharides in order to initiate successfully bacterial pneumonia in animals (38). A more recent demonstration of the involvement of cell surface elements in disease comes from studies of the relative virulence of *Escherichia coli* strains. *E. coli* is a weakly virulent

[1] Supported in part by a grant from the United States Atomic Energy Commission No. AT(11-1)-1426.

[2] Predoctoral Fellow of the United States Department of Health, Education and Welfare under Title IV of the National Defense Education Act.

organism, but some strains are more virulent than others. Two *E. coli* mutants have been isolated which are characterized by defective O-antigenic polysaccharides (55), polysaccharides which are located on the cell surface. The capacity of these mutant strains to kill mice is strikingly different from that of the parental strain which is 1000 times as virulent as one of the mutants and 100 times as virulent as the other.

The lethality of the various *E. coli* strains upon intraperitoneal injection in mice is correlated with a strain's ability to resist phagocytosis and to persist in the peritoneal cavity. The parent strain is known to resist phagocytosis by macrophages *in vivo* and by polymorphonuclear leukocytes *in vitro*. The mutants do not, and the mutant strain more deficient in O-antigenic polysaccharide is the more susceptible to phagocytosis and less virulent. Medearis et al. have pointed out that the effect on phagocytosis of an alteration in the composition of one portion of the cell wall polysaccharide reflects, at least in part, the importance of an interaction between the surface of the bacterial cell and the surface of the macrophage or polymorphonuclear leukocyte (55). In our opinion, these results reflect, in an elegant manner, the involvement of the cell surface in the infective process.

Bacterial viruses recognize their hosts through an interaction between molecules on the tail fibers of the bacteriophage and the O-antigenic polysaccharides of the bacterial cell surface. No laboratory has contributed more to the data establishing the role of the O-antigen in the recognition process than that of Robbins and his co-workers (18, 50, 67) who have observed that the structure of the normal O-antigenic polysaccharide of *Salmonella anatum* is modified when the bacterium is lysogenic for the bacteriophage ε^{15}. Such bacteria contain the viral genome in a non-replicating state. The O-antigen of non-lysogenic (normal) *S. anatum* cells contains the repeating unit 1,4-galactosyl-α-1,6-mannosyl-α-1,4-rhamnose. The repeating unit of the O-antigenic polysaccharide of a bacterium lysogenic for ε^{15} is the same. The galactosyl-1,6-mannose linkage, however, is altered from the α- to the β-configuration, and the O-acetyl substituent, which in the repeating unit of the normal antigen is attached to the C-6 position of the galactose residue, is absent in the lysogenic strain.

It is of particular interest that the strain of *S. anatum* which is lysogenic for ε^{15} exhibits an altered susceptibility to further bacteriophage infection. Viruses which are normally able to infect *S. anatum,* including ε^{15} itself, cannot attach themselves to the surface of the lysogenic bacterium. Thus, ε^{15} infection alters the resistance of the host bacterium to attack by other viruses through a structural modification of a bacterial cell wall polysaccharide. The O-antigenic polysaccharide, then, is a critical determinant in the infective process.

Cell surface structures also play an important role in other cellular recognition processes (80). These processes are believed to be involved in controlling such cellular functions as DNA replication (61). Viral or chemical transformation of normal cells to the neoplastic state is accompanied by a structural alteration of the cell surface which results in the loss of contact inhibition (1, 79). A wheat germ component which specifically agglutinates virally transformed neoplastic cells has been described (4, 5, 6). This agglutinin has been isolated and found to be a glycoprotein (19, 20, 32). In addition it has been demonstrated that the tumor-specific surface structure which interacts with the agglutinin contains N-acetylglucosamine.

Most, if not all, neoplastic mammalian cells appear to have N-acetylglucosamine as a constituent of their surface glycolipids; few normal cells possess this surface determinant (19, 20). The agglutination of neoplastic cells by the wheat germ glycoprotein can be reversed by treatment with N-acetylglucosamine as would be anticipated if a glycolipid containing this hexosamine were the critical surface determinant for the agglutination of these cells. By such serological techniques, the transformation from the normal to a neoplastic state of a wide variety of animal cell types has been shown to be accompanied by an alteration of the antigenic glycolipids of the cell surface. Such observations suggest that specific surface structures play an important role in the conversion to or the maintenance of the neoplastic state. Thus, one sees that across the biological spectrum cell surfaces are intimately involved in pathogenesis.

Evidence from Plant Systems

The universal ability of plant pathogens to produce polysaccharide-degrading enzymes.—There can be no more compelling reason to study cell wall polysaccharide-degrading enzymes than the observation that every microbial plant pathogen examined has shown the ability to produce such enzymes. Not only are these enzymes universally produced by pathogens, but the variety of such enzymes, in terms of their substrate specificities, is staggering. Enzymes have been found which cleave virtually every glycosidic linkage known to occur in wall polysaccharides. A given linkage is often attacked by more than one enzyme.

No class of polysaccharide-degrading enzymes has been studied more than the pectic enzymes (16). There are enzymes which attack only the ends of galacturonide-containing polymers and enzymes which cleave such polymers at random internal intervals. There are enzymes which attack only methyl esterified galacturonosyl linkages and enzymes which attack only galacturonosyl residues with free carboxyl groups. Some pectic enzymes cleave glycosidic bonds between galacturonosyl residues through a hydrolytic mechanism, while others open the same linkage through an eliminative mechanism. The opportunities for differential control of polysaccharide degradation are further expanded by an organism's ability to produce more than one protein with the same enzymatic activity (isozymes).

What is the value of an organism's having evolved a variety of enzymes to cleave a single type of glycosidic linkage? One advantage is that the production of each enzyme can be controlled individually. For example, the synthesis of such enzymes might be induced or repressed in response to different stimuli, thereby increasing the likelihood that at least one enzyme would be produced under a variety of environmental conditions. The control of polysaccharide-degrading enzyme production is, in our opinion, the key to the role played by saccharides in a plant's resistance to microbial disease.

Another widely studied group of pathogen-produced enzymes is the cellulases (52, 58, 65). Here, as in the case of the pectic enzymes, considerable variation in the substrate specificity and in the structure of these proteins is observed (14, 64, 68, 72, 77).

Plant pathogens produce many other degradative enzymes. These include arabinosidases, xylosidases, mannosidases, galactosidases, and glucosidases (81). In fact, virtually every component of the primary cell wall of plants, including the proteins, is susceptible to attack.

Environmental control of the production of polysaccharide-degrading enzymes by plant pathogens.—The environment of a pathogenic microorganism not infrequently determines whether the pathogen produces a particular enzyme. This determination of enzyme production may be controlled by a variety of mechanisms. *Fusarium oxysporum* f. *lycopersici* secretes both pectin esterase and polygalacturonase into the culture medium when pectin is the carbon source, but produces only pectin esterase when pectin is replaced by glucose (76). Data of this type suggest one of three mechanisms: (*a*) both enzymes are produced without specific induction and polygalacturonase synthesis is repressed by glucose; (*b*) pectin esterase is secreted without induction and polygalacturonase synthesis is induced by pectin but not by glucose; or (*c*) synthesis of both enzymes is induced by pectin but only pectin esterase synthesis is induced by glucose. Irrespective of how this is accomplished, it is clear that in *F. oxysporum* f. *lycopersici* the synthesis of each enzyme is controlled in a different manner.

The fungus *Rhizoctonia solani* (isolate R-B) secretes both an endo-polygalacturonase and a pectate lyase (12). In extracts of bean hypocotyls infected with this pathogen, endo-polygalacturonase activity is predominant. When the same organism is cultured in bean hypocotyl medium, the lyase activity predominates. This system provides an interesting example of host control of the synthesis of pathogen-produced degradative enzymes.

Differential control of the synthesis of endo- and exo-polygalacturonases has also been observed in three species of *Botrytis* (37). *B. allii* produces both endo- and exo-polygalacturonases when grown on detached onion leaves or on leaves of intact onion plants, but this fungus fails to produce either enzyme when grown in potato-dextrose broth. *B. cinerea* and *B. squamosa* on the other hand, produce endo-polygalacturonase in all three environments but secrete significant amounts of exo-polygalacturonase only when grown on detached leaves.

Hancock has since established the differential control of pectate lyase and exo-polygalacturonase synthesis in *Colletotrichum trifolii* (34). When this fungus is grown in a strongly buffered medium containing polygalacturonic acid as carbon source, exo-polygalacturonase is secreted into the culture medium 5 to 7 days prior to the appearance of pectate lyase.

Differential control mechanisms have been demonstrated for pectic-degrading enzyme production in *Fusarium solani* f. *phaseoli* (11). When grown in culture on potato-pectin or potato-glucose-pectin media, this fungus secretes three pectic-degrading enzymes: a pectate hydrolase, a pectin hydrolase, and a pectate lyase. Only the lyase is produced, however, when the pathogen is grown on autoclaved bean hypocotyls or roots.

An extracellular xylan-degrading enzyme system produced by the grape pathogen *Diplodia viticola* is also under environmental control (70). This fungus secretes xylosidase when grown on surface-sterilized or autoclaved grapes, or in liquid culture containing either xylan, cellulose, xylose, or arabinose, but not when glucose or cellobiose is the carbon source.

Corticium rolfsii is stimulated by araban or by bran extract to secrete large amounts of α-L-arabinofuranosidase in culture (43). Growth of the fungus in a medium containing pectin or arabinose results in the secretion of intermediate amounts of this enzyme, while growth in a medium containing either xylose, glucose, galactose, or sucrose yields relatively small amounts of α-L-arabinofuranosidase.

In an interesting series of papers (40, 41, 45), Horton & Keen have reported their studies of the control of endo-polygalacturonase and of cellulase synthesis in *Pyrenochaeta terrestris.* They have observed that cellulase synthesis in this fungus is controlled by a combination of induction and repression (40), a mechanism which provides unique sensitivity to environmental change. Cellulase synthesis in this pathogen is induced by cellulose, but there is an optimal concentration above which cellulase synthesis is repressed. When the cellulose concentration exceeds the optimal level, the rate of glucose release from cellulose exceeds the rate at which glucose is utilized by the fungus. When the accumulation of released glucose exceeds a concentration of 0.0005 *M,* cellulase synthesis is repressed. Endo-polygalacturonase synthesis is induced in *P. terrestris* (45) by galacturonic acid, pectin, polygalacturonic acid, mucic acid, tartonic acid, and dulcitol. When this organism is grown in a pectin-containing medium, endo-polygalacturonase synthesis is further stimulated by hexose supplements as high in concentration as 0.005 *M*. The rate of endo-polygalacturonase synthesis is diminished only by hexose concentrations of 0.05 *M* or above (41).

Colletotrichum lindemuthianum is capable of producing a variety of polysaccharide-degrading enzymes (24, 26). When the alpha or gamma strain of this fungus is grown in a medium containing galactose as the sole carbon source, large quantities of α-galactosidase are secreted. In the presence of either glucose or xylose, α-galactosidase synthesis is repressed. β-Glucosidase appears to be produced constitutively in the alpha strain of *C. lindemuthianum,* while β-xylosidase is induced when either the alpha or gamma strain of the fungus is grown in culture with xylose as the sole carbon source. The various strains of *C. lindemuthianum,* which are characterized by their differential virulence toward certain *Phaseolus vulgaris* varieties, appear to have different systems for controlling degradative enzyme synthesis.

The results described above demonstrate that carbohydrates in the environment of plant pathogens regulate the production of polysaccharide-degrading enzymes. In the host plant, a pathogen is surrounded by carbohydrates. Small changes in the composition, structure, or accessibility of the host's carbohydrates can alter qualitatively and quantitatively the array of polysaccharide-degrading enzymes secreted by an invading pathogen.

Polysaccharide-degrading enzymes in diseased tissues.—Polysaccharide-degrading enzyme activities have frequently been observed in extracts of infected plant tissues, enzyme activities which are not found, or which are found to be very much lower, in healthy tissues. Much of the research in this area has focused on the pectic-degrading enzymes. Bateman & Millar (16) have recently reviewed the studies dealing with this group of enzymes. Those papers discussed by them will not be considered here. Several reports of the presence of one or more pectic-degrading enzymes in extracts of infected plant tissues have recently appeared (7, 8, 11, 33, 36, 60).

In considering the role of pectic-degrading enzymes in the maceration of plant tissue, Bateman & Millar (16) have concluded "that endo-pectic glycosidases and lyases must be considered as important agents in tissue maceration and that, in many instances, they can account *in toto* for the macerating phenomenon." This conclusion has found support in the observation (22) that the macerating factor produced by the fungus *Sclerotinia fructigena* is most likely a pectin lyase and not the α-L-arabinofuranosidase which had earlier been implicated in this process (21). There is evidence that a pectate lyase is the macerating factor secreted by the soft-rot bacterium *Erwinia aroideae* (23). Other evidence indicates that the macerating enzymes produced by the bacterium *Erwinia carotovora* and the fungus *Pythium aphanidermatum* are pectate lyases (73). The macerating enzyme produced by *Rhizopus* spp. appears to be an endo-polygalacturonase (71) as does the purified macerating enzyme from *Sclerotium rolfsii* (13).

Horton & Keen (41) have studied the endo-polygalacturonase and cellulase found in extracts of *Pyrenochaeta terrestris*-infected onion roots. The synthesis of these pathogen-produced enzymes in culture is repressed by the addition of glucose (41). The investigators have shown that, upon removal of the cotyledons from young onion seedlings infected with *P. terrestris,* the sugar content of the seedlings is reduced by approximately one-fourth. This is accompanied by an increase in both endo-polygalacturonase and cellulase activities and an acceleration of disease development. In contrast, disease development is retarded and enzyme accumulation is not observed when tissue glucose levels are increased by spraying infected plants with a solution of glucose or maleic hydrazide. These observations suggest that free sugars as well as polysaccharide components can participate in the regulation of polysaccharide-degrading enzyme production. Recently Patil & Dimond (62) have extended this line of evidence by demonstrating that the addition of glucose to the cut stems of *F. oxysporum* f. *lycopersici*-infected tomato plants reduces both the rate of symptom development and the level of vascular system polygalacturonase activity (62). Glucose also represses the synthesis of polygalacturonase by this pathogen in culture. The hypothesis that free sugars are involved in the regulation of plant disease development has been extensively documented (39).

The presence of xylosidase in diseased plant tissue was first reported by Strobel (70). Hancock has since reported the presence of this enzyme in sunflower hypocotyls infected with *Sclerotinia sclerotiorum* (35) and has confirmed an earlier report (21) of arabinosidase production in tissues infected with this fungus. Van Etten & Bateman (13, 74) have demonstrated the presence, in extracts of *Sclerotium rolfsii*-induced bean hypocotyl lesions, of enzymes which degrade xylan, galactan, galactomannan, cellulose, and polygalacturonic acid. Van Etten, Maxwell & Bateman (75) have observed endo-polygalacturonase and cellulase in bean hypocotyl lesions caused by *Rhizoctonia solani* (75). Extracts of these lesions have more recently been shown to degrade xylan, galactan, galactomannan, and araban in addition to polygalacturonic acid and carboxymethylcellulose (17). One of the most interesting observations of this investigation is the discovery of polysaccharide-degrading enzymes in the lesion as soon as infection is detectable.

The observations discussed in the preceding section have established that microbial plant pathogens produce an array of polysaccharide-degrading enzymes in culture, enzymes capable of degrading every polysaccharide of the cell wall. The investigations considered in this section make it equally clear that such pathogens do, in fact, secrete a vast array of degradative enzymes during the infective process. Moreover, these enzymes are found at the earliest detectable stages of infection, suggesting that an infection may be initiated only when polysaccharide-degrading enzymes are produced. If polysaccharide-degrading enzymes are required for successful infection, then such enzymes would be expected to degrade plant cell walls. Enzymatic degradation of cell walls will be considered in the next section.

Cell wall degradation by pathogen-produced enzymes.—The presence of pathogen-produced polysaccharide-degrading enzymes in lesions is not proof per se that the host's wall polysaccharides are being degraded. Little additional evidence is required, however, to conclude that pectic enzymes degrade cell walls, since maceration is observed when tissue is treated with purified pectic enzymes (see above).

After having detected endo-pectate lyase in *Fusarium solani* f. *cucurbitae*-infected squash hypocotyl tissue, Hancock (36) found that half of the galacturonic acid-containing polymers had been removed from the walls of infected host tissue. He also observed that the degree of polymerization of the residual pectic substances in infected tissue is approximately one-half that of pectic substances extracted from healthy tissue. Earlier, Hancock (33) had found that sunflower stem tissue infected with *S. sclerotiorum* contains large amounts of polygalacturonase and that the galacturonide content of infected tissue is only one-fourth that of healthy tissue. He had also observed (35) that walls from sunflower stems infected with *S. sclerotiorum* have lost most of their arabinosides and galactosides. The xylose content of this tissue is not significantly reduced, although an active xylosidase is present. This enzyme degrades both solubilized sunflower xylan and a commercially available xylan. This is an interesting example of an enzyme's being able to degrade soluble model substrates but failing to degrade cell wall components of similar composition. Such observations may indicate the involvement of a complex enzyme system of the type described for the degradation of cellulose (68). In the case of cellulose degradation, the presence of an additional factor (the C-1 component) is necessary to render native cellulose susceptible to attack by β-1,4-glucosidase (the C-X enzyme).

Hancock's observation that certain enzymes are able to degrade soluble substrates but not cell wall components is not unique. Enzymes secreted by *C. lindemuthianum* degrade the glucose-, xylose-, galactose-, arabinose-, and mannose-containing soluble extracellular polysaccharides produced by cultured sycamore (*Acer pseudoplatanus*) cells (54). These same enzymes, however, degrade only the glucose- and mannose-containing polymers of isolated sycamore cell walls, and have little or no ability to degrade cell walls isolated from barley, corn, or wheat.

When a pathogen produces enzymes able to degrade cell walls, such enzymes are able to affect similarly cell walls isolated from both resistant and susceptible varieties of the host. English & Albersheim have observed that, when *C. lindemuthianum* is grown in culture, an α-galactosidase is secreted which extracts galactose at the same rate from cell walls isolated from both resistant and susceptible varieties of bean (24). This provides experimental justification for the hypothesis that it is the control of polysaccharide-degrading enzyme production, rather than the activity of the enzymes produced, which regulates virulence in the pathogen and resistance in the host.

Alpha-galactosidase by itself, however, is not sufficient to degrade the galactose-containing polymers of bean cell walls. Although this enzyme readily hydrolyzes *p*-nitrophenyl-α-D-galactoside, an additional factor is required for the removal of galactose from isolated cell walls (25). The first indication of such a requirement was detected in experiments of the type summarized in the data of Figure 1. In such an experiment, *C. lindemuthianum* is grown in shake culture with cell walls isolated from 8-day-old Red Kidney bean hypocotyls serving as the sole carbon source. A portion of the extracellular fluid is assayed daily for its ability to hydrolyze *p*-nitrophenyl-α-D-galactoside. Relative α-galactosidase activity is indicated by the dashed line in Figure 1. A second portion of each sample is assayed for its ability to degrade cell walls isolated from Red Kidney bean hypocotyls. The acid hydrolyzable galactose remaining in treated hypocotyl

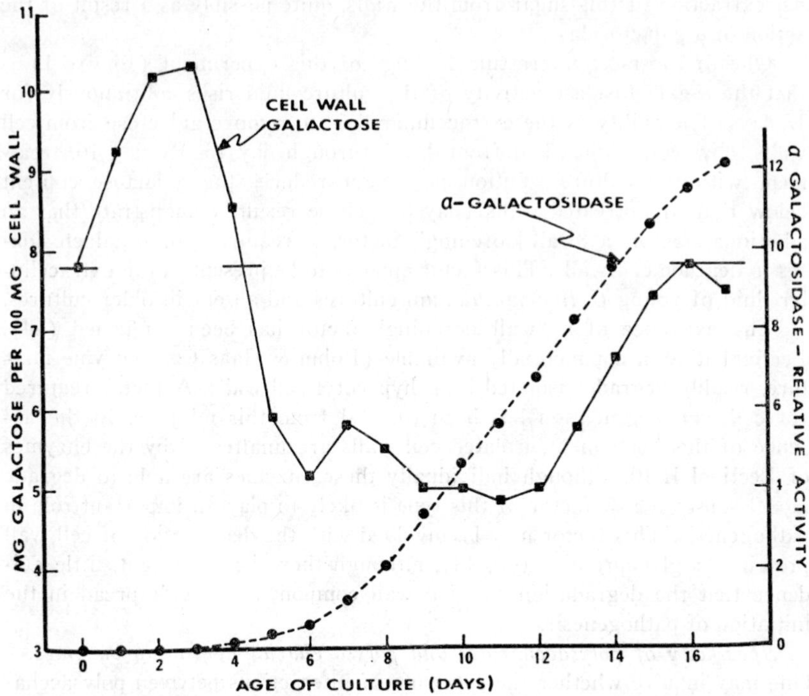

FIG. 1. The ability of enzymes in the culture medium of *C. lindemuthianum* to hydrolyze *p*-nitrophenyl-α-galactoside (dashed line) and to alter the galactose content of cell walls isolated from 8-day-old Red Kidney bean hypocotyl tissues (solid line) followed over a 17-day period. See text for explanation.

cell walls is determined gas chromatographically (3). The ability of *C. lindemuthianum* extracellular enzymes to alter the galactose content of hypocotyl cell walls is indicated by the solid line of Figure 1. It should be recognized that the increased galactose yields obtained from cell walls treated with the extracellular fluid of 1- to 5-day-old cultures imply that the enzymes in the culture fluid, although unable to extract galactose from the walls, render galactose-containing polymers more susceptible to acid hydrolysis. Increased sugar yields upon acid hydrolysis of cell walls following treatment of the walls with extracellular fluid from young *C. lindemuthianum* cultures have been observed repeatedly.

After 4 days of growth, the enzymatic activity of *C. lindemuthianum* culture fluid is altered, and the yield of acid hydrolyzable galactose following culture fluid treatment of walls begins to decrease. The yield of galactose from hypocotyl cell walls decreases just as the α-galactosidase activity appears in the extracellular fluid. The decreased galactose yield represents a net extraction of this sugar from the walls, quite possibly as a result of the action of α-galactosidase.

One of the most interesting findings of this experiment (Figure 1) is that the α-galactosidase activity of the culture fluid rises continuously for 17 days. The ability of the extracellular fluid to remove galactose from cell walls, however, is maximal from day 5 through day 13. By day 16, treatment with the culture solution no longer reduces the galactose content below that of untreated walls (day 0). These results demonstrate that an additional factor, a "wall-loosening" factor, is required for α-galactosidases to degrade cell walls. This factor appears to be present in the extracellular fluid of young *C. lindemuthianum* cultures and absent in older cultures.

The existence of a "wall-loosening" factor has been confirmed (44). Pectinol R-10, a commercially available (Rohm & Haas Co.) enzyme mixture, readily degrades isolated bean hypocotyl cell walls. A factor required for cell wall degradation has been isolated from this mixture. In the absence of this component, isolated cell walls are unaffected by the enzymes of Pectinol R-10, although individually these enzymes are able to degrade model substrates. A factor of this type is likely to play an important role in pathogenesis. This factor may be involved with the degradation of cell wall proteins or glycoproteins (46, 48), although there is, at present, little evidence that the degradation of such wall components is widespread in the initiation of pathogenesis.

Specificity of polysaccharides and polysaccharide-degrading enzymes.—One may inquire whether the spectrum of interactions between polysaccharide-degrading enzymes and the carbohydrates of plants is sufficiently broad to account for the variety of interactions observed between plants and pathogens. Can the proposed hypothesis explain the observation that one strain of a microbial pathogen can infect plant variety A but not plant variety B, while a second strain of the same pathogen can infect variety B but not variety A? This section will summarize the evidence that the interactions between polysaccharide-degrading enzymes and plant carbohydrates are sufficiently sensitive to account for these phenomena. The mechanisms of these interactions must now be considered.

The first point to be made is that sugars and sugar-containing molecules are uniquely sensitive to specific recognition by proteins. There is in all of biology no molecular interaction more specific, more sensitive to structural alteration, than the recognition of saccharides by proteins. The ability of proteins to discriminate between similar saccharides is exemplified by the chemistry of the blood-group substances (78). The structural difference upon which the distinction between the A, B, and O blood-group substances rests, resides solely in terminal nonreducing residues of glycoproteins on the erythrocyte surface. In persons carrying just the A gene, these terminal residues are N-acetylgalactosamine. Those with only the B gene have galactose in place of N-acetylgalactosamine, while in those carrying neither gene (type O) the terminal residue is absent. The difference between A and B phenotypes is based on whether a hydroxyl group or an N-acetyl group is present at the C-2 position of a galactosyl residue.

Antibodies against erythrocyte surface saccharides differentiate between blood groups. This is the classic example of a protein's distinguishing between slightly different sugar-containing molecules. Indeed, the recognition in this case is considerably more sensitive to minor structural differences than any which has been observed involving antibody recognition of proteins or nucleic acids, since no structural modification of a protein or nucleic acid so slight as the substitution of galactose for N-acetylgalactosamine has been shown to result in a completely altered serological response.

The O-antigen system of *Salmonella* offers another excellent example of specific polysaccharide-protein interaction. The genus *Salmonella* encompasses hundreds of serotypes which are believed to differ phenotypically only in the structure of their O-antigens (51). The polysaccharides determining the O-antigenic behavior of *Salmonella* serotypes are composed of repeating units which contain up to four different sugars. There are a number of *Salmonella* serotypes the O-antigens of which are composed of repeating units containing identical sequences of sugars, but which differ only in the nature of the linkages between sugar molecules. Structural differences between O-antigens are sufficient to define hundreds of specific reactions between these molecules and antibodies. In higher plants, the spectrum of sugars and sugar derivatives of which wall polysaccharides are composed (2) is considerably broader than it is in *Salmonella.* As a result, the specificity or the variety of molecular structures which plant walls display to pathogens is unlimited.

The specific nature of the interactions between enzyme proteins and their substrates is well documented. It is not so generally appreciated, however, that, among biopolymer degrading systems, it is those acting on polysaccharides which exhibit the greatest degree of substrate specificity. Virtually all nucleic acids are degraded by phosphodiesterase. Almost all proteins are subject to degradation by a given peptidase. Polysaccharide-degrading enzymes, on the other hand, exhibit rather circumscribed specificities. An exo-β-1,3-glucanase, for example, has little ability to degrade β-1,4- or β-1,6-glucans, while a β-1,4-glucanase does not hydrolyze α-1,4-glucans or β-1-4-galactans (66). These specificities reflect the monosaccharides connected by the glycosidic linkage, as well as both the configuration (α or β) of and the carbon atom involved in the glycosidic bond.

In recognizing the components of a polysaccharide, degradative enzymes respond to the presence of such substituents as an amino group at carbon atom 2, a carboxyl group at carbon atom 6, methyl ether groups, and esters of both hydroxyl and carboxyl groups, as well as to the stereochemistry and oxidation state of a particular carbon atom. Such specificities as these imply that minute alterations in the structure of a polysaccharide can dramatically affect its susceptibility to degradation by a particular enzyme. Since polysaccharides appear to be composed of repeating units or to contain recurring sugar linkages, a minor structural change could be so magnified as to become a major factor in determining a polymer's susceptibility to degradation.

If plant cell wall polysaccharides are involved in disease resistance, these polymers must possess specific structures, and the biosynthesis of each must be controlled. If wall composition were random rather than being carefully determined, it would be difficult to conceive of specific resistance or susceptibility based on the presence, absence, or alteration of one or more of these macromolecules. Recently, a major effort has been undertaken to ascertain the degree of exactness inherent in the composition of plant cell walls (56, 57). The results of these studies have demonstrated that the polysaccharide compositions of plant cell walls are precisely determined.

Cell wall material from each of the various morphological parts (roots, hypocotyls, first internodes, and primary leaves) of a plant has a characteristic sugar composition. The wall polysaccharide compositions of mature plant parts from varieties within a given species are essentially identical. Differences in the sugar composition are observed in cell walls prepared from different species of the same genus, as well as from species of different genera. Cell wall compositions are more similar in closely related plants than in those more distantly related (56).

The composition of plant cell wall polysaccharides changes dramatically during seedling development. Large differences occur between developmental patterns of the cell walls in various parts of a seedling. The general patterns of change in polysaccharide composition, however, are similar for walls of analogous organs among the varieties of a species. Nevertheless, small but significant differences in the rates of change in wall composition have been detected between varieties of the same species (57). These results have established the dynamic nature of the cell wall during growth, as well as the quantitative and qualitative exactness with which the biosynthesis of plant cell walls is genetically directed.

The use of cultured cells has permitted a study of the effects of various media on cell wall composition, and has made possible an assessment of the degree to which metabolic abrogation of the primary genetic controls of wall composition could be achieved (56). The relative amount of monosaccharides released from the walls of suspension-cultured sycamore cells by acid hydrolysis is dependent upon the carbohydrate used as carbon source. Perhaps of greater significance than the changes in the relative sugar composition of the walls is the variation in the total yield of monosaccharides. Such yields range from 24 per cent of wall weight for cells grown in mannose culture to 62 per cent for cells grown in sucrose culture. Since wall composition is essentially invariant within a particular tissue of the intact plant, it seems reasonable to assume that substrates for wall synthesis are drawn from a rather specific mixture of metabolites. Hence, when the substrate pool is altered, as it is in the case of suspension-cultured cells, such metabolic control mechanisms as catabolite repression may influence the composition of the wall.

The results of the sycamore cell experiments revealed an additional fact of considerable importance. These cells grow satisfactorily despite their possessing walls of differing compositions (56). Thus, while the composition of cell walls in plants is carefully controlled, precise wall composition may not be of great importance for many metabolic functions of the plant. Hence, new varieties which exhibit differential disease resistance could have altered walls without greatly affecting the metabolic welfare of the plant. It is unlikely that any other cell organelle could be altered to this degree without a fatal result. Consequently, it is difficult to imagine any other organelle's possessing the potential of the wall in developing new types of resistance.

The correlation between ability to produce polysaccharide-degrading enzymes and virulence.—If a correlation could be made between the production of polysaccharide-degrading enzymes by a microbial pathogen and the organism's virulence, a critical role for such enzymes in the infective process would be clearly indicated. The experimental approach to making such a correlation is more difficult than one might suppose. One must select for study an enzyme which is essential for successful infection in the particular host-pathogen combination being considered. In many instances, this constraint has been overlooked, as it is difficult to determine whether a particular enzyme is critical for disease development.

A second complication is that pathogens frequently produce degradative enzymes when grown under one set of conditions, but fail to do so in a different environment. This has been demonstrated repeatedly (see above). In order to obviate the difficulties engendered by environmental variability, one should attempt to correlate virulence with a pathogen's production in the host of an essential degradative enzyme. To conduct such an experiment is all but impossible, however, since the failure of an avirulent pathogen to produce an enzyme may reflect a lack of pathogen growth rather than reduced production of a particular enzyme. The best compromise is to measure production of the essential enzyme when the pathogen is grown in culture under conditions which approximate the environment of the pathogen in the host. This precaution is difficult to observe. A lack of correlation between virulence and the production of an enzyme by a pathogen in culture (53, 59, 60) is not definitive, since either the enzyme studied may not be essential to the infective process, or the growth medium may stimulate or suppress production of the enzyme in a different way from that of the host. With such limitations in mind, it is interesting to note some of the impressive correlations which have been made.

When a suspension of the soft rot bacterium, *Pseudomonas marginalis,* is irradiated with ultraviolet light, a majority of the surviving bacteria have lost both their virulence toward the test hosts, chicory and lettuce, and their ability to produce pectic-degrading enzymes (29). In these studies, 10 out of 17 single-colony isolates tested had lost their pathogenicity, and all "five isolates chosen at random from among the ten avirulent isolates failed in repeated tests to produce pectolytic enzymes in culture." Unless there was some hidden selection for loss of pathogenicity in the isolation procedures, the yield of such mutants, 10 of 17 survivors tested, is unreasonably high. This question notwithstanding, the correlation between pectic-degrading enzyme production and virulence in *P. marginalis* is impressive.

Garber et al. (30) have investigated the production of degradative enzymes by several strains of *Penicillium italicum* and *Penicillium digitatum*. They have observed that orange tissue rotted by either of these species contains enzymes which degrade polygalacturonic acid and carboxymethylcellulose. On the other hand, necrotic or less seriously affected tissue, resulting from infection by avirulent mutants of these species, has carboxymethylcellulose-degrading activity, but lacks the ability to degrade polygalacturonic acid. These results provide another interesting correlation between virulence and a pathogen's ability to produce a specific degradative enzyme.

Using six species of *Verticillium,* Leal & Villanueva (49) have studied the production of pectic-degrading enzymes by forty strains of this fungus. High pectic-degrading activity was characteristic of all the pathogenic strains examined, while nonpathogenic strains had no such activity. Similarly, a close correlation between polygalacturonase activity and virulence has been found for three *Colletotrichum falcatum* strains (69). In addition, Kelman & Cowling (47) have observed that filtered media from cultures of virulent *Pseudomonas solanacearum* isolates exhibit higher cellulase activities than do media from weakly parasitic or avirulent strains of this bacterium. A correlation between the virulence of *C. lindemuthianum* isolates and the ability of these isolates to produce α-galactosidase has been demonstrated (24). In these experiments, several isolates of the same *C. lindemuthianum* strains were shown to exhibit differential virulence. When isolates possessing differential virulence were grown under identical conditions, virulence was observed to be correlated with α-galactosidase secretion.

The problems encountered in attempting to correlate enzyme production with virulence are illustrated by comparing observations from two laboratories regarding virulence and pectic enzyme production in the fungus *R. solani.* Barker & Walker (9) have reported a correlation between isolate virulence and polygalacturonase production. Papavizas & Ayers (59), however, have been unable to demonstrate in this fungus a relationship between isolate virulence and ability to produce this enzyme. A major difference between the two investigations was that Barker & Walker obtained their isolates (presumably heterokaryons) directly from infected host plants, while Papavizas & Ayers used single-basidiospore (homokaryon) isolates. Regardless of the difficulty of making significant correlations, data obtained for a variety of pathogens indicate that lack of ability to produce a degradative enzyme may account for avirulence.

Direct evidence for the involvement of cell wall polysaccharides and polysaccharide-degrading enzymes in disease resistance.—The development of resistance in hypocotyls of maturing Red Kidney beans to infection by *R. solani* is correlated with the conversion in the hypocotyl cell walls of pectin to calcium pectate (15). Calcium pectate is resistant to the polygalacturonase secreted by this fungus (10).

R. solani secretes a variety of polysaccharide-degrading enzymes when grown in culture on hypocotyl cell walls from either 4-day or 20-day-old Red Kidney bean plants as sole carbon source. The enzymes in filtered growth medium from either culture can degrade hypocotyl cell walls isolated from susceptible 4-day-old Red Kidney beans, but can degrade only slightly hypocotyl walls from resistant 20-day-old plants. Moreover, extracts prepared from the earliest detectable lesions of *R. solani*-infected plants can degrade hypocotyl walls from 4-day-old Red Kidney beans, but can degrade only slightly hypocotyl walls from 20-day-old plants. Following treatment with extracts of immature lesions, walls isolated from 4-day-old plants are observed to have lost 59 per cent of the arabinose, 19 per cent of the xylose, 72 per cent of the galactose, and 23 per cent of the glucose normally recovered upon acid hydrolysis. Forty-three per cent of the neutral sugars normally recovered from the walls of 4-day-old plants are removed by the action of lesion extract enzymes, while only 4 per cent of the neutral sugars are removed from the walls of 20-day-old plants (17).

A similar phenomenon has been observed in the ability of the extracellular enzymes of *C. lindemuthianum* cultures to degrade hypocotyl cell walls isolated from susceptible 5-day-old bean plants, but not hypocotyl walls from resistant 18-day-old plants (24). Hypocotyls of 5-day-old bean plants are susceptible to infection by this pathogen; hypocotyls of 18-day-old plants are not. Of the neutral sugars removed from the hypocotyl walls of susceptible bean plants upon treatment with the enzymes secreted by either *C. lindemuthianum* or *R. solani,* more than half is galactose. Van Etten & Bateman (74) have observed the secretion of galactan-degrading enzymes by a third bean pathogen, *S. rolfsii.* The results of these three investigations suggest that degradation of galactose-containing polymers in bean hypocotyl walls is essential for pathogenesis.

English & Albersheim (24) have examined resistance of Red Kidney, Pinto, and Small White beans to infection by the alpha, beta, and gamma strains of *C. lindemuthianum.* When grown in shake culture with hypocotyl cell walls as the carbon source, each of the three fungal strains secretes an equivalent but relatively small amount of β-galactosidase and of β-xylosidase. The secretion of these two enzymes is not dependent upon the bean variety from which the walls had been isolated. However, each of the *C. lindemuthianum* strains secretes a greater amount of α-galactosidase when grown on hypocotyl walls from a susceptible bean variety than when grown on walls from a resistant variety. For example, the beta strain secretes 20 times more α-galactosidase when grown in suspension culture on hypocotyl walls isolated from susceptible Red Kidney bean plants than is secreted when this strain is grown on walls isolated from resistant Pinto beans, and 5 times more α-galactosidase than is secreted when grown on walls isolated from slightly susceptible Small White beans. The alpha strain, on the other hand, produces more α-galactosidase when grown on walls isolated from Pinto (susceptible to the alpha strain) hypocotyls, than it does when grown on walls isolated from Red Kidney (resistant to the alpha strain) hypocotyls. Such results constitute a direct demonstration of the control by cell walls isolated from a host plant of polysaccharide-degrading enzyme production by a pathogen.

What, Then Might Be the Mechanisms Underlying Resistance and Virulence?

A hypothetical mechanism for the induction of pathogenesis, a mechanism based on molecular interactions, which accounts for a number of observations concerning resistance in plants and virulence in pathogens will now be considered. Once again, it must be stressed that this hypothesis deals only with initiation of infection by microorganisms. It concerns only those factors present in the host when it first comes in contact with the pathogen which determine whether infection will be successful. The plant's responses, e.g., phytoalexin production, hypersensitive reaction, and other changes in host metabolism, following its recognition of the pathogen's presence are important resistance mechanisms but are not germane to this discussion.

This proposal is based on the data accumulated during years of breeding plants for resistance. Flor (28) was the first to study carefully the genetics of both components of a host-pathogen system; his contributions remain the dominant influence on research in this field. Flor recognized that, for each gene determining resistance in flax, there is a specific and related gene determining virulence in flax rust (*Melampsora lini*). This is the case not only for flax and flax rust, but for a wide variety of host-pathogen combinations (27, 63).

The intent of this article has been to present the evidence in support of the following hypotheses: (*a*) Molecular interactions between the carbohydrate constituents of a host and the polysaccharide-degrading enzymes produced by a pathogen account for the inherent resistance of plants to most microorganisms. (*b*) These interactions account equally well for the rare instances in which a microorganism successfully infects a plant. Such interactions will be used to account, theoretically, for the genetic observations relating to pathogenesis in the *Phaseolus vulgaris: Colletotrichum lindemuthianum* system.

Resistance in the true bean, *P. vulgaris* to *C. lindemuthianum* is inherited as a dominant trait (82). The manner in which virulence is inherited in *C. lindemuthianum* has not been established because the fungus lacks an available sexual stage. In most pathogens, virulence is inherited as a recessive trait (27, 82). The inheritance of virulence in *C. lindemuthianum* both as a recessive and as a dominant trait will be considered.

The alpha, beta, gamma, and delta strains of *C. lindemuthianum* are distinguished by their differential virulence toward certain *P. vulgaris* varieties. Goth & Zaumeyer (31) have reported the susceptibility of a large number of *P. vulgaris* varieties to the four *C. lindemuthianum* strains. The responses for representative varieties are summarized in Table I. Genes are assigned to the *P. vulgaris* varieties and the *C. lindemuthianum* strains in a manner which provides the simplest gene-for-gene relationship (63). It is interesting that no *C. lindemuthianum* strain is assigned fewer than two genes for virulence.

There are eight recognized groups of *P. vulgaris* varieties, each group of which possesses a unique set of genes phenotypically dominant for *C. lindemuthianum* resistance. One group of beans, represented in Table I by Cornell 49–242, has been assigned the genotype Cl_4. The varieties in this group must possess at least one resistance gene not present in other varieties, since all other varieties are susceptible to infection by at least one *C. lindemuthianum* strain. The available data fail to indicate whether any of the other resistance genes (Cl_1, Cl_2, and Cl_3) are present in Cornell 49–242.

The use of Table I is best demonstrated by example. Any strain of *C. lindemuthianum* possessing virulence gene "1" (i.e., the alpha, beta, and delta strains) is virulent toward those bean varieties which possess either no resistance gene (e.g., Bountiful) or only resistance gene Cl_1 (e.g., Top

TABLE I

FOUR STRAINS OF *C. LINDEMUTHIANUM* DISTINGUISHED BY DIFFERENT RESPONSES OF *P. VULGARIS* VARIETIES POSSESSING VARIOUS COMBINATIONS OF RESISTANCE GENES

P. vulgaris variety	Dominant resistance genes in *P. vulgaris*	*C. lindemuthianum* strain			
		alpha 1, 2[a]	*beta* 1, 3	*gamma* 2, 3	*delta* 1, 2, 3
Bountiful	—	S[b]	S	S	S
Top Crop	Cl_1	S	S	—[c]	S
Genefer Market	Cl_2	S	—	S	S
Red Kidney	Cl_3	—	S	S	S
Small White	Cl_1, Cl_2	S	—	—	S
Emerson 51–2	Cl_1, Cl_3	—	S	—	S
Perry Marrow	Cl_2, Cl_3	—	—	S	S
Sanilac	Cl_1, Cl_2, Cl_3	—	—	—	S
Cornell 49–242	Cl_4	—	—	—	—

[a] Virulence genes in the fungus.
[b] A susceptible response.
[c] A resistant response.

Crop). The alpha strain possesses virulence gene "2" in addition to gene "1." Therefore, the alpha strain successfully infects any bean variety not possessing a resistance gene other than Cl_1 and Cl_2. Bean genotypes susceptible to attack by the alpha strains include varieties represented by Bountiful, Top Crop, Genefer Market, and Small White.

In order to demonstrate how carbohydrate constituents might impart resistance to a plant and how the production of polysaccharide-degrading enzymes might lead to pathogen virulence, theoretical functions are assigned to two *P. vulgaris* resistance genes and to two *C. lindemuthianum* virulence genes. Assume that *C. lindemuthianum* is virulent only when it produces large amounts of α-galactosidase. When virulence is inherited as a recessive trait, it is postulated that sufficient α-galactosidase is produced to result in a virulent reaction. Only when the synthesis of this enzyme is repressed is α-galactosidase secretion insufficient for the establishment of infection. The effector, a small molecule which combines with the product of a regulator gene to yield active repression of enzyme synthesis, is, for the purposes of this illustration, glucose. Thus, a plant will be resistant when it provides the pathogen sufficient glucose to repress α-galactosidase synthesis. Conversely, the pathogen will be virulent unless it obtains from a host sufficient quantities of glucose to repress α-galactosidase synthesis.

Resistance gene Cl_2 is assigned the function of encoding an enzyme which adds glucose side chains to a polysaccharide in the cell wall. These glucosyl residues are attached through α-glycosidic linkages. Resistance gene Cl_3 encodes an enzyme which adds glucose side chains attached through β-glycosidic linkages. Thus, Genefer Market cell walls would contain α-glucosides, Red Kidney cell walls β-glucosides, and Perry Marrow cell walls both α- and β-glucosides.

In a pathogen in which virulence is inherited as a recessive trait, virulence reflects the absence of a specific gene product. Let virulence gene "2" prevent α-glucosidase synthesis, and let virulence gene "3" prevent β-glucosidase synthesis. Thus, the alpha strain produces β-glucosidase, and the beta strain produces α-glucosidase, while the gamma and delta strains produce neither of these enzymes.

The responses predicted by this hypothetical system are consistent with those which have been observed experimentally (Table I). For example, the alpha strain, which fails to produce α-glucosidase, cannot release glucose from the cell walls of Genefer Market. As a result, when the alpha strain attacks Genefer Market, the synthesis of α-galactosidase is not repressed, and infection can occur. The alpha strain, although unable to produce α-glucosidase, does produce β-glucosidase. This enzyme hydrolyzes the β-glucosides of Red Kidney cell walls. Therefore, the synthesis of α-galactosidase is repressed when the alpha strain attacks Red Kidney, and the absence of α-galactosidase precludes the establishment of infection. Perry Marrow, the walls of which contain α- and β-glucosides, is resistant to both the alpha and beta strains. However, Perry Marrow is susceptible to the gamma and delta strains, since neither of these strains is able to synthesize either α- or β-glucosidase. By assigning functions to resistance genes Cl_1 and Cl_4 and to virulence gene "1," one can account for all the observed host-pathogen responses.

Now consider the case in which virulence in *C. lindemuthianum* is inherited as a dominant trait. Again, assume that virulence requires production of large amounts of α-galactosidase. In this case, the requisite enzyme is postulated to be an exo-α-galactosidase. Therefore, for significant degradation of galactan to occur, this substrate must be free of interfering side chains. In this instance, glucose does not act as a co-repressor of α-galactosidase synthesis. Resistance gene Cl_2 again encodes an enzyme that catalyzes the addition of α-glucosides to the wall, and resistance gene Cl_3 encodes an enzyme which adds β-glucosides. In this case, however, it is postulated that either the α- or the β-glucosides, or both, are added to the galactan which must be degraded by the pathogen-produced exo-α-galactosidase if a virulent response is to occur. When pathogen virulence is inherited as a dominant trait, virulence reflects the presence of an essential gene product. In this case, virulence gene "2" encodes α-glucosidase and virulence gene "3," β-glucosidase.

Again, the predicted responses are consistent with those which have been observed (Table I). The α-glucosidase produced by the alpha strain removes the glucose side chains from the wall galactan of Genefer Market, rendering the galactan susceptible to degradation by exo-α-galactosidase. This leads to infection of Genefer Market by the alpha strain. The α-glucosidase of the alpha strain cannot remove the β-glucosyl residues from the wall galactan of Red Kidney, leaving the galactan immune to exo-α-galactosidase attack, and the Red Kidney plant resistant to infection by the alpha strain. Perry Marrow, which contains α- and β-glucosyl residues attached to the galactan, is resistant to both the alpha and beta strains but is susceptible to both the gamma and delta strains, each of which secretes a mixture of α- and β-glucosidases.

In the model under consideration, plants with differing wall structures are considered distinct varieties, much as O-antigen structure distinguishes serotypes of the genus *Salmonella.* It is predicted that plants with structurally different wall polysaccharides are, in many instances, differentially resistant to pathogen infection, just as bacteria with slightly differing O-antigenic polysaccharides often differ in resistance to phage infection. It would be expected, and has been observed (56, 57), that the more closely related are two plants, the more similar are their wall polysaccharides. Among plant varieties within a given species, analogous cells have walls with identical or very similar sugar compositions. The wall polysaccharides of such closely related plants are likely to differ only in the nature of the glycosidic linkages. Even these small structural differences, however, are sufficient to cause differing interactions with pathogen-produced enzymes. Finally, if strains of a pathogen arise because of their abilities to infect plants which differ only in wall polysaccharide structure, it follows that the pathogen strains could differ only in their abilities to interact with wall polysaccharides of the hosts.

The number of possible variations in plant cell wall polysaccharide structures is very large (2). This variability is made possible by the number of polymer constituents and by the variety of linkages through which they are joined. There are, in total, about ten common neutral sugar and uronic acid constituents of plant cell wall polysaccharides, and an approximately equal number of such sugar derivatives as ethers and esters. The monomeric units of polysaccharides are interconnected through α- or β-glycosidic linkages, and these linkages can be to any of three or four carbon atoms of the adjoining sugar. Branching of wall polysaccharides is the rule rather than the exception. Thus, there is no limit to the number of structures which plant cell walls can present to pathogen proteins. In the future, resistant varieties may be selected on the basis of specific wall polysaccharide characteristics. The wall structures for which to select could be determined by studying carbohydrate effects on the control of polysacharide-degrading enzyme production in the pathogen.

Disaccharides are as likely to act as specific effectors in the control of pathogen enzyme synthesis as the monosaccharide glucose which is used in the example presented above. In *E. coli*, β-galactosidase synthesis is induced by the disaccharide lactose, and is even more efficiently induced by the disaccharide analogue isopropylthiogalactoside which is not metabolized by *E. coli* (42). Partial enzymatic degradation of plant cell wall polysaccharides produces a vast array of disaccharides. These disaccharides could act as highly specific effectors for the induction and repression of polysaccharide-degrading enzyme synthesis by pathogens. Exciting possibilities for the control of plant diseases may lie in the application to plants of specific, poorly metabolized analogues of effectors, effectors which control production by pathogens of specific polysaccharide-degrading enzymes.

Acknowledgment

The authors wish to acknowledge gratefully the contributions of Dora Kelling in the preparation of this article.

LITERATURE CITED

1. Abercrombie, M., Ambrose, E. J. The surface properties of cancer cells: a review. *Cancer Res.*, **22**, 525–48 (1962)
2. Albersheim, P. Biogenesis of the cell wall. In *Plant Biochemistry*, 298–321 (Bonner, J., Vorner, J. E., Eds., Academic Press, New York, 1054 pp., 1965)
3. Albersheim, P., Nevins, D. J., English, P. D., Karr, A. A method for the analysis of sugars in plant cell-wall polysaccharides by gas-liquid chromatography. *Carbohydrate Res.*, **5**, 340–45 (1967)
4. Aub, J. C., Sanford, B. H., Cote, M. N. Studies on reactivity of tumor and normal cells to a wheat germ agglutinin. *Proc. Natl. Acad. Sci. U.S.*, **54**, 396–99 (1965)
5. Aub, J. C., Sanford, B. H., Wang, L. Reactions of normal and leukemic cell surfaces to a wheat germ agglutinin. *Proc. Natl. Acad. Sci. U.S.*, **54**, 400–02 (1965)
6. Aub, J. C., Tieslau, C., Lankester, A. Reactions of normal and tumor cell surfaces to enzymes. I. Wheat-germ lipase and associated muco-polysaccharides. *Proc. Natl. Acad. Sci. U.S.*, **50**, 613–19 (1963)
7. Ayers, W. A., Papavizas, G. C. An exocellular pectolytic enzyme of *Aphanomyces euteiches*. *Phytopathology*, **55**, 249–53 (1965)
8. Ayers, W. A., Papavizas, G. C., Diem, A. F. Polygalacturonate transeliminase and polygalacturonase production by *Rhizoctonia solani*. *Phytopathology*, **56**, 1006–11 (1966)
9. Barker, K. R., Walker, J. C. Relationship of pectolytic and cellulytic enzyme production by strains of *Pellicularia filamentosa* to their pathogenicity. *Phytopathology*, **52**, 1119–25 (1962)
10. Bateman, D. F. An induced mechanism of tissue resistance to polygalacturonase in *Rhizoctonia*-infected hypocotyls of bean. *Phytopathology*, **54**, 438–45 (1964)
11. Bateman, D. F. Hydrolytic and trans-eliminative degradation of pectic substances by extracellular enzymes of *Fusarium solani* f. *phaseoli*. *Phytopathology*, **56**, 238–44 (1966)
12. Bateman, D. F. Alteration of cell wall components during pathogenesis by *Rhizoctonia solani*. In *The Dynamic Role of Molecular Constituents in Plant-Parasitic Interaction*, 58–79 (Mirocha, C. J., Uritani, I., Eds., The American Phytopathological Society, St. Paul, Minn., 372 pp., 1967)
13. Bateman, D. F. The enzymatic maceration of plant tissue. *Neth. J. Plant Pathol.*, **74**, Suppl., 67–80 (1968)
14. Bateman, D. F. Some characteristics of the cellulase system produced by *Sclerotium rolfsii* Sacc. *Phytopathology*, **59**, 37–42 (1969)
15. Bateman, D. F., Lumsden, R. D. Relation of calcium content and nature of the pectic substances in bean hypocotyls of different ages to susceptibility to an isolate of *Rhizoctonia solani*. *Phytopathology*, **55**, 734–38 (1965)
16. Bateman, D. F., Millar, R. L. Pectic enzymes in tissue degradation. *Ann. Rev. Phytopathol.*, **4**, 119–46 (1966)
17. Bateman, D. F., Van Etten, H. D., English, P. D., Nevins, D. J., Albersheim, P. Susceptibility to enzymatic degradation of cell walls from bean plants resistant and susceptible to *Rhizoctonia solani* Kühn. *Plant Physiol.*, **44**, 641–48 (1969)
18. Bray, D., Robbins, P. W. Mechanism of ε^{15} conversion studied with bacteriophage mutants. *J. Mol. Biol.*, **30**, 457–75 (1967)
19. Burger, M. M. Isolation of a receptor complex for a tumuor specific agglutinin from the neoplastic cell surface. *Nature*, **219**, 499–500 (1968)
20. Burger, M. M., Goldberg, A. R. Identification of a tumor-specific determinant on neoplastic cell surfaces. *Proc. Natl. Acad. Sci. U.S.*, **57**, 359–66 (1967)
21. Byrde, R. J. W., Fielding, A. H. An extracellular α-L-arabinofuranosidase secreted by *Sclerotinia fructigena*. *Nature*, **205**, 390–91 (1965)
22. Byrde, R. J. W., Fielding, A. H. Pectin methyl-*trans*-eliminase as the maceration factor of *Sclerotinia fructigena* and its significance in brown rot of apple. *J. Gen. Microbiol.*, **52**, 287–97 (1968)

23. Dean, M., Wood, R. K. S. Cell wall degradation by a pectic transeliminase. *Nature,* **214,** 408–10 (1967)
24. English, P. D., Albersheim, P. Host-pathogen interactions. I. A. correlation between α-galactosidase production and virulence. *Plant Physiol.,* **44,** 217–24 (1969)
25. English, P. D., Albersheim, P. (Unpublished data)
26. English, P. D., Jurale, B., Albersheim, P. (Unpublished data)
27. Fincham, J. R. S., Day, P. R. Genetics of pathogenicity. *Fungal Genetics,* 2nd ed. 257–73 (F. A. Davis Co., Philadelphia, 1965)
28. Flor, H. H. The complementary genic systems in flax and flax rust. *Advan. Genet.,* **8,** 29–53 (1956)
29. Friedman, B. A., Ceponis, M. J. Effect of ultraviolet light on pectolytic enzyme production and pathogenicity of *Pseudomonas. Science,* **129,** 720–21 (1959)
30. Garber, E. D., Beraha, L., Shaeffer, S. G. Genetics of phytopathogenic fungi. XIII. Pectolytic and cellulolytic enzymes of three phytopathogenic *Penicillia. Botan. Gaz.,* **126,** 36–40 (1965)
31. Goth, R. W., Zaumeyer, W. J. Reactions of bean varieties to four races of anthracnose. *Plant Disease Reptr.,* **49,** 815–18 (1965)
32. Hakomori, S., Murakami, W. T. Glycolipids of hamster fibroblasts and derived malignant-transformed cell lines. *Proc. Natl. Acad. Sci. U.S.,* **59,** 254–61 (1968)
33. Hancock, J. G. Degradation of pectic substances associated with pathogenesis by *Sclerotinia sclerotiorum* in sunflower and tomato stems. *Phytopathology,* **56,** 975–79 (1966)
34. Hancock, J. G. Pectate lyase production by *Colletotrichum trifolii* in relation to changes in pH. *Phytopathology,* **56,** 1112–13 (1966)
35. Hancock, J. G. Hemicellulose degradation in sunflower hypocotyls infected with *Sclerotinia sclerotiorum. Phytopathology,* **57,** 203–06 (1967)
36. Hancock, J. G. Degradation of pectic substances during pathogenesis by *Fusarium solani* f. sp. *cucurbitae. Phytopathology,* **53,** 62–69 (1968)
37. Hancock, J. G., Millar, R. L., Lorbeer, J. W. Pectolytic and cellulolytic enzymes produced by *Botrytis allii, B. cinerea,* and *B. squainosa in vitro* and *in vivo. Phytopathology,* **54,** 928–31 (1964)
38. Hayes, W. *The Genetics of Bacteria and their Viruses* (John Wiley & Sons, New York, 740 pp., 1965)
39. Horsfall, J. G., Dimond, A. E. Interactions of tissue sugar, growth substances, and disease susceptibility. *Z. Pflanzenbau. Pflanzenshutz,* **64,** 415–21 (1957)
40. Horton, J. C., Keen, N. T. Regulation of induced cellulase synthesis in *Pyrenochaeta terrestris* gorenz et al. by utilizable carbon compounds. *Can. J. Microbiol.,* **12,** 209–20 (1966)
41. Horton, J. C., Keen, N. T. Sugar repression of endopolygalacturonase and cellulase synthesis during pathogenesis by *Pyrenochaeta terrestris* as a resistance mechanism in onion pink root. *Phytopathology,* **56,** 908–16 (1966)
42. Jacob, F., Monod, J. Genetic regulatory mechanisms in the synthesis of proteins. *J. Mol. Biol.,* **3,** 318–56 (1961)
43. Kaji, A., Yoshihara, M. (Personal communication)
44. Karr, A., Albersheim, P. (Unpublished data)
45. Keen, N. T., Horton, J. C. Induction and repression of endopolygalacturonase synthesis by *Pyrenochaeta terrestris. Can. J. Microbiol.,* **12,** 443–53 (1966)
46. Keen, N. T., Williams, P. H., Walker, J. C. Protease of *Pseudomonas lachrymans* in relation to cucumber angular leaf spot. *Phytopathology,* **57,** 263–71 (1967)
47. Kelman, A., Cowling, E. Cellulase of *Pseudomonas solanacearum* in relation to pathogenesis. *Phytopathology,* **55,** 148–55 (1965)
48. Lamport, D. T. A. Hydroxyproline-*O*-glycosidic linkage of the plant cell wall glycoprotein extensin. *Nature,* **216,** 1322–24 (1967)
49. Leal, J. A., Villanueva, J. R. Lack of pectic enzyme production by non-pathogenic species of *Verticillium. Nature,* **195,** 1328–29, (1962)
50. Losick, R., Robbins, P. W. Mechanism of ε^{15} conversion studied with a bacterial mutant. *J. Mol. Biol.,* **30,** 445–55 (1967)
51. Lüderitz, O., Staub, A. M., Westphal, O. Immunochemistry of O and R

antigens of *Salmonella* and related *Enterobacteriaceae. Bacteriol. Rev.*, **30,** 193–255 (1966)

52. Mandels, M., Reese, E. T. Inhibition of celluases. *Ann. Rev. Phytopathol.*, **3,** 85–102 (1965)
53. Mann, B. Role of pectic enzymes in the *Fusarium* wilt syndrome of tomato. *Trans. Brit. Mycol. Soc.*, **45,** 169–78 (1962)
54. McNab, J. M., Nevins, D. J., Albersheim, P. Differential resistance of cell walls of *Acer pseudoplatanus, Triticum vulgare, Hordeum vulgare,* and *Zea mays* to polysaccharide-degrading enzymes. *Phytopathology,* **57,** 625–31 (1967)
55. Medearis, D. N., Camitta, B. M., Heath, E. C. Cell wall composition and virulence in *Escherichia coli. J. Exptl. Med.*, **128,** 399–414 (1968)
56. Nevins, D. J., English, P. D., Albersheim, P. The specific nature of plant cell wall polysaccharides. *Plant Physiol.*, **42,** 900–06 (1967)
57. Nevins, D. J., English, P. D., Albersheim, P. Changes in cell wall polysaccharides associated with growth. *Plant Physiol.*, **43,** 914–22 (1968)
58. Norkrans, B. Degradation of cellulose. *Ann. Rev. Phytopathol.*, **1,** 325–50 (1963)
59. Papavizas, G. C., Ayers, W. A. Virulence, host range, and pectolytic enzymes of single-basidiospore isolates of *Rhizoctonia praticola* and *Rhizoctonia solani. Phytopathology,* **55,** 111–16 (1965)
60. Papavizas, G. C., Ayers, W. A. Polygalacturonate trans-eliminase production by *Fusarium oxysporum* and *Fusarium solani. Phytopathology,* **56,** 1269–73 (1966)
61. Pardee, A. B. Cell Division, and a hypothesis of cancer. *Natl. Cancer Inst. Monograph,* **14,** 7 (1964)
62. Patil, S. S., Dimond, A. E. Repression of polygalacturonase synthesis in *Fusarium oxysporum* f. sp. *lycopersici* by sugars and its effect on symptom reduction in infected tomato plants. *Phytopathology,* **58,** 676–82 (1968)
63. Person, C. Gene-for gene relationships in host:parasite systems. *Can. J. Botany,* **37,** 1101–30 (1959)
64. Pettersson, G., Porath, J. Studies on cellulolytic enzymes. II. Multiplicity of the cellulytic enzymes of *Polyporus versicolor. Biochim. Biophys. Acta,* **67,** 9–15 (1963)
65. Reese, E. T., Ed. *Advances in Enzymic Hydrolysis of Cellulose and Related Materials* (The Macmillan Company, New York, 290 pp., 1963)
66. Reese, E. T., Maguire, A. H., Parrish, F. W. Glucosidases and exo-glucanases. *Can. J. Biochem.*, **46,** 25–34 (1968)
67. Robbins, P. W., Uchida, T. Chemical and macromolecular structure of O-antigens from *Salmonella anatum* strains carrying mutants of bacteriophage ε^{15}. *J. Biol. Chem.*, **240,** 375–83 (1965)
68. Selby, K., Maitland, C. C. The cellulase of *Trichoderma viride.* Separation of the components involved in the solubilization of cotton. *Biochem. J.*, **104,** 716–24 (1967)
69. Singh, G. P., Husain, A. Relation of hydrolytic enzyme activity with the virulence of strains of *Colletotrichum falcatum. Phytopathology,* **54,** 1100–01 (1964)
70. Strobel, G. A. A xylanase system produced by *Diplodia viticola. Phytopathology,* **53,** 592–96 (1963)
71. Suzuki, H., Abe, T., Urade, M., Nisizawa, K., Kuroda, A. Nature of the macerating enzymes from *Rhizopus* sp. *J. Ferment. Technol.*, **45,** 73–85 (1967)
72. Tomita, Y., Suzuki, H., Nisizawa, K. Chromatographic patterns of cellulase components of *Trichoderma viride* grown on the synthetic and natural media. *J. Ferment. Technol.*, **46,** 701–10 (1968)
73. Turner, M. T., Bateman, D. F. Maceration of plant tissues susceptible and resistant to soft-rot pathogens by enzymes from compatible host-pathogen combinations. *Phytopathology,* **58,** 1509–15 (1968)
74. Van Etten, H. D., Bateman, D. F. Enzymatic degradation of galactan, galactomannan, and xylan by *Sclerotium rolfsii* Sacc. (Unpublished data)
75. Van Etten, H. D., Maxwell, D. P., Bateman, D. F. Lesion maturation, fungal development, and distribution of endopolygalacturonase and cellulase in *Rhizoctonia*-infected bean hypocotyl tissues. *Phytopathology,* **57,** 121–26 (1967)

76. Waggoner, P. E., Dimond, A. E. Production and role of extracellular pectic enzymes of *Fusarium oxysporum* f. *lycopersici*. *Phytopathology,* **45,** 79–87 (1955)
77. Wakabayashi, K., Kanda, T., Nisizawa, K. Separation of two cellulase components from a culture filtrate of *Irpex lacteus* and some of their properties. *J. Ferment. Technol.* **44,** 669–81 (1966)
78. Watkins, W. M. Blood-group substances. *Science,* **152,** 172–81 (1966)
79. Weiss, L. The mammalian tissue cell surface. In *Biochem. Soc. Symp., No. 22,* 32–50 (1963)
80. Weiss, L. *The Cell Periphery, Metastasis and Other Contact Phenomena* (John Wiley & Sons, Inc., New York, 388 pp., 1967)
81. Wood, R. K. S. *Physiological Plant Pathology* (Blackwell Scientific Publications, Oxford, 570 pp., 1967)
82. Zaumeyer, W. J., Thomas, H. R. *A Monographic Study of Bean Diseases and Methods for Their Control.* Tech. Bull., No. 868, U.S.D.A. (1957)

Copyright 1969. All rights reserved

THE ROLE OF PHENOLICS IN HOST RESPONSE TO INFECTION

By Tsune Kosuge
Department of Plant Pathology, University of California
Davis, California

Introduction

Numerous reports have been published on the appearance and accumulation of phenolic compounds in plant tissue in response to infection. Many of these compounds are toxic to certain fungi and bacteria, and others inactivate viruses *in vitro*. A case has been made that a particular compound or an enzyme confers resistance to plant pathogens in some instances, but the idea is not fully accepted.

Space limitations do not permit an exhaustive review of all papers on this subject. Such does not seem necessary, however, since many of the compounds involved have not been identified conclusively, nor has their relevance to host-parasite interaction been clearly established. Many deserving papers also are not cited because of the lack of space. Moreover, there are a number of recent reviews covering information pertinent to the present subject matter (26, 27, 37, 39, 86–88, 130, 133, 158). Other reviews are cited which are pertinent to the particular subject being discussed and serve as points of departure for each section.

Biosynthesis of Phenolic Compounds

Important mechanisms undoubtedly involved in controlling biosynthesis and metabolism of phenolic compounds in higher plants are (*a*) the feedback inhibition or activation of enzyme activity by intermediates and end products of the metabolic pathway; and (*b*) the repression or induction of enzyme synthesis (183). Evidence exists that such mechanisms function during the response of plants to infection by pathogens. Discussion of the biosynthesis of phenolic compounds will therefore emphasize those investigations which deal with one or both of these mechanisms.

The shikimic acid pathway for aromatic biosynthesis.—The shikimic acid pathway elucidated first in bacteria appears to function in higher plants for synthesis of phenolic compounds which arise from phenylalanine and tyrosine (22, 45–47, 113). The initial reaction of this pathway involves the condensation of phosphoenol pyruvate and erythrose-4 phosphate (22, 113). In addition, phosphoenol pyruvate, in its reaction with shikimate-5-phosphate to yield enolpyruvylshikimate-5-phosphate, provides the carbons for the side chain of phenylpropane compounds. Although relatively little is known about regulation of the carbohydrate metabolism in higher plants

(49), particularly in relation to plant disease (169), factors which influence the activity of the pathways which yield erythrose phosphate and phosphoenol pyruvate would obviously have a direct influence upon the synthesis of phenolic compounds.

Studies on microorganisms (91) reveal that several of the enzymes of the shikimic acid pathway are regulatory enzymes. In higher plants, dehydroshikimate reductase from mung bean is inhibited by vanillin, *p*-hydroxybenzoic acid, catechol (6), cinnamic acid, ferulic acid (36), and other phenolic compounds. Pea seedlings contain a chorismate mutase which is inhibited by L-phenylalanine and L-tyrosine (25). Two isozymes of chorismate mutase were isolated from etiolated mung bean seedlings (175). One isozyme which compromises the major part of the chorismate mutase activity in extracts from the plant is inhibited by L-phenylalanine and L-tyrosine, and activated by L-trytophan. The second isozyme is unaffected by these amino acids. The first isozyme therefore appears to be associated with regulation of phenylalanine and tyrosine synthesis in etiolated seedlings of this plant. Loss of regulatory properties of allosteric enzymes involved in aromatic biosynthesis could account in part for increased accumulation of phenolic compounds in diseased plant tissue.

Deamination of aromatic amino acids.—The conversion of L-phenylalanine to *trans*-cinnamic acid, catalyzed by phenylalanine ammonia-lyase (phenylalanine deaminase), provides the phenylpropane skeleton for hydroxylated cinnamic acid derivatives such as caffeic acid, the B ring and 3-carbon bridge of flavonoid compounds such as pisatin (53, 55), and the phenylpropane building blocks for lignin formation (13, 131). The enzyme is widely distributed in plants. Its activity may be regulated *in vivo* in part by *trans*-cinnamate and *p*-coumarate, which are strong inhibitors of the enzyme (63, 85). The enzyme occurs preformed in high concentration in lignifying tissue (131), but appears to be lacking in sweet potato roots. Mechanical injury or colonization by *Ceratocystis fimbriata* results in rapid increase in its activity (104). *Monilinia fructicola* spore suspensions and $CuCl_2$ increased phenylalanine ammonia-lyase activity in pea pods 10- to 12-fold over that in water-treated pods. Application of actinomycin D with *M. fructicola* spore suspensions increased lyase activity 8- to 9-fold over that obtained with the spore suspensions alone (57). Lyase activity increased in excised bean axes (171), gherkins (35), and potato tuber discs (181, 182). In the last tissue, the appearance of lyase activity was stimulated over 3-fold by exposure of the discs to white light (181). The close relation between the activity of the enzyme and phenolic biosynthesis is indicated by studies with sweet potato root tissue (105) and potato tuber slices (181). In pea pods, factors which stimulate pisatin formation cause a rapid appearance of phenylalanine ammonia-lyase activity (57), indicating a close relation of the enzyme to pisatin synthesis.

Tyrosine ammonia-lyase, which catalyzes the conversion of L-tyrosine to *p*-coumaric acid (4-hydroxy-cinnamic acid), exists in high concentrations in grasses in contrast to very low concentrations in dicotyledonous plants (22, 113). In Gramineae, this enzyme and phenylalanine ammonia-lyase provide the phenylpropane carbon skeleton for the synthesis of flavonoids, phenolic phenylpropanes, and lignin (22, 113). The enzyme is synthesized in small amounts in sweet potato root discs in response to colonization by *C. fimbriata* or mechanical injury (104). The enzyme has not been studied intensively in connection with host-parasite interactions, but it may be important in phenolic biosynthesis in host-parasite interactions involving cereal plants.

An enzyme that catalyzes the conversion of 3,4-dihydroxyphenylalanine to caffeic acid has been reported to exist in large amounts in dandelion and in barley and in small amounts in broad bean (96). Little information is available regarding this enzyme, but its possible function in phenolic synthesis in infected plant tissue deserves attention.

Hydroxylation of phenylpropanes.—An enzyme, cinnamate hydroxylase, that catalyzes the conversion of cinnamic acid to *p*-coumaric acid has been purified from spinach (111) and from epicotyls of pea seedlings (134). Light and wounding of plant tissue stimulate production of the enzyme (4). Cinnamate hydroxylase activity is important to phytoalexin synthesis since *p*-coumaric acid appears to be the point where the phenylpropane unit is incorporated into pisatin and phaseollin (53, 84). Phenylalanine is readily converted to tyrosine by phenylalanine hydroxylase in mammals, but ^{14}C-tracer experiments (22, 113) indicate that the enzyme is lacking or at least of little significance in higher plants. Therefore, a report of a highly active phenylalanine hydroxylase from spinach awaits confirmation (112). An enzyme of the so-called phenolase complex appears to function as a hydroxylase that converts monophenols to diphenols (59, 62, 137, 168). Hydroxylase activity undoubtedly plays an important function in the rapid synthesis of dihydric phenols in discs of potato tubers and sweet potato roots (59, 105, 181).

Glycoside and ester formation.—Enzymes exist which utilize uridine diphosphate glucose and convert phenolic compounds to glucosides and glucose esters (126, 163). Such reactions are considered detoxifying mechanisms since administration of phenolics to plants results in rapid conversion to these glucose derivatives. Many phenolic compounds occur in plants as depsides but the manner in which the latter compounds are synthesized is unknown.

The acetate pathway for phenolic biosynthesis.—Head-to-tail condensation of acetate units appears to be involved in the synthesis of ring A of the isoflavone, pisatin, (53, 55, 113) and 3-methyl-6-methoxy-8-hydroxy dihydroisocoumarin (6-methoxy mellein) (21). Probably, malonyl coenzyme A and acetyl coenzyme A are involved as intermediates and the mechanism of synthesis may resemble that of fatty acid synthesis (109, 113). Enzymes involved in the formation of malonyl and acetyl CoA have been characterized in plants (109), but no intermediates leading to the formation of phenolics from these compounds have been demonstrated.

Stimulation and interaction of pathways in the production of phenolic compounds.—During infection of many plant tissues, it is evident from the accumulated products that terpene synthesis via acetate and mevalonate is stimulated (2). Frequently this is accompanied by a simultaneous increase in synthesis of phenylpropane phenolic compounds (106, 161). Similarly, both the acetate-phenol and shikimic acid pathways contribute to the synthesis of pisatin (55, 56).

In tobacco tissue infected by *Pseudomonas solanacearum,* scopoletin and IAA increased markedly (143). Pronounced increases in the amounts of phenylalanine and tryptophan and significant, but less impressive increases in the amount of tyrosine and dihydroxyphenylalanine were observed within 24 hr of inoculation and before disease symptoms appeared (123). Since concentrations of most of the other amino acids decreased during this period, it appeared that the increase in concentration of the aromatic amino acids was not the result of protein breakdown but was due to specific synthesis of these compounds. Particularly significant are the large increases in concentration of phenylalanine and tryptophan which are the precursors, respectively, of scopoletin and IAA. In diseased tobacco tissue, there was a marked increase in activity of phenylalanine ammonia-lyase which provides the cinnamic acid moiety for the synthesis of scopoletin (123). In potato tubers inoculated with *Phytophthora infestans* synthesis of chlorogenic acid was more rapid in resistant than in susceptible tissue (121). Infection of wheat with rust increased synthesis of phenylalanine and tyrosine. The phenomenon was more pronounced in the susceptible interaction than in the resistant combination (44, 129).

Growth substances and phenolic biosynthesis.—Some investigators propose that IAA and other growth substances directly stimulate the production of phenolic compounds in host-parasite interactions. Others, however, believe that the accumulation of phenolic compounds parallels the increased synthesis of IAA because the compounds originate from a common biosynthetic pathway. Both views find support. Thus, the studies on *P. solanacearum* in tobacco (123, 141, 143) indicate that the accumulation of tryptophan precedes accumulation of IAA in infected tissue. In this situation, general stimulation of aromatic amino acid synthesis in response to infection may account for the accumulation of IAA and scopoletin, and the increased production of the former probably does not cause accumulation of the latter. On the other hand, tobacco cuttings to which was administered a solution of IAA contained greater amounts of scopoletin than did untreated cuttings (143). Similarly, 2,4-dichlorophenoxyacetic acid stimulated accumulation of scopoletin in tobacco (33).

The relationship between the ethylene-induced (15) and *C. fimbriata* stimulated (21) production of 6-methoxymellein in carrot roots seems to have been clarified (18, 19). When carrot slices were inoculated with certain isolates of *C. fimbriata,* active ethylene evolution began within 24 hr and continued for 72 hr (18). Accumulation of 6-methoxymellein began in discs 48 hr after inoculation with isolates which effectively induced ethylene production (19). Uninoculated discs evolved little ethylene, but when these discs were exposed to ethylene, 6-methoxymellein accumulation became rapid within 24 hr.

A strong correlation between ethylene production and 6-methoxymellein accumulation also has been established (19). Thus, colonization by certain isolates of *C. fimbriata* appears to initiate ethylene production and the latter, in turn, triggers the production of 6-methoxymellein. An isolate of *Helminthosporium carbonum* yielded similar results. There appears to be no direct relation between the amounts of ethylene produced in culture and those produced in host tissue by the fungus (18). Both the host tissue and the fungus probably synthesize ethylene.

Chlorogenic acid also accumulates in carrot root discs inoculated with *C. fimbriata* (86). In contrast to 6-methoxymellein, chlorogenic acid accumulation seems to occur without a lag period (86). Ethylene also stimulates synthesis of phenylalanine ammonia lyase and accumulation of chlorogenic acid in sweet potato root tissue (72).

Possible mechanisms of the stimulated phenolic biosynthesis in host-parasite interactions.—Pisatin formation was stimulated by heavy metal ions and by *p*-chloromercuribenzoate, iodoacetate, fluoride, azide, cyanide, and thioglycollate. Perhaps sulfhydryl enzymes were inactivated, thus disrupting normal flavonoid metabolism (124). It was proposed that fungal metabolites interfere with normal isoflavonoid synthesis and cause a shift toward the synthesis of pisatin during infection of peas (124).

The effect of inhibitors of protein synthesis on pisatin production has been tested. When a low concentration of cycloheximide was used, pisatin production was greatly enhanced. Actinomycin D greatly stimulated production of pisatin at low concentration but was inhibitory at high concentrations. Other microbial metabolites including puromycin, gliotoxin, phytoactin B, and mitomycin C stimulated synthesis of pisatin (140). Treating pea pods with *M. fructicola* spore suspensions or $CuCl_2$ caused a several-fold increase in phenylalanine ammonia-lyase activity and pisatin. Both increases could be inhibited by cycloheximide (57, 140).

The proposal that the pea genome contains an operon with an adjacent operator site and a polycistronic structural gene where enzymes for pisatin production are coded has been used to explain the phytoalexin response (58). The activity of the operon is regulated by a regulator gene which continuously produces a specific repressor. In healthy pea pods, this repressor occupies the operator site and prevents expression of the genes for pisatin production. Both synthesis and catabolism of the repressor occur. Therefore, if repressor synthesis is inhibited, the quantity of repressor is reduced. As a result, the operator site is freed of repressor, genes for pisatin production are "turned on," and pisatin synthesis ensues. Certain microbial metabolites (for example, cycloheximide, actinomycin D, or substances present in spore suspensions) in low concentrations derepress the pisatin operon by selectively inhibiting synthesis of its repressor. Thus, pisatin production occurs. At high concentrations, selective inhibition is lost, protein synthesis in general is inhibited, and pisatin synthesis is reduced. Cations such as Cu^{+2} and Hg^{+2}, which also stimulate pisatin production, may complex with the repressor and thus derepress the pisatin operon (58). This is in contrast to the proposal that these cations inhibit sulfhydryl enzymes and redirect flavonoid synthesis to the production of pisatin (124).

It appears that the effect of the chemicals in stimulating pisatin production is not to realign an actively functioning pathway, but to activate a new pathway which begins with the conversion of phenylalanine to *trans*-cinnamic acid.

Allen (3) suggested that many of the infection-induced responses are general reactions of the plant for repair of damage. Hence, parasitic and mechanically inflicted injury often initiated the same response in plants although the triggering action is not the same. It therefore seems pertinent to discuss recent studies of the injury-induced rapid increase in phenylalanine ammonia-lyase activity that occurs in certain plant tissues (35, 104, 171, 181). If *de novo* synthesis of the enzyme is involved in these cases inhibitors of protein synthesis should prevent the increase in lyase activity. This was tested using sweet potato root slices. The increase in lyase activity was strongly suppressed by actinomycin D at 10 to 20 μmg/ml and by blasticidin S at 0.02 to 0.1 μg/ml (105). An increase in lyase activity was also effectively suppressed by other inhibitors of protein synthesis: puromycin, *p*-fluorophenylalanine, the respiratory inhibitor antimycin A; and an uncoupler of oxidative phosphorylation, 2,4-dinitrophenol. Cycloheximide at $10 \mu M$ completely inhibited the increase in lyase activity in potato tuber slices, but benzimidazole, inhibitory at 5×10^{-3} *M,* stimulated lyase production at 10^{-4} *M* (181, 182). Cycloheximide (5 μg/ml) and actinomycin D (25 μg/ml) effectively inhibited the development of lyase activity in excised bean axes (171). Thus, the results agree with the idea that *de novo* synthesis of phenylalanine ammonia-lyase is involved, and support the suggestion that synthesis of the enzyme in uninjured potato tuber tissue is strongly repressed (181). Injury, such as that which occurs when the tuber is sliced, derepresses lyase synthesis. Both mechanical and parasitic injury seem to derepress synthesis of this enzyme in sweet potato roots, but parasitic injury seems to intensify the derepression (104). Light also derepresses the enzyme in potato tuber slices (181), but apparently has no effect on lyase synthesis in excised bean axes (171).

Interesting observations have been made on the turnover or inactivation of phenylalanine ammonia-lyase in plant tissue. Lyase activity reached a peak within 15 to 24 hr of excision and then declined to 50 per cent (35, 104, 171, 182). Similarly, lyase activity reached a peak after 24 hr in *M. fructicola*-treated pea pods, and after 48 hr in $CuCl_2$-treated pods. In these cases, however, the peak in activity was followed by a rapid decline to a level less than 10 per cent of the maximum (57). The decline in lyase activity in potato tuber discs (182) and bean axes (171) was virtually eliminated when cycloheximide or actinomycin D was added to the tissue just prior to the expected drop in enzyme activity. Thus the disappearance of lyase activity in these tissues may depend upon the production of a protein, possibly a proteolytic enzyme that destroys lyase, or a polypeptide which inhibits lyase activity, or an enzyme that modifies the active site of lyase (63). Whatever the mechanism involved, the observations provide an example of another means for regulating lyase activity. The phenomenon of rapid lyase synthesis followed by the synthesis of a lyase-inactivating system, termed sequential induction (181, 182), emphasizes the dynamic nature of the injury-induced response in potato tuber tissue and reveals another mechanism by which the production of phenolic compounds can be regulated.

Of relevance to host-parasite interaction is the fact that most of the compounds used in the above-mentioned studies were microbial and plant metabolites. They were effective in concentrations that would be expected to occur under physiologic conditions. They not only have the potential to regulate the production of a particular enzyme in host tissue, but also appear to possess the capacity to regulate the activity of that enzyme.

PHENOL-OXIDIZING ENZYMES

Peroxidase.—In the presence of H_2O_2, this enzyme carries out the oxidation of a variety of compounds, including aliphatic and aromatic amines and phenolic compounds. It catalyzes a H_2O_2-dependent oxidative polymerization of phenylpropane compounds, yielding lignin-like substances (13, 153). Peroxidase also shows catalase activity by its capacity to catalyze the conversion of H_2O_2 to H_2O and O_2 (138). Although H_2O_2 is added in assays for peroxidase activity, peroxidase preparations catalyze a variety of reactions which do not require its addition to the reaction mixture. (In these cases, catalytic amounts of H_2O_2 might be generated by the enzyme system.) For example, preparations of crystalline horseradish peroxidase catalyze a pyridoxal-dependent oxidative decarboxylation of amino acids which yields the amide of the acid having one less carbon atom than the substrate (102). Bean preparations with peroxidase activity catalyze the conversion of oxalacetate to malonate (148). Crystalline horseradish peroxidase catalyzes the oxidation of NADH and NADPH (1, 48), and acts as a hydroxylase in the presence of dihydroxyfumaric acid (138). These preparations also act as an IAA oxidase by catalyzing the conversion of IAA to 3-hydroxymethyl oxindole (68). Peroxidase also appears to be involved in the formation of ethylene from methional and α-keto methylthiobutyric acid (97, 177, 178).

The enzyme is widely distributed in higher plants. Within cells of nonphotosynthetic tissue, it appears to be associated with particulate components as well as the soluble fraction (54). Isozymes of the enzyme occur (146, 152), and the pattern of distribution and quantity of multiple forms often vary (74, 146). Peroxidase isozymes are known to increase in number as a result of disease or injury (77, 152). Shannon (147) isolated seven peroxidase isozymes from horseradish. Although differing in catalytic properties and electrophoretic mobility, all had the capacity to oxidize oxalacetate in the absence of H_2O_2, and all possessed the capacity to oxidize *o*-anisidine in the presence of H_2O_2. However, isozymes which possessed the highest specific activity toward oxalacetate showed the lowest specific activity for *o*-anisidine peroxidation. All of the isozymes showed activity toward IAA. However, in tobacco tissue, purified fractions with high IAA oxidase activity showed very low peroxidase activity towards polyphenols (144). The fractions could contain an apoperoxidase (150) which lacks peroxidase activity but still functions as an IAA oxidase.

Phenolase.—This well-known copper protein catalyzes the oxidation of mono- and ortho-dihydroxy-phenolic compounds to yield quinones. The enzymes occurs in multiple forms that differ in substrate specificity (75, 76). Identity and tissue localization of phenolase substrates have been determined in some cases (94), but generally this aspect of substrate-enzyme compartmentalization has been ignored. Reports have indicated that greater activity of the enzyme occurs in virus-infected tissue (76, 116), tissue colonized by fungi and bacteria (70, 71, 83), and in mechanically injured tissue (70).

Latent forms of the enzyme exist in broad bean leaves, and repeated $(NH_4)_2SO_4$ treatment of broad bean extracts has resulted in a 45-fold increase in phenolase activity (80, 81). The latent phenolase was also activated by treatment of broad bean extracts with acid or alkali (80), with various anionic wetting agents (81, 128), with pectic substances, and with carboxymethylcellulose (31). In many plants, the enzyme occurs bound to chloroplasts, mitochondria, and other particulate fractions (34, 60). Triton X-100 has been used to remove the enzyme from particulate fractions (60, 170).

It seems likely that some of the observed increases in phenolase activity reported in host-parasite interaction are not the result of *de novo* synthesis but come from increased solubilization of phenolase from particulate material or activation of latent phenolase. In other instances, however, *de novo* synthesis of the enzyme appears to be involved (71). Synthesis of phenolase might be stimulated by ethylene (72).

Mechanisms and Modulation of Biological Activity of Phenolics

Possible reactions of peroxidase in host-parasite interactions.—Although the enzyme has been studied in connection with the oxidation of phenolic compounds, other aspects related to its catalytic activities are potentially important to phytopathological studies. An interesting finding is that the enzyme catalyzes the conversion of IAA to 3-hydroxymethyl oxindole (68). The latter, in turn, dehydrates nonenzymically, yielding 3-methylene oxindole (165).

Methylene oxindole is bacteriostatic to *E. coli* and at very low concentrations is stimulatory to certain bacteria (156, 164). The compound is a strong inhibitor of sulfhydryl enzymes and desensitizes certain bacterial regulatory enzymes to feedback inhibition. It also possesses considerable plant growth-promoting activity (164). This suggests that the growth-promoting activity of IAA is due to its conversion to methylene oxindole by peroxidase. Obviously, this idea is contrary to the long-held notion that the enzyme destroys the growth-promoting activity of IAA. Moreover, *E. coli* and peas possess enzyme systems which reduce the biologically active methylene oxidole to the inactive methyloxindole. Thus, the reductase could help regulate the quantity of 3-methylene oxindole (156, 157). The chemical reactivity of methylene oxindole might not explain all the plant responses attributed to IAA, but being biologically active it could interact with phenolic metabolism in infected tissue.

Peroxidase appears to be involved in the formation of lignin in higher plants (13, 153, 154). Potentially at least, lignin formation can act as a barrier to further spread of the invading pathogen (66). A lignin-like substance was isolated from radish roots inoculated with *Alternaria japonica* (5). The amount of lignin-like material extracted was 5 to 10 times as great from inoculated roots as from healthy tissue and boiled-root controls. The extent to which peroxidase activity is enhanced in the inoculated tissue was not determined. Since the fungus used to induce lignin formation normally attacks leaves and not roots of the radish plant, the significance of the induced lignin synthesis in disease resistance in this situation is unclear. It is possible that operation of this defense mechanism in the roots limits the organism to the foliage.

Many investigators have noted a close correlation between disease resistance and peroxidase (40). Peroxidase activity of different organs of the potato plant was correlated positively with resistance of *Phytophthora infestans* (40). Peroxidase activity increased more than ten times as much in rust-resistant wheat varieties as in a susceptible variety. Peroxidase also stimulated germination of self-inhibited wheat rust uredospores (95). The latter effect was greater if both peroxidase and peroxide were added. The nature of these effects is unknown, but it seems possible that the enzyme converts some constituent of the culture medium to a product which functions as the stimulator of germination. Alternatively, destruction of an inhibitor by peroxide and peroxidase might be involved.

Heat-killed cells of *Pseudomonas tabaci* injected into tobacco leaves increased peroxidase activity and resistance to subsequent inoculation with the bacterium (93). After 4 days, two new isozymes of peroxidase were detected in half-leaves injected with heat-killed bacteria, but half-leaves receiving water injections yielded one new isozyme. Half-leaves remained symptomless after 4 days if treated with a solution of peroxidase and then inoculated with the bacterium. However, half-leaves were not protected by injection with water or with a boiled peroxidase preparation.

The possible relationship of reducing power and virulence of pathogens is frequently mentioned (92). If there is a relationship, the capacity of peroxidase to destroy reducing power (e.g., NADH) might account for the correlation of high peroxidase activity of plants and resistance to certain pathogens.

Modulation of peroxidase activity by phenols.—Phenolic compounds possibly regulate the growth-promoting activity of IAA by virtue of their effects on peroxidase actvity. Since the *in vivo* aspects of the problem have been discussed (149), the following discussion of phenols and peroxidase is limited to studies carried out at the enzyme level.

Chlorogenic acid at 5×10^{-6} *M* completely inhibits the oxidative decarboxylation of amino acids by peroxidase (102). Caffeic acid causes 96 per cent inhibition at the same concentration. Phenol at this concentration stimulates the reaction more than fivefold. The oxidation of reduced pyridine nucleotides by peroxidase requires a monophenol, whereas the *o*-dihydroxyphenol, catechol, at 7.6×10^{-6} *M* inhibits the reaction 85 per cent (1). The peroxidase-catalyzed formation of ethylene from methional is stimulated as much as 15-fold by 20 μM phenol (177, 178).

The oxidation of IAA by pineapple preparations is strongly stimulated by 10^{-5} *M* *p*-coumaric acid (52). Chlorogenic acid at 3×10^{-7} *M* is inhibitory, as are caffeic acid, dihydroxyphenylalanine, and dihydroxyphenylpropionic acid at 10^{-6} *M*. Ferulic acid is stimulatory at 5×10^{-6} *M* but inhibitory at 2×10^{-5} *M*. Similarly, scopoletin is stimulatory at 2×10^{-4} *M* but inhibitory at 10^{-3} *M*. In contrast to the quantities needed to inhibit pineapple preparations, inhibition of IAA oxidation by tobacco preparations is complete when scopoletin is used in concentrations from 1 to 9 μg/ml (142). As scopoletin accumulates in tobacco tissue infected by *P. solanacearum,* it effectively inhibits IAA oxidase (142). It was suggested that inhibition of IAA oxidase resulted in reduced catabolism of IAA during pathogenesis and thus accounts for accumulation of the latter in host tissue.

Mono- and ortho-dihydroxy-phenols are widely distributed in plants, generally as glycosides, sugar esters, and depsides (126, 163). Upon conversion to these forms, the acids can change considerably in activity as peroxidase inhibitors or activators. For example, glucoside formation of *p*-coumaric acid catalyzed by UDPG glucosyl transferase (126) would abolish its stimulatory effect since the phenolic group is involved in linkage to glucose. On the other hand, ester and depside formation, which involve the carboxyl group of *p*-coumaric acid, would not diminish its effectiveness as a peroxidase activator. Other enzymes involved in the metabolism of phenolic compounds also might regulate peroxidase activity. Thus a hydroxylase can convert the strong activator, *p*-coumaric acid, to the potent inhibitor, caffeic acid, thereby helping regulate the relative amounts of the two compounds. *o*-Methyltransferase (42) can convert the potent inhibitor, caffeic acid, to ferulic acid, which at low concentration acts as a peroxidase activator and at high concentrations performs as an inhibitor. Thus, the possible interactions are numerous and the interplay of the various phenolic compounds that influence peroxidase activity deserves further attention in studies of host-parasite interaction.

Effect of ethylene on the production of peroxidase.—Ethylene enhanced peroxidase activity in sweet potato root tissue (72, 155). This event paralleled increased resistance to colonization by *C. fimbriata.* Upon exposure to ethylene, root tissue from a susceptible variety of sweet potato became resistant to infection by *C. fimbriata* (155). Similarly, inoculation of sweet potato tissue with certain nonpathogenic strains of *C. fimbriata* induced resistance to subsequent colonization by pathogenic isolates of the fungus. The capacity of nonpathogenic isolates to induce resistance was correlated positively with the capacity to accumulate high levels of ethylene in the atmosphere above sweet potato tissue (155). In contrast, others found that ethylene did not retard development of sweet potato roots by this fungus (18). Mechanical injury (slicing) of sweet potato and carrot root tissue induced slight evolution of ethylene lasting up to about 48 hr (72). Infection by *C. fimbriata,* however, resulted in a high rate of ethylene production lasting over several days. The involvement of peroxidase in ethylene production, discussed previously, and the stimulation of peroxidase synthesis by ethylene suggests that an interesting interplay between the gas and the enzyme might function in injured storage tissue such as roots of carrot and sweet potato.

Little information is available on events concerning substrates and enzymes for ethylene production during early stages after injury. The earliest detectable event in sweet potato is the evolution of ethylene in amounts sufficient to stimulate peroxidase synthesis (72). After the first 6 hr following injury, production of ethylene decreases gradually and the rate of peroxidase synthesis declines. Upon exposure to an exogenous source of ethylene, the accelerated increase in peroxidase activity is re-established.

Reactivity of products formed by phenolase.—Inactivation of phenolase occurs during reaction with *o*-diphenols, such as chlorogenic acid and catechol. Under certain assay conditions, inactivation of the enzyme is complete with 20 min (73). In other cases inactivation is so rapid that it appears as if a particular compound is not being utilized as a substrate (43). The exact mechanism of the inactivation is unknown, but the rapidity of inactivation makes difficult the exact measurement of phenolase activity. Mayer, Harel & Ben-Shaul (101) compared three methods (manometry, spectrophotometry, and polarography) commonly used for phenolase assay, and concluded that polarographic measurements were the most accurate.

The inhibitory activity of the phenolase-polyphenol system is generally attributed to the reactivity of the quinone which the system generates. Antimicrobial activity of quinones is attributed to reaction with proteins or intracellular amino acids, alteration of the cellular redox potential, interference with cofactor and enzyme synthesis, and inhibition of specific enzyme systems (69, 172). Quinones may inhibit enzymes by (*a*) complexing with metal ions which participate in catalysis; (*b*) reaction with sulfhydryl groups by 1,4-addition; (*c*) oxidation of sulfhydryl groups; (*d*) reaction with the substrate or a cofactor; (*e*) production of H_2O_2 during oxidation of polyphenols; (*f*) nonspecific binding to enzymes through the aromatic ring; and (*g*) competition with the substrate (172).

The phenolase-polyphenol system has also been studied in connection with the inactivation of extracellular enzymes produced by pathogens (14, 31, 89) and darkening of host tissue during lesion formation (117, 133). Enzymically generated quinones reacted with N-terminal primary amino groups, aliphatic amino groups, secondary amines, and thiol groups in amino acids (98). Only thiol-containing compounds and aromatic amines reacted with melanogenic quinones derived from 3,4-dihydroxyphenylalanine. From those studies it is inferred that inactivation of an enzyme by enzymically generated quinones would involve addition to an amino or thiol group on the protein essential for catalytic activity.

In high concentration, phenolase catalyzed the oxidation of tyrosine groups on proteins (23, 24, 179). The enzyme catalyzed the oxidation of tyrosyl groups on insulin and caused lysis of rat red blood cells (23, 24). Perhaps phenolase catalyzes the oxidation of tyrosyl residues in membranes of the red blood cells, thereby causing lysis by disrupting membrane permeability (23). It will be of interest to determine if phenolase can affect the membranes of plants and microorganisms.

The phenolase-phenolic compound reaction appears to inactivate plant viruses *in vitro* and possibly in necrosing tissue where virus, phenolase, and phenolase substrates mix. Since inhibitors such as diethyldithiocarbamate can reduce or prevent inactivation of plant viruses *in vitro,* it has been proposed that the quinone generated by the phenolase system inactivates the virus by 1,4-addition to the protein coat. It was reasoned, therefore, that certain completely substituted *ortho* quinones are not likely to participate in such addition reactions, and consequently should be ineffective as virus inactivators (107). Tetrachloro-orthoquinone, however, caused complete inactivation of Tulare apple mosaic virus (107). Partially substituted quinones were less effective in inactivating the virus. Moreover, treatment of the virus with other oxidants induced marked changes in the protein coat, as revealed by alterations in ultraviolet absorption spectra and serological reactivity, though it did not inactivate the virus completely. Those investigators suggested that quinone inactivation of the virus occurs as a result of oxidation of the viral nucleic acid. Not all plant viruses are inactivated by quinones: possibly their ribinucleic acid is not accessible to the oxidant (107). However, there is a report that infectious ribonucleic acid from tobacco mosaic virus was not inactivated by exposure to oxidized chlorogenic acid (103).

Several investigators have shown that products formed by oxidation of *o*-dihydric phenols inactivate enzymes including hydrolytic enzymes produced by plant pathogenic fungi (14, 120). Patil & Dimond (120) proposed that inhibition of *Verticillium albo-atrum* polygalacturonase by the phenolase-phenol system was due to 1,4-addition of quinones to an amino or imino group on the enzyme. However, since the half-lives of *o*-quinone and the quinone of dihydroxyphenylalanine are 0.9 and 0.07 min, respectively (7), rapid addition to proteins or self polymerization would be expected to occur. Inactivation of the polygalacturonase should therefore have occurred more rapidly if it involved 1,4-addition to a functional group essential for catalytic activity.

Effective inactivation of pectinase by products of phenolase might not occur in host tissue since these highly reactive compounds would be free to react with other proteins and amino acids of host or pathogen origin. Highly active polygalacturonase preparations have been obtained from lesions on bean hypocotyls infected with *Rhizoctonia solani* (8). The lesions were in advanced stages of necrosis and contained high phenolase and peroxidase activity (100).

Modulation of phenolase activity.—An NADH and NADPH-coupled quinone reductase has been purified from spinach and other plants (176). Some investigators attach importance to the enzyme as a means of reducing and thereby detoxifying quinones generated by the action of phenolase. However, rapid reduction of *o*-quinones with NADH occurs in the absence of the reductase (176). Other reductants such as ascorbate also reduce *o*-quinones nonenzymically (73). The importance of quinone reductase to reverse the quinone-forming activity of phenolase, therefore, is debatable. Pectinase produced by *Erwinia carotovora* stimulated the production of phenolase in potato tuber tissue (92). Oxidized phenols, formed by phenolase, produced a zone of blackened tissue which functioned as an infection barrier. As the bacterium spread in the tuber tissue, however, the formation of the zone of dark tissue was progressively suppressed. Presumably the bacteria produced a system which reduced the oxidized phenols and prevented them from forming the infection barrier. Earlier, the relationship of reducing power and pathogenicity was investigated by Kelman (79), who used a triphenyl tetrazolium medium to select virulent isolates of *P. solanacearum.* However, colonies which strongly reduced the dye were either weakly virulent or avirulent, whereas those that weakly reduced the dye were highly virulent. Thus, in the case, lack of reducing power was associated with virulence. The resistance of potato tuber to *P. infestans* during the early stages of infection is characterized by increased phenolase activity and simultaneous inhibition of dehydrogenase activity of the host (132). Susceptibility to infection is associated with an increase in dehydrogenase activity of the host during invasion (64).

The involvement of β-glycosidases and phenolase in the response of plants to infection has received considerable attention. Phloretin occurs bound as a β-glucoside (phloridzin) in apple leaves. Upon injury to leaf tissue during invasion by *Venturia inaequalis,* a host β-glucosidase hydrolyzes phloridzin and yields phloretin (115). The latter is oxidized to yield products that arrest the development of the invading pathogen. Since tissue collapse must occur to permit the β-glucosidase-phenolase system to function, resistance may depend upon the rate of tissue collapse in the host-parasite interaction and not necessarily upon the amounts of the substrate and enzymes in host tissue (115). Apple leaf extracts oxidize both phloridzin and phloretin (115, 127). The first oxidation product of phloridzin appears to be 3-hydroxyphloridzin which is then oxidized to its *o*-quinone (127). Because phloridzin oxidation products are toxic to *Venturia inaequalis,* Raa (127) suggested that these materials represent a defense mechanism of the host against the fungus.

Antibiotic activity of various parts of pear trees to *Erwinia amylovora* is associated with arbutin and a high β-glucosidase content of those tissues (67). The enzyme catalyzes the hydrolysis of arbutin to yield hydroquinone. The latter is toxic to the organism, possibly as a semiquinone which may be formed by phenolase action or by autoxidation (67). Furthermore, hydroquinone is oxidized to form quinones which polymerize and lose toxicity. Thus, the interplay of phenolase and β-glucosidase affects hydroquinone concentration in pear tissue and is undoubtedly a factor that affects susceptibility of pear tissue to fire blight (67a). The balance between appearance and disappearance of hydroquinone in the arbutin-hydroquinone system also appears to be a factor that controls intervarietal resistance of pears to *E. amylovora* (151). Arbutin need not be hydrolyzed to hydroquinone for enzymic oxidation to occur. Leaves of *Pyrus* species contain two independent pathways of arbutin oxidation. One of these converts arbutin to 3,4-dihydroxyphenyl-β-D-glucoside, and the latter to a quinone which polymerizes to form melanin-like pigments (62).

Production and destruction of indole-3-acetic acid by phenolase systems.—Interactions in incubation mixtures containing phenolase, tryptophan, and an *ortho* dihydroxy phenol result in the production of IAA (51, 94, 99). Reactants present in such reaction mixtures should produce complexes between IAA and phenols (90). Although it is possible that conditions favoring such interactions could exist in infected tissue, a positive correlation between IAA formation and phenolase activity could not be demonstrated in potato tuber tissue (41). IAA was destroyed when incubated with phenolase and catechol (12) although no destruction occurred upon incubation with chlorogenic acid and phenolase (41).

Toxicity of phenols.—Information on the rate at which pisatin, phaseollin, and 6-methoxy mellein are formed and on their toxicity to various fungi has been reviewed (86–88). Therefore, only recent results are discussed here. In carrot root slices inoculated with *C. fimbriata,* the rate of accumulation of 6-methoxy mellein lagged during the first 48 hr after inoculation (86). Thereafter, the compound accumulated rapidly approximating 8 mg/g fresh weight of carrot root per 24 hr. When incorporated into potato-dextrose agar at 10^{-3} M (0.2 mg/ml), the compound inhibited the growth of *C. fimbriata* 98 per cent (65). Thus, the amount of 6-methoxy mellein that accumulated in carrot root discs inoculated with the organism was well above the concentration needed to inhibit the growth of the organism in culture. On the other hand, certain *C. fimbriata* isolates readily colonize carrot root discs (19). These isolates also induced greater accumulation of 6-methoxy mellein and greater production of ethylene. Nevertheless, such isolates were inhibited by 6-methoxy mellein incorporated into potato-dextrose agar (PDA). Thus it appears that the capacity of the compound to inhibit growth of the fungus in culture does not relate to its capacity to inhibit growth of the organism in carrot tissue. Moreover, since carrot root discs are now used for the selective isolation of *C. fimbriata* (108), the proposed function of 6-methoxy mellein as a resistance factor appears to be in doubt.

In bean hypocotyls infected with *Rhizoctonia solani,* phaseollin, absent in healthy hypocotyl tissue, accumulated to 0.5 mg/g fresh weight in young lesions and to 2.5 mg/g fresh weight in mature lesions (125). These concentrations are well above the ED_{50} of 13 to 25 μg/ml for the compound in PDA and suggest that phaseollin helps limit development of the fungus in necrosing hypocotyl tissue. Two strains of *Aschochyta pisi* were tested on several pea cultivars susceptible and semi-resistant to the organism (29). Generally, there was an inverse correlation between disease severity and pisatin concentration in endocarp diffusates, indicating a role for pisatin in limiting growth of the fungus. At 1 per cent O_2 tension, concentrations of pisatin in pea-endocarp diffusates and phaseollin in bean-pod diffusates were less than their ED_{50} values for *M. fructicola* and the fungus grew well on the pods. At higher O_2 tensions, yields of pisatin and phaseollin increased and paralleled inhibition of fungal growth (30). Perhaps the differential oxygen requirements of the fungus for growth and of the host for phytoalexin production could be important in relation to host reactions as in root-fungal interactions (30). It will be of interest to extend these studies to such host-parasite combinations.

The mechanism of the activating effect of monophenols on peroxidase may be associated with the capacity to participate in the oxidation-reduction reaction during catalysis (1). Inhibition of the enzyme by *o*-dihydroxyphenols might be due to complexing with the heme iron, thus preventing oxidation of the substrate. Inhibition appears to be noncompetitive in this case (102). Phenolic compounds are known to inhibit other enzymes. Thus, chlorogenic acid inhibits potato tuber phosphorylase (139), transaminases (11), and dihydroxyphenylalanine (DOPA) decarboxylase (61). Ferulate and *p*-coumarate noncompetitively inhibit apple phenolase (145). Phloretin and phloridzin inhibit glucose-6-phosphate phosphohydrolase from rat liver mitochondria (180). Thus phenolic compounds may posses the potential to inhibit essential enzymes of plants or plant pathogens, and studies to explore this possibility should be made. The situation where necrotic lesions on potato tubers inoculated with *P. infestans* yielded phenolic compounds which uncoupled oxidative phosphorylation in tuber mitochondria (118) may reflect this possibility. In general, phenolic compounds are more toxic to plant pathogens in their quinone forms.

Little is known about the mechanism of toxicity of pisatin and other phytoalexins toward plant pathogens. Several compounds related to pisatin (e.g., coumestrol, formononetin, biochanin A) are known to increase in diseased plant tissue and are toxic to mammals (10).

Catabolism of toxic phenolic compounds by host and pathogens.—Little is known about the metabolism of the phenolic phytoalexins. No reports have appeared on the turnover (or lack of it) of phenolic phytoalexins in plants other than those which indicate that pisatin is metabolically inert in the pea plant (28, 56). Studies of ethylene effects on 6-methoxy mellein accumulation in carrot root slices suggest that the compound is metabolically active in this tissue. This is indicated by the drop in 6-methoxy mellein content which occurs in the slices when exposure to ethylene is stopped (19).

Certain plant pathogenic fungi degrade pisatin readily (166). For example, *Fusarium solani* and *F. oxysporum* (114), *Mycosphaerella pinoides, F. oxysporum* f. *pisi,* and *Cladosporium cucumerinum* (32) apparently possess this capacity whereas *M. fructigena* and *Colletotrichum lindemuthianum* do not (32). The degradation mechanism is unknown, but possession of this capacity presumably permits the fungus to overcome the resistance brought about by pisatin (166). *A. pisi,* which is relatively insensitive to pisatin (28), apparently accumulates but does not metabolize the chemical.

Phenolics and Hypersensitivity

Rapid tissue necrosis and death of cells (hypersensitivity) are common in interactions between plants and certain pathogens (83, 100, 110, 119). This phenomenon in connection with bacterial plant pathogens has been discussed in detail (83). The onset of the hypersensitive reaction is accompanied by increased activity of oxidative enzymes, increased synthesis of phenolic compounds (100, 129, 132, 160), and darkening of necrosing tissue (133). During the invasion of potato tuber cells by *P. infestans,* about 30 per cent of the cells died within 20 min of penetration by the hyphae of an incompatible race of the fungus but remained alive 2 days or longer when penetrated by hyphae of a compatible race (158, 159). Despite cell death, hyphal growth of both races continued at the same rate for at least 2.5 hr after cell penetration. With the incompatible-fungus hypersensitive-host combination, subsequent darkening of the cellular consituents occurred, and accelerated synthesis of phenols was observed in a zone 10 to 15 cells deep surrounding the point of infection (135). Hyphal development was arrested. Thin tuber slices from the hypersensitive variety were susceptible to heavy inoculation with the incompatible race, but regained resistance when treated with solutions of chlorogenic acid, caffeic acid, catechol, guaiacol, and hydroquinone (136). Similarly treated tuber discs from a susceptible variety showed little resistance to the incompatible race. Phenolic compounds appear to contribute to resistance of the tuber to infection by the incompatible race of the fungus. A second factor, present in the resistant but absent in the susceptible tissue appears to convert phenolic compounds to a form more toxic to the fungus. Nonphenolic compounds, such as rishitin, also appear to contribute to resistance to the incompatible race of the fungus (162). Lesion maturation in bean hypocotyls infected by *R. solani,* was accompanied by an increase in peroxidase, catalase, cytochrome oxidase, and phenolase activity, particularly during the mid and latter stages of lesion formation (100). This activity presumably resulted from "degenerative metabolism" associated with death of the host cells.

If the hypersensitive reaction results from activity of oxidative enzymes, it should be inhibited by reducing compounds either applied externally or generated within the host tissue. Thus, fewer TMV lesions were formed on *N. glutinosa* leaves which had been infiltrated with ascorbic acid before inoculation (38, 119). Results were similar with tobacco necrosis virus on *N. tabacum* var. Samsun. Infected *N. glutinosa* leaves consumed 50 per cent more ascorbate than did healthy leaves. The system that consumed ascorbate was not identified, but the findings imply that quinones generated during lesion formation accounted for the greater consumption of ascorbate in infected tissue. Continuous exposure to ascorbate was required for lesion formation to be reduced because the compound was rapidly consumed by the plant (38). Lesions which formed on ascorbate-treated *N. glutinosa* leaves were larger but less pigmented than those that formed on untreated leaves (119). It was suggested that the increased reducing power detected in cells surrounding TMV-induced lesions on *N. glutinosa* leaves hinders movement of quinones from the infection court (38). Lesion size, therefore, should have been restricted by the enhanced reducing power supplied by ascorbate (119), but such was not the case. On Pinto hean leaves inoculated with TMV, sites of lesion formation were detected as fluorescing spots several hours before necrosis of tissue (103). Ascorbate concentration increased after inoculation and subsequently decreased rapidly due to oxidation to dehydroascorbate. However, rapid ascorbate oxidation occurred long after the area of lesion formation was delimited, indicating no relation of ascorbate oxidation to restriction of lesion size.

In general, it appears that increased activity of certain metabolic pathways and alterations in others occur in necrosing tissue and in cells surrounding those undergoing lesion formation (9, 135, 160, 161). At all stages of lesion maturation in the *R. solani*-bean hypocotyl combination, respiration increased in infected tissue and in uninfected tissue below the lesions (9). The pentose phosphate shunt appeared to be the predominant pathway for glucose catabolism in the lesion itself and in the cells below the developing lesion. This pathway would provide erythrose 4-phosphate and NADPH needed for the synthesis of phenolic compounds. Browning of potato tuber discs induced by treatment with chlorogenic and ascorbic acid was accompanied by a shift in glucose catabolism from the Embden-Meyerhoff to the pentose phosphate pathway (161). Enhanced reduction of triphenyl tetrazolium around TMV-induced lesions on *N. glutinosa* was attributed to increased dehydrogenase activity in cells surrounding those undergoing necrosis (38). Increased activity of the pentose phosphate shunt in these cells could provide this reducing power in the form of glucose-6-phosphate dehydrogenase and NADPH. In accord with its proposed function *in vivo* the latter would also be utilized for increased synthetic activity that surrounds the lesion.

TMV causes local lesions on *N. glutinosa* at 25° C but infects the plant systemically at 36° C. When plants infected systemically at 36° C were returned to 25° C, the virus-infected portions rapidly necrosed and the entire plant quickly collapsed (78). One of the earliest detectable changes in cell ultrastructure in plants undergoing systemic collapse is rupture of membranes of various cell organelles such as chloroplast membranes (17). Results were similar during lesion formation in various virus-infected plants (16, 173, 174).

The involvement of phenolase and phenolic compounds appears to be secondary to the event that triggers necrosis. The triggering event might involve a change in membrane structure leading to altered membrane permeability (50, 83, 174) and loss of cell compartmentalization. Once this occurs, regulation of cellular metabolism would be lost, phenolase could be activated, and this and other oxidative and hydrolytic enzymes and their substrates, previously separated by compartmentalization, would come together. The resulting degenerative activities would then lead to the formation of lesions.

Concluding Remarks

A review of papers on this subject revealed a number of instances where techniques are misused. For example, many investigators apparently assume that published assay procedures for measurement of an enzyme from one organism are optimum for assaying the enzyme from another. That is a faulty assumption since the properties of an enzyme frequently vary considerably from source to source. Some investigators have taken measurements after long incubation periods although the reaction is obviously no longer linear with time. In such cases, true reaction rates are not obtained. Also, as noted earlier, certain enzymes, such as phenolase, undergo rapid inactivation during reaction. Long incubations cannot be used and calculation of true enzyme activity requires correction for loss of activity. The latter has been done by certain investigators (167). There are also instances where investigators have used potent enzyme preparations, producing rates of reaction too rapid to measure accurately. A simple dilution of the enzyme preparation would have solved the problem. In addition, enzymes with broad substrate specificity were encountered frequently although only a single substrate was used to assay activity. Such enzymes, depending upon their source, frequently show unusual differences in activity toward substrates. There is also a marked tendency for investigators to neglect to identify the product of a reaction being investigated. This procedure is important where enzyme assays depend on techniques that are relatively nonspecific for the product.

One must always bear in mind that enzyme assays are (and should be) conducted under optimum conditions for quantitating the amount of enzyme present in a preparation. Such information reveals the potential amount of enzyme activity available, though it does not necessarily reveal the manner in which it functions *in vivo*. Actual activity *in vivo* depends on the availability and concentration of substrate and cofactor, the temperature and pH of the environment, the presence of competing enzymes, and numerous other factors.

Many of the compounds and enzymes described in this report have been investigated for toxicity to plant pathogens. However, evidence is generally lacking to establish their role unequivocally as disease-resistance factors. Moreover, with few exceptions, investigators have given little consideration to other possible functions that phenolic compounds might have in host parasite interactions. Perhaps the most interesting outcome of the studies reported here relates to protein changes in the host as a result of interaction with the pathogen. This information might eventually be applicable to numerous interactions that do not involve phenolic compounds. As far as phytoalexins are concerned, more studies are required on systems that most closely approximate the natural host-parasite interaction. The *A. pisi*-pisatin-pea system and the *R. solani*-phaseollin-bean interaction deserve further investigation. However, best suited in this regard might be a system of *Phytophthora megasperma* var. *sojae* and soy bean, studied by Klarman & Gerdemann (82) and Paxton & Chamberlain (20, 122). A crucial step in the progress of these studies will be the identification of the phytoalexins involved.

Hopefully, future work in this area will emphasize the mechanism of action of the compounds proposed to function as phytoalexins. Elucidation of pathways and characterization of enzymes leading to synthesis of the phytoalexins are awaited. Knowledge of genetic control over the enzymes of this pathway is urgently needed and will prove or disprove ideas set forth on the induction of pisatin biosynthesis. Further attention should be paid to elucidating the mechanism underlying the observed increase in phenolase and peroxidase in infected tissue. Rigorous proof is needed to establish the identity of many of the phenolic compounds alleged to be involved in defense responses to infection. More studies are needed on the mechanism of action of peroxidase, if indeed it is involved in defense reactions of the host to the potential pathogen.

It is hoped that investigators, although confronted by numerous pitfalls, are not discouraged from this area of research. It remains one of the most challenging aspects of the host-parasite interaction. Discoveries in the future should contribute significantly to the nature of the host-parasite interaction and to the understanding of responses evoked in plants by plant pathogens.

LITERATURE CITED

1. Akazawa, T., Conn, E. E. The oxidation of reduced pyridine nucleotides by peroxidase. *J. Biol. Chem.*, **232,** 403–15 (1958)
2. Akazawa, T., Uritani, I., Akazawa, Y. Biosynthesis of Ipomeamarone. I. The incorporation of acetate-2-C^{14} and mevalonate-2-C^{14} into ipomeamarone. *Arch. Biochem. Biophys.*, **99,** 52–59 (1962)
3. Allen, P. J. Physiology and biochemistry of defense. In *Plant Pathology,* **1,** 435–67 (Horsfall, J. G., Dimond, A. E., Eds., Academic Press, New York, 674 pp., 1959)
4. Amrhein, N., Zenk, M. H. Induction of cinnamic acid 4-hydroxylase by light and wounding. *Naturwissenschaften,* **55,** 394 (1968)
5. Asada, Y., Matsumoto, I. Formation of lignin in the root tissues of Japanese radish affected by *Alternaria japonica. Phytopathology,* **57,** 1339–43 (1967)
6. Balinsky, D., Davies, D. D. Aromatic biosynthesis in higher plants. 2. Mode of attachment of shikimic acid and dehydroshikimic acid to dehydroshikimic reductase. *Biochem. J.*, **80,** 296–300 (1961)
7. Ball, E. G., Chen, T. T. Studies on oxidation-reduction. XX. Epinephrine and related compounds. *J. Biol. Chem.*, **102,** 691–719 (1933)
8. Bateman, D. F. Pectolytic activities of culture filtrates of *Rhizoctonia solani* and extracts of *Rhizoctonia*-infected tissues of bean. *Phytopathology,* **53,** 197–204 (1963)
9. Bateman, D. F., Daly, J. M. The respiratory pattern of *Rhizoctonia*-infected bean hypocotyls in relation to lesion maturation. *Phytopathology,* **57,** 127–31 (1967)
10. Bickhoff, E. M. Flavonoid estrogens in plants. *Am. Perfumer Cosmet.*, **83,** 59–62 (1968)
11. Braunstein, E. A. Transamination and the integrative functions of the dicarboxylic acids in nitrogen metabolism. In *Advances in Protein Chemistry,* **3,** 1–52b, (Anson, M. L., Edsall, J. T., Eds., Academic Press, New York, 524 pp., 1947)
12. Briggs, W. R., Ray, P. M. An auxin inactivation system involving tyrosinase. *Plant Physiol.*, **31,** 165–67 (1956)
13. Brown, S. A. Lignins. *Ann. Rev. Plant Physiol.*, **17,** 223–44 (1966)
14. Byrde, R. J. W. Natural inhibitors of fungal enzymes and toxins in disease resistance. In *Perspectives of Biochemical Plant Pathology,* (Conn. Agr. Expt. Sta. Bull., **663,** 31–41 (Rich, S., Ed., New Haven, Conn., 191 pp., 1963)
15. Carlton, B. C., Peterson, P. E., Tolbert, N. E. Effects of ethylene and oxygen on production of a bitter compound by carrot roots. *Plant Physiol.*, **36,** 550–52 (1961)
16. Carroll, T. W. Lesion development and distribution of tobacco mosaic virus in *Datura stramonium. Phytopathology,* **56,** 1348–53 (1966)
17. Carroll, T. W., Kosuge, T. Changes in structure of chloroplasts accompanying necrosis of tobacco leaves systematically infected with tobacco mosaic virus. *Phytopathology* (In press)
18. Chalutz, E., DeVay, J. E. Production of ethylene in vitro and in vivo by *Ceratocystis fimbriata* in relation to disease development. *Phytopathology* (In press)
19. Chalutz, E., DeVay, J. E., Maxie, E. C. Ethylene-induced isocoumarin formation in carrot root tissue. *Plant Physiol.* (In press)
20. Chamberlain, D. W., Paxton, J. D. Protection of soybean plants by phytoalexin. *Phytopathology,* **58,** 1349–50 (1968)
21. Condon, P., Kuć, J., Draudt, H. N. Production of 3-methyl-6-methoxy-8-hydroxy-3,4-dihydroisocoumarin by carrot root tissue. *Phytopathol-*

ogy, **53,** 1244–50 (1963)

22. Conn, E. E. Enzymology of phenolic biosynthesis. In *Biochemistry of Phenolic Compounds,* 399–435 (Harborne, J. B., Ed., Academic Press, New York, 618 pp., 1964)
23. Cory, J. G. Evidence for a role of tyrosyl residues in cell membrane permeability. *J. Biol. Chem.,* **242,** 218–21 (1967)
24. Cory, J. G., Bigelow, C. C., Frieden, E. Oxidation of insulin by tyrosinase. *Biochemistry,* **1,** 419–22 (1962)
25. Cotton, R. G. H., Gibson, F. The biosynthesis of phenylalanine and tyrosine in the pea (*Pisum sativum*): chorismate mutase. *Biochim. Biophys. Acta,* **156,** 187–89 (1968)
26. Cruickshank, I. A. M. Phytoalexins. *Ann. Rev. Phytopathol.,* **1,** 351–74 (1963)
27. Cruickshank, I. A. M., Perrin, D. R. Pathological function of phenolic compounds in plants. In *Biochemistry of Phenolic Compounds,* 511–44 (Harborne, J. B., Ed., Academic Press, New York, 618 pp., 1964)
28. Cruickshank, I. A. M., Perrin, D. R. Studies on Phytoalexins VIII. The effect of some further factors on the formation, stability and localization of pisatin *in vivo. Australian J. Biol. Sci.,* **18,** 817–28 (1965)
29. Cruickshank, I. A. M., Perrin, D. R. Studies on phytoalexins IX. Pisatin formation by cultivars of *Pisum sativum* L. and several other *Pisum* species. *Australian J. Biol. Sci.,* **18,** 829–35 (1965)
30. Cruickshank, I. A. M., Perrin, D. R. Studies on phytoalexins. X. Effect of oxygen tension on the biosynthesis of pisatin and phaseollin. *Phytopathol. Z.,* **60,** 335–42 (1967)
31. Deverall, B. J., Wood, R. K. S. Chocolate spot of beans (*Vicia faba* L.)—interactions between phenolase of host and pectic enzymes of the pathogen. *Ann. Appl. Biol.,* **49,** 473–87 (1961)
32. DeWit-Elshove, A. Breakdown of pisatin by some fungi pathogenic to *Pisum sativum. Neth. J. Plant Pathol.,* **74,** 44–47 (1968)
33. Dieterman, L. J., Lin, C. Y., Rohrbaugh, L., Thiesfeld, V., Wender, S. H. Identification and quantitative determination of scopolin and scopoletin in tobacco plants treated with 2,4-dichlorophenoxyacetic acid. *Anal. Biochem.,* **9,** 139–45 (1964)
34. Drawert, F., Gebbing, H. Bound phenoloxidase and its release from plant cell structures. *Naturwissenschaften,* **54,** 226–27 (1967)
35. Engelsma, G. Effect of cycloheximide on the inactivation of phenylalanine diaminase in gherkin seedlings. *Naturwissenschaften,* **54,** 319–20 (1967)
36. Farkas, G. L. A comparison of biochemical changes associated with virus-induced lesion formation and mechanical injury. In *Biochemical Regulation in Diseased Plants or Injury,* 123–35 (Hirai, T., Ed., Kyoritsu Printing Co., Ltd., Tokyo, 351 pp., 1968)
37. Farkas, G. L., Király, Z. Role of phenolic compounds in the physiology of plant diseases and disease resistance. *Phytopathol. Z.,* **44,** 105–50 (1962)
38. Farkas, G. L., Király, Z., Solymosy, F. Role of oxidative metabolism in the localization of plant viruses. *Virology,* **12,** 408–21 (1960)
39. Farkas, G. L., Solymosy, F. Host metabolism and symptom production in virus-infected plants. *Phytopathol. Z.,* **53,** 85–93 (1965)
40. Fehrmann, H., Dimond, A. E. Peroxidase activity and *Phytophthora* resistance in different organs of the potato plant. *Phytopathology,* **57,** 69–72 (1967)
41. Fehrmann, H., Dimond, A. E. Studies on auxins in the *Phytophthora* disease of the potato tuber. I. Role of indole-acetic acid in pathogenesis. *Phytopathol. Z.,* **59,** 83–100 (1967)
42. Finkle, B. J., Masri, M. S. Methylation of polyhydroxy aromatic compounds by pampas grass *O*-methyltransferase. *Biochim. Biophys. Acta,* **85,** 167–69 (1964)
43. Fling, M., Horowitz, N. H., Heinemann, S. F. The isolation and properties of crystalline tyrosinase from *Neurospora. J. Biol. Chem.,* **238,** 2045–53 (1963)

44. Fuchs, A., Rohringer, R., Samborski, D. J. Metabolism of aromatic compounds in healthy and rust-infected primary leaves of wheat. II. Studies with L-phenylalanine-U-^{14}C, L-tyrosine-U-^{14}C, and ferulate-U-^{14}C. *Can. J. Botany,* **45,** 2137–53 (1967)
45. Gamborg, O. L. Aromatic metabolism in plants. III. Quinate dehydrogenase from mung bean cell suspension cultures. *Biochim. Biophys. Acta,* **128,** 483–91 (1966)
46. Gamborg, O. L. Aromatic metabolism in plants. II. Enzymes of the shikimate pathway in suspension cultures of plant cells. *Can. J. Biochem.,* **44,** 791–99 (1966)
47. Gamborg, O. L., Keeley, F. W. Aromatic metabolism in plants. I. A study of the prephenate dehydrogenase from bean plants. *Biochim. Biophys. Acta,* **115,** 65–72 (1966)
48. Gamborg, O. L., Wetter, L. R., Neish, A. C. The role of plant phenolic compounds in the oxidation of reduced diphosphopyridine nucleotide by peroxidase. *Can. J. Biochem. Physiol.,* **39,** 1113–24 (1961)
49. Gibbs, M. Carbohydrates: their role in plant metabolism and nutrition. In *Plant Physiology,* Vol. IVB, 3–115 (Steward, F. C., Ed., Academic Press, New York, 599 pp., 1966)
50. Goodman, R. N. The hypersensitive reaction in tobacco: a reflection of changes in host cell permeability. *Phytopathology,* **58,** 872–73 (1968)
51. Gordon, S. A., Paleg, L. G. Formation of auxin from tryptophan through action of polyphenols. *Plant Physiol.,* **36,** 838–45 (1961)
52. Gortner, W. A., Kent, M. J. The coenzyme requirement and enzyme inhibitors of pineapple indoleacetic acid oxidase. *J. Biol. Chem.,* **233,** 731–35 (1958)
53. Grisebach, H., Barz, W., Hahlbrock, K., Kellner, S., Patschke, L. Recent investigations on the biosynthesis of flavonoids. In *Biosynthesis of Aromatic Compounds,* 25–36 (Billek, G., Ed., Pergamon Press, New York, 142 pp., 1966)
54. Hackett, D. P., Ragland, T. E. Oxidation of menadiol by fractions isolated from non-photosynthetic plant tissues. *Plant Physiol.,* **37,** 656–62 (1962)
55. Hadwiger, L. A. The biosynthesis of pisatin. *Phytochemistry,* **5,** 523–25 (1966)
56. Hadwiger, L. A. Changes in phenylalanine metabolism associated with pisatin production. *Phytopathology,* **57,** 1258–59 (1967)
57. Hadwiger, L. A. Changes in plant metabolism associated with phytoalexin production. *Neth. J. Plant Pathol.,* **74,** suppl. 1, 163–69 (1968)
58. Hadwiger, L. A., Schwochau, M. E. Host resistance responses—an induction hypothesis. *Phytopathology,* **59,** 223–27 (1969)
59. Hanson, K. R., Zucker, M. The biosynthesis of chlorogenic acid and related conjugates of the hydroxycinnamic acids: chromatographic separation and characterization. *J. Biol. Chem.,* **238,** 1105–15 (1963)
60. Harel, E., Mayer, A. M., Shain, Y. Purification and multiplicity of catechol oxidase from apple chloroplasts. *Phytochemistry,* **4,** 783–90 (1965)
61. Hartman, W. J., Akawie, R. I., Clark, W. G. Competitive inhibition of 3,4-dihydroxyphenylalanine (DOPA) decarboxylase *in vitro. J. Biol. Chem.,* **216,** 507–29 (1955)
62. Hattori, S., Sato, M. The oxidation of arbutin by isolated chloroplasts of arbutin-containing plants. *Phytochemistry,* **2,** 385–95 (1963)
63. Havir, E. A., Hanson, K. R. L-Phenylalanine ammonia-lyase. II. Mechanism and kinetic properties of the enzyme from potato tubers. *Biochemistry,* **7,** 1904–14 (1968)
64. Held, A., Kedar, N., Birk, Y. Dehydrogenase activity of potato tuber tissue infected with *Phytophthora infestans. Phytopathology,* **55,** 970–76 (1965)
65. Herndon, B. A., Kuć, J., Williams, E. B. The role of 3-methyl-6-methoxy-8-hydroxy-3,4-dihydroisocoumarin in the resistance of carrot root to *Ceratocystis fimbriata* and *Thielaviopsis basicola. Phytopathology,* **56,** 187–91 (1966)
66. Hijwegen, T. Lignification, a possible mechanism of active resistance against pathogens. *Neth. J. Plant Pathol.,* **69,** 314–17 (1963)
67. Hildebrand, D. C., Schroth, M. N.

Antibiotic activity of pear leaves against *Erwinia amylovora* and its relation to β-glucosidase. *Phytopathology,* **54,** 59–63 (1964)

67a. Hildebrand, D. C., Schroth, M. N. Arbutin-hydroquinone complex in pear as a factor in fire blight development. *Phytopathology,* **54,** 640–45 (1964)

68. Hinman, R. L., Lang, J. Peroxidase-catalyzed oxidation of indole-3-acetic acid. *Biochemistry,* **4,** 144–58 (1965)

69. Hoffman-Ostenhof, O. Enzyme inhibition by quinones. In *Metabolic Inhibitors,* Vol. II, Chap. 22, 145–59 (Hochster, R. M., and Quastel, J. H., Eds., Academic Press, New York, 753 pp., 1963)

70. Hyodo, H., Uritani, I. σ-Diphenol oxidases in sweet potato infected by the black rot fungus. *J. Biochem. (Japan),* **57,** 161–66 (1965)

71. Hyodo, H., Uritani, I. The inhibitory effect of some antibiotics on increase in σ-diphenol oxidase activity during incubation of sliced sweet potato tissue. *Agr. Biol. Chem.,* **30,** 1083–86 (1966)

72. Imaseki, H., Asahi, T., Uritani, I. Investigations on the possible inducers of metabolic changes in injured plant tissues. In *Biochemical Regulation in Diseased Plants or Injury,* 189–201 (Hirai, T., Ed., Kyoritsu Printing Co., Ltd., Tokyo, 351 pp., 1968)

73. Ingraham, L. L. Reaction inactivation of polyphenol oxidase: catechol and oxygen dependence. *J. Am. Chem. Soc.,* **77,** 2875–76 (1955)

74. Ivanova, T. M., Danydova, M. A., Rubin, B. A. Mitochondrial peroxidase and its possible role in oxidative processes. *Biokhimiya,* **31,** 1167–73 (1966)

75. Jolley, R. L., Jr., Mason, H. S. The multiple forms of mushroom tyrosinase-interconversion. *J. Biol. Chem.,* **240,** PC 1489-91 (1965)

76. John, V. T., Weintraub, M. Phenolase activity in *Nicotiana glutinosa* infected with tobacco mosaic virus. *Phytopathology,* **57,** 154–58 (1967)

77. Kanazawa, Y., Shichi, H., Uritani, I. Biosynthesis of peroxidases in slices of black rot-infected sweet potato roots. *Agr. Biol. Chem.,* **29,** 840–47 (1965)

78. Kassanis, B. Some effects of high temperature on the susceptibility of plants to infection with viruses. *Ann. Appl. Biol.,* **39,** 358–69 (1952)

79. Kelman, A. The relationship of pathogenicity in *Pseudomonas solanacearum* to colony appearance on a tetrazolium medium. *Phytopathology,* **44,** 693–95 (1954)

80. Kenten, R. H. Latent phenolase in extracts of broad-bean (*Vicia faba* L.) leaves. 1. Activation by acid and alkali. *Biochem. J.,* **67,** 300–7 (1957)

81. Kenten, R. H. Latent phenolase in extracts of broad-bean (*Vicia faba* L.) leaves. 2. Activation by anionic wetting agents. *Biochem. J.,* **68,** 244–51 (1958)

82. Klarman, W. L., Gerdemann, J. W. Resistance of soybeans to three *Phytophthora* species due to the production of a phytoalexin. *Phytopathology,* **53,** 1317–20 (1963)

83. Klement, Z., Goodman, R. N. The hypersensitive reaction to infection by bacterial plant pathogens. *Ann. Rev. Phytopathol.,* **5,** 17–44 (1967)

84. Kosuge, T. Possible routes of biosynthesis of phenolics in diseased tissue of higher plants. In *Phenolics in Normal and Diseased Fruits and Vegetables,* 83–102 (Runeckles, V. C., Ed., Imperial Tobacco Co., Montreal, 102 pp., 1964)

85. Koukol, J., Conn, E. E. The metabolism of aromatic compounds in higher plants. IV. Purification and properties of the phenylalanine deaminase of *Hordeum vulgare. J. Biol. Chem.,* **236,** 2692–98 (1961)

86. Kuć, J. Phenolic compounds and disease resistance in plants. In *Phenolics in Normal and Diseased Fruits and Vegetables,* 63–81 (Runeckles, V. C., Ed., Imperial Tobacco Co., Montreal, 102 pp., 1964)

87. Kuć, J. Resistance of plants to infectious agents. *Ann. Rev. Microbiol.,* **20,** 337–70 (1966)

88. Kuć, J. Biochemical control of disease resistance in plants. *World Rev. Pest Control,* **7,** 42–55 (1968)

89. Kuć, J., Williams, E. B., Maconkin, M. A., Ginzel, J., Ross, A. F.,

Freedman, L. J. Factors in the resistance of apple to *Botryosphaeria ribis*. *Phytopathology,* **57,** 38–42 (1967)

90. Leopold, A. C., Plummer, T. H. Auxin-phenol complexes. *Plant Physiol.,* **36,** 589–92 (1961)
91. Lingens, F. The biosynthesis of aromatic amino acids and its regulation. *Angew. Chem.* (Intern. Edition), **7,** 350–60 (1968)
92. Lovrekovich, L., Lovrekovich, H., Stahmann, M. A. Inhibition of phenol oxidation by *Erwinia carotovora* in potato tuber tissue and its significance in disease resistance. *Phytopathology,* **57,** 737–42 (1967)
93. Lovrekovich, L., Lovrekovich, H., Stahmann, M. A. The importance of peroxidase in the wildfire disease. *Phytopathology,* **58,** 193–98 (1968)
94. Mace, M. E. Phenols and their involvement in fusarium wilt pathogenesis. In *Phenolics in Normal and Diseased Fruits and Vegetables,* 13–19 (Runeckles, V. C., Ed., Imperial Tobacco Co., Montreal, 102 pp., 1964)
95. Macko, V., Woodbury, W., Stahmann, M. A. The effect of peroxidase on the germination and growth of mycelium of *Puccinia graminis* f. sp. *tritici*. *Phytopathology,* **58,** 1250–54 (1968)
96. Macleod, N. J., Pridham, J. B. Deamination of β-(3,4-dihydroxyphenyl)-L-alanine by plants. *Biochem. J.,* **88,** 45P–46P (1963)
97. Mapson, L. W., Mead, A. Biosynthesis of ethylene. Dual nature of cofactor required for the enzymic production of ethylene from methional. *Biochem. J.,* **108,** 875–81 (1968)
98. Mason, H. S., Peterson, E. W. Melanoproteins. I. Reactions between enzyme-generated quinones and amino acids. *Biochim. Biophys. Acta,* **111,** 134–46 (1965)
99. Matta, A., Gentile, I. A. The relation between polyphenoloxidase activity and ability to produce indoleacetic acid in *Fusarium*-infected tomato plants. *Neth. J. Plant Pathol.,* **74,** suppl. 1, 47–51 (1968)
100. Maxwell, D. P., Bateman, D. F. Changes in the activities of some oxidases in extracts of *Rhizoctonia*-infected bean hypocotyls in relation to lesion maturation. *Phytopathology,* **57,** 132–36 (1967)
101. Mayer, A. M., Harel, E., Ben-Shaul, R. Assay of catechol oxidase—a critical comparison of methods. *Phytochemistry,* **5,** 783–89 (1966)
102. Mazelis, M. The pyridoxal phosphate-dependent oxidative decarboxylation of methionine by peroxidase. I. Characteristics and properties of the reaction. *J. Biol. Chem.,* **237,** 104–8 (1962)
103. Milo, G. E., Jr., Santilli, V. Changes in the ascorbate concentration of pinto bean leaves accompanying the formation of TMV-induced local lesions. *Virology,* **31,** 197–206 (1967)
104. Minamikawa, T., Uritani, I. Phenylalanine deaminase and tyrosine deaminase in sliced or black rot-infected sweet potato roots. *Arch. Biochem. Biophys.,* **108,** 573–74 (1964)
105. Minamikawa, T., Uritani, I. Phenylalanine ammonia-lyase in sliced sweet potato roots. Effect of antibiotics on the enzyme formation and its relation to the polyphenol biosynthesis. *Agr. Biol. Chem.,* **29,** 1021–26 (1965)
106. Minamikawa, T., Uritani, I. Phenylalanine ammonia-lyase in sliced sweet potato roots. *J. Biochem.,* **57,** 678–88 (1965)
107. Mink, G. I., Huisman, O., Saksena, K. N. Oxidative inactivation of Tulare apple mosaic virus. *Virology,* **29,** 437–43 (1966)
108. Moller, W. J., DeVay, J. E. Carrot as a species-selective isolation medium for *Ceratocystis fimbriata*. *Phytopathology,* **58,** 123–24 (1968)
109. Mudd, J. B. Fat metabolism in plants. *Ann. Rev. Plant Physiol.,* **18,** 229–52 (1967)
110. Müller, K. O. Hypersensitivity. In *Plant Pathology,* Vol. 1, 469–519 (Horsfall, J. G., Dimond, A. E., Eds., Academic Press, New York, 674 pp., 1959)
111. Nair, P. M., Vining, L. C. Cinnamic acid hydroxylase in spinach. *Phytochemistry,* **4,** 161–68 (1965)
112. Nair, P. M., Vining, L. C. Phenylalanine hydroxylase from spinach leaves. *Phytochemistry,* **4,** 401–11 (1965)
113. Neish, A. C. Major pathways of bio-

synthesis of phenols. In *Biochemistry of Phenolic Compounds,* 295–359 (Harborne, J. B., Ed., Academic Press, N.Y., 618 pp., 1964)

114. Nonaka, F. Inactivation of pisatin by pathogenic fungi. *Saga Daigaku Nogaku Iho,* **24,** 109–21 (1967)

115. Noveroske, R. L., Williams, E. B., Kuć, J. β-Glycosidase and phenoloxidase in apple leaves and their possible relation to resistance to *Venturia inaequalis. Phytopathology,* **54,** 98–103 (1964)

116. Nye, T. G., Hampton, R. E. Biochemical effects of tobacco etch virus infection on tobacco leaf tissue. II. Polyphenol oxidase activity in subcellular fractions. *Phytochemistry,* **5,** 1187–89 (1966)

117. Overeem, J. C., Sijpesteijn, A. K. The formation of perylenequinones in etiolated cucumber seedlings infected with *Cladosporium cucumerinum. Phytochemistry,* **6,** 99–105 (1967)

118. Ozeretskovskaya, O. L., Metlitskii, L. V., Chalenko, G. I. Role of energy exchange in hypersensitive reactions. *Tr. Vses. Nauch. Issled. Inst. Zashch. Rast.,* **26,** 103–7 (1966)

119. Parish, C. L., Zaitlin, M., Siegel, A. A study of necrotic lesion formation by tobacco mosaic virus. *Virology,* **26,** 413–18 (1963)

120. Patil, S. S., Dimond, A. E. Inhibition of *Verticillium* polygalacturonase by oxidation products of polyphenols. *Phytopathology,* **57,** 492–96 (1967)

121. Patil, S. S., Zucker, M., Dimond, A. E. Biosynthesis of chlorogenic acid in potato roots resistant and susceptible to *Verticillium alboatrum. Phytopathology,* **56,** 971–74 (1966)

122. Paxton, J. D., Chamberlain, D. W. Acquired local resistance of soybean plants to *Phytophthora* spp. *Phytopathology,* **57,** 352–53 (1967)

123. Pegg, G. F., Sequeira, L. Stimulation of aromatic biosynthesis in tobacco plants infected by *Pseudomonas solanacearum. Phytopathology,* **58,** 476–83 (1968)

124. Perrin, D. R., Cruickshank, I. A. M. Studies on phytoalexins VII. Chemical stimulation of pisatin formation in *Pisum sativum* L. *Australian J. Biol. Sci.,* **18,** 803–16 (1965)

125. Pierre, R. E., Bateman, D. F. Induction and distribution of phytoalexins in *Rhizoctonia*-infected bean hypocotyls. *Phytopathology,* **57,** 1154–60 (1967)

126. Pridham, J. B. Low molecular weight phenols in higher plants. *Ann. Rev. Plant Physiol.,* **16,** 13–36 (1965)

127. Raa, J. Polyphenols and natural resistance of apple leaves against *Venturia inaequalis. Neth. J. Plant Pathol.,* **74,** suppl. 1, 37–45 (1968)

128. Robb, D. A., Mapson, L. W., Swain, T. Activation of the latent tyrosinase of broad bean. *Nature,* **201,** 503–4 (1964)

129. Rohringer, R., Fuchs, A., Lunderstädt, J., Samborski, D. J. Metabolism of aromatic compounds in healthy and rust-infected primary leaves of wheat. I. Studies with $^{14}CO_2$, quinate-U-^{14}C, and shikimate-U-^{14}C as precursors. *Can. J. Botany,* **45,** 863–89 (1967)

130. Rohringer, R., Samborski, D. J. Aromatic compounds in the host-parasite interaction. *Ann. Rev. Phytopathol.,* **5,** 77–86 (1967)

131. Rubery, P. H., Northcote, D. H. Site of phenylalanine ammonia-lyase activity and synthesis of lignin during xylem differentiation. *Nature,* **219,** 1230–34 (1968)

132. Rubin, B. A., Aksenova, V. A. Participation of polyphenolase system in defense reactions of potato against *Phytophthora infestans. Biokhimiya,* **22,** 191–97 (1957) (English translation)

133. Rubin, B. A., Artsikhovskaya, E. V. Biochemistry of pathological darkening of plant tissues. *Ann. Rev. Phytopathol.,* **2,** 157–78 (1964)

134. Russel, D. W., Conn, E. E. The cinnamic acid 4-hydroxylase of pea seedlings. *Arch. Biochem. Biophys.,* **122,** 256–58 (1967)

135. Sakai, R., Tomiyama, K., Ishizaka, N., Sato, N. Phenol metabolism in relation to disease resistance of potato tubers. 3. Phenol metabolism in tissue neighboring the necrogenous infection. *Ann. Phytopathol. Soc. Japan,* **33,** 216–22 (1967)

136. Sakuma, T., Tomiyama, K. The role of phenolic compounds in the resistance of potato tuber tissue to infection by *Phytophthora infes-*

tans. Ann. Phytopathol. Soc. Japan, **33,** 48–58 (1967)

137. Sato, M. Metabolism of phenolic substances by the chloroplasts. III. Phenolase as an enzyme concerning the formation of esculetin. *Phytochemistry,* **6,** 1363–73 (1967)
138. Saunders, B. C., Holmes-Siedle, A. G., Stark, B. P. *Peroxidase* (Buttersworth, Washington, D. C., 271 pp., 1964)
139. Schwimmer, S. Influence of polyphenols and potato components on potato phosphorylase. *J. Biol. Chem.,* **232,** 715–21 (1958)
140. Schwochau, M. E., Hadwiger, L. A. Stimulation of pisatin production in *Pisum sativum* by Actinomycin D and other compounds. *Arch. Biochem. Biophys.,* **126,** 731–33 (1968)
141. Sequeira, L. Growth regulators in plant disease. *Ann. Rev. Phytopathol.,* **1,** 5–30 (1963)
142. Sequeira, L. Inhibition of indoleacetic acid oxidase in tobacco plants infected by *Pseudomonas solanacearum. Phytopathology,* **54,** 1078–83 (1964)
143. Sequeira, L., Kelman, A. The accumulation of growth substances in plants infected by *Pseudomonas solanacearum. Phytopathology,* **52,** 439–48 (1962)
144. Sequeira, L., Mineo, L. Partial purification and kinetics of indoleacetic acid oxidase from tobacco roots. *Plant Physiol.,* **41,** 1200–8 (1966)
145. Shannon, C. T., Pratt, D. E. Apple polyphenol oxidase activity in relation to various phenolic compounds. *J. Food Sci.,* **32,** 479–83 (1967)
146. Shannon, L. M. Plant Isozymes. *Ann. Rev. Plant Physiol.,* **19,** 187–210 (1968)
147. Shannon, L. M. Physical and catalytic properties of peroxidase isoenzymes. In *Biochemical Regulation in Diseased Plants or Injury,* 181–88 (Hirai, T., Ed., Kyoritsu Printing Co., Ltd., Tokyo, 351 pp., 1968)
148. Shannon, L. M., deVellis, J., Lew, J. Y. Malonic acid biosynthesis in bush bean roots. II. Purification and properties of enzyme catalyzing oxidative decarboxylation of oxaloacetate. *Plant Physiol.,* **38,** 691–97 (1963)
149. Shantz, E. M. Chemistry of naturally-occurring growth-regulating substances. *Ann. Rev. Plant Physiol.,* **17,** 409–38 (1966)
150. Siegel, B. Z., Galston, A. W. Indoleacetic acid oxidase activity of apoperoxidase. *Science,* **157,** 1557–59 (1967)
151. Smale, B. C., Keil, H. L. A biochemical study of the intervarietal resistance of *Pyrus communis* to fire blight. *Phytochemistry,* **5,** 1113–20 (1966)
152. Solymosy, F., Szirmai, J., Beczner, L., Farkas, G. L., Changes in peroxidase—isozyme patterns induced by virus infection. *Virology,* **32,** 117–21 (1967)
153. Stafford, H. A. Comparison of lignin-like products found naturally or induced in tissues of Phleum, Elodea, and Coleus, and in a paper peroxidase system. *Plant Physiol.,* **39,** 350–60 (1964)
154. Stafford, H. A. Factors controlling the synthesis of natural and induced lignins in Phleum and Elodea. *Plant Physiol.,* **40,** 844–51 (1965)
155. Stahmann, M. A., Clare, B. G., Woodbury, W. Increased disease resistance and enzyme activity induced by ethylene and ethylene production by black rot infected sweet potato tissue. *Plant Physiol.,* **41,** 1505–12 (1966)
156. Still, C. C., Fukuyama, T. T., Moyed, H. S. Inhibitory oxidation products of indole-3-acetic acid. Mechanism of action and route of detoxification. *J. Biol. Chem.,* **240,** 2612–18 (1965)
157. Still, C. C., Olivier, C. C., Moyed, H. S. Inhibitory oxidation products of indole-3-acetic acid: enzymic formation and detoxification by pea seedlings. *Science,* **149,** 1249–51 (1965)
158. Tomiyama, K. Physiology and biochemistry of disease resistance of plants. *Ann. Rev. Phytopathol.,* **1,** 295–324 (1963)
159. Tomiyama, K. Further observation on the time requirement for hypersensitive cell death of potatoes infected by *Phytophthora infestans* and its relation to metabolic activity. *Phytopathol. Z.,* **58,** 367–78 (1967)
160. Tomiyama, K., Ishizaka, N. Phenol

metabolism in relation to disease resistance of potato tuber. II. Exponential equation relating levels of phenols or phenolic enzymes to distance from locus of injury of plant tissue. *Plant Cell Physiol.*, **8,** 217–20 (1967)

161. Tomiyama, K., Sakai, R., Otani, Y., Takemori, T. Phenol metabolism in relation to disease resistance of potato tuber. I. Activities of phenol oxidase and some enzymes related to glycolysis as affected by chlorogenic acid treatment. *Plant Cell Physiol.*, **8,** 1–13 (1967)
162. Tomiyama, K., Sakuma, T., Ishizaka, N., Sato, N., Katsui, N., Takasugi, M., Masamune, T. A new antifungal substance isolated from resistant potato tuber tissue infected by pathogens. *Phytopathology,* **58,** 115–16 (1968)
163. Towers, G. H. N. Metabolism of phenolics in higher plants and micro-organisms. In *Biochemistry of Phenolic Compounds,* 249–94 (Harborne, J. B., Ed., Academic Press, N.Y., 618 pp., 1964)
164. Tuli, V., Moyed, H. S. Desensitization of regulatory enzymes by a metabolite of plant auxin. *J. Biol. Chem.*, **241,** 4564–66 (1966)
165. Tuli, V., Moyed, H. S. Inhibitory oxidation products of indole-3-acetic acid: 3-hydroxymethyloxindole and 3-methyleneoxindole as plant metabolites. *Plant Physiol.*, **42,** 425–30 (1967)
166. Uehara, K. Relationship between the host specificity of pathogen and phytoalexin. *Ann. Phytopathol. Soc. Japan,* **24,** 103–10 (1964)
167. Van Kammen, A., Brouwer, D. Increase of polyphenoloxidase activity by a local virus infection in uninoculated parts of leaves. *Virology,* **22,** 9–14 (1964)
168. Vaughan, P. F. T., Butt, V. S. The role of *σ*-dihydric phenols in the hydroxylation of *p*-coumaric acid. *Biochem. J.*, **107,** 7P-8P (1968)
169. Verleur, J. D. Regulation of carbohydrate and respiratory metabolism in fungal diseased plants. In *Biochemical Regulation in Diseased Plants or Injury,* 275–85 (Hirai, T., Ed., Kyoritsu Printing Co., Ltd., Tokyo, 351 pp., 1968)
170. Walker, J. R. L., Hulme, A. C. Studies on the enzymic browning of apple. III. Purification of apple phenolase. *Phytochemistry,* **5,** 259–62 (1966)
171. Walton, D. C., Sondheimer, E. Effects of abscisin II on phenylalanine ammonia-lyase activity in excised bean axes. *Plant Physiol.*, **43,** 467–69 (1968)
172. Webb, J. L. Quinones. *Enzymes and Metabolic Inhibitors,* Vol. 3, 421–594 (Academic Press, New York, 1028 pp., 1966)
173. Weintraub, M., Ragetli, H. W. An electron microscope study of tobacco mosaic virus lesions in *Nicotiana glutinosa* L. *J. Cell Biol.*, **23,** 499–509 (1964)
174. Wheeler, H., Hanchey, P. Permeability phenomena in plant disease. *Ann. Rev. Phytopathol.*, **6,** 331–50 (1968)
175. Woodin, T., Kosuge, T. Evidence for isozymes of chlorismate mutase in etiolated mung bean seedlings. *Plant Physiol.*, **43,** S-47 (1968)
176. Wosilait, W. D., Nason, A., Terrell, A. J. Pyridine nucleotide-quinone reductase. II. Role in electron transport. *J. Biol. Chem.*, **206,** 271–282 (1954)
177. Yang, S. F. Biosynthesis of ethylene. Ethylene formation from methional by horseradish peroxidase. *Arch. Biochem. Biophys.*, **122,** 481–87 (1967)
178. Yang, S. F. Biosynthesis of ethylene. In *Biochemistry and Physiology of Plant Growth Regulators* (Wightman, F., Setterfield, G., Eds., Runze Press, Ottawa, In press)
179. Yasunobu, K. T., Peterson, E. W., Mason, H. S. The oxidation of tyrosine-containing peptides by tyrosinase. *J. Biol. Chem.*, **234,** 3291–95 (1959)
180. Zerr, C., Novoa, W. B. The inhibition of glucose-6-phosphatase by phlorizin and structurally related compounds. *Biochem. Biophys. Res. Commun.*, **32,** 129–133 (1968)
181. Zucker, M. Induction of phenylalanine deaminase by light and its relation to chlorogenic acid synthesis in potato tuber tissue. *Plant Physiol.*, **40,** 779–84 (1965)
182. Zucker, M. Sequential induction of phenylalanine ammonia-lyase and

a lyase-inactivating system in potato tuber disks. *Plant Physiol.*, **43,** 365–74 (1968)

183. Zucker, M., Hanson, K. R., Sondheimer, E. The regulation of phenolic biosynthesis and the metabolic roles of phenolic compounds in plants. In *Phenolic Compounds and Metabolic Regulation,* 68–93 (Finkle, B. J., Runeckles, V. C., Eds., Appleton-Century-Crofts, New York, 157 pp., 1967)

Copyright 1969. All rights reserved

RESISTANCE IN MALUS TO VENTURIA INAEQUALIS[1]

By E. B. Williams and Joseph Kuć

Departments of Botany and Plant Pathology, and Biochemistry
Purdue University, Lafayette, Indiana

Introduction

Scab, caused by *Venturia inaequalis* (Cke.) Wint., has been one of the most serious and difficult diseases of apple to control. Because of the necessity of careful timing of application and difficulty in eradication, fungicides are frequently not effective. The best means of control appears to depend on the development of resistant varieties. Knight (69) has compiled an excellent bibliography of research on pome fruit breeding and genetics up to 1960.

In this review, we will discuss research dealing with types and sources of resistance in the host, pertinent facts about the fungus, and the present status of knowledge on the nature of the host-parasite interaction.

The Host

Several historical reviews (16, 30, 32, 74, 143) are available on the initial attempts in apple improvement. The first recorded attempt (143) was in France in 1683 by Venette, who sowed open pollinated seed with the aim of securing new varieties. The crossing method was established in 1796 in Germany by Diel, who developed most of the techniques of emasculation of flowers, collection of pollen, and transfer of pollen to stigmas. During the latter part of the 18th Century, Thomas Knight of England conducted hybridizing experiments with apple (32). Fruit breeding in state institutions was first performed in Germany about 1880 (143).

Early attempts at apple improvement did not involve disease resistance. However, by the end of the 19th Century, researchers were becoming aware of the possibility of developing scab resistant varieties. Aderhold (1), in Germany, presented data on the susceptibility of 160 apple varieties. While none showed absolute immunity, 11, including Antonovka, proved to be markedly resistant. Wallace (122), in the USA, doubted the probability of finding and maintaining resistant varieties. Numerous reports (24, 38, 41, 74, 90, 91) on the resistance levels of certain commercial varieties were published during the next 4 decades.

In 1934 and 1935, Rudloff & Schmidt (97, 98) contended that no exist-

[1] Purdue University AES Journal Paper No. 3593. Certain of the unpublished studies reported were supported by USDA Cooperative Agreement 12-14-100-5570(34).

ing commercial high quality variety possessed satisfactory resistance to scab under all conditions. They established, however, that four small-fruited *Malus* species, *M. atropurpurea, M. micromalus, M. spectabilis* and *M. spectabilis Kaido,* did possess this type of resistance as they were either entirely free of scab or showed very light infection during field inoculation tests. The commercial apple Antonovka also possessed a high level of resistance. These workers believed that cross-breeding presented the most likely solution for developing scab resistant varieties and proceeded to initiate a suitable program. They demonstrated that it was possible to transmit the original level of scab resistance, but they were discouraged in that factors controlling small fruit size seemed to be dominant. Consequently, further breeding attempts in Schmidt's program (102–106) were primarily confined to the use of larger fruited varieties such as Antonovka and Ernst Bosch as parents. Schmidt's program was interrupted by World War II. However, some of the Antonovka source selections developed by him are currently used in breeding programs in Germany (142, 143), Sweden (30), the USA (75, 108, 112), and Canada (118).

Malus is a poor choice as a genetic tool. The haploid number of chromosomes is 17. This number probably developed from the basic number of seven in the Rosaceae by the duplication of four chromosomes and the triplication of three (32, 82). Except for scab resistance in certain small fruited species (42, 43, 112), intensified anthocyanin production in *M. pumila Niedzwetzkyana* (77), and possibly extra petals and pendulous growth habit (99), most factors in *Malus* that have been studied (14-18, 68, 123) have been shown to be inherited quantitatively. As the apple is essentially self sterile (18, 32), it is difficult to self in order to obtain inbred parental lines with homozygous factors.

Genes for resistance—Two types of scab resistance in *Malus* are available for use by the breeder. These are quantitative or multiple factor and qualitative or single major factor determinants. The former type is sensitive to the effect of environment, host conditioning, or both, while the latter is expressed under all conditions, usually as single dominant genes. Both types may condition field immunity in that no macroscopic evidence of infection is present, or resistance may be expressed as a reduced number and size of sporulating lesions. In general, there is good correlation between leaf and fruit resistance (110, 131).

The low to moderate level of scab resistance exhibited by many commercial varieties (1, 30, 38, 46, 54, 55, 70, 74, 100) is apparently of the multiple factor type. This resistance usually results in fewer and smaller sporulating lesions when massed inoculum is used, and in the formation of nonsporulating necrotic flecks when selected single isolates are used. The value in utilizing the resistance present in existing commercial varieties in breeding programs for the development of scab immune varieties has been discussed (138, 139). The major problem encountered in such an approach is the necessity of growing and screening large numbers of seedlings to secure a few showing the desired level of immunity. In spite of this, it was suggested that multiple factor resistance may be more valuable in the long run than the relatively easy-to-use qualitative type. Consequently, this approach is being applied in scab resistance breeding in Europe (30, 46, 95, 139, 143).

Hough (41) classified his breeding material with respect to field reaction. Classes of scab infection were arbitrarily designated, ranging from a rating of 1 for little or no infection to 5 for severe infection with heavy defoliation. He found that the majority of orchard varieties, when intercrossed, yielded progeny of average (class 4) susceptibility with no individuals in class 1. Certain varieties, including Jonathan, Jeffers, and Duchess, when crossed, gave rise to individuals with above average resistance. The distribution of individuals in these progenies suggests multiple factor inheritance. There was also evidence of cytoplasmic influence in expression of scab resistance.

Hough (41) also found six small fruited selections of Asiatic *Malus* species that exhibited field immunity to scab. These included: *M. atrosanguinea* 804, *M. floribunda* 821, *M. prunifolia* 19651, *M. ringo, M. toringo* (dwarf spreading), and *M. zumi*. All had been collected by Crandall (16). Progeny segregation of one (*M. floribunda* 821) approximated a 1:1 ratio (41, 42), suggesting that either resistance in this selection is controlled by a single qualitative gene or by a block of closely linked quantitative genes. Further studies with these selections and others established definite reaction classes under greenhouse conditions for each source of resistance (42, 113). These classes are as follows: 0, signifying no macroscopic evidence of infection; 1, pin-point pits and no sporulation; 2, irregular chlorotic or necrotic lesions and no sporulation; 3, few restricted sporulating lesions; and 4, extensive abundantly sporulating lesions. Since the original reaction classes were described, another (class M) has been added. This is described as a mixture of necrotic, nonsporulating and sparsely sporulating lesions. This same method of scoring plants is used in our program. In our system, only class 4 has been considered as field susceptible. The class 1 or pin-point reaction is considered a hypersensitive response (113). The host epidermal cells below the infection peg collapse within 40 to 72 hr and shortly thereafter the fungus is killed (25). In contrast, the other reaction types are not expressed until 3 to 12 days after inoculation and the fungus remains viable for as long as 21 days.

In 1945, a cooperative program of scab resistance breeding was initiated by Purdue University and the University of Illinois. Rutgers University was included in 1950. Similar projects were initiated at the New York Experiment Station, Geneva, and at the Central Experimental Farm, Ottawa, Canada. Emphasis was on resistance present in the small fruited Asiatic species. Many articles are available (20, 33, 41, 50, 75, 79, 80, 101, 108–110, 112, 113, 116, 117, 119, 120, 137, 139, 141), discussing the sources of these selections, and, with many, their response to the scab organism.

A large pool of scab resistant gene sources exists in *Malus* (Table I).

TABLE I

APPLE SPECIES AND VARIETIES USED AS SOURCES OF RESISTANCE TO SCAB

Source	*Source*
I. Monogenic A. Resistant to all known races (21, 43, 111, 113, 116) *M. atrosanguinea* 804 (3 type) *M. baccata jackii* *Jonsib* *M. floribunda* 821 *M. micromalus* (3 type) *M. prunifolia* 19651 *M. prunifolia microcarpa* 782–26 *M. prunifolia xanthocarpa* 691–25 *M. A.* 4 *M. A.* 8 *M. A.* 16 *M. A.* 1255 *M. pumila* R12740–7A non-differential *Antonovka* selection PI 172612	B. Susceptible to certain races (80, 115, 116, 132) *M. baccata* vars. Dolgo, Alexis, Bittercrab *Geneva* (*M. pumila* var. niedzwetzkyana o. p.) *M. pumila* R12740-7A race 2 differential *M. pumila* R12740–7A race 4 differential *M. atrosanguinea* 804 (pit type) *M. micromalus* (pit type)
II. Multigenic A. Field resistant to all known races (116) *Antonovka* 1½ lb and certain other selections of *Antonovka* *M. baccata* (selected seedlings) *M. sargenti* 843 *M. sieboldii* 2982–22 *M. toringo* 852 *M. zumi calocarpa*	B. Partially resistant to all known races (116) Certain commercial varieties such as Jonathan, Golden Delicious, etc.
III. Other resistance sources[a] A. All commercial varieties adequately tested (resistant only to certain naturally occurring strains) (9, 10) B. Crabapple varieties (5, 24, 79, 80)	C. Old Irish and English varieties (74, 131) D. Selected European varieties (46, 100)

[a] Segregation data insufficient to determine mode of inheritance. Most are probably multigenic.

In the initial crosses of Wolf River × *M. atrosanguinea* 804 and Wolf River × *M. micromalus,* both class 1 (pit) and class 3 reactions were obtained with each cross (111). Segregation data indicated that the resistance of each was controlled by a single dominant gene. With the discovery of race 5 of the fungus (132), it was determined that the "pit" gene in *M. micromalus* was masking the effect of another closely linked gene which conditioned an M reaction. This gene is at a different locus than the 3 type discovered in the original Wolf River × *M. micromalus* cross.

M. floribunda 821, when class 4 only is considered susceptible, gave a sharp 1:1 resistant/susceptible ratio when crossed in quantity with susceptible selections. This ratio was maintained through several modified backcrosses (43). This interpretation is at variance with that expressed by the workers in Ottawa (118). They considered sporulation, however slight, as evidence of susceptibility. Eight to 32 per cent of their seedlings were resistant. While our ratios with *M. floribunda* 821 remained constant, there was a change in reaction type through the backcross generations. The original *M. floribunda* 821 resistance was expressed as a class 1 reaction. Selections $F_2$26829–2–2 and $F_2$26830–2, obtained from sibbing F_1's of 821 × Rome Beauty, gave a class 2 reaction (113). In the modified backcross progeny, the resistant reaction classes range from 2 to M to 3 (131). This suggests that the original level of resistance was not due to a single qualitative gene, but was due either to a group of rather closely linked quantitative genes or to a class 3 reaction qualitative gene, closely linked with one or more quantitative genes.

The resistance of the original R12740–7A selection was expressed primarily as a class 2 reaction (113). In an effort to explain segregation ratios, a tentative genotype scheme was set up (21) in which resistance was controlled by two qualitative and three quantitative genes, all in a heterozygous condition. With the discovery of race 2 (115) and race 4 (116), it was determined that there are at least three qualitative genes and an undetermined number of quantitative genes in the original R12740–7A.

den Boer & Collins (24) classified 300 species and varieties of *Malus* with reference to field response to scab. These followed closely the response noted by Hough (41) for comparable selections. Nichols (83, 84) also catalogued the field resistance to scab of flowering crabapples. Sixty of more than 300 named cultivars were resistant. Greenhouse inoculations (131) with grouped inoculum of 40 conidial isolates of races 1, 2, 3, and 4 were made on 34 named flowering crabapples. Of the 12 showing resistance, four, Arnold, Liset, Prairie Rose and Schiedecker, exhibited a hypersensitive pit reaction. Schiedecker is susceptible in an ornamental planting at Vincennes, Indiana (131). Bagga & Boone (5) reported the results of testing 41 crabapple cultivars with a single fungal isolate. They found that resistance to this isolate in 25 selections was due to a single dominant gene, in 11 selections to two genes, and in five selections to three genes. At least seven of these, *M. spectabilis* variety Alba, Aldenham, Hopa, Katherine, Morden 457,

Dolgo and Henry F. DuPont, were susceptible in our greenhouse tests when inoculated with the combined inoculum of 20 isolates (131).

Of the 13 Irish varieties listed by Lamb (74) as being scab free, seven were tested in our greenhouse inoculations, using the combined inoculum from 20 isolates. Six of these varieties, Ballyfatten, Greasy Pippin, Green Chisel, Keegans Crab, Red Brandy and Striped Brandy, exhibited moderate to high levels of resistance (131). These have remained scab free for 12 years in our field plantings. Four English varieties, Early Victoria, Potts Seedling, Grenadier and Yarlington Mill, also were resistant in our greenhouse inoculations and field plantings.

Scab gene relationships—A program was initiated in 1955 to determine the relationship of scab resistance genes and the results were published (22, 23, 133, 134, 136). Symbols were designated to identify the different gene loci with ten of the qualitative genes being located at the V_f (*M. floribunda* 821) locus, and two at the V_m (*M. micromalus* pit) locus. Quite possibly most of the ten allelic genes at the V_f locus may be the same. Most of the resistance source selections containing these genes came originally from the Arnold Arboretum. Some were distributed as seed. However, one, Hansen's baccata no. 1, was brought to South Dakota as seed from a collection in Siberia (80). With the discovery of race 5 (132), and the subsequent finding that class 1 selections of both *M. micromalus* and *M. atrosanguinea* 804 are susceptible to this race, we can assume that the same gene is present in both. Three other loci, V_b (Hansen's baccata no. 2), V_{bj} (*M. baccata jackii*), and V_r (*M. pumila* R12740-7A), were identified with a single gene pair at each.

Although the number of qualitative scab resistance genes available from the Asiatic species is less than previously thought, there may be an additional number of this type from the flowering crabapple cultivars. Based on the reaction class incited with grouped inoculum, the resistance from some seems as complete as that of the Asiatic species. It is likely that in some instances the resistance will be the same, since the ornamental Asiatic species are probably the ancestors of many of these.

Apparently, a large pool of qualitative and quantitative genes is available for use. Ideally, the best program for developing durable scab resistant varieties will involve the incorporation of both types of resistance in the final selections.

The Pathogen

The literature dealing with nutrition, genetics, and pathogenicity has been reviewed (10, 49, 50, 116). We will cover only the research which applies directly to the eventual host-parasite interaction.

The apple scab pathogen, *V. inaequalis,* is an ascomycete, belonging to the Venturiaceae. The asexual stage is usually listed as *Fusicladium dendriticum* (Wallr.) Fckl. The development of the ascocarp has been described (29, 58) as have been the nuclear phenomena associated with the ascus and with the asexual stage of the fungus (3). The vegetative mycelium, conidiophores and conidia are uninucleate (3) which greatly facilitates the use of this organism in genetic studies. Normally, the haploid chromosome number in *V. inaequalis* is seven (19, 47).

Using the eight ascosporic lines isolated in serial order from a single ascus, Keitt & Palmiter (56, 57) demonstrated that the sexual stage pairings could be manipulated *in vitro*. They concluded that *V. inaequalis* is heterothallic in the sense that the isolates studied fell into two sterility groups, being hermaphroditic, self-sterile, intra-group sterile, and inter-group fertile. Sexual compatibility was investigated further by Keitt & Langford (52), who found that only two classes of sexual compatibility exist for *V. inaequalis*. Other investigations (114, 135) indicate that sexual compatibility is controlled by single gene pairs, whose locus was studied with respect to its centromere.

Infection process—V. inaequalis is representative of a group of fungi characterized by a distinctive type of parasitism (89). The invading mycelium is limited to a position between the cuticle and outer epidermal cell walls where it exists as a stroma consisting of short, thick-walled cells. The fungus, nevertheless, is intimately associated with the underlying host cells in that it derives its nourishment from them and also exerts profound influence on them.

On both susceptible and resistant hosts, spore germination, appressorial formation, and cuticle penetration are similar (8, 25, 89, 113). During germination of the spore, a short germ tube normally develops which forms an appressorium. The appressoria are apparently held fast to the leaf surface by means of a mucilaginous sheath. Direct cuticular penetration is accomplished by means of a minute infection hypha which develops through a thin walled pore-like area about 2 μ in diameter in the lower surface of the appressorium. As soon as the infection hypha reaches the epidermal cell wall, it flattens into an irregularly shaped primary hypha. With a susceptible host, the mycelium continues to develop and forms a stroma, forcing the cuticle from the epidermal cell walls. Within 10 to 14 days, conidiophores are formed, piercing or splitting the cuticular layer, and conidia develop. At this time, macroscopic symptoms are expressed as small olivacious spots which enlarge rapidly. Accompanying this effect, depletion of plastids and vacuolation occur in the palisade cells. Within 2 to 3 weeks, the leaf tissue underlying the fungus stroma becomes chlorotic to necrotic, resulting eventually in the collapse of the entire infected area.

Modifications of the effect on host tissue depend upon the type of resistance involved. With the hypersensitive pit type (113), penetration occurs in the normal way, but within 36 to 48 hr the epidermal cells underlying the point of penetration collapse. The cells become filled with a granular deposition of material (131), and immediately beneath these cells the membranes of plastids in the palisade cells are apparently altered with a loss of definition in the plastids. There is also an accumulation of darkly staining spheroid bodies in the cytoplasm of the palisade cells. Correlated with the collapse of epidermal cells, fungal growth is inhibited and, with some combinations, the fungus is killed (25). Numerous fungi not pathogenic to apple are also able to induce the formation of pits, indicating that the hypersensitive response is not specific (31). With reaction 2 type resistance, macroscopic symptoms are not expressed until 6 to 9 days after inoculation at which time the infected area becomes necrotic, without sporulation. However, the fungus remains viable for as long as 21 days in these lesions (25).

Only the young, rapidly expanding leaves are susceptible to scab. Also, with resistant hosts, only those leaves respond with macroscopic symptoms. Approximately the same percentage of spores germinated, and appressoria formed on leaves of all ages of a *Malus* selection with hypersensitive resistance (8). However, the hypersensitive response occurred only on the youngest four to five leaves.

Physiologic specialization—Palmiter (91) demonstrated conclusively the existence of pathogenic strains of *V. inaequalis* as had been suggested by others (81, 122, 124). He also confirmed the observation (2) that infection by *V. inaequalis* occurred only in the genus *Malus*. Subsequent investigations by others (4, 37, 48, 52, 53, 57, 102, 115, 116, 132) extended the differential host range within *Malus*. These studies indicate that *V. inaequalis* is made up of many strains that differ to varying degrees in their physiologic characters. At present, five races have been named (115, 116, 132).

Inheritance of pathogenicity—Using commercial apple varieties inoculated with monosporic lines of *V. inaequalis*, Keitt & Langford (51, 52) demonstrated the "lesion" and "fleck" types of host-parasite response. The lesion type was expressed as abundantly sporulating spots; while the fleck reaction was expressed as chlorotic or necrotic spots with little or no sporulation. The factors in the fungus initiating this response were found to be inherited primarily in a 1:1 ratio. Based on this study and others (9, 54), seven pathogenicity gene pairs (*p*-1^{+}/*p*-1 through *p*-7^{+}/7) were identified and studied with reference to their centromeres. Bagga & Boone (4), using ornamental crabapple cultivars, identified six pathogenicity gene pairs (*p*-14^{+}/*p*-14 through *p*-19^{+}/19). From the host-resistance gene pools present in *Malus* species, Williams (130) and Williams & Shay (135) identified six gene pairs (*p*-8^{+}/p-8 through *p*-13/*p*-13′) which condition either an increase or a reduction in pathogenicity.

Two of the 19 gene pairs, *p*-12′ and *p*-13′, induce a reduced necrotic fleck on *M. sikkimensis* instead of the normal resistant reaction ordinarily obtained on this selection. These genes were shown to be inherited independently (130, 135). Of the 19 gene pairs, only three were found to be located close enough to their respective centromeres to be mapped: *p*-4^{+}/*p*-4 (13 crossover units), *p*-12/*p*-12′ (22.7 units), and *p*-17^{+}/*p*-17 (26.2 units). Three of the gene pairs, *p*-12/*p*-12′, *p*-8^{+}/*p*-8, and *p*-9^{+}/*p*-9, formed a loose linkage group (130, 135). The linkage relationships of gene pairs studied by the three groups of investigators were not correlated.

Host-Pathogen Interaction

Penetration of host and fungal nutrition—The mechanism by which the infection peg penetrates the cuticle is not known. Wittshire (140) suggested that the fungus produces a substance which dissolves the cuticle and aids in penetration. Nusbaum & Keitt (89) found no evidence to support this. It is possible that penetration is effected by means of mechanical pressure, or the action of hydrolytic enzymes, or both. The possibility of selective or differential channels, possibly pectin in nature, cannot be overlooked.

The chemical nature of the stratum in which the fungus exists beneath the cuticle and above the epidermal cell walls is not clearly defined. Probable components include proteins, lipids, pectic substances, hemicelluloses and cellulose. The fungus can obtain nutrients in this stratum by solubilizing components, by absorbing nutrients diffusing into the stratum from underlying cells, or by a combination of both. Franke (28) has demonstrated the existence of cytoplasmic links (called ectodesmata) which extend through outer epidermal walls. These in *Malus* could provide intimate contact between host cytoplasm and fungus and serve as a bridge with underlying host tissue.

V. inaequalis grows on a mineral medium containing either glucose, cellobiose, dextrin, fructose, maltose, mannitol, mannose, melibiose, raffinose, or sucrose as the sole source of carbon (76). Potassium nitrate and ammonium sulfate are satisfactory sources of nitrogen, though growth in liquid culture is increased when the inorganic sources of nitrogen are replaced by casein hydrolyzate (92). Fothergill & Ashcroft (27) verified growth of the fungus in a liquid medium containing mineral salts and glucose, but noted that growth was markedly enhanced by the addition of vitamins. Thiamine had the strongest growth-stimulating effect of the vitamins tested, but folic acid, nicotinic acid, pyridoxine, and ascorbic acid also stimulated growth. The fungus produces extracellular pectinolytic enzymes and can utilize pectin, calcium pectinate, and calcium pectate as sources of carbon (40, 76, 93).

The production of extracellular cellulolytic enzymes is uncertain. Raa (93) and Leben & Keitt (76) reported that the fungus made little or no growth with cellulose as sole source of carbon. Holowczak, Kuc' & Williams, (40) however, used six isolates of the fungus representing three races and demonstrated good growth with 10 per cent w/v cellulose.

As an aid in studying the effect on virulence of a broad spectrum of changes related to genetics and nutrition, a large number of biochemical nutritional mutants of the fungus were induced (12). Of the many biochemical mutants tested, those requiring biotin, inositol, nicotinic acid, pantothenic acid, or reduced sulfur were virulent; whereas others requiring choline, riboflavin, purines, pyrimidines, arginine, histidine, methionine, or proline were avirulent (11). Virulence could be wholly or in part restored to six of the eight avirulent biochemical mutants by supplementing the required substances to the surfaces of inoculated apple leaves during the incubation period (67). The avirulent mutants germinated, penetrated the leaf and established themselves, but required nutrient supplementation for subsequent development and symptom expression. The virulent mutants obtained a sufficient quantity of the required nutrient for growth, development and symptom expression in the area beneath the cuticle and above the epidermis. Avirulent mutants grew in apple leaf homogenates, indicating that required substances are either present in the leaf but not available at the site of fungal growth, or are in forms not utilizable by the fungus.

The work with mutants supplied valuable information concerning the nutrition of the fungus and the availability of nutrients at the site of fungal growth. However, since wild type strains of the fungus grow on inorganic salts plus glucose, and since external supplements of amino acids on apple leaves inoculated with wild-type lines of *V. inaequalis* have no effect on resistance, it is doubtful that this work has any direct bearing on the nature of disease resistance. A possibility remains that specific nutrients are not important for growth but as initiators of metabolic changes in the fungus or for the production of extracellular products that determine susceptibility or resistance.

Toxins produced by V. inaequalis—A heat-stable metabolite which causes collapse of leaf tissue of susceptible and resistant *Malus* selections is produced in deionized water by germinating spores of *V. inaequalis* (85, 88). The symptoms were identical to those produced on resistant varieties by the pathogen. Deionized water in which ungerminated spores were kept for the same length of time as germinated spores did not cause collapse of leaf tissue. However, no consistent correlation could be established between the susceptibility and resistance of host selections and their reaction to the metabolite. Hignett & Kirkham (35) reported the production of a series of nondialysable pigmented proteins by the fungus when cultured in a malt medium. These proteins were precipitated from culture filtrates by alcohol and purified by gel filtration on Sephadex G-100. At least three major components were found in the complex mixture. A scab-susceptible variety injected with some or all of the isolated proteins showed interveinal leaf desiccation and upward curling of leaf tips and edges within 24 hr followed by interveinal necrosis. General wilting of shoots did not occur. At similar dosages, an apple variety showing field resistance was unaffected, although rapid interveinal necrosis could be caused within 5 hr by raising the dosage. Leaf expansion in the susceptible variety was inhibited by injection of a protein fraction at dosages below those necessary to cause damage. A "field-immune" variety was unaffected under similar conditions. The development of scab lesions in inoculated varieties was stimulated by injection of protein fractions at nonphytotoxic dosages. The leaf area covered by lesions was frequently double that observed in the control. Scab development was not observed in the field-immune variety after injection of the protein fractions. This work suggested that protein produced by *V. inaequalis* has a selective toxic effect on susceptible hosts at high dosage and a stimulatory effect on scab development at low dosage.

The work of Raa (93) may be in conflict with the above. *V. inaequalis* was grown in shake culture in an inorganic medium containing vitamin free casamino acids and 2 per cent glucose. After 20 to 30 days at 24° C, the culture medium was filtered and its toxicity towards apple leaves was determined. In another experiment, washed conidia were suspended in distilled water containing 0.5 per cent glucose and the suspension was stirred vigorously at 18 to 20° C. After 1, 2, and 3 days, samples of this suspension were centrifuged, and the toxicity of the supernatant to apple leaves was determined by placing freshly cut leaves into small flasks containing the culture filtrates. The flasks were kept at 18 to 20° C, and the rate of wilting was taken as an indication of toxicity. Forty-two resistant and 70 susceptible plants from the cross Antonovka 34–20 × Golden Delicious differed in their sensitivity to culture filtrates of a pathogenic isolate. These filtrates caused rapid wilting of the leaves of resistant plants after 1 hr whereas the leaves of susceptible plants kept their turgor. Drops of undiluted culture filtrate added twice daily to the surface of young leaves caused necrosis in resistant plants after 2 to 3 days and in susceptible plants after 7 to 8 days. Culture filtrates of an avirulent, nonsporulating isolate of the fungus caused wilting in considerable numbers of susceptible seedlings. Thus, it would appear that both the nonsporulating and avirulent isolate and the virulent isolate of *V. inaequalis* form toxins, but the toxins of the former act less specifically than those of the latter.

Further work (36, 64) suggests a different role for extracellular proteins produced by *V. inaequalis* from that suggested by Raa. Fungal melanoprotein isolated from culture filtrates of *V. inaequalis* had biological activity when injected into test plants. Solute transport in xylem of healthy plants was interrupted after injection of melanoproteins into the petioles, and tracer solutes were contained within vascular tissues in test leaves. Similar distribution patterns of tracer solutes were observed in leaves after inoculation with the fungus. The patterns changed as the lesions matured, with tracer compounds accumulating at the lesion sites. The percentage of leaf coverage by lesions was increased by application of melanoprotein with spore inoculum to susceptible plants. These effects were not duplicated by numerous other proteins. The authors suggest that the pathogen redirects host metabolism in favor of the developing lesion by keeping nutrients within the leaf. Thus, a mechanism for susceptibility is suggested, explaining the availability of nutrients to the fungus in at least the latter stages of its development. Injection of melanoproteins at concentrations higher than that required to localize solutes in the leaf vein caused cupping of the leaves followed by interveinal desiccation and necrosis. Data are not presented to show a differential effect of melanoprotein on various susceptible and resistant hosts.

It is possible that both groups (36, 64, 93) were working with the same extracellular products but obtaining different physiological effects because of differences in the concentration of the extracellular product applied or infused into the plant. Raa's work (93) suggests that the collapse of cells in resistant hosts triggers a series of metabolic events which produce metabolites that inhibit extracellular enzymes produced by the fungus and thereby inhibit its growth.

Metabolic change associated with host-fungus interaction—Work in this area has been concerned chiefly with the metabolism of phenolic compounds found in the apple leaf and the activities of β-glycosidase and phenoloxidase. Phenolics found in apple leaves, fruit, bark, and roots include phloridzin, leucoanthocyanins, epicatechins, catechins, quercetin, cyanidin, 3-hydroxyphloridzin, *p*-coumarylglucose, kaempferol, *p*-coumarylquinic acid, isochlorogenic acid, and chlorogenic acid (13, 45, 96, 125–129). The principal phenolics of leaves and fruits, respectively, are phloridzin and chlorogenic acid. Aside from metabolic changes occurring to phloridzin, little is known concerning the qualitiative or quantitative changes of other phenolic compounds following inoculation. Also, the location of the phenolic compounds in the leaf has not been established, and little or nothing is known concerning the metabolic pathways operative in the pathogen.

A fraction containing water-soluble constituents, extracted from apple leaves with ethyl acetate, and its separate main constituents were strongly inhibitory to the growth and sporulation of the fungus (59). An inorganic source of nitrogen in the medium increased and an organic source decreased inhibitory action. Water-soluble fractions from a susceptible and a resistant apple variety injected into leaves of a susceptible variety markedly reduced the area of leaf tissue covered by lesions. The effect of the fraction was to reduce susceptibility of apple leaves but it did not change the disease reaction of the leaf tissue, i.e., did not make a susceptible leaf resistant. Protection was reduced or lost by injecting urea with the fraction. It was suggested that organic sources of nitrogen aided in the degradation of phenols in the fraction, either by the plant or fungus, into nonfungitoxic products.

Subsequently, it was found that cinnamic acid was highly inhibitory to spore germination and growth of *V. inaequalis* and reduced susceptibility of the host when injected into apple shoots (62). The presence of cinnamic acid in untreated host tissues was not reported. In general, hydroxylation of cinnamic acid reduced its toxicity. *o*-Hydroxycinnamic acid was the most toxic of the hydroxy derivatives tested. The quantity of hydroxy derivatives of cinnamic acid in apple leaves, other than chlorogenic acid, isochlorogenic acid, and *p*-coumarylquinic, has not been reported. Further studies of the major components on the phenolic fractions of resistant and less resistant apple and pear varieties (60) suggested no qualitative differences related to resistance. There were suggestions, however, of a possible correlation between quantitative differences and resistance. The existence of an optimal balance between phenolic and nitrogen content for development of the pathogen in a host-parasite interaction was presented (61). It was suggested (26) that phenols in the host are important in the physiology of the host-pathogen interaction, but resistance to *V. inaequalis* was not necessarily due to compounds or a mixture of compounds unique for resistant plants.

A fluorescent phenolic compound increased in concentration in the peel of fruits and leaves of several apple selections infected with either *V. inaequalis* or *Podosphaera leucotricha* (6). Diseased leaves of Jonathan, Gallia Beauty, Golden Delicious, Geneva, and 384-1 contained a higher concentration of the fluorescent compound than healthy leaves. In the race differential selections Geneva and 384-1, the concentration was higher in the susceptible host-parasite combination than in the resistant combination. The increase appeared to be localized in and around infected areas and continues from 40 hr to 23 days after inoculation. In uninoculated tissues, the compound occurred in trace quantities with some increase with age, but concentrations were always higher in infected tissues of corresponding age. The response is not specific to *V. inaequalis* since *P. leucotricha* also induced production of the fluorescent compound. Mechanical injury failed to increase the concentration of the compound, and it could not be detected in chromatograms of culture filtrates or extracts of *V. inaequalis*. Though concentrations were higher in susceptible-reacting tissues than in resistant ones, concentrations per unit infected area may be higher in resistant-reacting tissues since whole leaves were extracted and resistant lesions are much smaller than susceptible ones. The increased concentration of the fluorescent compound as early as 40 hr after inoculation, prior to any symptom expression on susceptible or resistant hosts, indicates the compound arises due to a physiological interaction between host and pathogen and not merely as a result of gross injury to the host.

Recently Hunter & Kirkham (44) investigated the fluorescent compounds found in the leaves of a susceptible apple variety and a variety that is resistant under British orchard conditions. A larger number of compounds which fluoresce bright blue in the presence of ammonia vapor were apparent in extracts of the resistant as compared to the susceptible variety. Two of these fluorescent compounds were separated from the bulk of the leaf extract and were shown to inhibit the growth of *V. inaequalis in vitro* and decrease susceptibility when injected *in vivo*. The same compounds increased in the leaves of the resistant variety within a short time after inoculation.

Since phloridzin is the most abundant glycoside found in the leaves of apple, most of the studies of the chemical changes following infection have centered around changes occurring in this compound and its various degradation products. Barnes & Williams (7) showed that phloridzin and phloretic acid stimulated and phloretin inhibited growth of *V. inaequalis in vitro*. They hypothesized that phloridzin was metabolized to phloretin, phloroglucinol, phloretic acid, *p*-hydroxybenzoic acid, and protocatechuic acid. Since growth stimulation by phloretic acid occurs at lower concentrations than that by phloridzin, and since phloretic acid is lacking in culture filtrates containing phloridzin it appears that phloretic acid may be utilized by the fungus and be directly responsible for the stimulating property of phloridzin. In culture, *V. inaequalis* degraded phloridzin to phloretin, phloroglucinol, phloretic acid, *p*-hydroxybenzoic acid, and glucose when an additional carbon source was added (40). It could not utilize phloridzin or phloretin as the sole carbon source.

Noveroske (85) studied the nature and role of host constituents associated with the host-pathogen interaction. Leaf washings from heavily inoculated leaves of the hypersensitive selection *M. atrosanguinea* 333-9 inhibited spore germination of *V. inaequalis;* whereas washings from uninoculated leaves of the same selection and leaves of selection 384-1, exhibiting sporulating lesions, did not. *V. inaequalis* germinates on the cuticle of susceptible and resistant hosts and, with the exception of hosts with hypersensitive resistance, makes considerable growth beneath the cuticle. Consequently, growth rather than spore germination would be a far better assay for the presence of fungitoxic materials in apple leaf; however, the slow growth of the fungus makes it difficult to study the effect of unstable inhibitors for other than a fungicidal effect.

Extracts of unoxidized leaf tissue of eight different apple selections were more inhibitory to spore germination when assayed in the light than in the dark (88). Solutions of phloridzin and extracts of unoxidized tissue held in the light were toxic when assayed in the dark, demonstrating that the effect of light was on components of the solutions rather than the fungus. It was suggested that toxicity in extracts of unoxidized tissue resulted from the autoxidation of phloridzin. Toxicity in extracts from oxidized plant tissue did not change when assayed in the light or dark. Toxicity to spore germination of leaf extracts also varied with the length of time tissue autolyzed prior to extraction. Inhibition increased and then decreased, phloridzin decreased, and phloretin first increased, then decreased, paralleling changes in inhibition. Autolyzing leaves in the presence of the reducing agent sodium metabisulfite reduced toxicity, phloretin accumulated and poridzin hydrolysis was unaffected. Autolysis in the presence of the β-glycosidase inhibitor glucono-1,5-lactone prevented phloretin accumulation, reduced toxicity levels about 50 per cent, and reduced the rate of phloridzin hydrolysis.

The conclusion that both phloridzin and phloretin are oxidized by enzymes in apple leaves (88) was substantiated by further experiments (86). Intermediates, suggested to be the dihydroxy analogues of phloridzin and phloretin, were detected when enzymes of apple leaves were incubated with phloridzin. No correlation existed between phloridzin content, activity of β-glycosidase or phenoloxidase, and resistance with eight selections of apple. Since a potential to restrict *V. inaequalis* appeared to exist in resistant as well as susceptible hosts, the mechanism of resistance was suggested to reside in the sensitivity of host tissue to metabolites produced by the fungus. A diffusate has been obtained from germinating spores of the fungus which caused collapse of leaf tissue (85, 88); however, collapse of both susceptible and resistant tissue was observed.

Raa (93) and Raa & Overeem (94) confirmed the findings of Noveroske and co-workers (85, 86, 88) that: (*a*) there was no relation between phloridzin content of the host and resistance; (*b*) phenoloxidase and β-glycosidase were present in resistant and susceptible varieties; (*c*) phloridzin was converted to phloretin by leaf homogenates; and (*d*) both phloridzin and phloretin were oxidized by leaf homogenates. They further confirmed the work of Holowczak, Kuć & Williams (40), indicating that phloretin, but not phloridzin, inhibited the growth of *V. inaequalis*. They disagree with Noveroske and co-workers in that oxidation products of phloretin are the principal inhibitors of the fungus. They present evidence that oxidation products of phloridzin rather than phloretin are of prime importance in the conversions of phloridzin catalyzed by phenoloxidase and β-glycosidase. In incubation mixtures of phloridzin with acetone powders of apple leaves, phloridzin was shown to be oxidized to 3-hydroxyphloridzin and further to the correspondong *o*-quinone. Optimum pH for both reactions is between 4 and 5 with a sharp maximum. 3-Hydroxyphloridzin does not accumulate in the incubation mixture unless reducing substances are present during oxidation. The *o*-quinone of phloridzin is extremely reactive and undergoes spontaneous oxidative coupling reactions with the phloroglucinol nucleus of phloretin. Thus the phloretin level decreases as phloridzin is oxidized. The *o*-quinone can also react with amino groups of amino acids. Simultaneous with oxidation, enzymatic hydrolysis of phloridzin takes place with the formation of glucose and phloretin. Phloretin is oxidized, however, at only 7 to 14 per cent of the rate of phloridzin. The oxidation product of phloretin is 3-hydroxyphloretin.

The report by Noveroske, Kuć & Williams (86) that phloridzin was quantitatively transformed into phloretin before appreciable oxidation took place may be explained by a difference in the preparation of the crude enzymes. As suggested by Raa (93), it appears that the relative proportion of β-glucosidase to phenoloxidase in the enzyme preparations employed by these workers was higher than that used by Raa. The reason for the discrepancies may be that Noveroske's group studied the transformation reactions of phloridzin in dilute aqueous extracts of homogenized leaves; whereas Raa used undiluted preparations. Since the β-glucosidase from various sources dissolves readily in water (121), whereas the extraction of the polyphenoloxidase occurs more slowly (34), it is possible that the ratio of phenoloxidase to β-glucosidase is higher in undiluted leaf preparations than in the extracts used by Noveroske's group. A scheme for the transformation reactions of phloridzin in the presence of apple leaf enzymes is presented by Raa (93) and Raa & Overeem (94).

The importance of phloridzin and its transformations as reported by both Raa's and Noveroske's group has a bearing on the disease resistance problem only if the toxic oxidation products can be shown to occur in resistant but not in susceptible hosts. Since, apparently, host cell collapse is associated with these transformations, a metabolite produced by the fungus capable of causing the collapse of cells in resistant but not in susceptible varieties is suggested. It is uncertain, however, whether the collapse of host cells is due to such a metabolite or rather is due to the formation of oxidation products of phloridzin and phloretin. It is entirely possible that an extra-cellular metabolite produced by the fungus may cause a loss of compartmentation of enzymes and substrate in the epidermal cells initiating the mixing of phloridzin, phenoloxidase, and β-glycosidase, which leads to the formation of fungitoxic and phytotoxic oxidation products and host cell collapse. It is also uncertain if the fungus is restricted prior to cell collapse and phloridzin transformations.

Control of disease resistance and susceptibility.—Kirkham (59) found that water-soluble plant phenols may be strongly inhibitory to the growth and sporulation of *V. inaequalis,* depending on the amount and source of nitrogen in the medium. He infused apple leaves with the phenolic fraction extracted from Cox's Orange and found that this reduced the amount of leaf surface covered by scab lesions. The infusion of urea with the phenolic fraction negated this effect, and urea alone increased the amount of leaf area covered by lesions. Infusions produced their maximum effect on leaves approaching the mature resistant stage. In subsequent experiments (26, 60–63, 65, 66), it was reported that cinnamic acid, *o*-coumaric acid, and esters of *o*-coumaric acid inhibited the growth and sporulation of the fungus *in vitro* and reduced the amount of leaf area covered by lesions *in vivo.* The isobutyl esters of *o*-coumaric and cinnamic acids showed systemic activity. In other experiments (73), phenylthiourea, a potent inhibitor of phenoloxidase, was infused through leaf petiols into actively growing shoots of potted apple trees resistant to race 1 but susceptible to race 2. The compound conditioned resistance to both races. Phenylthiourea prevented growth of the fungus at the concentrations used for infusion. Another polyphenoloxidase inhibitor, 4-chlororesorcinol, fed at levels not appreciably inhibitory to the fungus *in vitro* was found to induce sporulation in lesions on a resistant host (87). The 4-chlororesorcinol infused into the plant is readily converted to 1-0-(β-D-glucopyranosyl)-4-chlororesorcinol (107). The infusion of 4-chlororesorcinol into shoots markedly inhibited peroxidase and polyphenoloxidase but did not affect ascorbic acid oxidase and β-glucosidase (107). At concentrations as low as 8×10^{-8} *M*,4-chlororesorcinol inhibited peroxidase activity. The effect of the phenylthiourea can probably be accounted for solely on the basis of its fungitoxicity. The reversal of resistance by 4-chlororesorcinol may be due to its inhibitory activity on peroxidase, or polyphenoloxidase, or both; implying that an oxidation product is important in the disease resistance mechanism. This supports the work of

Raa and Noveroske dealing with oxidation products of phloridzin and phloretin.

In studies on the chemotherapeutic effect of amino acids on susceptibility of apple varieties to scab, Kuć et al., (72) reported that the D- and DL-isomers of phenylalanine affected the scab resistance of seven apple selections. A solution of the amino acid was infused into growing shoots, and leaves above points of infusion were then inoculated. The number and size of sporulating lesions were markedly reduced on five commercial varieties, Golden Delicious, McIntosh, Jonared, Yellow Transparent, and Gallia Beauty, infused with D- or DL-phenylalanine. These commercial varieties are susceptible to races 1, 2, and 3 of *V. inaequalis* and infusion did not result in a resistant reaction. After infusion with D- or DL-phenylalanine, selections of R12740-7A, susceptible to race 2 and resistant to races 1 and 3, and Geneva, susceptible to race 3 and resistant to races 1 and 2, were resistant to all three races. The L- or naturally occurring isomer of phenylalanine did not alter the disease reaction or severity of symptoms of the seven selections tested. Solutions of D-alanine, DL-α-aminobutyric acid, and to a lesser extent D- and DL-leucine had an effect similar to that of D- or DL-phelylalanine, whereas DL- or L-alanine, L-leucine, DL-, D- of L-tyrosine, DL-, D-, or L-dihydroxyphenylalanine, and DL- or L-cysteine hydrochloride had no effect. The DL-, D-, and L-isomers of dihydroxyphenylalanine were inhibitory *in vitro* to several isolates of race 2; however, the other amino acids were not appreciably inhibitory *in vitro* at concentration equal to or greater than those used for infusion.

In further studies, leaves of 384-1 infused with 0.03 *M* α-aminoisobutyric acid (AIB) exhibited pitting after inoculation with virulent strains of *V. inaequalis* (78). Leaves of three commercial varieties infused with the compound and inoculated exhibited either no reaction or small nonsporulating necrotic lesions. Pitting was also evident on inoculated leaves of 384-1 and three commercial varieties infused with D-α-amino-*n*-butyric acid. L-α-Amino-*n*-butyric acid, DL-β-amino-*n*-butyric acid, butyric acid, DL-β-aminoisobutyric acid, isobutyric acid, and ammonium nitrate had little or no effect on the disease reaction. The acids were not fungitoxic at concentrations up to 0.1 *M*, and AIB up to 0.4 *M*, *in vitro*. The pathogen grown on media containing 0.03 *M* AIB did not lose virulence. Physiological studies with AIB-1-C^{14} showed that the fungus decarboxylated 0.4 per cent of the compound in 2 weeks and host tissue decarboxylated 1 per cent in 48 hr. The host-parasite combination produced acidic and neutral breakdown products to the extent of 0.5 to 1.1 per cent in 6 days. There was no stimulation of AIB breakdown by the host-parasite combination and no evidence that the small amounts of breakdown products were involved in the induced resistance. Glycine or the L-forms of alanine, valine, leucine, isoleucine, serine, or threonine did not alter the host reaction, nor did they reverse the chemotherapeutic action of AIB when fed in combination with it.

In light of the implication of phloridzin and phloretin in the disease resistance of apple to *V. inaequalis,* it is interesting that Holowczak, Kuć & Williams (39) reported that C^{14}-labeled DL-phenylalanine caused the accumulation of labeled phenolic compounds in two apple varieties with differential resistance. Two of these phenolics were identified as phloretin and phloretic acid. No such accumulation occurred in a susceptible variety infused with either DL- or L-isomers. When the DL- or L-forms of labeled phenylalanine were fed to the three varieties tested, large amounts of label were found in phloridzin. Label was restricted to the β-ring of the phloretic acid portion of the molecule. The accumulation of labeled phenolics in tissues of varieties with differential resistance fed DL-phenylalanine may explain the induced resistance. In the susceptible variety, phenolics did not accumulate and the D- or DL-isomers of phenylalanine did not induce a resistant reaction.

In light of the universality of resistance and the apparent high degree of metabolic adaptation necessary for susceptibility, it is interesting that most reports in the literature describe a shift from susceptibility to resistance, but not from resistance to susceptibility. This suggests that a disruption of the normal metabolism of the host by introducing a "stress" factor can disrupt the metabolic balance necessary for susceptibility. The infusion of chemicals into apple leaves has provided information useful in disease control, but has shed little light on the mechanisms for disease resistance other than the generalization that abnormal metabolites, probably capable of inducing metabolic stress, induce resistance. This adds evidence to the concept that the genetic information for disease resistance is often present but not expressed even in highly susceptible hosts (71).

Discussion

The genetic regulation of metabolism or structure of apple and *V. inaequalis* which determines susceptibility or resistance is unknown. Indeed, the reason for containment of the fungus in the resistant host is uncertain. However, several conclusions concerning the interaction appear indisputable. The existence of quantitative and specific qualitative genes for resistance in the host and many genes for pathogenicity in the fungus eliminate the possibility of a unique host factor or unique fungal factor as determinants of susceptibility or resistance in all host-pathogen interactions. The multiplicity of genetic control and its specificity strongly support the concept that susceptibility or resistance is determined by contributions of both host and pathogen. The existence of differentially resistant selections, the ability to induce resistance in susceptible hosts, and the containment of the fungus in defined lesions of susceptible varieties argue for the existence of genetic information for resistance in most if not all hosts.

Resistance or susceptibility may depend on the expression of structural genes by regulatory metabolites, activating or inactivating repressor molecules, giving rise to the appropriate metabolic environment. If a host metabolite is responsible for containment of the fungus, it is not necessary to postulate a different host metabolite for each resistant host-parasite interaction. Specificity may be determined by the ability to induce a specific metabolic environment leading to an inhibitor common to resistance. Thus, phloridzin is common in apple, but its oxidation and hydrolysis may be determined by specific effects on cell permeability.

The suggestions of Raa (93) concerning genetic control serve as an orientation for thought on the problem. He postulates genes for avirulence in the pathogen and genes for resistance in the host, leading to a common metabolite responsible for fungus containment. An equally plausable explanation is that genes for virulence repress the expression of resistance genes in the host. This concept would be consistent with existing views concerning the inheritance of genes for virulence and the universality of resistance, with a high degree of metabolic compatibility between host and pathogen being most often characteristic of susceptibility.

LITERATURE CITED

1. Aderhold, R. Ein Beitrag zur Frage der Empfanglichkeit der Apfelsorten für *Fusicladium dendriticum* (Wallr.) Fckl. und deren Beziehungen zum Wetter. *Arb. K. Biol. Anst. Landw. Forstw.,* **2,** 560–66 (1902)
2. Aderhold, R. Kann das *Fusicladium* von *Crataegus* and von *Sorbus*-Arten auf den Apfelbaum ubergehen? *Arb. K. Biol. Anst. Landw. Forstw.,* **3,** 436–39 (1903)
3. Backus, E. J., Keitt, G. W. Some nuclear phenomena in *Venturia inaegualis. Bull. Torrey Bot. Club,* **67,** 765–70 (1940)
4. Bagga, H. S., Boone, D. M. Genes in *Venturia inaequalis* controlling pathogenicity to crabapples. *Phytopathology,* **58,** 1176–82 (1968)
5. Bagga, H. S., Boone, D. M. Inheritance of resistance to *Venturia inaequalis* in crabapples. *Phytopathology,* **58,** 1183–87 (1968)
6. Barnes, E. H., Williams, E. B. A biochemical response of apple tissue to fungus infection. *Phytopathology,* **50,** 844–46 (1960)
7. Barnes, E. H., Williams, E. B. The role of phloridzin in the host-parasite physiology of the apple scab disease. *Can. J. Microbiol.,* **7,** 525–34 (1961)
8. Biehn, W. L., Williams, E. B., Kuć, J. Resistance of mature leaves of *Malus atrosanguinea* 804 to *Venturia inaequalis* and *Helminthosporium carbonum. Phytopathology,* **56,** 588–89 (1966)
9. Boone, D. M., Keitt, G. W. *Venturia inaequalis* (Cke.) Wint. XII. Genes controlling pathogenicity of wild type lines. *Phytopathology,* **47,** 403–9 (1957)
10. Boone, D. M., Keitt, G. W. Genetics and nutrition of *Venturia inaequalis* in relation to its pathogenicity. *Recent Advances in Botany. Rept. Intern. Bot. Congr., 9th, Montreal, Can., Aug. 19–29,* 498–502 (1959)
11. Boone, D. M., Kline, D. M., Keitt, G. W. *Venturia inaequalis* (Cke.) Wint. XIII. Pathogenicity of induced biochemical mutants. *Am. J. Botany,* **44,** 791–96 (1957)
12. Boone, D. M., Stauffer, J. F., Stahmann, M. A., Keitt, G. W. *Venturia inaequalis* (Cke.) Wint. VII. Induction of mutants for studies on genetics, nutrition, and pathogenicity. *Am. J. Botany,* **43,** 199–204 (1956)
13. Bradfield, A. E., Flood, A. E., Hulme, A. C., Williams, A. H. Chlorogenic acids in fruit trees. *Nature,* **170,** 168 (1952)
14. Brown, A. G. The inheritance of mildew resistance in progenies of the cultivated apple. *Euphytica,* **8,** 81–88 (1959)
15. Brown, A. G. The inheritance of shape, size and season of ripening in progenies of the cultivated

apple. *Euphytica,* **9,** 327–37 (1960)

16. Crandall, C. S. Apple breeding at the University of Illinois. *Illinois Agr. Expt. Sta. Bull.,* **275,** 341–600 (1926)
17. Crane, M. B., Lawrence, W. J. C. Genetical studies in cultivated apples. *J. Genet.,* **28,** 265–96 (1933)
18. Crane, M. B., Lawrence, W. J. C. *The Genetics of Garden Plants* (Macmillan & Co. Ltd., London, 301 pp., 1952)
19. Day, P. R., Boone, D. M., Keitt, G. W. *Venturia inaequalis* (Cke.) Wint. XI. The chromosome number. *Am. J. Botany,* **43,** 835–38 (1956)
20. Dayton, D. F. Progress in breeding scab immune apples. *Trans. Illinois State Hort. Soc.,* **92,** 54–56 (1958)
21. Dayton, D. F., Shay, J. R., Hough, L. F. Apple scab resistance from R12740-7A, a Russian apple. *Proc. Am. Soc. Hort. Sci.,* **62,** 334–40 (1953)
22. Dayton, D. F., Williams, E. B. Independent genes in *Malus* for resistance to *Venturia inaequalis. Proc. Am. Soc. Hort. Sci.,* **92,** 89–94 (1968)
23. Dayton, D. F., Williams, E. B., Shay, J. R. Gene pools for apple scab resistance. *Proc. Intern. Hort. Congr., 17th., College Park, Maryland, Aug. 15–20,* 13 (1966)
24. den Boer, A. F., Collins, W. H. Apple scab resistance in ornamental crab apples. *Trans. Iowa Hort. Soc.,* **86,** 44–64 (1952)
25. Enochs, Nettie J. *Histological Studies on the Development of* Venturia inaequalis *(Cke.) Wint. in Susceptible and Resistant Selections of Malus* (Master's thesis, Purdue Univ., Lafayette, Indiana, 1964)
26. Flood, A. E., Kirkham, D. S. The effect of some phenolic compounds on the growth and sporulation of two *Venturia* species. In *Phenolics in Plants in Health and Disease,* 81–85 (Pridham, J. B., Ed., Pergamon Press, New York, 131 pp., 1960)
27. Fothergill, P. G., Ashcroft, R. The nutritional requirements of *Venturia inaequalis. J. Gen. Microbiol.,* **12,** 387–95 (1955)
28. Franke, W. Ectodesmata and foliar adsorption. *Am. J. Botany,* **48,** 683–91 (1961)
29. Frey, C. N. The cytology and physiology of *Venturia inaequalis* (Cooke) Wint. *Trans. Wisconsin Acad. Sci., Arts and Letters,* **21,** 303–43 (1924)
30. Granhall, I. Kan äpfelskorven bemästras genom växforadling? *Sverig. Pomol. Foren Arsskr.,* **54,** 17–28 (1953)
31. Grijseels, A. J., Williams, E. B., Kuć, J. Hypersensitive response in selections of *Malus* to fungi nonpathogenic to apple. *Phytopathology,* **54,** 1152–54 (1964)
32. Hall, A. D., Crane, M. B. *The Apple* (Martin Hopkinson, Ltd., London, 235 pp., 1933)
33. Hansen, N. E. Plant introductions, 1895–1927. *South Dakota Expt. Sta. Bull.,* **224,** 1–64 (1927)
34. Harel, E., Mayer, A. M., Shain, Y. Purification and multiplicity of catechol oxidase from apple chloroplasts. *Phytochemistry,* **4,** 783–90 (1965)
35. Hignett, R. C., Kirkham, D. S. Factors concerned in host specificity of *Venturia inaequalis* (Cke.) Wint. *Biochem. J.,* **97,** 34 (1965)
36. Hignett, R. C., Kirkham, D. S. The role of extracellular melanoproteins of *Venturia inaequalis* in host susceptibility. *J. Gen. Microbiol.,* **48,** 269–75 (1967)
37. Hirst, J. M., Storey, I. F., Ward, W. C., Wilcox, H. J. The origin of apple scab epidemics in the Wisbeck area in 1953 and 1954. *Plant Pathol.,* **4,** 91–96 (1955)
38. Hockey, J. E., Eidt, C. C. Resistance of apple seedlings to scab. *Sci. Agr.,* **24,** 542–50 (1944)
39. Holowczak, J., Kuć, J., Williams, E. B. Metabolism of DL- and L-phenylalanine in *Malus* related to susceptibility and resistance to *Venturia inaequalis. Phytopathology,* **52,** 699–703 (1962)
40. Holowczak, J., Kuć, J., Williams, E. B. Metabolism *in vitro* of phloridzin and other compounds by *Venturia inaequalis. Phytopathology,* **52,** 1019–23 (1962)
41. Hough, L. F. A survey of the scab resistance of the foliage on seedlings in selected apple progenies. *Proc. Am. Soc. Hort. Sci.,* **44,** 260–72 (1944)
42. Hough, L. F., Shay, J. R. Breeding for scab resistant apples. *Phytopathology,* **39,** 10 (1949) (Abstr.)
43. Hough, L. F., Shay, J. R., Dayton, D. F. Apple scab resistance from *Malus floribunda* Sieb. *Proc. Am.*

Soc. Hort. Sci., **62,** 341–47 (1953)

44. Hunter, L. D., Kirkham, D. S. Active resistance of apples to scab. *Intern. Congr. Plant Pathol., 1st, London, July,* 93 (1968) (Abstr.)
45. Hutchinson, A., Taper, C. D., Towers, G. H. N. Studies of phloridzin in *Malus. Can. J. Biochem. Biophys.*, **37,** 901–10 (1959)
46. Isaev, S. I. (Some results of forty years apple breeding.) *Sel'skohoz. Biol.*, **11,** 170–82 (1967). Taken from *Plant Breed. Abstr.*, **38,** 146 (1968)
47. Julien, J. B. Cytological studies of *Venturia inaequalis. Can. J. Botany,* **36,** 607–13 (1958)
48. Julien, J. B., Spangelo, L. P. S. Physiological races of *Venturia inaequalis. Can. J. Plant Sci.*, **37,** 102–07 (1957)
49. Keitt, G. W. Inheritance of pathogenicity in *Venturia inaequalis* (Cke.) Wint. *Am. Naturalist,* **86,** 373–90 (1952)
50. Keitt, G. W., Boone, D. M., Shay, J. R. Genetics and nutritional controls of host-parasite interactions in apple scab. In *Plant Pathology, Problems and Progress, 1908–1958,* 156–67 (Univ. Wisconsin Press, Madison, Wis., 588 pp., 1959)
51. Keitt, G. W., Langford, M. H. A preliminary report on genetic studies on pathogenicity and the nature of saltation in *Venturia inaequalis. Phytopathology,* **31,** 1142 (1941) (Abstr.)
52. Keitt, G. W., Langford, M. H. *Venturia inaequalis* (Cke.) Wint. I. A groundwork for genetic studies. *Am. J. Botany,* **28,** 805–20 (1941)
53. Keitt, G. W., Langford, M. H., Shay, J. R. *Venturia inaequalis* (Cke.) Wint. II. Genetic studies on pathogenicity and certain mutant characters. *Am. J. Botany,* **30,** 491–500 (1943)
54. Keitt, G. W., Leben, C., Shay, J. R. *Venturia inaequalis* (Cke.) Wint. IV. Further studies on the inheritance of pathogenicity. *Am. J. Botany,* **35,** 334–36 (1948)
55. Keitt, G. W., Nusbaum, C. J. Cytological studies of the parasitism of two monoconidial isolates of *Venturia inaequalis* on the leaves of susceptible and resistant apple varieties. *Phytopathology,* **26,** 97–98 (1936) (Abstr.)
56. Keitt, G. W., Palmiter, D. H. Heterothallism in *Venturia inaequalis. Science,* **85,** 498 (1937)
57. Keitt, G. W., Pamiter, D. H. Heterothallism and variability in *Venturia inaequalis. Am. J. Botany,* **25,** 338–45 (1938)
58. Killian, K. Uber die Sexualitat von *Venturia inaequalis* (Cooke) Aderh. *Z. Botan.*, **9,** 353–98 (1917)
59. Kirkham, D. S. Significance of the ratio between the water soluble aromatic and nitrogen constituents of apple and pear in the host-parasite relationships of *Venturia* species. *Nature,* **173,** 690–91 (1954)
60. Kirkham, D. S. Studies on the significance of polyphenolic host metabolites in the nutrition of *Venturia inaequalis* and *Venturia pirina. J. Gen. Microbiol.*, **17,** 120–34 (1957)
61. Kirkham, D. S. The significance of polyphenolic metabolites of apple and pear in the host relations of *Venturia inaequalis* and *Venturia pirina. J. Gen. Microbiol.*, **17,** 491–504 (1957)
62. Kirkham, D. S., Flood, A. E. Inhibition of *Venturia* spp. by analogues of host metabolites. *Nature,* **179,** 422–23 (1956)
63. Kirkham, D. S., Flood, A. E. Some effects of respiration inhibitors and *o*-coumaric acid on the inhibition of sporulation in *Venturia inaequalis. J. Gen. Microbiol.*, **32,** 123–29 (1963)
64. Kirkham, D. S., Hignett, R. C. Control of host susceptibility and solute transport by metabolites of *Venturia inaequalis. Nature,* **212,** 211 (1966)
65. Kirkham, D. S., Hunter, L. D. Systemic antifungal activity of isobutyl *o*-coumarate in apple. *Nature,* **201,** 638–39 (1964)
66. Kirkham, D. S., Hunter, L. D. Studies of the *in vivo* activity of esters of *o*-coumaric and cinnamic acids against apple scab. *Ann. Appl. Biol.*, **55,** 359–71 (1965)
67. Kline, D. M., Boone, D. M., Keitt, G. W. *Venturia inaequalis* (Cke.) Wint. XIV. Nutritional control of pathogenicity of certain induced biochemical mutants. *Am. J. Botany,* **44,** 797–803 (1957)
68. Klein, L. G. The inheritance of certain fruit characters in the apple. *Proc. Am. Soc. Hort. Sci.*, **72,** 1–14 (1958)
69. Knight, R. L. Abstract bibliography of fruit breeding and genetics to

1960. *Malus* and *Pyrus*. *Commonwealth Bureau Horticulture and Plantation Crops. Tech. Comm.*, **29,** 535 pp. (1963) (Abstr.)

70. Kotecov, P. The susceptibility of apple varieties to scab, *Venturia inaequalis* (Cooke) Aderhold. *Nauchni Tr. Vissh Selskostopanshi Inst. Agron. Fak.*, **15,** 333–36 (1966). Taken from *Plant Breed. Abstr.*, **37,** 369 (1967)
71. Kuć, J. Biochemical control of disease resistance in plants. *World Rev. Pest Control,* **7,** 42–55 (1968)
72. Kuć, J., Barnes, E. H., Daftsios, A., Williams, E. B. The effect of amino acids on susceptibility of apple varieties to scab. *Phytopathology,* **49,** 313–15 (1959)
73. Kuć, J., Williams, E. B., Shay, J. R. Increase of resistance to apple scab following injection of host with phenylthiourea and D-phenylalanine. *Phytopathology,* **47,** 21–22 (1957) (Abstr.)
74. Lamb, J. G. D. The apple in Ireland; its history and varieties. *Econ. Proc. Roy. Dublin Soc.*, **4,** 1–63 (1951)
75. Lamb, R. C., Hamilton, J. M. Breeding scab resistant apples. *N. Y. State Agr. Expt. Sta., Geneva, N.Y., Farm Rept.*, **22,** 15 (1956)
76. Leben, C., Keitt, G. W. *Venturia inaequalis* (Cke.) Wint. V. The influence of carbon and nitrogen sources and vitamins on growth *in vitro*. *Am. J. Botany,* **35,** 337–43 (1948)
77. Lewis, D., Crane, M. B. Genetical studies in apples. *J. Genet.*, **37,** 119–28 (1938)
78. MacLennan, D. H., Kuć, J., Williams, E. B. Chemotherapy of the apple scab disease with butyric acid derivatives. *Phytopathology,* **53,** 1261–66 (1963)
79. McCrory, S. A. Preliminary evaluation and description of domestic and introduced fruit plants. *South Dakota Agr. Expt. Sta. Bull.*, **471,** 1–39 (1958)
80. McCrory, S. A., Shay, J. R. Apple scab resistance survey of South Dakota apple varieties and breeding stocks. *Plant Disease Reptr.*, **35,** 433–34 (1951)
81. Morris, H. E. A contribution to our knowledge of apple scab. *Montana Agr. Expt. Sta. Bull.*, **96,** 69–102 (1914)
82. Nebel, B. R. Recent findings in cytology of fruits (Cytology of *Pyrus* III). *Proc. Am. Soc. Hort. Sci.*, **27,** 406–10 (1930)
83. Nichols, L. P. Some observations on the susceptibility of flowering crabapples to scab. *Arborist News,* **27,** 75–76 (1962)
84. Nichols, L. P. Observations in 1961 and 1962 on the relative susceptibility of cultivars of ornamental crabapples to scab, cedar rust and powdery mildew. *Plant Disease Reptr.*, **47,** 311–14 (1963)
85. Noveroske, R. L. *Phloridzin—Its Metabolism and Role in Conditioning Resistance in* Malus *to Venturia inaegualis* (Doctoral thesis, Purdue University, Lafayette, Indiana, 107 pp., 1962)
86. Noveroske, R. L., Kuć, J., Williams, E. B. Oxidation of phloridzin and phloretin related to resistance of *Malus* to *Venturia inaequalis*. *Phytopathology,* **54,** 92–97 (1964)
87. Noveroske, R. L., Williams, E. B., Kuć, J. Breakdown of host resistance to *Venturia inaequalis* by a polyphenoloxidase inhibitor. *Phytopathology,* **52,** 23 (1962) (Abstr.)
88. Noveroske, R. L., Williams, E. B., Kuć, J. β-Glycosidase and phenoloxidase in apple leaves and their possible relation to resistance to *Venturia inaequalis*. *Phytopathology,* **54,** 98–103 (1964)
89. Nusbaum, C. J., Keitt, G. W. A cytological study of host-parasite relations of *Venturia inaequalis* on apple leaves. *J. Agr. Res.*, **56,** 595–618 (1938)
90. Palmiter, D. H. Variability of *Venturia inaequalis* in cultural characters and host relations. *Phytopathology,* **22,** 21 (1932) (Abstr.)
91. Palmiter, D. H. Variability in monoconidial cultures of *Venturia inaequalis*. *Phytopathology,* **24,** 22–47 (1934)
92. Pelletier, R. L., Keitt, G. W. *Venturia inaequalis* (Cke.) Wint. VI. Amino acids as sources of nitrogen. *Am. J. Botany,* **41,** 362–71 (1954)
93. Raa, J. *Natural Resistance of Apple Plants* to Venturia inaequalis (Doctoral thesis, University of Utrecht, Utrecht, the Netherlands, 100 pp., 1968)
94. Raa, J., Overeem, J. C. Transformation reactions of phloridzin in the presence of apple leaf enzymes. *Phytochemistry,* **7,** 721–31 (1968)

95. Remy, P., Decourtye, L. État d'avancement des recherches sur l'amelioration du pommier pour la resistance à la tavelure. *Ann. Amelior. Plantes,* **12,** 219–44 (1962)

96. Richmond, D. V., Martin, J. T. Studies on plant cuticles. III. The composition of cuticles of apple leaves and fruits. *Ann. Appl. Biol.,* **47,** 583–92 (1959)

97. Rudloff, C. F., Schmidt, M. *Venturia inaequalis* (Cooke) Aderh. II. Zur Zuchtung schorfwiderstandsfahiger Apfelsorten. *Zuchter,* **6,** 288–94 (1934)

98. Rudloff, C. F., Schmidt, M. Der Erreger des Apfelschorfes *Venturia inaequalis* (Cooke) Aderh. Grundlagen und Moglichkeiten für senie Bekämpfung auf Zuchterischen Wage II. *Zuchter,* **7,** 65–74 (1935)

99. Sampson, D. R., Cameron, D. F. Inheritance of bronze foliage, extra petals and pendulous habit in ornamental crabapples. *Proc. Am. Soc. Hort. Sci.,* **86,** 717–22 (1965)

100. Sarasola, Maria D. C. Resistencia a la sarna del manzano en la Argentina. *Rev. Invest. Agricolas,* **15,** 673–81 (1961)

101. Sax, K. The cytogenetics of facultative apomixis in *Malus* species. *J. Arnold Arboretum,* **40,** 289–97 (1959)

102. Schmidt, M. *Venturia inaequalis* (Cooke) Aderhold. VI. Zur Fage nach den Vorkommen physiologischer spezialisierter Rassen bein Erreger des Apfelschorfes. Erste Mitteilung. *Gartenbauwissenschaft,* **10,** 478–99 (1936)

103. Schmidt, M. *Venturia inaequalis* (Cooke) Aberhold. VI. Zur Frage phologie und Physiologie der Widerstandsfahigkeit gegen den Erreger des Apfelschorfes. *Gartenbauwissenshaft,* **11,** 221–30 (1937)

104. Schmidt, M. *Venturia inaequalis* (Cooke) Aderhold. VIII. Weitere Untersuchungen zur Zuchtung schorfwiderstandsfahiger Apfelsorten. *Zuchter,* **10,** 280–91 (1938)

105. Schmidt, M. *Venturia inaequalis* (Cooke) Aderhold. IX. Fünfjahrige Freilandbeobachtungen über den Schorfbefall von Apfelsorten. *Gartenbauwissenschaft,* **13,** 567–86 (1939)

106. Schmidt, M. Untersuchungen über den Biologie von *Venturia inaequalis* in Zusammenhang mit der Züchtung schorfwiderstandfahiger Apfelsorten. *Gartenbauwissenschaft,* **18,** 19–20 (1942)

107. Settle, Eileen J. *Biosynthesis of 1-o(β-D-Glucopyranosyl)-4-Chlororesorcinol and the Effect of 4-Chlororesorcinol on Peroxidase, Polyphenoloxidase, Ascorbic Acid Oxidase and β-Glucosidase in* Malus atrosanguinea (Master's thesis, Purdue University, Lafayette, Indiana, 46 pp., 1965)

108. Shay, J. R. Breeding vegetable and fruit crops for resistance to disease. Biological and chemical control of plant and animal pests. *Am. Assoc. Advan. Sci.,* **61,** 229–44 (1960)

109. Shay, J. R. Need for extensive disease testing of apple breeding lines. *Proc. Intern. Hort. Congr., 17th, College Park, Maryland, Aug. 15–20,* 27–30 (1966)

110. Shay, J. R., Brown, A. G. Problems in disease resistance in certain of the tree fruits. *Recent Advances in Botany. Rept. Intern. Bot. Congr., 9th, Montreal, Can., Aug. 19–29,* 448–51 (1959)

111. Shay, J. R., Dayton, D. F., Hough, L. F. Apple scab resistance from a number of *Malus* species. *Proc. Am. Soc. Hort. Sci.,* **62,** 348–56 (1953)

112. Shay, J. R., Dayton, D. F., Hough, L. F., Williams, E. B., Janick, J. Apple scab resistance. *Rept. Intern. Hort. Congr., 14th, Wageningen, the Netherlands,* 733–39 (1955)

113. Shay, J. R., Hough, L. F. Evaluation of apple scab resistance in selections of *Malus. Am. J. Botany,* **39,** 288–97 (1952)

114. Shay, J. R., Keitt, G. W. The inheritance of certain mutant characters in *Venturia inaequalis. J. Agr. Res.,* **70,** 31–41 (1945)

115. Shay, J. R., Williams, E. B. Identification of three physiologic races of *Venturia inaequalis. Phytopathology,* **46,** 190–93 (1956)

116. Shay, J. R., Williams, E. B., Janick, J. Disease resistance in apple and pear. *Proc. Am. Soc. Hort. Sci.,* **80,** 97–104 (1962)

117. Spangelo, L. P. S., Julien, J. B. Breeding apples for resistance to scab. *Can. Dept. Agr. Res. for Farmers,* **8,** 3–4 (1963)

118. Spangelo, L. P. S., Julien, J. B., Racicot, H. N., Blair, D. S. Breeding apples for resistance to scab.

Can. J. Agr. Sci., **36**, 329–38 (1956)

119. Van Eseltine, G. P. Notes on the species of apples. II. The Japanese flowering crabapples of the *Sieboldii* group and their hybrids. *N.Y. State Agr. Expt. Sta., Geneva, N.Y., Tech. Bull.*, **214**, 1–21 (1933)
120. Van Eseltine, G. P. Flowering Crabapples. *Morris Arboretum Bull.*, **1**, 103–5 (1937)
121. Viebel, S. β-Glycosidase. In *The Enzymes*. Vol. **1**, 583–620 (Summer, J. B., Myrback, K., Eds., Academic Press, New York, 724 pp., 1950)
122. Wallace, E. Scab disease of apples. *N.Y.* (Cornell) *Agr. Expt. Sta. Bull.*, **335**, 545–92 (1913)
123. Wellington, R. An experiment in breeding apples. II. *N.Y. Agr. Expt. Sta., Geneva, N.Y., Tech. Bull.*, **106**, 1–149 (1924)
124. Wiesmann, R. Untersuchungen über Apfel- und Birnschorfpilz *Fusicladium pirinum* (Lib.) Fckl. sowie die Schorfanfälligkeit einzelner Apfel- und Birnsorten. *Landwirtsch. Jahrb. Schweiz*, **35**, 109–56 (1931)
125. Williams, A. H. Phenolic substances in pear-apple hybrids. *Nature*, **175**, 213 (1955)
126. Williams, A. H. A new dihydrochalkone from *Malus* species. *Chem. & Ind.* London, 1306 (1956)
127. Williams, A. H. The dihydrochalkone of *Malus* species. *Chem. & Ind.* London, 934–35 (1960)
128. Williams, A. H. The distribution of phenolic compounds in apple and pear trees. In *Phenolics in Plants in Health and Disease*, 3–7 (Pridham, J. B., Ed., Pergamon Press, New York, 131 pp., 1960)
129. Williams, A. H. Dihydrochalkones of *Malus* species. *J. Chem. Soc.*, **4**, 33–36 (1961)
130. Williams, E. B. The inheritance of pathogenicity and certain other characters in *Venturia inaequalis* (Cke.) Wint. (Doctoral thesis, Purdue University, Lafayette, Indiana, 82 pp., 1954)
131. Williams, E. B. (Unpublished data)
132. Williams, E. B., Brown, A. G. A new physiologic race of *Venturia inaequalis*, incitant of apple scab. *Plant Disease Reptr.*, **52**, 799–801 (1968)
133. Williams, E. B., Dayton, D. F. Four additional sources of the V_f locus for *Malus* scab resistance. *Proc. Am. Soc. Hort. Sci.*, **92**, 95–98 (1968)
134. Williams, E. B., Dayton, D. F., Shay, J. R. Allelic genes in *Malus* for resistance to *Venturia inaequalis*. *Proc. Am. Soc. Hort. Sci.*, **88**, 52–56 (1966)
135. Williams, E. B., Shay, J. R. The relationship of genes for pathogenicity and certain other characters in *Venturia inaequalis* (Cke.) Wint. *Genetics*, **42**, 704–11 (1957)
136. Williams, E. B., Shay, J. R., Dayton, D. F. Allelism of genes in *Malus* for resistance to *Venturia inaequalis*. *Proc. Intern. Botan. Congr., 10th, Edinburgh, Scotland, Aug. 1–11*, 86–87 (1964)
137. Williams, E. B., Shay, J. R., Janick, J. Progress report on the apple scab resistance breeding project. *Trans. Indiana Hort. Soc.* **102**, 43–44 (1963)
138. Williams, W. Department of Plant Breeding. Breeding new varieties. *Rept. John Innes Hort. Inst.*, **47**, 5–11 (1956)
139. Williams, W., Brown, A. G. Breeding new varieties of fruit trees. *Endeavour*, **19**, 147–55 (1960)
140. Wiltshire, S. P. Infection and immunity studies on the apple and pear scab fungi (*Venturia inaequalis* and *V. pirina*). *Ann. Appl. Biol.*, **1**, 335–50 (1915)
141. Wyman, D. Where some of the crabapples come from. *N.Y. Botan. Garden J.*, **8**, 122–24 (1958)
142. Zwintzscher, M. Uber die Widerstandsfahigkeit von F_2 Bastarden des Apfels gegenüber dem Schorferreger, *Venturia inaequalis* (Cooke) Aderhold. *Gartenbauwissenschaft*, **19**, 22–35 (1954)
143. Zwintzscher, M. Fruit breeding in Western Europe. *Proc. Intern. Hort. Congr., 17th, College Park, Maryland, Aug. 15–20*, 31–39 (1966)

Copyright 1969. All rights reserved

MODES OF INFECTION AND SPREAD OF FOMES ANNOSUS

By Charles S. Hodges

Principal Plant Pathologist, Forest Service
United States Department of Agriculture
Research Triangle Park, North Carolina

Introduction

The root and butt rot caused by *Fomes annosus* is one of the most economically important diseases of conifers in the north temperate zone of the world. Sinclair (112) lists almost 150 species of trees, including some angiosperms, on which *F. annosus* has been reported. Species of *Abies, Juniperus, Larix, Picea, Pinus, Pseudotsuga,* and *Tsuga* are the major hosts. On most species of *Pinus* and *Juniperus, F. annosus* causes primarily a root rot and kills trees within several months to years after infection. The major exception is *Pinus strobus,* which is subject to typical butt rot as well as root rot, and trees usually remain alive for several years after infection unless windthrown. Spruces, true firs, larches, hemlocks, and Douglas fir may be killed by *F. annosus* when young, but later are primarily damaged by butt rot which may extend for several meters up the stem.

Koenigs' (59) bibliography lists 438 references on *F. annosus.* Since then, more than 300 additional papers have been published. This review will not attempt to evaluate all aspects of the *F. annosus* problem. Instead, attention will be focused only on the aspects of spread and infection which are important in devising controls for this disease.

Regardless of host and type of disease produced, spread of *F. annosus* is accomplished in two major ways: (*a*) by spores, and (*b*) by growth of mycelium across contacts between healthy and infected roots. Spores are responsible for initiation of new infection centers in previously uninfected stands whereas growth of mycelium at points of root contacts or grafts results in the enlargement of established infection centers. Factors relating to these two types of spread will be discussed separately.

Spread By Spores

Through the freshly cut stump surface.—In 1950, Rishbeth (100) observed that in every site where the disease developed, *F. annosus* was present in stumps created through thinning or harvesting the previous crop. Subsequently (101), in probably the most significant contribution to the infection biology of *F. annosus,* he demonstrated conclusively that freshly cut stump surfaces could be infected by airborne basidiospores. The fungus then colonized the stump and its root system and spread to nearby trees through root contact or grafts. Infection by this means was demonstrated by inoculating freshly cut stump surfaces with basidiospores and following progressive colonization of the stump by the fungus. Rishbeth also found that the fungus could be isolated from naturally infected stumps within a month after thinning, and later could be found growing in lateral roots away from the stump. These observations are consistent with infection of the cut surface by airborne spores. Further evidence for stump infection was obtained by painting or creosoting the stump surface immediately after felling. This materially reduced the percentage of stumps infected by *F. annosus.* A positive correlation between amount of stump infection and subsequent infection of residual trees was shown later (104). Earlier workers (26, 43, 88) had noted the relationship of infected stumps in the spread to residual trees but they did not determine how stumps were initially infected.

The infection of stump surfaces by spores of *F. annosus* has been substantiated indirectly in Europe, Canada, and the United States by use of chemical and biological stump-surface protectants, (7, 23, 24, 71, 78, 105, 112, 124, and others), and this means of spread is recognized as the primary way *F. annosus* enters a previously healthy stand.

The freshly cut stump surface is highly selective for *F. annosus* (71). However, this selectivity deceases with time after felling (17, 18, 101, 108, 125). Usually, little infection takes place more than 2 weeks after felling. This is attributed primarily to increasing fungal competition after that time (101, 108). Meredith (72) attempted to inoculate stumps already infected with other microorganisms but was unsuccessful, except for stumps containing blue stain fungi. Infections in these stumps were associated with small areas of wood apparently free of stain.

The period of highest stump infection corresponds primarily to the period of greatest spore production by the fungus. In Europe, Canada, and the northern United States, some spores are produced in most seasons of the year (10, 22, 40, 58, 71, 98, 101, 104, 112, 113, 125), except during periods of extreme cold or drought. Maximum production of spores usually occurs during the late summer or fall in these areas (58, 98, 112, 125). Orlos & Tawaroska (89), however, reported the greatest rate of spore deposition in Poland from May to September, with almost no deposition during other periods of the year. Some spores may be produced at temperatures below 0° C (40, 70) and disseminated even when there is 75 cm of snow on the ground (98, 113).

The rate of spore production by *F. annosus* in Europe, Canada, and the northern United States is directly correlated with temperature (112, 125), but may (112, 125) or may not (40, 98, 109a) be correlated with rainfall prior to trapping.

In the southeastern and central United States, greatest spore production occurs during fall, winter, and spring; few, if any, spores are produced during the summer (13, 25, 109a). Ross (109a) found that sporulation and temperature are inversely correlated in southernmost states. The estimated mean average temperature and the estimated maximum temperature 1 week prior to periods when no spores were trapped, averaged 28° C and 38° C, respectively. In the upper southern states where high and low temperature limits are not extreme, spores are produced in abundance throughout the year and spore production is not correlated with temperature.

Spore release by *F. annosus* apparently follows a diurnal pattern. Wood (122) measured spore release at 2-hr intervals for a 40-day period from September through November. He found that maximum release of spores occurred around midnight, and minimum release around noon. Hourly fluctuations in spore release were correlated with relative humidity and temperature 60 cm above the forest floor. Sinclair (112), on the other hand, found spore deposition lowest at night, increasing during the morning and rising sharply in the afternoon. The discrepancy between these two reports may be due to differences in trapping methods. Wood collected spores directly beneath sporophores, whereas Sinclair sampled the spores in the atmosphere.

A positive correlation is not always found between the number of spores produced and the degree of stump infection. Rishbeth (101) showed that the periods of maximum and minimum percentages of stumps infected varied from year to year. He believed that infection was influenced by the relative number of spores produced by competing microorganisms, especially *Peniophora gigantea,* rather than on the actual number of *F. annosus* spores produced. This hypothesis was confirmed by Meredith (72), who found that inoculation of stumps with a mixture of spores of *F. annosus* and *P. gigantea* resulted in only 23 per cent of the stump area being colonized, compared to 80 per cent when inoculated with *F. annosus* alone. Climatic factors may also influence infection of the stump surface. In the southeastern United States, temperatures at the stump surface often are high enough to kill mycelium of *F. annosus* in wood during the summer months (37, 107). Stump infection would not occur under these conditions, even if spores were present. Other unexplained factors may be involved in lack of infection of freshly cut stump surfaces when spores are present. Stumps of Douglas fir are highly susceptible to *F. annosus* when cut in the spring but not during the fall, even under favorable climatic conditions (18).

Once the stump surface becomes infected with *F. annosus,* the fungus grows rapidly down the stump and into the roots. A growth rate of about 1 m per year has been reported by most workers (12, 35, 63, 71, 72, 101, 117). Hodges (45), however, reported the growth rate of *F. annosus* through roots of slash pine stumps in South Carolina to be almost 2 m after 12 months, considerably greater than growth rates previously reported. Unpublished studies by the writer indicate that the rate of growth of *F. annosus* in stump roots is related to the length of time the root cambium remains alive. In the southern United States this varies from 2 to 6 months, depending on the size of the stump, the season in which the tree was cut, and probably the host vigor. After the cambium dies, the roots are colonized by saprophytic organisms in the soil or root surface and this usually blocks the continued growth of *F. annosus.* Lack of complete root colonization may not be an important factor in the spread of *F. annosus* to surrounding trees because roots of these trees often make contact with infected stump roots near the stump or with the stump itself. Roots of stumps and of residual trees are commonly in close contact for a distance of a meter or more.

The amount of initial stump infection is not necessarily correlated with the number of trees infected in the residual stand. Even if *F. annosus* thoroughly colonizes the stump and its root system initially, it is often replaced by other microorganisms, especially by *P. gigantea* and *Trichoderma* spp. (12, 45, 63, 72). In addition, certain biological, chemical, and physical soil factors may prevent spread of *F. annosus* from infected stumps to residual trees (33, 86, 91, 96, 97, 101, and numerous others).

Under certain conditions, *F. annosus* can survive for many years in the roots of stumps and dead trees. In England, Low & Gladman (69) and Rishbeth (101) found the fungus in stumps 44 years and 25 years, respectively, after the trees were felled. In New Hampshire, Miller (73) found sporophores of *F. annosus* on a pine stump cut more than 40 years before. Stumps decompose much more rapidly in the southern United States (65), and except in very large stumps, *F. annosus* probably does not survive for more than 5 to 10 years. The fungus may survive longer in heavily resinous wood than in nonresinous wood (6).

Through dead roots.—Prior to the establishment by Rishbeth (101) of the role of the stump in the initiation of new infection centers, most workers believed that trees were infected by spores germinating on the surface of dead or moribund roots. The fungus then colonized the dead portion and finally entered the living tissues of the root. Hiley (44) showed that dead roots could be infected when he succeeded in inoculating roots which had been killed by immersion in boiling water. He also observed in the field that a large percentage of tree roots die, especially deep roots in soil previously used for agriculture. He attributed this to poor aeration. It was in these deep roots, especially the tap root and vertically growing lateral roots, that he found what he believed to be most of the original points of infection. He believed these dead roots became infected by basidiospores or conidia coming into contact with them, or by mycleium growing in the soil, even though he had been unable to demonstrate that mycelium could grow in nonsterile soil. Other reports (2, 20, 40, 43, 52, 53, 60, 66, 67, 86, 94, 110) stressed the fact that dead roots, principally the tap root or other vertically growing roots beneath the base of the tree, were the primary point of infection and were always more decayed than the laterals. On the basis of these observations, the planting of Norway spruce so that no tap roots will develop has been suggested to reduce infection (53). Møller (76) also felt that older, partly or wholly inactive wood was most likely to be infected, but he believed shallow lateral roots were infected more often than had been suggested previously.

The observations that tap and vertical lateral roots are often decayed more than shallow laterals are valid. It seems unlikely that these roots were dead before infection by *F. annosus* occurred, however. Rishbeth (100) has clearly shown that *F. annosus* rarely invades wood already occupied by other fungi. Similar results have been obtained by the writer.

There are several possible explanations for the observations that the tap and other vertical roots usually are the first to become infected. Perhaps the competition from other microorganisms at the root surface diminishes with depth (100). This lack of competition furnishes more favorable conditions for infection, either by root contact or by spores that filter down through the soil. Another possibility is that *F. annosus* can tolerate relatively low oxygen tensions (39, 126), which would also give it an advantage in soils where aeration is poor. If many of these deep roots are weakened by poor aeration or other factors, as has been suggested (44, 66, 110), the fungus would colonize and decay such roots faster than vigorous roots. An analogy might be made between such weakened roots and roots of suppressed trees where *F. annosus* makes more rapid growth than it does through roots of vigorous trees (74, 102, 117).

Through roots of stumps and living trees.—Although dead roots are probably not infected by spores of *F. annosus* in the soil, there is considerable evidence that roots of freshly cut stumps and possibly even living trees may become infected in this way. This latter type of infection may be responsible for infection centers which occur in stands where no trees have been cut and stumps are not available for infection by airborne spores.

Rishbeth (101) first demonstrated the presence of viable spores in the soil by using soil collected in infested areas to inoculate freshly cut stump surfaces. Stump infection occurred when soil taken from a depth of 50 to 90 cm was placed on the stump surface, indicating that some spores were washed deeply into the soil. Rishbeth believed it was possible for these spores to germinate on the surface of stump roots and infect them, but he did not consider it likely because he was unable to obtain mycelial growth of *F. annosus* on unsterilized root pieces in the laboratory (101) or on wounded or nonwounded roots (102) after inoculation with spores.

Molin (75), however, was able to wash spores of *F. annosus* through a column of sand 20 to 40 cm thick where they germinated on a block of wood, and he postulated that localized infections in the field may arise as a result of spore infection. Kuhlman (61), by use of a selective medium, recovered conidia of *F. annosus* for up to 10 months after adding them to the soil. He also demonstrated the presence of basidiospores in soils in infection centers and disease-free areas.

Wallis (117), Jorgensen (55), Kuhlman (62), and Punter (92) successfully inoculated roots of newly cut stumps with suspensions of conidia or basidiospores. Percentage infection in all cases was higher if the roots were wounded before inoculation. Kuhlman (62) obtained infection by applying only 44 conidia to an area of 2.5 sq cm on nonwounded roots.

Little experimental evidence is available for direct root infection under field conditions. Hendrix & Kuhlman (42) found that approximately 7 per cent of 700 severed but otherwise undisturbed roots of slash pine became infected with *F. annosus* from naturally available inoculum. Sixty-five per cent of the infected roots became infected from the severed end and may have become infected from airborne spores at the time they were severed. The remainder were infected some distance qway from the severed end and could only have become infected from the soil. Two of 442 unsevered roots, from dominant trees, also were infected. Although the exact source of inoculum was not known, spores were suspected.

Gibbs (35) repeated the experiments of Hendrix & Kuhlman (42) but was unable to recover *F. annosus* in severed roots of Scots pine. He indicated that this species is moderately deep rooted and the opportunity for chance infection may be smaller than for slash pine. Gibbs also found evidence that direct root infection appears likely only on heavily suppressed trees. He inoculated wounds on both severed and intact roots of dominant and suppressed Scots pine with basidiospores. After 8 months, percentage infection was 0 in intact and 60 in the severed roots of dominant trees, whereas it was 50 in intact and 25 in the severed roots of suppressed trees. Average growth from the inoculation point in infected roots was similar and varied from 2.4 to 2.9 cm with a maximum growth of 17 cm in one root where development was associated with bark beetle attack. The average growth rate of the fungus in roots infected by spores was only 0.35 cm/month compared to the growth rate of 4.1 cm/month which Gibbs obtained by inoculation of roots with wood colonized by *F. annosus*. Such differences in average growth rate might be expected, however, because actual root infection probably takes place sooner in the roots inoculated with wood.

The only other report of roots infected by means other than through the stump surface was by Hodges (45), who found several roots of stumps treated with borax to be infected with *F. annosus* at distances from the stump too great to have resulted from stump-surface infection.

Based on both personal research and data available from the literature, infection of freshly cut stump surfaces by airborne basidiospores is the major means for the initiation of new infection centers. However, assessment of the relative importance of direct infection of roots by spores in the soil is much more difficult. Although artificial inoculation studies furnish ample evidence for direct root infection, only Hendrix & Kuhlman (42) and Hodges (45) have reported its occurrence under natural conditions. In these studies, less than 10 per cent of the roots sampled became infected directly. The percentage of roots colonized on stumps infected through the cut surface may also be low in some cases, even though the entire area of the stump surface was colonized initially (12, 45, 62). No data are available on the fate of *F. annosus* in roots infected directly except for a relatively short period of time after infection. Following inoculation on the root surface, the infection process appears to be rather slow (35, 62). Once the fungus is established, however, growth rate should be equal to that in roots that were infected from stumps. It should also be noted that the first appearance of *F. annosus* in roots of stumps infected through the cut surface may be several months following surface infection, depending upon stump height and temperature. For this reason, colonization of the root resulting from direct infection may occur sooner than if infection occurred via the stump surface.

Because of the possibility of direct infection of stump roots, protection of the cut surface of the stump with chemicals may not always give adequate control of the disease. Some stump protectants, like creosote and borax, tend to keep the stump alive for a considerable time after felling and may increase the possibility of direct root infection (45).

Through trunk wounds.—The ability of *F. annosus* to infect through above-ground wounds is dependent on tree species. Pines are seldom infected in this manner. Rishbeth (102) reported one instance where *F. annosus* may have entered pruning wounds on a suppressed tree and eventually reached the root system. However, his attempts to inoculate wounds or freshly cut branch-ends of Scots pine were unsuccessful. He believed that pine must be weakened almost to the point of death before infection can occur through stem wounds. Witcher (120), on the other hand, found that a number of infection centers in an unthinned plantation of slash pine developed where the trees were pruned to remove branch galls of *Cronartium fusiforme.* The pruning wounds were 2 to 8 cm in diameter and 10 to 90 cm above ground. An adjacent unpruned stand was free of infection.

Meredith (71) inoculated the surfaces of freshly severed members of double stemmed trees forked just above the ground level, but no infection took place; eight of ten stumps of single trees were infected. Meredith associated the lack of infection of stumps of the forked trees to copious resin flow. He compared these results with what might be expected with wounds on living trees. These results are similar to those noted by the writer on stumps of loblolly pine connected by natural grafts to nearby trees. These stumps remained alive for more than 3 years and were free of infection from *F. annosus* and other fungi, even though almost all nongrafted stumps nearby were colonized by *F. annosus.*

The effect of copious resin flow in reducing infection is supported by the work of Froelich (33a). He found no infection of turpentine faces on slash pine in areas where basidiospore inoculum in the air had been high over an extended period of time.

Most pines are apparently resistant to butt rot caused by *F. annosus.* The same factors associated with resistance to stem decay may be responsible for preventing infection of stem wounds.

Norway spruce, one of the most important hosts of *F. annosus* in Europe, is very susceptible to annosus butt rot. However, there is little evidence in the literature that this species is more susceptible to trunk wound infection than species of pine. Vakin (116) and Day (21) reported that *F. annosus* sometimes develops after injuries to the roots and stems of Norway spruce. Braun (14, 15), however, did not isolate *F. annosus* from decay found beneath wounds made by deer on stems of this species. Yde-Andersen (123) found that up to 25 per cent of the Norway spruce in some stands was damaged by logging; however, none of the decay and discoloration associated with these wounds was caused by *F. annosus*. Zycha (127) believed that infection through wounds on standing timber is of minor importance because they are quickly covered by resin.

Western hemlock is very susceptible to infection through trunk wounds by *F. annosus* (16, 27, 28, 50, 99, 118). In a study by Rhoads & Wright (99) of more than 600 scars, sunscald lesions, and broken tops on 198 western hemlocks, decay caused by *F. annosus* was found in connection with 119 scars, five sunscald lesions, and three broken tops (that showed decay). *F. annosus* surpassed any other fungus in frequency of association with injuries; the volume of decay by this fungus exceeded that of all other fungi combined. Conversely, of 36 Sitka spruce trees which had a total of 104 scars, one sunscald lesion, and four broken tops, *F. annosus* was isolated from beneath only one scar. However, infection of Sitka spruce through wounds has been reported by Bier et al. (8). Hunt & Krueger (50) found that of 54 wounds on western hemlock where decay was evident or suspected, *F. annosus* occurred in 30 per cent and accounted for 80 per cent of the decay volume. The fungus occurred beneath wounds in hemlock stands of all ages. Logging injuries were responsible for most of the wounds.

No experiments have been conducted to determine the ability of *F. annosus* to infect trunk wounds on different tree species except those reported by Rishbeth (102) and Meredith (71). The paucity of reports in the literature would indicate that trunk wounds are not important infection courts on species other than western hemlock. There are no data to indicate why western hemlock should be more susceptible to wound infection by *F. annosus* than other conifers.

Role of conidia.—In addition to basidiospores, *F. annosus* also produces a conidial stage that is conspicuous in culture and on infected wood incubated under conditions of high humidity. Although inconspicuous in nature, they have been found in underground cavities (48), on pieces of infected wood (52, 57), between fragments of bark on pine roots (56), on the surface of infected stumps (78, 104), and in tunnels made by various insects (3, 4, 101). Rishbeth (101, 104) & Hiley (44) believe conidia rarely occur in nature. However, Morris & Knox (78) observed conidia on discs cut from the top of stumps resulting from a thinning 4 months earlier. When these discs were incubated for 12 hr at 75° F in a closed polyethylene bag, numerous additional conidiophores and conidia were found on the upper stump surface, indicating that conidia form rapidly under certain conditions and may serve as an important source of inoculum.

There is ample evidence that conidia germinate and result in colonization of woody substrates in the field. They have been used by numerous workers to inoculate successfully freshly cut stump surfaces (12, 45, 63, 112, and others). Hüppel (51) inoculated seedlings of Scots pine and Norway spruce under sterile conditions with a conidial suspension of *F. annosus* which killed the seedlings. Kuhlman (62) obtained infection of stump roots in the field using only 44 conidia placed on the bark of an uninjured root.

There is some indication that conidia may not be as effective an inoculum as basidiospores. Punter (92) inoculated stump roots in the field with basidiospores but conidia failed to cause infection under the same conditions. Kuhlman et al. (64) found that growth rate of the fungus through stumps was faster with basidiospore inoculum than with conidia, even though fewer viable basidiospores than conidia were used.

Conidia can remain viable for periods up to 1 year under certain conditions (44, 61). Ross (109) found that conidia also were most resistant to high temperatures than were basidiospores. Thermal inactivation of conidia required from 90 to 120 min at 45° C, whereas basidiospores were killed after 30 min at this temperature.

Because the conidial stage is so inconspicuous in nature and because conidia cannot be distinguished morphologically from basidiospores, it is difficult to evaluate their role in the spread of *F. annosus.* They do develop if the relative humidity is sufficiently high, however, and can initiate infection. They may be especially important in spread of the fungus by insects.

Spread by Mycelium

Through and along roots.—F. annosus spreads from infected tree or stump roots to healthy roots at points of root contact or grafts. In pine, Rishbeth (102) observe that the mycelium of the fungus passes directly from the bark of infected roots onto that of living roots. In small roots up to about 3.0 cm in diameter, the wood is invaded soon after the bark. The distal end of the root dies and is rapidly invaded by other microorganisms. The fungus continues to grow toward the tree in the live portion of the root, invading all tissues at about the same rate. On larger roots, *F. annosus* colonizes the bark before invading the woody tissues, and progresses through the root until it reaches the bole. The tree dies after a variable period following the girdling of the tree at the root collar. Crown symptoms may or may not appear, depending upon the number of roots killed before the tree itself dies or is windthrown. Except for white pine (43) and red pine (112), *F. annosus* normally does not extensively invade the heartwood of living pines. Further spread of the fungus to the lateral roots on the side of the tree opposite the point of initial infection enables the fungus to pass to living roots of adjacent trees, which are in turn killed. The infection center thus gradually increases in diameter.

The average rate of spread of the fungus in individual infection centers varies among stands. In England, Rishbeth (104) reported average rates of spread of about 0.5 m per year in two stands and about 1.0 m per year in a third. The first infections of living roots in contact with dead stump roots containing *F. annosus* were seen 1.5 to 2 years after felling. Sinclair (112) observed in New York that the average rate of spread in a red pine plantation was about 0.6 m per year during an 8 year period. In Massachusetts, Fowler (30) reported a somewhat more rapid average rate of spread of 1.7 and 2.1 m per year in infection centers 10 and 8 years old, respectively. Hodges (46) reported death of trees up to 2 m away from stumps inoculated 10 months earlier. Pine generally becomes less susceptible with age, and infection centers finally stabilize (46, 81, 95, 102).

Inoculation of living pine roots with *F. annosus* has generally resulted in growth of the fungus at rates of less than 2 cm per month (11, 35, 88, 102, 112, 114), although Miller & Kelman (74) obtained growth rates averaging 5.2 cm per month in roots of dominant loblolly pine.

The rapid spread of *F. annosus* in infection centers cannot be explained by reported rates of growth in wood of roots of stumps, or of living trees. Rishbeth (102) observed growth of *F. annosus* mycelium on the surface of pine roots up to 1 m beyond the point of invasion of living tissue. The extent of this superficial growth varied with the soil type, being greater on roots in alkaline than in acid soils. Rishbeth attributed this to the scarcity of competing organisms in alkaline soils. However, Gibbs (34) was unable to explain the differential growth of *F. annosus* on roots in different soils either in terms of the populations of antagonistic fungi or in terms of a direct pH effect on the antibiotics produced by such fungi. Hodges (46) also was unable to observe visible growth of mycelium on the surface of roots in acid soil, but the fungus was isolated from bark scales at points ahead of infection in wood. Although no data are available on the rate of growth of *F. annosus* on the root surface, it is considerably faster than through roots of living trees, as evidenced by its presence ahead of the fungus in wood. Faster surface growth, affording the opportunity for multiple infections, could account for the rapid spread of the fungus through the stand.

Growth of *F. annosus* in roots of older spruces, firs, and other species susceptible to butt rot is somewhat different from growth in the roots of pines (103). Relatively small roots usually are the first to be infected. All tissues of roots under 2 cm diameter are colonized , but in larger roots the fungus is confined to the central xylem. Resin exudation and infiltration of infected wood is seldom noted, except in larch (44). Once established in the center of the root, the fungus progresses rapidly toward the bole. Gibbs (36) reported growth of 75 to 100 cm in 18 months in roots of Sitka spruce. The position of the rot column in the bole depends on the number of roots infected. It is central if several roots are infected; but if only one root is infected, it occurs immediately above the infected root. Rate of spread in the heartwood varies with species but averages from 10 to 70 cm per year (36, 52, 80, 83-85, 95, 103).

Because older trees seldom die, few data are available on the rate of spread of *F. annosus* in individual infection centers in spruce and fir stands. Paludan (90) found that, 3.5 years after thinning, *F. annosus* could not be found in Norway spruce in rows adjacent to a row of heavily infected stumps (rows 1.2 m apart). After 5 to 7 years, *F. annosus* was isolated from 31 and 42 per cent of the trees in the rows on either side of the stump row. In some plots, *F. annosus* had reached the second row from the stump row after 8 years. A few trees in the third row from the stump row (3.6 m) were infected after 10 years. There is no indication in the literature that species susceptible to heart rot become more resistant with increasing age, or that infection centers finally stabilize.

Resin exudation and infiltration of root and stem wood is associated with infection of pine by *F. annosus* (41, 44, 46, 88, 111, 112, 121), and is believed responsible for the relatively slow growth of *F. annosus* in living trees. Rishbeth (102) observed that growth of *F. annosus* in roots of trees on acid soils was less than on alkaline soils and related this to greater resin production in the former situation. When he inoculated roots of Scots and Corsican pines, Sitka spruce, and Douglas fir with *F. annosus,* Gibbs (36) found that the degree of infection and growth rate of the fungus in the root wood was least in the two pine species. Gibbs thought this was related to the greater production of resin in pine than in spruce and Douglas fir. He also postulated that the greater severity of *F. annosus* on dry sites might be related to resin production because it depends on the water regime in the tree which, in turn, is dependent on the water regime in the soil. This was recently verified by Lorio & Hodges (68), who found that dirunal patterns of oleoresin exudation pressure in loblolly pine were related to changes in soil and atmospheric moisture. Gibbs (36) also pointed out that more vigorous trees produce the most resin, and the relationship between host vigor and resistance might be explained on this basis.

Bega & Tarry (6) found that in pure culture, resin had no toxic effects on growth or sporulation of *F. annosus* and that at the optimum temperature for growth of *F. annosus* (24° C), all isolates tested showed increased growth rate and increased sporulation when resin was added to the medium. They also observed, as did Shain (111), that the fungus penetrates resin-impregnated wood, although at a greatly reduced rate. Ross (109a) found that spores of *F. annosus* would germinate on resin only if adjacent to organic debris. Once germination had occurred, however, the fungus would grow over the surface of the resin. Conversely, Cobb et al. (19) found that volatile constituents of crude oleoresin, collected from ponderosa pine, significantly reduced growth of *F. annosus,* as well as changed colony characteristics and sporulation. Heptane exhibited the greatest effect on growth of any of the components tested; *F. annosus* ceased to grow after 1 or 2 days' exposure to this compound. Similar results were obtained when the compounds were added to the culture medium. Effects of vapors of these compounds on the germination of spores of *F. annosus* were not as marked, but germ-tube elongation was greatly reduced, especially by heptane. The different results obtained by Bega & Tarry (6) are probably due to loss of toxic volatile components in their system (19).

Unpublished data of the writer indicate that the degree of resin exudation from roots of loblolly, slash, and longleaf pines is directly correlated with susceptibility to *F. annosus.* Percentage of infection, rate of growth, and resin exudation in inoculated roots of dominant longleaf pine, were considerably less than in roots of slash and loblolly pines.

These reports indicate that copious exudations of resin and probably other compounds are induced in response to infection of roots and stems of most pine species. There is some controversy, however, whether resin itself is toxic to *F. annosus.* Slower growth of *F. annosus* through resin-impregnated tissue may be due to other compounds, such as phenols, or to physical factors. It is also possible that the chemical composition of the resin may be more important than the amount of resin produced in response to infection (19).

There is some question as to whether *F. annosus* can infect roots through uninjured bark at points of root contact. Hiley (44) was unable to inoculate roots with infected wood blocks unless the roots had been wounded first. Likewise, Braun (14) found infection to occur through wounds, lenticels, and cracks in the bark which extended into the inner bark, but indicated the fungus was unable to penetrate into the root if the "tannin bark layers", which the fungus cannot attack, were intact. Most reports (88, 102, 112, 117, 121), however, indicate that *F. annosus* can infect through intact bark.

Growth through soil.—Attempts to demonstrate growth of *F. annosus* mycelium through nonsterile soil and litter have met with varying success. Francke-Grossman (32) obtained mycelial growth through nonsterile soil by using a decoy block of Sitka spruce treated with ferric sulphate. Release of volatile products of fatty acids by the decoy block apparently influenced the fungus to grow toward the block from infected wood buried in the soil. The fungus did not grow through all soils and no growth took place through humus, probably because of the presence of large numbers of antagonistic microorganisms. Incubation at temperatures of 8 to 14° C favored the growth of the fungus. The author suggested that mycelium of *F. annosus* in roots of stumps may be induced to grow through soil toward stimuli such as dying roots of conifers. Mussell (79), using freshly cut stemwood sections of loblolly pine as traps, isolated *F. annosus* from soil at distances up to 30 cm from infected stumps. Rishbeth (100) obtained limited mycelial growth through acid and alkaline soil from a depth of 15 to 30 cm, but growth through litter, humus, and soil from a depth of 8 cm was negligible. He was unable to observe growth of mycelium in the soil in the field, however, even from heavily infected roots. Negrutskii (82) was able to grow *F. annosus* on nonsterile forest litter but reported increased growth on sterilized litter. He also obtained infection by covering wounded roots of living trees with soil from within an infection center. However, no infection took place when soil from beyond the limits of the infection center was used. Others (9, 29, 67, 93, 106) have postulated that the fungus could spread in the humus layer from tree to tree.

Most workers (9, 14, 44, 47, 75, 97, 115) have been unable to demonstrate growth of *F. annosus* through nonsterile soil. Growth in sterile soil or litter is easily demonstrated, indicating that antagonistic microorganisms are responsible for the lack of growth in these substrates under natural conditions (14, 44, 94, 96, 97, 100, 115, and others). Low populations of microorganisms also may allow the growth of *F. annosus* in nonsterile soil in the deeper horizons (100). In some soils, however, inhibitors of mycelial growth can be found to depths of 1.3 m (75).

The fruiting of *F. annosus* in litter without any discernible connection of the sporophore to infected roots has been reported (1). Others (14, 41, 101), however, found such fruiting bodies not developing from mycelium growing freely in the litter, but attached to small rootlets, some as small as 1/16 in. All fruiting bodies occurring in the litter examined by the writer were also attached to small roots.

It would appear unlikely that *F. annosus* mycelium grows more than a few cm through nonsterile humus and litter because of the presence of high populations of antagonistic microorganisms in such soils. Growth through nonsterile soils where competition is low may be possible, but this is likely only in the lower soil layers. Close contact appears to be necessary for the spread of the fungus from root to root.

Other Modes of Spread

Insects.—Trees and stumps infected with *F. annosus* are often colonized by different insects (5, 31, 56). Although no fungus-vector relationship has been demonstrated, there appears to be ample opportunity for insects to transmit spores of *F. annosus* to previously uninfected trees or stumps. The conidial stage of *F. annosus* has been found in galleries of different bark beetles in pine trees and stumps (3, 4, 101, 119). Weidensaul (119) isolated *F. annosus* from both on and in 17 per cent of adult *Dendroctonus terebrans* he sampled. Nuorteva & Laine (87) recovered spores of *F. annosus* up to 5 days after individuals of *Blastophagus piniperdi, Hylastes brunneus, Rhizophagus ferrugineus,* and *Hylobius abietis* were placed for a short time on pure cultures of the fungus.

If some degree of insect transmission of *F. annosus* does occur in nature, as seems probable, conidia produced in insect galleries on infected trees, particularly in galleries of bark beetles, would be the most likely means of transmission. Conidia have been shown to survive in soil for several months (61), and up to a year in dried cultures (44), and thus should easily survive insect flights to new substrates. Although the dry conidia of *F. annosus* would not adhere to the body of insects as readily as fungi with sticky spores, the spores can survive in the gut (119) and can be recovered from fecal pellets (87).

Freshly cut stumps, highly selective for *F. annosus,* would be an excellent substrate for *F. annosus* following invasion by infested insects. Stump infection in this manner, as well as by direct root infection, might make chemical stump protection less effective.

Burrowing animals.—Hartig (41) suggested that mice and possibly other burrowing animals carry spores of *F. annosus* on their fur and rub them onto healthy roots where they germinate and root infection follows. This means of dissemination was also suggested by Huet (49) and Mook & Eno (77), but no substantiating experimental data were presented. In the southern United States, sporophores of *F. annosus* are commonly found attached to roots over pine mouse tunnels, and these animals undoubtedly carry spores of the fungus on their fur. Whether they subsequently deposit these spores on susceptible root tissue remains to be demonstrated.

Miscellaneous.—Gram & Rostrum (38) and Jorgensen (54) suggested that many of the infection centers in shelter belts of *Crataegus* sp., *Sorbus intermedia,* and Sitka spruce in Denmark originated from nearby fence posts made of stems infected with *F. annosus.* In Britain, sporophores of *F. annosus* have been found on infected stakes 30 years after they were put in the ground (69).

Jorgensen (55) observed that the occurrence of *F. annosus* in Ontario closely corresponds to the area of early reforestation with seedlings imported from Europe, and indicated that *F. annosus* may have been introduced to Ontario by spores on seedlings. He showed that viable spores could be recovered from seedlings bundles 8 weeks after application of spores. However, roots of seedlings inoculated with spores and subsequently planted did not become infected after 2 years, whether they were wounded or not. Punter (92) later concluded that introduction of *F. annosus* on seedling stocks was highly unlikely because he was also unable to obtain infection of seedlings by inoculation with basidiospores.

Summary

Long-distance spread of *F. annosus* is accomplished primarily by the aerial dissemination of spores. Basidiospores probably constitute the majority of the air spora, but conidia are also produced by the fungus under conditions of high humidity. In most parts of Europe, Canada, and the northern United States, basidiospores are produced during all seasons of the year, except during periods of extreme cold or drought. Highest spore production in these areas occurs in spring and fall. In the southernmost United States, spore discharge is greatest during the fall, winter, and spring, and few, if any, spores are released during the summer.

Infection of freshly cut stump surfaces by airborne spores is probably the most common means of entrance of *F. annosus* into a previously uninfected thinned stand. These cut surfaces are highly selective for the spores of *F. annosus* during the first 2 weeks after cutting, but the degree of infection decreases rapidly after that time. This decrease is primarily due to colonization of stumps by other microorganisms, especially *P. gigantea.*

Growth of the fungus through the stump and its roots proceeds at a rate of approximately 1 to 2 m per year. In many stumps, the fungus is replaced by other microorganisms, particularly *P. gigantea* and *Trichoderma* spp., and thus some degree of biological control is achieved.

The fungus spreads from infected stump roots to adjacent healthy trees at points of root contact or graft. Dying trees may first be noted from 1 to 3 years following thinning. From the slow rate of growth of *F. annosus* through inoculated roots of pine, growth of the fungus on the surface of roots, which may result in multiple infections ahead of the part of the root colonized, is probably responsible for the relatively rapid enlargement of infection centers. Spread in most centers averages one or more meters per year. Most of these centers, at least in pine, appear to stabilize when the trees are 30 or 35 years old. Growth of the fungus in roots of conifers subject to heart rot is somewhat more rapid than in trees of resistant species.

Roots of live stumps can also become infected by airborne spores which pass down through the soil and germinate on the root surface. Invasion of the root by the fungus then follows. Wounds are not required for such infections but may increase the chance for their occurrence. The relative importance of direct root infection versus root infection via the stump surface is difficult to assess, but the latter is undoubtedly the most common means of stump infection. The possibility of direct root infection and spread of the fungus to stumps by insects may negate somewhat the effectiveness of stump protectants, and some stump infection may take place even when protectants are used. Some protectants, like creosote and borax, may increase the likelihood of direct root infection by keeping the stump alive longer, thereby preventing its invasion by soilborne organisms antagonistic to *F. annosus*.

Direct root infection may also occur on roots of suppressed or otherwise weakened trees in unthinned stands where no stumps are present. The extent of colonization of these roots, and the probability of spread to adjacent trees, depends on how fast they are invaded by saprophytic organisms. Infection of roots of vigorous trees and dead roots colonized by other microorganisms is unlikely to be initiated by spores of *F. annosus*.

There is some disagreement about whether *F. annosus* can grow through nonsterile litter or soil. Experimental substantiation of this is meager; most workers have been able to demonstrate only little if any growth through this substrate. Because *F. annosus* shows little competitive ability against many of the microorganisms found in litter and soil, it appears unlikely that growth through these substrates plays an important role in spread of the fungus, except possibly in soils low in antagonistic microorganisms, or where inoculum is massive.

Colonization of above-ground wounds by airborne basidiospores on species of pine and Norway spruce is apparently negligible. They are a very important means of entry in western hemlock, however, and may sometimes occur in Sitka spruce and other species.

Trees weakened by *F. annosus* are commonly invaded by bark beetles and other insects. Conidia of the fungus can develop in insect galleries and on the surface of insects, and could possibly be transported to uninfected trees or stumps.

Infected fence posts and spores on the surface of seedling roots and on small burrowing animals also have been implicated in the spread of *F. annosus,* but these means of spread are considered relatively unimportant.

LITERATURE CITED

1. Anon. Root rots. *U. S. Dept. Agr., Forest Serv., Northeast Forest Expt. Sta. Ann. Rept. 1959,* 45–48 (1960)
2. Anderson, M. L. Heart rot in conifers. *Trans. Scot. Arbor. Soc.,* **38,** 37–45 (1924)
3. Bakshi, B. K. Fungi associated with ambrosia beetles in Great Britain. *Trans. Brit. Mycol. Soc.,* **33,** 111–20 (1950)
4. Bakshi, B. K. *Oedocephalum lineatum* is a conidial stage of *Fomes annosus. Trans. Brit. Mycol. Soc.,* **35,** 195 (1952)
5. Beal, J. A., Massey, C. L. Bark beetles and ambrosia beetles (Coleoptera: Scolytidae): With special reference to species occurring in North Carolina. *Duke Univ., School of Forestry Bull. No. 10,* 178 pp. (1945)
6. Bega, R. V., Tarry, J. Influence of pine root oleoresins on *Fomes annosus. Phytopathology,* **56,** 870 (1966)
7. Berry, F. H. Treat stumps to prevent *Fomes annosus* in shortleaf pine plantations. *U. S. Dept. Agr., Forest Serv., Central States Forest Expt. Sta. Res. Note No. 34,* 4 pp. (1965)
8. Bier, J. E., Foster, R. E., Salisbury, P. J. Studies in forest pathology. IV. Decay of Sitka spruce on the Queen Charlotte Islands. *Can. Dept. Agr. Tech. Bull. No. 56,* 35 pp. (1946)
9. Bjorkman, E. Soil antibiotics acting against the root-rot fungus (*Polyporus annosus* Fr.). *Physiol. Plantarum,* **2,** 1–10 (1949)
10. Bjørnekaer, K. Studies on the biology of some Danish Polyporaceae with special reference to their spore discharge. *Friesia,* **2,** 1–41 (1938)
11. Boyce, J. S., Jr. *Fomes annosus* in white pine in North Carolina. *J. Forestry,* **60,** 553–57 (1962)
12. Boyce, J. S., Jr. Growth of *Fomes annosus* into slash pine stumps after top inoculation. *Plant Disease Reptr.,* **47,** 218–21 (1963)
13. Boyce, J. S., Jr. Colonization of pine stem sections by *Fomes annosus* and other fungi in two slash pine stands. *Plant Disease Reptr.,* **47,** 320–24 (1963)
14. Braun, H. J. Untersuchungen über den Wurzelschwamm *Fomes annosus* (Fr.) Cooke. *Forstwiss. Zentr.,* **77,** 65–88 (1958)
15. Braun, H. J. Zur Frage der Infektion von Schäl- und Schürfwunden durch den Wurzelschwamm *Fomes annosus* (Fr.) Cooke (*Trametes radiciperda* Hartig). *Allgem. Forst Jagd Ztg.,* **131,** 67–68 (1960)
16. Buckland, D. C., Foster, R. E., Nordin, V. J. Studies in forest pathology. VII. Decay in western hemlock and fir in the Franklin River Area, British Columbia. *Can. J. Res., Sect. C,* **27,** 312–31 (1949)
17. Cobb, F. W., Jr., Schmidt, R. A. Duration of susceptibility of eastern white pine stumps to *Fomes annosus. Phytopathology,* **54,** 1216–18 (1964)
18. Cobb, F. W., Jr., Barber, H. W., Jr. Susceptibility of freshly cut stumps of redwood, Douglas fir, and ponderosa pine to *Fomes annosus. Phytopathology,* **58,** 1551–57 (1968)
19. Cobb, F. W., Jr., Krstic, M., Zavarin, E., Barber, H. W., Jr. Inhibitory effects of volatile oleoresin components on *Fomes annosus* and four *Ceratocystis* species. *Phytopathology,* **58,** 1327–35 (1968)
20. Day, W. R. The penetration of conifer roots by *Fomes annosus. Quart. J. Forestry,* **42,** 99–101 (1948)
21. Day, W. R. Drought crack of conifers. *Gt. Brit. Forestry Comm., Forest Rec.,* **26,** 40 pp. (1954)
22. Dimitri, L. Untersuchungen über die Ausbreitung von *Fomes annosus* (Fr.) Cooke. *Phytopathol. Z.,* **48,** 349–69 (1963)

23. Driver, C. H. Effect of certain chemical treatments on colonization of slash pine stumps by *Fomes annosus*. *Plant Disease Reptr.*, **47,** 569–71 (1963)
24. Driver, C. H. Further data on borax as a control of surface infection of slash pine stumps by *Fomes annosus*. *Plant Disease Reptr.*, **47,** 1006–9 (1963)
25. Drummond, D. B., Bretz, T. W. Seasonal fluctuations of airborne inoculum of *Fomes annosus* in Missouri. *Phytopathology,* **57,** 340 (1967)
26. Dwyer, W. W., Jr. *Fomes annosus* in eastern redcedar in two piedmont forests. *J. Forestry,* **49,** 259–62 (1951)
27. Englerth, G. H. Decay of western hemlock in western Oregon and Washington. *Yale Univ., School Forestry Bull.,* **50,** 53 pp. (1942)
28. Englerth, G. H., Isaac, L. A. Decay of western hemlock following logging injury. *Timberman,* **45,** 34–35, 56 (1944)
29. Falck, R. Neue Mitteilungen über die Rotfäule. *Mitt. Forstwirtsch. Forstwissensch.,* **1,** 525 (1930)
30. Fowler, M. E. *Fomes annosus* in northeastern United States. In *Conference and Study Tour on Fomes annosus. Scotland, 1960,* 20–22, IUFRO, Firenze (1962)
31. Francke-Grossman, H. Feinde und Krankheiten der Sitkafichte auf norddeutschen Standorten. *Forst u. Holz,* **9,** 117–19 (1954)
32. Francke-Grossman, H. Under what conditions can *Fomes annosus* grow in nonsterilized soil? In *Conference and Study Tour on Fomes annosus. Scotland, 1960,* 22–28, IUFRO, Firenze (1962)
33. Froelich, R. C., Dell, T. R., Walkinshaw, C. H. Soil factors associated with *Fomes annosus* in the Gulf states. *Forest Sci.,* **12,** 356–61 (1966)
33a. Froelich, R. C. (Personal communication)
34. Gibbs, J. N. A study of the epiphytic growth habit of *Fomes annosus*. *Ann. Botany (London), N.S.,* **31,** 755–74 (1967)
35. Gibbs, J. N. The role of host vigor in the susceptibility of pines to *Fomes annosus*. *Ann. Botany (London), N.S.,* **31,** 803–15 (1967)
36. Gibbs, J. N. Resin and the resistance of conifers to *Fomes annosus*. *Ann. Botany (London), N.S.,* **32,** 649–65 (1968)
37. Gooding, G. V., Jr., Hodges, C. S., Jr., Ross, E. W. Effect of temperature on growth and survival of *Fomes annosus*. *Forest Sci.,* **12,** 325–33 (1966)
38. Gram, E., Rostrup, S. Survey of the diseases of cultivated agricultural plants in 1921. *Tidsskr. Planteavl.,* **28,** 185–246 (1922)
39. Gundersen, K. Growth of *Fomes annosus* under reduced oxygen pressure and the effect of carbon dioxide. *Nature,* **190,** 649 (1961)
40. Haraldstad, A. R. Investigations on *Fomes annosus* in Høylandskomplekset, south-western Norway. *Nytt Mag. Botany,* **9,** 175–98 (1961)
41. Hartig, R. *Die Zersetzungserscheinungen des Holzes* (Julius Springer, Berlin, 151 pp., 1878)
42. Hendrix, F. F., Kuhlman, E. G. Root infection of *Pinus elliottii* by *Fomes annosus*. *Nature,* **201,** 55–56 (1964)
43. Hepting, G. H., Downs, A. A. Root and butt rot in planted white pine at Biltmore, N.C. *J. Forestry,* **42,** 119–23 (1944)
44. Hiley, W. E. *Fungal Diseases of the Common Larch* (Clarendon Press, Oxford, 204 pp., 1919)
45. Hodges, C. S. Evaluation of stump treatment chemicals for control of *Fomes annosus*. In *Proc. Intern. Conf. Fomes annosus., 3rd, Aarhus, Denmark, 1968* (In press)
46. Hodges, C. S. *Fomes annosus* in the southern United States. In *Root Diseases and Soil-borne Pathogens* (Toussoun, T. A., Bega, R. V., Nelson, P. E., Eds., Univ. California Press, Berkeley, Calif., in press)
47. Hopffgarten, E. H. von. Beiträge zur Kenntnis der Stockfäule (*Trametes radiciperda*). *Phytopathol. Z.,* **6,** 1–48 (1933)
48. Hubert, E. E. Rootrots of the western white pine type. *Northwest Sci.,* **24,** 5–17 (1950)
49. Huet, M. La maladie du rond (*Polyporus annosus*). *Bull. Soc. Cent. For. Belg.* **43,** 349–71 (1936)
50. Hunt, J., Krueger, K. W. Decay associated with thinning wounds in young-growth western hemlock and Douglas-fir. *J. Forestry,* **60,**

336–40 (1962)

51. Hüpple, A. Inoculation of pine and spruce seedlings with conidia of *Fomes annosus* (Fr.) Cke. In *Proc. Intern. Conf. Fomes annosus., 3rd, Aarhus, Denmark, 1968* (In press)
52. Jorgensen, C. C., Lund, A., Treschow, C. Studies of the root-destroyer *Fomes annosus* (Fr.) Cke. *K. Vet-Hojsk. Aarsskr. 1939,* 71–128 (1939)
53. Jorgensen, C. A., Treschow, C. On the control of the agent of root rot *Fomes annosus* (Fr.) Cke. by superficial planting and the application of lime and phosphate. *Forstl. Forsogsv. i Danmark, Beret.,* **19,** 253–84 (1948)
54. Jorgensen, E. Trametes infection in shelter belts. *Dansk Skovfor. Tidsskr.,* **40,** 279–85 (1955)
55. Jorgensen, E. On the spread of *Fomes annosus* (Fr.) Cke. *Can. J. Botany,* **39,** 1437–45 (1961)
56. Jorgensen, E., Petersen, B. B. Attack of *Fomes annosus* (Fr.) Cke. and *Hylesinus piniperda* L. on *Pinus sylvestris* in the Djursland plantations. *Dansk Skovfor. Tidsskr.,* **36,** 453–79 (1951)
57. Kallio, T. Observations on *Fomes annosus* in spring 1967. *Metsät. Aikak.,* **84,** 228–30, 234 (1967)
58. Kallio, T. Distribution of *Fomes annosus* spores through the air in Finland. In *Proc. Intern. Conf. Fomes annosus, 3rd, Aarhus, Denmark, 1968* (In press)
59. Koenigs, J. W. *Fomes annosus.* A bibliography with subject index. *U. S. Dept. Agr., Forest Serv., Southern Forest Expt. Sta. Occas. Pap. No. 181,* 35 pp. (1960)
60. König, D. Über Rotfäulebestande und deren Behandlung. *Tharandter Forstl. Jahrb.,* **74,** 63–74 (1923)
61. Kuhlman, E. G. Survival of *Fomes annosus* spores in soil. *Phytopathology* (In press, 1969)
62. Kuhlman, E. G. Conidiospores levels necessary for stump root infection by *Fomes annosus. Phytopathology* (In press, 1969)
63. Kuhlman, E. G., Hendrix, F. F. Infection, growth rate, and competitive ability of *Fomes annosus* in inoculated *Pinus echinata* stumps. *Phytopathology,* **54,** 556–61 (1964)
64. Kuhlman, E. G., Hendrix, F. F., Jr., Hodges, C. S., Jr. Inoculation and colonization of stumps of *Pinus echinata* by *Fomes annosus. Phytopathology,* **52,** 739 (1962)
65. Kuhlman, E. G., Ross, E. W. Regeneration of pine on *Fomes annosus* infested sites in the southeastern United States. In *Proc. Intern. Conf. Fomes annosus, 3rd, Aarhus, Denmark, 1968* (In press)
66. Ladefoged, K. Research on the relations between felling grade, root death, and rot in Norway spruce. *Dansk Skovfor. Tidsskr.,* **44,** 5–53 (1959)
67. Lagerberg, T. Some aspects of care of standing timber and of wood. *Svenska Skogsvardsfor. Tidskr.,* **34,** 396–406 (1936)
68. Lorio, P. L., Jr., Hodges, J. D. Oleoresin exudation pressure and relative water content of inner bark as indicators of moisture stress in loblolly pines. *Forest Sci.,* **14,** 392–98 (1968)
69. Low, J. D., Gladman, R. J. *Fomes annosus* in Great Britain. An assessment of the situation in 1959. *Forest Rec.,* **41,** 22 pp. (1960)
70. Marite, H., Meyer, J. Contribution à l'étude de la propagation du *Fomes annosus* (Fr.) Cooke. *Parasitica,* **22,** 264–69 (1966)
71. Meredith, D. S. The infection of pine stumps by *Fomes annosus* and other fungi. *Ann. Botany (London), N.S.,* **23,** 455–76 (1959)
72. Meredith, D. S. Further observations on fungi inhabiting pine stumps. *Ann. Botany (London), N.S.,* **24,** 63–78 (1960)
73. Miller, O. K., Jr. The distribution of *Fomes annosus* (Fries) Karst. in New Hampshire red pine plantations and some observations on its biology. *Fox Forest Bull. No. 12,* 25 pp. (1960)
74. Miller, T., Kelman, A. Growth of *Fomes annosus* in roots of suppressed and dominant loblolly pines. *Forest Sci.,* **12,** 225–33 (1966)
75. Molin, N. The infection biology of *Fomes annosus. Statens Skogforskninst. Meddel.,* **47,** 36 pp. (1957)
76. Møller, C. M. Fresh investigations in Denmark on *Fomes annosus. Dansk Skovfor. Tiddskr.,* **24,** 433–54 (1939)
77. Mook, P. V., Eno, H. G. *Fomes*

Annosus. What it is and how to recognize it. *U. S. Dept. Agr., Forest Ser., Northeast Forest Expt. Sta. Paper No. 146,* 33 pp. (1961)

78. Morris, C. L., Knox, K. A. *Fomes annosus:* A report on the production of conidia in nature and other studies in Virginia. *Plant Disease Reptr.,* **46,** 340–41 (1962)
79. Mussell, H. W. *Mycelial Growth of Fomes Annosus Through Forest Soil in a Thinned Loblolly Pine Plantation* (Master's thesis, Duke Univ., Durham, N.C., 34 pp. 1964)
80. Negrutskii, S. F. *Fomes annosus* on *Juniperus sabina. Lesn. Hoz.,* **12,** 80 (1960)
81. Negrutskii, S. F. Some properties of the distribution of the root fungus in pine stands. *Lesn. Zh. Arkhangel'sk,* **4,** 35–38 (1961)
82. Negrutskii, S. F. On the sources of infection of the fungus *Fomitopsis annosa. J. Agr. Sci., Moscow,* **7,** 108–109 (1962)
83. Negrutskii, S. F. Some characteristics of the infection of *Pinus cembra* var. *sibirica* by *Fomes annosus. Izvest. Vysshikh. Uchebnykh Zavedenii Lesn. Zh.,* **6,** 22–26 (1963)
84. Negrutskii, S. F. The infection of larches with *Fomitopsis annosa* and its control. *Visnk. Sil's-Kohospodarskoyi Nauky,* **2,** 59–64 (1963)
85. Negrutskii, S. F. The character and features of infection by the fungus *Fomes annosus* in spruce-fir stands. *Nauch. Zap. Lugan. Sel.'-Khoz. Inst.,* **9,** 231–41 (1963)
86. Nissen, T. V. Actinomycetes antagonistic to *Polyporus annosus* Fr. *Experimentia,* **12,** 229–30 (1956)
87. Nuorteva, M., Laine, L. Über die Möglichkeiten der Insekten als Übertrager des Wurzelschwamms (*Fomes annosus* (Fr.) Cke.). *Ann. Ent. Fenn.,* **32,** 113–35 (1968)
88. Olson, A. J. A root disease of Jeffrey and ponderosa pine reproduction. *Phytopathology,* **31,** 1063–77 (1941)
89. Orlos, H., Tawarowski, I. Investigation on the dynamics of sporulation in certain species of Polyporaceae. *Prace Inst. Bad. Lesn.,* **319,** 203–26 (1967)
90. Paludan, F. Infection by *Fomes annosus* and its spread in young Norway spruce. *Forstl. Forsogzv. Danm.,* **30,** 19–47 (1966)
91. Powers, H. R., Jr., Boyce, J. S., Jr. *Fomes annosus* on slash pine in the Southeast. *Plant Disease Reptr.,* **45,** 306–07 (1961)
92. Punter, D. On *Fomes annosus* in eastern Canada. *Abstr. Intern. Congr. Plant Pathol., 1st., London, England,* 157 (1968)
93. Rennerfelt, E. Die Entwicklung von *Fomes annosus* Fr. bei Zusatz von Aneurin und verschiedenen Extrakten. *Svensk Bot. Tidskr.,* **38,** 153–63 (1944)
94. Rennerfelt, E. The occurrence and distribution of spruce butt rot. *Svenska Skogsvardsfor. Tidskr.,* **43,** 316–34 (1945)
95. Rennerfelt, E. On butt rot caused by *Polyporus (Fomes) annosus* in Sweden. Its distribution and mode of occurrence. *Statens Skogforskninst. Meddel.,* **35,** 1–88 (1947)
96. Rennerfelt, E. The effect of some antibiotic substances on the germination of the conidia of *Polyporus annosus. Acta Chem. Scand.,* **3,** 1343–49 (1949)
97. Rennerfelt, E. The effect of soil organisms on the development of *Polyporus annosus* (Fr.), the root rot fungus. *Oikos,* **1,** 65–78 (1949)
98. Reynolds, G., Wallis, G. W. Seasonal variation in spore deposition of *Fomes annosus* in coastal forests of British Columbia. *Bi-mon. Res. Notes,* **22,** 6–7 (1966)
99. Rhoads, A. S., Wright, E. *Fomes annosus* commonly a wound pathogen rather than a root parasite of western hemlock in western Oregon and Washington. *J. Forestry,* **44,** 1091–92 (1946)
100. Rishbeth, J. Observations on the biology of *Fomes annosus,* with particular reference to East Anglian pine plantations. I. The outbreak of disease and ecological status of the fungus. *Ann. Botany (London), N. S.,* **14,** 365–83 (1950)
101. Rishbeth, J. Observations on the biology of *Fomes annosus,* with particular reference to East Anglian pine plantations. II. Spore production, stump infection, and saprophytic activity in stumps. *Ann. Botany (London), N. S.,*

15, 1–22 (1951)

102. Rishbeth, J. Observations on the biology of *Fomes annosus* with particular reference to East Anglian pine plantations. III. Natural and experimental infection of pines, and some factors affecting severity of the disease. *Ann. Botany (London), N. S.,* **15,** 221–46 (1951)

103. Rishbeth, J. Butt rot by *Fomes annosus* Fr. in East Anglian conifer plantations and its relation to tree killing. *Forestry,* **24,** 114–20 (1951)

104. Rishbeth, J. Some further observations on *Fomes annosus* Fr. *Forestry,* **30,** 69–89 (1957)

105. Rishbeth, J. Stump protection against *Fomes annosus.* III. Inoculation with *Peniphora gigantea. Ann. Appl. Biol.,* **52,** 63–77 (1963)

106. Roll-Hansen, F. Studies in *Polyporus annosus* Fr., especially in respect to its occurrence in Norway south of the Dovre Fell. *Norske Skogsforsoksv. Meddel. NR,* **24,** 1–100 (1940)

107. Ross, E. W. Practical control of *Fomes annosus* in southern pines. In *XIV TUFRO Congr. Sect. 24, Munich, Germany,* pp. 321–24 (1967)

108. Ross, E. W. Duration of stump susceptibility of loblolly pine to infection by *Fomes annosus. Forest Sci.,* **14,** 206–11 (1968)

109. Ross, E. W. Thermal inactivation of conidia and basidiospores of *Fomes annosus* (Fr.) Karst. *Phytopathology* (In press, 1969)

109a. Ross, E. W. (Personal communication)

110. Ruzicka, J. On the decay of forest trees. *Czech. Acad. Agr. Bull.,* **4,** 8–9 (1928)

111. Shain, L. Resistance of sapwood in stems of loblolly pine to infection by *Fomes annosus. Phytopathology,* **57,** 1034–45 (1967)

112. Sinclair, W. A. Root- and butt-rot of conifers caused by *Fomes annosus,* with special reference to inoculum dispersal and control of the disease in New York. *Mem. Cornell Univ. Agr. Expt. Sta. No. 391,* 54 pp. (1964)

113. Stambaugh, W. J., Cobb, F. W., Schmidt, R. A., Krieger, F. C. Seasonal inoculum dispersal and white pine stump invasion by *Fomes annosus. Plant Disease Reptr.,* **46,** 194–98 (1962)

114. Towers, B., Stambaugh, W. J. The influence of induced soil moisture stress upon *Fomes annosus* root rot of loblolly pine. *Phytopathology,* **58,** 269–72 (1968)

115. Treschow, C. Zur Kultur von *Trametes* auf sterilisierten Waldhumus. *Zentr. Bakteriol.,* **2,** 104, 186–88 (1941)

116. Vakin, A. T. Die Herzfäule der Fichte in den Revieren des Rshevsky Forstes in Gouvernement Tver. *Mitt. Leningrad Forstinst.,* **35,** 105–54 (1927)

117. Wallis, G. W. Infection of Scots pine roots by *Fomes annosus. Can. J. Botany,* **39,** 109–21 (1961)

118. Wallis, G. W., Ginns, J. H., Jr. Annosus root rot in Douglas-fir and western hemlock. *Can. Dept. Fish., Forestry, Forest Pest Leaflet No. 15,* 7 pp. (1968)

119. Weidensaul, T. C. *Investigations of the Black Turpentine Beetle (Dendrotonus terebrans Hopkins) in Relation to Possible Transmission of Fomes Annosus (Fr.) Cke.* (Master's thesis, Duke Univ., Durham, N.C., 50 pp., 1963)

120. Witcher, W., Beach, R. E. *Fomes annosus* infection through pruned branches of slash pine. *Plant Disease Reptr.,* **46,** 64 (1962)

121. Woeste, U. Anatomische Untersuchungen über die Infektionswege einiger Wurzelpilze. *Phytopathol. Z.,* **26,** 225–72 (1956)

122. Wood, F. A. Patterns of basidiospores release by *Fomes annosus. Phytopathology,* **56,** 906–7 (1966)

123. Yde-Andersen, A. Heart rot in Norway spruce. *Dansk Skovfor. Tidsskr.,* **44,** 78–110 (1959)

124. Yde-Andersen, A. Stump treatment as a method of controlling *Fomes annosus. Dansk Skovfor. Tidsskr.,* **46,** 411–22 (1961)

125. Yde-Andersen, A. Seasonal incidence of stump infection in Norway spruce by air-borne *Fomes annosus* spores. *Forest Sci.,* **8,** 98–103 (1962)

126. Zycha, H. Über das Wachstum zweier holzzerstörrender Pilze und ihr Verhältnis zur Rohlensäure. *Zentr. Bakteriol.,* **117,** 223–44 (1937)

127. Zycha, H. Stand unserer Kenntnisse von der *Fomes annosus*—Rotfäule. *Forstarchiv,* **35,** 1–4 (1964)

Copyright 1969. All rights reserved

IRRIGATION AND PLANT DISEASES[1,2]

By J. Rotem and J. Palti

The Volcani Institute of Agricultural Research, Bet Dagan, and Extension Services, Ministry of Agriculture, Tel-Aviv, Israel

Introduction

In countries with long, rainless seasons, irrigation is the principal means of increasing food production. In addition, the practice of irrigation is rapidly expanding to areas with markedly irregular rainfall and thus one of the principal risks of farming is lessened. However, the water applied by irrigation supplies the moisture required by numerous pathogens as well as by their host crops. What makes the subject of irrigation so interesting to the student of crops and their diseases is that the farmer, by applying water, can influence the growth conditions of his crop more fundamentally than by any other means short of growing them under cover.

From the point of view of plant disease occurrence, irrigation occasionally makes it possible to avoid diseases altogether by growing crops out of season. Thus, as pointed out by Stakman & Harrar (93), potatoes and wheat are commonly grown in Mexico during the dry season under irrigation, to avoid destructive attacks of late blight and rust.

There are also cases in which irrigation renders the host more resistant to a given disease, mainly by increasing its vigor. More often, however, irrigation, by its effects on moisture and temperature conditions, will encourage epidemic development of diseases to a greater or lesser extent. In most of the countries in which irrigation is now practiced, diseases that used to be absent from crops grown without irrigation have become common ever since irrigation, especially in sprinkler form, has been applied.

The intensity of diseases in irrigated and unirrigated fields must, however, be considered against the background of the far-reaching changes that irrigation has introduced into the general practice of farming. Thus, the heavy investment involved makes the farmer plant his irrigated land with more valuable crops, often grown out-of-season and in improper rotation. In areas with moderate winters but rainless summers, year-round cultivation of many crops has thus become possible; as a result, diseases formerly limited to well-defined seasons can now develop throughout most or all of the year. In addition, whereas only sparse growth developed in the unirri-

[1] Contribution from the Volcani Institute of Agricultural Research, Bet Dagan, Israel. 1968 Series, No. 1459-E.

[2] A major portion of the studies on irrigation and plant disease in Israel has been carried out in a project sponsored by the United States Department of Agriculture (project No. A10-CR-69).

gated crop, a dense stand of plants develops under irrigation, with resultant changes in the microclimate.

However, neither irrigation in general, nor even sprinkling in particular, will invariably increase the incidence of plant diseases. The key to the effect of irrigation on disease development lies in the interaction between conditions created by irrigation, weather factors, and the specific natures of the pathogen and crop. It is, therefore, hardly surprising that even identical irrigation treatments applied to the same crop will not have the same effect on the development of a given disease under different sets of weather conditions. It follows that in a discussion of irrigation effects on disease, great care must be taken to indicate the specific conditions of weather and of crop growth, and that the ecological requirements of each pathogen and the differential effects of various irrigation techniques have to be taken into account.

Irrigation may so profoundly affect the vigor, rate of growth, overall development of the whole plant and its various parts, and the length of the crop's growth span, that many effects on proneness of the host to disease can be attributed to it only indirectly. Similarly, irrigation effects on soil microflora obviously have a bearing on the development of diseases caused by soil-inhabiting pathogens, on the survival of resting fungal bodies, and on other sources of inoculum in the soil.

The following survey of irrigation effects has been limited to fungal and bacterial diseases. Because of the greater difficulties in studying root pathogens and the complicating action of soil microorganisms, relatively less is known about the effect of irrigation on root diseases than on those of the shoot. The greater part of the following sections is, therefore, devoted to irrigation, and especially overhead sprinkling, in relation to air-borne diseases. Our review is based on the published literature, on personal communications from colleagues in various countries, and on the research and farming practice in Israel.

Soil-Borne Diseases

Although irrigation effects on air-borne diseases are conspicuous, the simultaneous effect of irrigation on the soil-borne pathogen and its host is of great significance also. The relation of soil-borne diseases to soil moisture has been reviewed by Garrett (37), who assembled a list of soil-borne diseases and the moisture levels that favor them. Studies published since Garrett's have confirmed his generalizations: high levels of moisture are mostly favorable to pathogens dependent on water for dissemination of zoospores, such as species of Phycomycetes, and many pathogens benefit from host susceptibility induced by high moisture. Other hosts, however, become more susceptible at low levels of soil moisture, which also favor strongly aerobic pathogens. Another aspect to be considered is the moisture effect on the microflora that compete with pathogens.

Aeration and soil moisture level.— The effects of soil moisture on the development of pathogens with different responses to aeration is best illustrated by reference to *Rhizoctonia solani* and *Pythium* spp. *R. solani,* which requires oxygen for growth, develops almost as well in dry as in wet, but aerated, soil (3). However, when aeration is deficient, development of *R. solani,* as reflected by pathogenicity to *Euphorbia pulcherrima,* lags behind *P. ultimum,* which is better adapted to low oxygen levels (1). But this does not imply that species of *Pythium* are restricted to poorly aerated soil. Frank (36) found that *Pythium* spp. affects peanuts more in abundantly irrigated soil when it is well aerated. The role of the host in these interactions can be illustrated by reference to some phytophthoral diseases. In citrus trees, which are sensitive to poor aeration, excessive irrigation inhibits growth and regeneration of roots and thus prevents the host from overcoming the effects of phytophthoral rots (95). On the other hand, conditions most favorable to *P. parasitica* in tobacco and to *P. drechsleri* in safflower include fluctuations of dry and wet conditions, since dryness increases host susceptibility, and wetness favors the pathogen (58, 105, 114). The fact that different levels of soil moisture are optimal to *Fusarium* spp. on various plants may be attributed mainly to the response of the host rather than to that of the pathogen (13, 15, 16, 27, 35, 87, 96).

Microflora.—The development of soil microflora under different irrigation management, and its effect on disease, has been studied in the case of *Streptomyces scabies* on potatoes. Developing best under dry conditions (88), common scab is largely prevented in fields irrigated to their full capacity (54, 55, 60). This has been attributed to the effect of irrigation on increased development of bacteria probably antagonistic to *S. scabies,* in wet soil, as against development of Actinomycetes in dry soil (56). On the other hand, Curl (22) found that, in fescues, larger numbers of Actinomycetes develop at high irrigation levels. It thus appears that the specific effect of the crop plant is also an important factor and this is the conclusion at which Curl arrived in his studies.

Preservation and dispersal of inoculum.—The preservation of inoculum of soil-borne pathogens in soils under irrigation has been studied intensively where attempts were made to eliminate such inoculum by flooding. This practice has been reported to be successful particularly in relation to *Sclerotinia sclerotiorum* (4, 66) and to *Fusarium oxysporum* f. *cubense* (97). A review of the effects of soil moisture on the survival of plant pathogens has been included in Sewell's (90) survey of the physical soil conditions affecting such pathogens.

Dispersal of inoculum of root and collar pathogens by irrigation water has been demonstrated in relation to several important diseases, either by propagules carried in water (52, 59), or by active spread of mycelium in the irrigated soil (113). Other aspects of dispersal in soil-borne pathogens were reviewed by Hirst (43).

The effects of irrigation on survival and dispersal of pathogens that persist on plant debris or in resting stages in the soil, but attack chiefly the above-ground parts of the plant, will be discussed in the sections dealing with these parts.

Air-Borne Diseases

Effects of soil moisture levels.—In the case of soil-borne diseases the level of soil moisture has often been stated to affect simultaneously the susceptibility of the host and the virulence of the pathogen. But as far as diseases of leaves, stems, and fruits are concerned, soil moisture effects are generally limited to the host. These effects are exerted in a variety of ways, the most important of which are as follows:

(*a*) Predisposition to pathogen attack by moisture levels prevailing prior to the actual time of such attack.

(*b*) Effects on the turgidity of host organs when they are about to be attacked. Where soil moisture is made available to crops by irrigation in warm and dry zones, additional effects to be taken into account are: (i) richness of shoot development, and the attendant microclimatic changes in the immediate vicinity of the plant; (ii) formation of periodic growth flushes especially susceptible to certain diseases; and (iii) extension of the growing period into seasons in which factors other than moisture render the host more susceptible to disease.

Soil moisture effects on predisposition of the host and on the turgidity of its organs have been reviewed by Yarwood (110). Leaf turgidity has been shown to favor infection of many hosts by a great variety of bacterial diseases (26, 73, 91, 98, 104). Among fungal diseases, *Peronospora tabacina* on tobacco (80), *Cercospora musae* on bananas (39), *Phytophthora infestans* (47) and *Alternaria porri* f.sp. *solani* (77) on potatoes, and *Colletotrichum lindemuthianum* on beans (47), have been described as favored by high turgidity.

There are, however, also a few records of diseases favored by low, or inhibited by high, turgidity of the leaf, e.g. *Botrytis* sp. on *Eucharis* sp. (6), *Erysiphe graminis* on wheat (74), and *Uromyces phaseoli* on beans (14).

The turgidity of a plant is conditioned by its ability to take up water, on the one hand, and by its rate of transpiration on the other hand. The importance of irrigation in this respect is obvious. Turgidity is generally highest during the hours of the night and early morning when most infection processes also take place. Excess moisture supply then frequently results in guttation, which in turn has been shown to favor the development of some diseases (30). This is apparently due to the host exudates, present in the drops of guttation water, which stimulate the pathogen. Duvdevani et al. found that overhead sprinkling increases guttation of cucumbers above the level obtaining on furrow-irrigated plots, and may thus favor *Pseudoperonospora cubensis* (28).

There are cases, however, in which infection caused by day-dispersed spores of low viability may succeed only in daytime; since leaf turgidity is then generally decreased, irrigation, which increases turgidity, favors the disease.

It is not yet clear how turgidity affects the proneness of the host to disease, whether only infection, or subsequent colonization and sporulation are also affected. In comparison to a flaccid leaf, other factors being equal, a turgid leaf tends to keep its stomata open, shows increased metabolic activities, but maintains its temperature at a lower level. The last factor is often critical in semi-arid conditions. The phytopathological effects of the other factors need to be studied.

While high soil moisture and pronounced turgidity of plants at the time of infection is thus seen to favor the great majority of air-borne diseases, low soil moisture has been shown to predispose many plants to the subsequent attack of pathogens (5, 18, 21, 29, 34, 65, 77, 80, 99, 101, 108). In a few cases, high levels of soil moisture have been described as favoring disease incidence directly, or not through a predisposing effect (48, 67, 106).

Only one reference was found to diseases not markedly affected by the level of soil moisture (17). Whether or not this is, as we suspect, a more frequent phenomenon not adequately covered by the literature, remains an open question.

Is the predisposing effect of low moisture stress of any practical importance?

The incidence of *Alternaria sesami* and *Corynespora cassiicola* on sesame (21), *Sclerotum bataticola* on sorghum (29), and *Hendersonula toruloidea* on walnuts (34), was clearly shown in field experiments to be reduced by additional irrigation.

As most other studies were carried out under laboratory conditions, their value in relation to field conditions is uncertain. In some cases, laboratory results were in contrast to field observations, where factors additional to host susceptibility induced by moisture stress governed the spread of diseases. Such cases include tobacco, which is more susceptible to *P. tabacina* under low moisture conditions, but is nevertheless attacked more severely in the field under sprinkling, which favors the pathogen (80).

Another aspect shown to be of importance in relation to the development of *Alternaria porri* f.sp. *solani* on potatoes and tomatoes in Israel, is the effect of the soil-moisture level on tuber or fruit formation: though both these hosts are predisposed to the disease by low moisture stress, their yield/foliage ratio affects their susceptibility even more strongly. Thus, well-watered plants, forming plenty of tubers or fruits, are eventually more susceptible than plants growing with a limited supply of water and giving low yields (77, 81).

Shoot development.—Under arid conditions, any kind of irrigation induces fuller shoot development, and this in its turn ensures more shade, lower temperatures, and longer periods of high moisture in the lower foliosphere and upper soil layers. Lower leaves and root collars are thus exposed to increased danger from the attack of the many fungi and bacteria favored by such conditions.

Those leaf fungi which prefer to attack the older lower leaves of field crops are particularly favored by irrigation. In Israel, this effect of irrigation is prominent in regard to *A. dauci* on carrots, and to *Cercospora personata* on peanuts. Some other fungi, such as *P. infestans,* for the attack of which increased moisture is more important than is the age of the host tissue, are also favored by the conditions of shade in fully developed crops. In southern Australia, greatly increased foliage growth made by vines under irrigation seems to induce a greater susceptibility to late summer infection of *Oidium,* due to the shading to the center of the vines (9).

Another effect of irrigation on leaf fungi is that due to the periodic formation of flushes of young growth, after application of water. Among the pathogens preferring such tender tissues, the downy mildew group is outstanding and profits most from this particular effect of irrigation. This has been observed in Israel, under all types of irrigation, with regard to the downy mildew of peas, alfalfa, cucumbers, and tobacco, and has been demonstrated experimentally with regard to *Sclerospora graminicola* on pearl millet (50) and to *Plasmopara viticola* on grapes (72).

Prolongation of the period over which organs susceptible to the vine mildew are formed has been pointed out as a cause of the increased incidence of *P. viticola* in irrigated vineyards in France (38).

Effects of irrigation on the microclimate of the foliosphere and on leaf surface conditions.—Before we proceed to discuss effects of irrigation on the pathogen, some description is required of the way irrigation acts on the microclimate of the foliosphere and on conditions at the leaf surface. Reliable data on this subject are scarce.

The cultural factors most markedly influencing irrigation effects in this respect are the size and shape of the irrigated area and the density of the crop. Simultaneous application of water to large, contiguous areas, as is sometimes done by flooding, may affect atmospheric humidities and temperatures for longer periods than irrigation of elongated strips; in the latter case, if the weather is dry, the inrush of air from adjoining dry areas will keep the microclimatic effects to a minimum. Stanhill (94) has aptly called this an "oasis" effect.

The density of the crop is of obvious relevance: Thus, in Israel, after sprinkling around noon on a summer day, the leaves of trellised tomatoes dried up within 45 min, while those of bananas in dense plantations remained wet for 2 to 3 hr and some banana organs well protected from wind and radiation remained wet for 24 hr (78, 79).

A distinction must, of course, be made between the effects of irrigation applied by various methods. In general, in dry countries, the effect exerted on the microclimate by surface irrigation is too weak to be significant from a phytopathological point of view (79, 83, 86). But in fields under sprinkler irrigation, considerable decreases of temperature and increases of relative humidity occur, at least during the time of irrigation and for various periods thereafter (70, 71, 79, 83, 100).

In some cases, sprinkling has also been reported to lengthen dew periods (71), but this was not found under other conditions (79, 83).

The effect of sprinkling on temperatures and relative humidities of the air is most pronounced under extreme macroclimatic conditions. In a series of experiments it was found that sprinkling effected the largest decreases in temperature (8° to 9° C) and increases in relative humidity (50 per cent) when the respective values for the unirrigated neighboring fields were 29° to 32° C and 12 to 25 per cent. However, when these values were 10° to 20° C and 80 to 90 per cent, sprinkling lowered temperature by about only 1° C and raised the relative humidity by no more than 2 to 3 per cent (79).

While these effects may be of some interest from a general, agrometeorological point of view, what matters primarily with regard to disease development are the actual temperature and moisture conditions on the leaf surface. In one study in Israel, leaf temperatures of potatoes sprinkled on a hot summer day were measured at various hours of the day. In the morning and at noon, when the air temperature was 24° to 29° C, and the leaf temperature in unirrigated plots was 30° to 36° C, the temperature of wet leaves dropped to 22° C. When irrigation was concluded, the leaf temperature began to rise but remained below that of unirrigated plots for the whole day. In plots irrigated in the evening, the temperature of wet leaves was only 4° C lower than the 22° C in the air (83).

The higher the air temperatures and the stronger the solar radiation, the more pronounced appears to be the effect of sprinkling in lowering leaf temperatures. This is of considerable significance in arid climates, where temperatures of dry leaves may often be marginal for infection, and the temperature drop due to irrigation may make all the difference to success or failure of the infection process on a hot day.

Production of air-borne inoculum and its dispersal.—Irrigation may affect production of air-borne spores by promoting sporulation in aboveground plant organs and by inducing spore discharge from resting bodies in plant or soil.

In specific cases, prolonged flooding, by its effects on atmospheric humidity around the lowest leaves, may promote sporulation. In other cases, sprinkling performed before or after natural dew fall, prolongs the period of leaf wetness and thus facilitates sporulation. Generally, however, day time irrigation does not promote sporulation, either because it is too brief or because many pathogens, such as some downy mildews, do not sporulate in light. In other cases, day temperatures, even when decreased by irrigation, are too high for sporulation.

Only sprinkling which is applied for very long periods can induce sporulation directly. Thus, in Jamaica, rains as well as under-trees sprinkling extending over 48 consecutive hours, resulted in heavy sporulation of *Deightoniella torulosa* on banana leaf trash. As a result, atmospheric spore concentration increased, reaching a peak 12 hr after sprinkling was concluded, and decreased rapidly as the plantation dried up. The sprinkled bananas were consequently more severely infected by "speckle" disease than were those grown under surface irrigation (63).

It has been established that wetting is essential for the discharge of spores from the fruiting bodies of many fungi (46). Irrigation, in this respect, has the same effect as rain. In Greece, sprinkling has been found to induce zoospore discharge from oospores of *P. viticola,* and the number of cases in which infections were recorded corresponded to the number of vineyards irrigated in April (111). In Australia, the release of ascospores of *Mycosphaerella pinodes* from debris left over from the previous year's pea crop is often affected by irrigation, assisted in some cases by rain (10)

A case in which the effect of sprinkling on spore release and dispersal is particularly pronounced is that of the bitter rot of apples (*Gloeosporium fructigenum*) in Israel. The spores of this fungus, formed in acervuli, are sticky, and under dry conditions they resist dispersal even by winds as strong as 20 km per hr. But as soon as the acervuli are wetted by overhead sprinkling, spores are spread. Apple orchards irrigated in this way are, therefore, affected by the disease, while those irrigated by under-tree sprinkling or surface irrigation are not (70).

The mechanism of spore dispersal by splashing has been studied by Gregory et al. (40). They found that the collision of falling drops with wet spore-bearing surfaces causes parts of the drops and of the surface water film to be incorporated in splash droplets that contain spores; the larger the incident drop, the greater the number of droplets formed. It has further been shown, by Hirst & Stedman (45), that splash dispersal is also possible from dry surfaces: The first drops that strike a dry spore-bearing surface cause spore dispersal by radial air movement in advance of the radially spreading splash, and by vibration.

Air-borne spores do not usually need water for their dispersal. However, dispersal occurring at the time leaves are covered by water is of obvious advantage to the pathogen. Faulwetter showed in 1917 (31) that raindrops are active agents in spore dispersal. Spore dispersal by rain or irrigation is considered to aid various diseases, including *Corynebacterium michiganense* in tomatoes (11), *Fusarium moniliforme* in bananas (102), *Pseudomonas phaseolicola* in beans (103), *Colletotrichum phomoides* in tomato (71), *P. infestans* in potatoes (44), and *Phytophthora* spp. in citrus (7). In dry regions, in which the moisture conditions suitable for infection of plants are often marginal, splash dispersal due to sprinkling is more important for the pathogen than is dispersal by rain in more humid areas. The role of sprinkling in simultaneously dispersing spores and enabling them to infect plants during a short 3½ hr period of water application on a hot dry day, was proved experimentally in Israel in relation to *P. infestans* in potatoes (83). The view that splash dispersal of spores by sprinkling is among its fundamental effects on plant diseases is accepted in relation to many diseases by many pathologists (9, 23, 42, 53, 64, 112).

The washing of inoculum from upper to lower leaves by sprinkling is undoubtedly responsible, in part, for greater infection of the latter. Washing spores from the shoot into the soil for infection of underground parts seems of lesser importance. However, this type of dispersal by sprinkling, certainly favors potato tuber infection by *P. infestans,* and has also been stated to aid their infection by *Alternaria solani* (32).

Survival of inoculum.—The effect of irrigation on the survival of fungi persisting on plant debris on or in the soil has been studied in Israel in relation to *Drechslera (Helminthosporium) teres* in barley (49), *A. porri* f.sp. *solani* in tomatoes and potatoes (76), and *A. dauci* in carrots (68). In all these cases, the pathogen is carried on debris through the rainless summer season, and subsequently produces inoculum for the infection of crops sown in the fall. It has been shown clearly that the length of persistence of the above pathogens in plant debris is inversely correlated to the amount of irrigation water applied to the soil, and persistence was longest in unirrigated soil. Irrigation in these cases may thus play an important role in diminishing the amount of inoculum surviving from one season to the next.

Techniques and management of irrigation and their effects on air-borne diseases.—How irrigation applied by the various techniques affects factors related to disease development is summarized in Table I. It is evident that

TABLE I

EFFECTS OF IRRIGATION TECHNIQUES ON FACTORS RELATED TO DISEASE DEVELOPMENT

Factor	Irrigation by: furrows	flooding	subsoil	trickling	overhead sprinkling
Percentage of soil surface wetted	20–25	80–90	0	20–30	100
Wetting of foliage and fruit	not wetted				wetted
Effect on tissue temperatures	negligible or none				pronounced
Splashing of soil and/or inoculum or both	no splashing				much splashing
Effect on fungicides	not washed off				partly washed off

sprinkling is by far the most favorable for most diseases. But in addition to the technique, pronounced effects are exerted also by the various factors that play a role in the management of irrigation: (*a*) the amount of water given at each application and the rate at which it is given (which is related to the time span of each application); (*b*) the intervals between successive irrigations; and (*c*) the time of day when irrigation is applied. These factors will be reviewed here chiefly in relation to sprinkling.

Sprinkling versus other irrigation techniques.—Sprinkling differs from all other methods of irrigation by wetting practically the entire surface of the soil and (except for under-tree sprinkling in orchards) of the crop, and by splashing about or washing off the inoculum and fungicidal deposits. It also has to be borne in mind that sprinkling is in many cases applied in smaller amounts and at shorter intervals than is sub-soil, furrow, or flood irrigation.

A long list of diseases has been reported to be favored by sprinkling as compared to other irrigation techniques. These include *P. infestans* (19, 84), *A. solani* (32, 41, 53, 64), *Pseudoperonospora cubensis* (28), the banana diseases caused by *Dothiorella gregaria,* species of *Fusarium* and *Diplodia natalensis* (78), and a great number of bacterial diseases (11, 57, 61).

In fruit crops, increased disease development on trees sprinkled overhead, as compared with those sprinkled under the tree, has been described in the cases of *P. cactorum,* which affected pears on the lower limbs of the tree only to the level wetted by sprinkling (51), and of *G. fructigenum* which affects apples in Israel (70). Fisher (33) has noted the deleterious effect of overhead sprinkling on citrus trees suffering from *Elsinoe fawcettii, Diaporthe citri,* and *Cercospora citri-grisea* in groves in Florida.

The amount, rate, and frequency of irrigation.—The limits within which the amount, rate, and frequency of irrigation can be varied evidently depend on characteristics of the crop, soil, and climate, as well as on the skill of the farmer and the equipment at his disposal.

Other things being equal, the frequency of successive sprinklings has the most pronounced effect on the development of leaf and fruit diseases. In potatoes infested with *P. infestans* in a semi-arid region of Israel, disease development was shown experimentally to be directly related to the intervals between sprinkling from 8 to 21 days (84). Frequent sprinklings have also been observed to accelerate development of *A. solani* on potatoes in Idaho (53), while extending irrigation intervals reduced the incidence of *Tranzschelia pruni-spinosae* on stone fruits in Australia (9).

Shorter irrigation intervals often go together with increased total amounts of water applied to the crop. This has been found to favor strawberry rots (8), and clover, and alfalfa diseases such as *Pseudoplea trifolii, Curvularia trifolii,* and *Pseudopeziza medicaginis* (24).

The effect of the rate at which water is applied (i.e., the amount per unit of time) has not been widely studied. In Israel, no difference in the development of *P. infestans* on potatoes was noticed when sprinkling was applied at the same frequencies and amounts, but at markedly differing rates (84). This may be the case with many pathogens that penetrate their hosts rapidly; however, pathogens that take a longer time for penetration, are likely to be favored by prolonged sprinkling, as has been proved for *Stemphylium botryosum* on tomatoes (79). The case of *Xanthomonas phaseoli* on beans probably belongs to this group (61).

Low rates of sprinkling usually involve small droplets, which disperse spores less than larger drops (40). This may to some extent compensate for adverse effects of prolonged sprinklings. Another aspect to be considered is that sprinklings at low rates are often applied at night when plants are covered with dew anyway, and such sprinklings may then be less favorable to disease than shorter daytime sprinklings at higher rates.

The length of the high moisture period produced by sprinkling in the foliosphere does not depend solely on the rate at which water is applied to a given plot. Especially when the plot sprinkled at any one time is an elongated strip, the sequence in which adjacent strips are sprinkled may be of great importance. If irrigation proceeds directly from one strip to the one next to it, plants on the borderline between the strips are certain to be sprinkled for the double length of time, and even plants further away from that border will be exposed to high humidity for extended periods of time. The only case of which we are aware where this has been tested experimentally, is that of a banana grove in Israel. Some blocks of this grove were sprinkled on their whole area, and tip rots, due mainly to *Dothiorella gregaria,* affected 3.3 fruits per bunch; in the other blocks to which irrigation was applied within 2 days on alternate rows in each one, as many as 5.7 fruits per bunch became infected (78).

It should be noted that the above statements about the effects of sprinkling do not apply to a number of powdery mildew diseases. These may be inhibited by sprinkling (2, 24, 28, 79), and may even be controlled by water application to the leaves (12, 25, 107). More details on this problem are presented in reviews by Yarwood (109) and by Schnathorst (89).

The effects of the time of day at which crops are sprinkled, on the development of diseases affecting them depend so much on interaction with other environmental factors, that they are best discussed in the section on such interaction.

Interaction between irrigation, weather, pathogen, and crop.—In the following, an attempt will be made to analyze overall effects of irrigation against the background of weather factors, and of the characteristics of pathogen and crop. It may perhaps seem premature to make such an attempt on the basis of the limited experimental data available, but some of the generalizations presented, speculative as they are, in part, may furnish the basis for a healthy discussion.

Among the factors that interact with irrigation under a given set of macroclimatic conditions, the following have been deemed important in the preceding sections: (*a*) the climate of the foliosphere, as determined largely by crop density, but also by topography, soil type, drainage, and other cultural factors; (*b*) supplementary sources of moisture, especially dew; and (*c*) the time of day at which spores are dispersed, their viability under marginal conditions of humidity and temperature, and the rate at which infection can be completed.

All these factors obviously interact with each other as well, but they will here be considered separately. Some of their interactions with irrigation, which here refers mainly to sprinkling, will be expressed in diagrams based, unfortunately, not so much on experimental facts as on hypothesis.

Sprinkling and crop density.—Where the macroclimate favors plant diseases, the microclimate of the field is of comparatively minor importance;

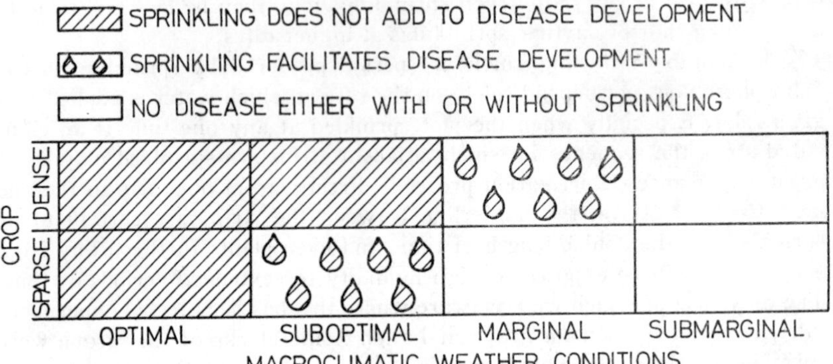

FIG. 1. Schematic representation of the effects of crop density and sprinkling on disease development under various macroclimatic conditions.

but conversely, as regional or seasonal conditions become progressively less favorable to the disease, the microclimate of the foliosphere assumes ever larger importance (82). This applies to the great majority of irrigated areas all over the world.

One of the factors profoundly affecting the microclimate in the crop is the density of foliage. This largely determines the rate at which moisture, whether derived from rain, dew, or irrigation, will evaporate. The effect thus exerted by crop density on disease development in a sprinkled field is depicted in Figure 1. This may be summed up by stating that under conditions highly favorable (optimal) or highly unfavorable (submarginal) to disease development, neither sprinkling nor density of the crop has an appreciable effect. However, where conditions are suboptimal or merely marginal for the pathogen, differential effects are apparent. Under suboptimal conditions the disease may develop even without sprinkling in dense crops, but needs the added moisture supplied by sprinkling to develop in sparse stands. Under marginal conditions the disease may fail to develop on sprinkled crops as long as the stand is sparse, and gain a foothold only as the foliage growth becomes dense. The above considerations are of fundamental value in forecasting disease outbreaks in drier climates, since they relate to the age (and, hence, density) at which the crop is likely to be attacked by certain fungi, such as *P. infestans* and downy mildews (69).

Sprinkling and spore dispersal.—Air-borne spores of many of the most important pathogens are formed at night, and released in the daytime. Although the amount of spores dispersed varies according to environmental and biotic factors, the diurnal periodicity of their release is quite constant in most fungal species. A survey of the literature suggests that in the semiarid zones, in which irrigation is most important, the spores of many species are released earlier in the day than in more temperate zones. Thus, in Israel, about 80 per cent of the spores of *A. porri* f.sp. *solani* are released during 2 to 3 hr, around 11 a.m. (75); those of many downy mildews, between 7 and 9 a.m., and under very dry conditions, even between 7 and 7:30 a.m.

Sprinkler irrigation affects this situation by precipitating spores from the air, splashing them about, and providing the moisture necessary for germination. As splash dispersal is conditioned by the presence of spores on the leaf surface, its highest rate will occur when irrigation is applied before these spores are released by natural means. Consequently, early morning sprinkling can potentially disperse by splashing spores of most fungi, while noon sprinkling will be effective chiefly for spores whose natural release occurs later. The evening and early night sprinklings will find few spores to disperse, while sprinklings performed just before morning will again meet spores to be splashed about. This has been found to be true for *P. infestans*

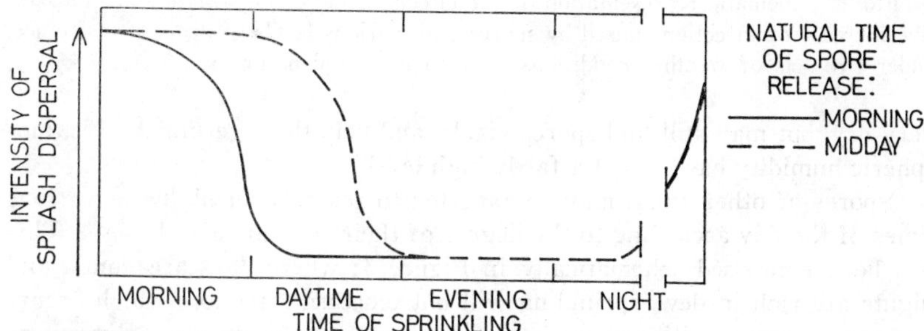

FIG. 2. Schematic representation of the effects of sprinkling applied at various times of the day, on the intensity of splash dispersal of spores released chiefly (*a*) in the morning and (*b*) at midday.

in Israel (83), but data for other fungi are hypothetical. This interplay of factors is presented schematically in Figure 2.

Sprinkling and viability of spores.—The viability of dispersed spores of various fungal species under dry conditions often furnishes the key to understanding the effects of sprinkling on the disease caused by these fungi. This will be illustrated here by reference to spores of two fungi; the extremely drought-resistant spores of *A. porri* f.sp. *solani* (76), and the highly drought-sensitive spores of *P. infestans* (20). Both of these pathogens form their spores at night, and they attack the same hosts, potatoes and tomatoes. But there the similarity ends: the *Alternaria* spores, dispersing at noon, are capable of surviving the dry hours of the day on which they are dispersed, and of many subsequent days. They will germinate as soon as dew falls, and sprinkling is, therefore, of little additional benefit to the disease where dew is present (76, 85). The sensitive spores of *P. infestans,* dispersing earlier in the morning, will not readily survive dry days. Surface irrigation will not help them in doing so, but sprinkling performed at or close to the time of dispersal will assist the freshly formed spores to infect hosts before losing their viability. Sprinkling applied later in the day may come too late to rescue spores if the preceding hours have been

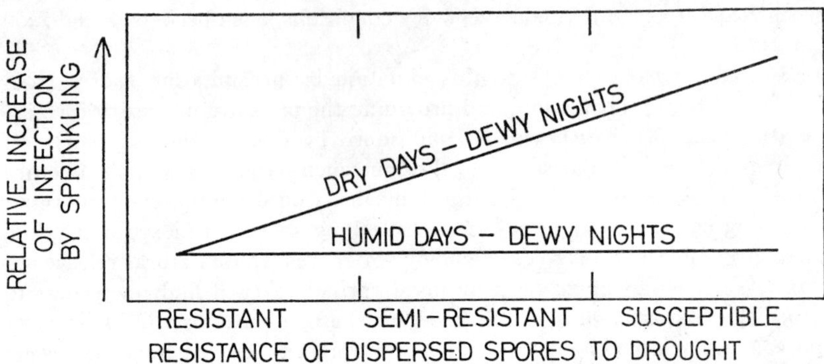

FIG. 3. Schematic representation of the effects of daytime sprinkling on the relative increase of infection caused by spores with various levels of drought resistance, under two sets of weather conditions (for further explanation, see text).

very dry, but may still find spores viable, and help them germinate, if atmospheric humidity has been at a fairly high level.

Spores of other fungi may be expected to react to sprinkling at various times of the day according to the degree of their resistance to drought. This has been expressed schematically in Figure 3: where days are humid and nights are rich in dew, sprinkling will not increase infection, which occurs no matter how sensitive or resistant the spores are to drought. But where dewy nights alternate with dry days, daytime sprinkling may greatly increase infection by sensitive spores which otherwise would not be able to withstand the dry day. To a lesser extent, sprinkling will increase infection caused by semi-resistant spores, and only infection by drought-resistant spores will not be increased, because it will be high anyway.

Sprinkling and the infection process.—Assuming that the spores of a pathogen have preserved their viability, the effect of sprinkling on its successful establishment in the host relates chiefly to the period over which the moisture required for penetration is furnished during the actual period of sprinkling and thereafter until the foliage dries. The rate of drying depends on the vapor pressure deficit of the air, on wind velocity, and on crop density. In one experiment in Israel, the rate of drying of tomato foliage after sprinkling, ranged from 5 min for outer leaves, under a cloudless sky, with strong wind, at 36° C and 16 per cent relative humidity to 4 hr for inner leaves, under a cloudy sky, with no wind, at 17° C and 86 per cent relative humidity. In the latter case, drops persisted until dewfall, and a continuous moisture period of 20 hr thus resulted from sprinkling and dew (79).

The extent to which sprinkling at various times of the day assists infection also depends greatly on the temperature level favoring spore germination and on the speed with which the infection process is completed. This has been depicted graphically in Figure 4 and is further explained in the legend to that figure.

Sprinkling and dew.—In rainless seasons, dew and sprinkling constitute the major sources of moisture on leaves, stems, and fruits. How these interact under various environmental conditions may again be demonstrated best by comparing their effect on those two contrasting types, *A. porri* f.sp. *solani* and *P. infestans*. This comparison is presented in Table II, and is based on recent studies (83–85).

The differences in behavior of the two pathogens under various dew and sprinkling combinations are, in our view, due chiefly to the above mentioned differences in the drought resistance of their spores. Those produced by *A. porri* f.sp. *solani* will remain viable under practically all the conditions under which potatoes and tomatoes are grown (76). In rainless periods the promotion, by sprinkling, of the disease caused by this fungus, is, therefore, limited to conditions where either dew is deficient or temperatures during

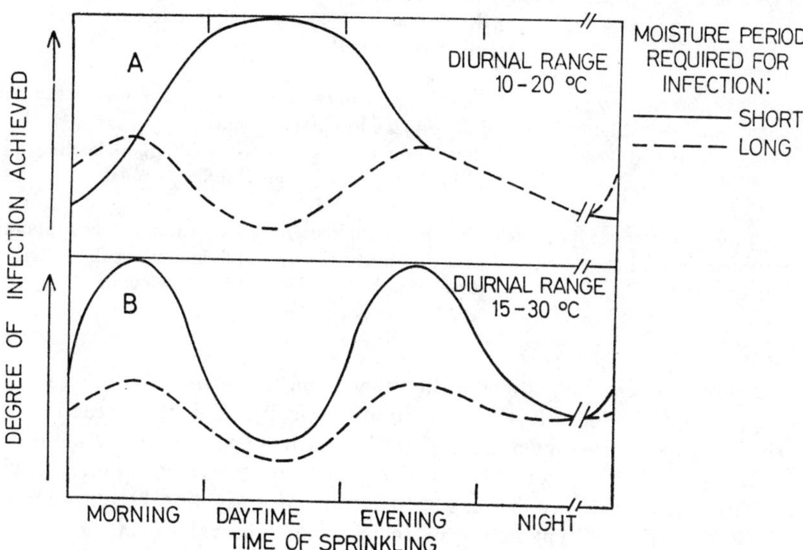

FIG. 4. Schematic representation of the effects exerted on infection by sprinkling, applied at various times of a dry day-dewy night cycle, at two diurnal temperature ranges, and in relation to pathogens requiring short or long moisture periods for completion of the infection process. The pathogens are assumed to produce spores viable throughout the day, and to have infection optima of 20° C. A. On days with a 10° to 20° daily range of minimum and maximum temperature, pathogens capable of completing infection within short moisture periods will achieve maximum infection under midday sprnkling; but pathogens requiring longer periods for infection can only utilize the extended moisture periods produced by dew combined with morning or evening sprinkling, and the amount of infection is then reduced by lower temperatures. B. On days with a 15 to 30° C daily range of temperature, even the pathogens requiring only short moisture periods find conditions in midday-sprinkled plots too hot, and will achieve better infection under morning or evening sprinkling; the pathogens requiring longer moisture periods will also take advantage of morning or evening sprinklings continuous with dew periods.

TABLE II

Interaction of Dew and Sprinkler Irrigation in Their Effect on *Alternaria porri* f.sp. *solani* and *Phytophthora infestans* on Potatoes and Tomatoes Under the Environmental Conditions Encountered in Various Locations in Israel

Environmental conditions	Development of *A. porri* f.sp. *solani*	Development of *P. infestans*
Completely arid and dewless oasis of the Negev desert	Slight development only under sprinklers	Complete absence
Daily atm. humidity minima 35 per cent or below, nights rich in dew, no rain, in N and W Negev	Dew alone suffices for infection and epidemic development; sprinkling is without effect	Dew alone inadequate; sprinkling required for blight outbreaks
Daily atm. humidity minima above 60 per cent, night rich in dew, no rain, in the coastal plain	Dew alone suffices for infection and epidemic development; sprinkling is without effect	Dew may suffice, but is materially assisted by sprinkling, in leading to blight outbreaks
Atm. humidity always high, dew plentiful, no rain, in the coastal plain (spring, fall)	Dew alone suffices for infection and epidemic development; sprinkling is without effect	Dew alone can support blight outbreaks; sprinkling irrelevant

the dew period are very low. On the other hand, for *P. infestans* and similar types of pathogens, daytime conditions critically affect their reaction to sprinkling; in the absence of rain, dew alone will support full infection only if fairly high daytime humidities enable spores to survive until dewfall. As humidity during the day decreases, the effect of sprinkling becomes more and more prominent, but in the extreme drought prevailing in the desert where nights lack dew, sprinkling alone cannot induce blight development.

Irrigation Management and Plant Disease Control

Plant disease control—apart from resistance breeding—consists largely in minimizing epidemics by cultural measures and in checking them by properly timed and effective applications of chemicals. Irrigation, and particularly sprinkling, is highly relevant to all aspects of such control.

Minimizing disease by irrigation management.—As outlined above, interaction of three groups of factors determines the time and scope of disease development in irrigated crops: (*a*) macroclimatic weather factors (rain, atmospheric humidity, temperature, radiation), through their effects on the climate of the foliosphere, on the formation of dew, and on the viability of air-borne spores; (*b*) pathogen and host characteristics, foremost among them the time of spore dispersal, drought resistance of spores, and speed of the infection process; and (*c*) the technique, frequency, and time of day chosen for water application. All these act against the background of crop density and of some other cultural factors. How can irrigation be managed to reduce risks of disease outbreaks in this setup?

The choice of the most suitable irrigation technique is no problem under completely arid conditions, in which irrigation water has no other sources of moisture with which to interact. But under semi-arid conditons, with little or no rain, the choice of surface irrigation instead of sprinkling may contribute more to disease control than does any chemical treatment. This has been stressed by Menzies (62) and by Snyder et al. (92), especially in relation to bacterial diseases of beans in the western United States, and has been found to be equally true in Israel for many, though not all diseases.

However, sprinkling has so many practical advantages wherever labor is in short supply or expensive, that farmers and pathologists have to learn to live with it and to minimize its adverse effects. This can be done chiefly in the following ways: (*a*) in dangerous cases sowing at distances and fertilizing at rates that will not result in excessive density of foliage; (*b*) judicious spacing of irrigations, according to crop and soil conditions, so as to irrigate at the longest possible intervals that will not induce moisture stress in the crop; (*c*) choosing the time of day in which sprinkling is least likely to enable the pathogen to complete the infection process; and (*d*) avoiding the sprinkling of adjacent strips of crops in quick succession, before the strip sprinkled first has had time to dry.

Prediction of air-borne diseases.—From the point of view of prediction, sprinkling has the great advantage over rain that its timing and length can be regulated; moreover, if properly timed, the length of time over which the effect of sprinkling extends beyond the period of actual application can be assessed accurately. This is much harder when moisture is provided by rain, and humid weather persists after the rain has ceased.

If the effect of sprinkling is thus viewed in relation to the various factors with which it interacts, it should be possible to predict with some precision when and where a given disease is likely to appear in irrigated crops, and where not. Prediction calendars of this type have, in fact, been worked out in Israel for *P. infestans,* several downy and powdery mildews, and some alternarial blights in various regions and seasons. This is not the forecasting of regional disease outbreaks in the usual sense, but an evaluation of all we know about the interaction of sprinkling, dew, crop density, and pathogen characteristics, to arrive at negative forecasts specifying where the disease will not appear in a field under certain irrigation managements.

Sprinkling and chemical disease control.—We have found no literature on this problem, though it is of paramount importance in most types of irrigated farming, chiefly for these reasons: (*a*) The high moisture period subsequent to each application of water constitutes the period of the greatest danger of infection by the majority of pathogens. Water congestion of host tissues and formation of flushes of young growth heighten the host's proneness to disease at this very time. (*b*) On heavy soils, application of fungicides by ground machinery is impracticable for a considerable number of days after each irrigation. (*c*) Where the crop has been sprinkled, fungicidal deposits deriving from previous applications are reduced at this crucial time.

It is to be hoped that the advent of systemic fungicides will resolve the dilemma as to whether fungicides should be applied before or after irrigation since everything, except the washing off of deposits by sprinkling, speaks in favor of the earlier application. Until such time, the time of fungicide application in relation to sprinkling must be decided in each case with due regard to the amount of unprotected growth made by the crop after the previous application of fungicides, the virulence of the pathogen, the extent to which field conditions favor rapid spread of disease, and the type of fungicidal formulation and equipment available for application.

LITERATURE CITED

1. Bateman, D. F. The effect of soil moisture upon development of poinsettia root rots. *Phytopathology,* **51,** 445–51 (1961)
2. Bennett, O. L. Irrigation of forage crops in eastern United States. *U.S. Dept. Agr. Prod. Res. Rept.,* **59,** 25 (1962)
3. Blair, I. D. Behaviour of the fungus *Rhizoctonia solani* Kühn in the soil. *Ann. Appl. Biol.,* **30,** 118–27 (1943)
4. Brooks, A. N. Control of sclerotiniose of celery on Florida muck. (Abstr.) *Phytopathology,* **30,** 703 (1940)
5. Brooks, C., Fisher, D. F. Irrigation experiments on apple-spot diseases. *J. Agr. Res.,* **12,** 109–37 (1918)
6. Brown, W., Harvey, C. C. Studies in the physiology of parasitism. X. On the entrance of parasitic fungi into the host plant. *Ann. Botany (London),* **41,** 643–62 (1927)
7. Butler, E. J., Jones, S. G. *Plant Pathology* (Macmillan & Co., London, 979 pp., 1949)
8. Cannell, G. H., Voth, V., Bringhurst, R. S., Proebsting, E. L. The influence of irrigation levels and application methods, polyethylene mulch, and nitrogen fertilization on strawberry production in southern California. *Proc. Am. Soc. Hort. Sci.,* **78,** 281–91 (1961)
9. Carter, M. V. (Personal communication, 1965)
10. Carter, M. V., Moller, W. J. Factors affecting the survival and dissemination of *Mycosphaerella pinodes* (Berk. & Blox.) Vestergr. in South Australian irrigated pea fields. *Australian J. Agr. Res.,* **12,** 878–88 (1961)
11. Cass Smith, W. P., Goos, O. M. Bacterial canker of tomatoes. *J. Dept. Agr. W. Australia,* **23,** 147–56 (1946)
12. Cherewick, W. J. Studies on the biology of *Erysiphe graminis* DC. *Can. J. Res., C,* **22,** 52–86 (1944)
13. Clayton, E. E. The relation of soil moisture to the Fusarium wilt of the tomato. *Am. J. Botany,* **10,** 133–47 (1923)
14. Cohen, M. Increased resistance to bean rust associated with water infiltration. (Abstr.) *Phytopathology,* **41,** 937 (1951)
15. Colhoun, J., Park, D. Fusarium diseases of cereals. I. Infection of wheat plants, with particular reference to the effects of soil moisture and temperature on seedling infection. *Trans. Brit. Mycol. Soc.,* **47,** 559–72 (1964)
16. Colhoun, J., Taylor, G. S., Tomlinson, R. Fusarium diseases of cereals. II. Infection of seedlings by *F. culmorum* and *F. avenaceum* in relation to environmental factors. *Trans. Brit. Mycol. Soc.,* **51,** 397–404 (1968)
17. Couch, H. B., Bedford, Ellis R. Fusarium blight of turfgrasses. *Phytopathology,* **56,** 781–86 (1966)
18. Couch, H. B., Bloom, J. R. Influence of environment on diseases of turfgrasses. II. Effect of nutrition, Ph, and soil moisture on Sclerotinia dollar spot. *Phytopathology,* **50,** 761–63 (1960)
19. Cox, A. E., Large, E. C. Potato

blight epidemics. *Handbook U.S. Dept. Agr., No. 174,* 230 pp. (1960)

20. Crosier, W. Studies in the biology of *Phytophthora infestans* (Mont.) DeBary. *Mem. 155 N.Y. (Cornell) Agr. Exptl. Sta.* (1934)
21. Culp, T. W., Thomas, C. A. Alternaria and Corynespora blights of sesame in Mississippi. *Plant Disease Reptr.,* **48,** 608–9 (1964)
22. Curl, E. A. Influence of sprinkler irrigation and four forage crops on populations of soil microorganisms including those antagonistic to *Sclerotium rolfsii* Sacc. *Plant Disease Reptr.,* **45,** 517–21 (1961)
23. Curl, E. A. (Personal communication, 1965)
24. Curl, E. A., Weaver, H. A. Diseases of forage crops under sprinkler irrigation in the Southeast. *Plant Disease Reptr.,* **42,** 637–44 (1958)
25. Delp, C. J. Effect of temperature and humidity on the grape powdery mildew fungus. *Phytopathology,* **44,** 615–26 (1954)
26. Diachun, S., Valleau, W. D., Johnson, E. M. Invasion of water-soaked tobacco leaves by bacteria, solutions, and tobacco mosaic virus. *Phytopathology,* **34,** 250–53 (1944)
27. Dickson, J. G. Influence of soil temperature and moisture on the development of the seedling-blight of wheat and corn caused by *Gibberella saubinetii. J. Agr. Res.,* **23,** 837–69 (1923)
28. Duvdevani, S., Reichert, I., Palti, J. The development of downy and powdery mildew of cucumbers as related to dew and other environmental factors. *Palest. J. Botany, Rehovot ser.,* **5,** 127–51 (1946)
29. Edmunds, L. K., Voigt, R. L., Carasso, F. M. Use of Arizona climate to induce charcoal rot in grain sorghum. *Plant Disease Reptr.,* **48,** 300–2 (1964)
30. Endo, R. M., Amacher, R. H. Influence of guttation fluid on infection structures of *Helminthosporium sorokinianum. Phytopathology,* **54,** 1327–34 (1964)
31. Faulwetter, R. C. Wind blown rain, a factor in disease dissemination. *J. Agr. Res.,* **10,** 639–48 (1917)
32. Fedderson, H. D. Target spot of potatoes. *Leaflet Australian Dept. Agr.,* **3678,** 10 pp. (1962)
33. Fisher, F. E. *Disease Control Problems with Overhead Irrigation.* (Presented at 6th Intern. Congr. Plant Protec., Vienna, Abstracta, **685,** 1965)
34. Foot, J. H., Hendrickson, A. H., Wilson, E. E. Walnut branch wilt. *Calif. Agr.,* **9(10),** 11 (1955)
35. Foster, R. E., Walker, J. C. Predisposition of tomato to Fusarium wilt. *J. Agr. Res.,* **74,** 165–85 (1947)
36. Frank, Z. R. Effect of irrigation procedure on Pythium rot of groundnut pods. *Plant Disease Reptr.,* **15,** 414–16 (1967)
37. Garrett, S. D. Soil conditions and the root-infecting fungi. *Biol. Rev.,* **13,** 159–85 (1938)
38. Gaudineau, M. *France rapport nationale* (le partie). (7th Congr. Intern. de la Vigne et du Vin, 272–93, 1953)
39. Goos, R. D., Tschirch, M. Greenhouse studies on the Cercospora leaf spot of banana. *Trans. Brit. Mycol. Soc.,* **46,** 321–30 (1963)
40. Gregory, P. H., Guthrie, E. J., Bunce, M. E. Experiments on splash dispersal of fungus spores. *J. Gen. Microbiol.,* **20,** 328–54 (1959)
41. Guthrie, J. W. Early blight of potatoes in southeastern Idaho. *Plant Disease Reptr.,* **42,** 246 (1958)
42. Guthrie, J. W. (Personal communication, 1965)
43. Hirst, J. M. Dispersal of soil microorganisms. In *Ecology of Soil-Borne Plant Pathogens,* 69–81 (Baker, K. F., Snyder, W. C., Eds., Univ. of California Press, Berkeley, Calif., 591 pp., 1965)
44. Hirst, J. M., Stedman, O. J. The epidemiology of *Phytophthora infestans.* II. The source of inoculum. *Ann. Appl. Biol.,* **48,** 489–517 (1960)
45. Hirst, J. M., Stedman, O. J. Dry liberation of fungus spores by raindrops. *J. Gen. Microbiol.,* **33,** 335–44 (1963)
46. Ingold, C. T. Dispersal by air and water. In *Plant Pathology,* **3,** 137–66 (Horsfall, J. G., Dimond, A. E., Eds., Academic Press, New York & London, 675 pp., 1960)
47. Johnson, J. Water-congestion and fungus parasitism. *Phytopathology,* **37,** 403–17 (1947)
48. Kendrick, J. B., Jr., Walker, J. C.

Predisposition of tomato to bacterial canker. *J. Agr. Res.,* **77,** 169–86 (1948)

49. Kenneth, R. G. *Aspects of the taxonomy, biology and epidemiology of Pyrenophora teres Drechsl. (Drechslera teres Sacc. Shoemaker), the causal agent of net blotch disease of barley* (Doctoral thesis, Hebrew Univ. of Jerusalem, 1960)
50. Kenneth, R. G. Studies on downy mildew diseases caused by *Sclerospora graminicola* (Sacc.) Shroet and *S. sorghi* Weston & Uppal. *Scripta Hierosolymitana,* **18,** 143–70 (1966)
51. Kienholz, J. R. Phytophthora rot of pears under sprinkler irrigation at Hood River, Oregon. *Plant Disease Reptr.,* **30,** 30–31 (1946)
52. Klotz, L. J.. Wong, Po-Ping, De Wolfe, T. A. Survey of irrigation water for the presence of *Phytophthora* spp. pathogenic to citrus. *Plant Disease Reptr.,* **43,** 830–32 (1959)
53. Knutson, K. W. (Personal communication, 1965)
54. Lapwood, D. H. The effects of soil moisture at the time potato tubers are forming on the incidence of common scab (*Streptomyces scabies.*) *Ann. Appl. Biol.,* **58,** 447–56 (1966)
55. Lapwood, D. H., Lewis, B. G. Observations on the timing of irrigation and the incidence of potato common scab (*Streptomyces scabies.*) *Plant Pathol.,* **16,** 131–35 (1967)
56. Lewis, B. G. Effects of irrigation on potato common scab. (Abstr.) *Trans. Brit. Mycol. Soc.,* **47,** 302 (1964)
57. Mackie, W. W., Snyder, W. C., Smith, F. L. Production in California of snap-bean seed free from blight and anthracnose. *Bull. Univ. Calif. Agr. Expt. Sta.,* 689 (1945)
58. McCarter, S. M. Effect of soil moisture and soil temperature on black shank disease development in tobacco. *Phytopathology,* **57,** 691–95 (1967)
59. McIntosh, D. L. The occurrence of *Phytophthora* spp. in irrigation systems in British Columbia. *Can. J. Botany.* **44,** 1591–96 (1966)
60. McKee, R. K. Effect of soil moisture on incidence of potato scab. *European Potato J.,* **11,** 111–16 (1968)
61. Menzies, J. D. Effect of sprinkler irrigation in an arid climate on the spread of bacterial diseases of beans. *Phytopathology,* **44,** 553–56 (1954)
62. Menzies, J. D. Plant diseases related to irrigation. In *Irrigation of Agricultural Lands, Agronomy Series,* **11,** 1058–64 (Hagan, R. M., Haise, H. R., Edminster, T. W., Eds., Madison, Wis., USA, 1967)
63. Meredith, D. S., Fruit-spot ("speckle") of Jamaican bananas caused by *Deightoniella torulosa* (Syd.) Ellis. III. Spore formation, liberation, and dispersal. *Trans. Brit. Mycol. Soc.,* **44,** 391–405 (1961)
64. Moller, W. J. (Personal communication, 1965)
65. Moore, L. D., Couch, H. B., Bloom, J. R. Influence of environment on diseases of turf grasses. III. Effect of nutrition, pH, soil temperature, air temperature, and soil moisture on Pythium blight of highland bentgrass. *Phytopathology,* **53,** 53–57 (1963)
66. Moore, W. D. Flooding as a means of destroying the sclerotia of *Sclerotinia sclerotiorum. Phytopathology,* **39,** 920–27 (1949)
67. Muse, R. R., Couch, H. B. Influence of environment on diseases of turfgrasses. IV. Effect of nutrition and soil moisture on Corticum red thread of creeping red fescues. *Phytopathology,* **55,** 507–10 (1965)
68. Netzer, D., Kenneth, R. G. Studies on persistence and transmission of *Alternaria dauci* (Kühn) Groves et Skolko (the causal agent of carrot leaf blight) under semiarid conditions in Israel. *Ann. Appl. Biol.* (In press)
69. Palti, J., Rotem, J. *Spray Warnings for the Control of Seasonal Disease Outbreaks in Field Crops Under Semi-arid Conditions* (Abstr. Intern. Congr. Plant Pathol., 1st, 143, London, July 14 to 26, 1968)
70. Pappo, S. *Apple Bitter Rot caused by Gloeosporium fructigenum (Berk.)* (Doctoral thesis, Hebrew Univ. of Jerusalem, 134 pp., 1965)
71. Raniere, L. C., Crossan, D. F. The

influence of overhead irrigation and microclimate on *Colletotrichum phomoides*. *Phytopathology*, **49,** 72–74 (1959)

72. Reichert, I., Hershenzon, Zahara, Bental, A. (Unpublished data, 1953)
73. Riker, A. J. Studies on the influence of environment on infection by certain bacterial plant parasites. (Abstr.) *Phytopathology*, **19,** 96 (1929)
74. Riviera, V. Cryptogamic epidemics and the environmental factors that determine them. *Intern. Rev. Sci. Pract., Agr.*, N. S. **ii,** 604–9 (1924) [after *Rev. Appl. Mycol.*, **4,** 108, 1925 (Abstr.)]
75. Rotem, J. The effect of weather on dispersal of *Alternaria* spores in a semi-arid region of Israel. *Phytopathology*, **54,** 628–32 (1964)
76. Rotem, J. Thermoxerophytic properties of *Alternaria porri* f.sp. *solani*. *Phytopathology*, **58,** 1284–87 (1968)
77. Rotem, J. An effect of soil moisture on the incidence of early blight on potato and tomato. *Israel J. Agr. Res.* (In press)
78. Rotem, J., Chorin, Mathilda. Grove aeration and mode of irrigation as factors in the development of Dothiorella rot of banana fruits. *Israel J. Agr. Res.*, **11,** 189–92 (1961)
79. Rotem, J., Cohen, Y. The relationship between mode of irrigation and severity of tomato foliage diseases in Israel. *Plant Disease Reptr.*, **50,** 635–39 (1966)
80. Rotem, J., Cohen, Y., Spiegel, Sarah. Effect of soil moisture on the predisposition of tobacco to *Peronospora tabacina*. *Plant Disease Reptr.*, **52,** 310–13 (1968)
81. Rotem, J., Feldman, S. The relation between the ratio of yield to foliage and the incidence of early blight in potato and tomato. *Israel J. Agr. Res.*, **15,** 115–22 (1965)
82. Rotem, J., Palti J. Epidemiology of some field crop diseases under semi-arid conditions, with special reference to irrigation effects. *Abstr. 1st Intern. Congr. Plant Pathol.*, **167,** London, **July 14** to 26 (1968)
83. Rotem, J., Palti, J., Lomas, J. The effects of morning, midday, and evening sprinkling on late blight development on potatoes grown in various seasons (In preparation)
84. Rotem, J., Palti, J., Rawitz, E. Effect of irrigation method and frequency on development of *Phytophthora infestans* on potatoes under arid conditions. *Plant Disease Reptr.*, **46,** 145–49 (1962)
85. Rotem, J., Reichert, I. Dew = a principal moisture factor enabling early blight epidemics in a semi-arid region of Israel. *Plant Disease Reptr.*, **48,** 211–15 (1964)
86. Rotem, J., Shoham, Z., Lomas, J., Palti, J. A contribution to the epidemiology of apple scab in Israel. *Israel J. Botany*, **15,** 176–81 (1966)
87. Ryker, T. C. Fusarium yellows of celery. *Phytopathology*, **25,** 578–600 (1935)
88. Sanford, G. B. The relation of soil moisture to the development of common scab of potato. *Phytopathology*, **13,** 231–6 (1923)
89. Schnathorst, W. C. Environmental relationships in the powdery mildews. *Ann. Rev. Phytopath.*, **3,** 343–66 (1965)
90. Sewell, G. W. F. The effect of altered physical conditions of soil on biological control. In *Ecology of Soil-Borne Plant Pathogens*, 479–94 (Baker, K. F., Snyder, W. C., Eds., Univ. of California Press, Berkeley, Calif., 591 pp., 1965)
91. Shaw, L. Intercellular humidity in relation to fire-blight susceptibility in apple and pear. *Mem. Cornell Univ. Agr. Exptl. Sta.*, **181,** 40 pp. (1935)
92. Snyder, W. C., Grogan, R. G., Bardin, R., Schroth, M. N. Overhead irrigation encourages wet-weather plant diseases. *Calif Agr.*, **19(5),** 11 (1965)
93. Stakman, E. C., Harrar, J. G. Principles of plant pathology. (The Ronald Press Co., New York, 581 pp., 1957)
94. Stanhill, G. The concept of potential evapotranspiration in arid zones. *Proc. Intern. Symp. Methodology of Plant Ecophysiology, Montpellier, France, April 1962, UNESCO Arid zone series 109–17*
95. Stolzy, L. H., Latey, J., Klotz, L. J., Labanauskas, C. K. Water and aeration as factors in root decay

of *Citrus sinensis. Phytopathology,* **55,** 270–75 (1965)

96. Stover, R. H. The effect of soil-moisture on the growth and survival of *Fusarium oxysporum* f. *cubense* in the laboratory. *Phytopathology,* **43,** 499–504 (1953)
97. Stover, R. H. Flood-fallowing for eradication of *Fusarium oxysporum* f. *cubense*: II. Some factors involved in fungus survival. *Soil Sci.,* **77,** 401–14 (1954)
98. Thomas, H. E., Ark, P. A. Nectar and rain in relation to fire blight. *Phytopathology,* **24,** 682–85 (1934)
99. Towers, B. Effect of induced soil-moisture stress on the growth of *Fomes annosus* in inoculated loblolly pine stumps. *Plant Disease Reptr.,* **50,** 747–49 (1966)
100. Van den Bring, C., and Carolus, R. L. Removal of atmospheric stresses from plants by overhead sprinkler irrigation. *Quart. Bull. Mich.,* **47,** 358–63 (1965)
101. Volk, A. Einflüsse des Bodens, der Luft und des Lichtes auf die Empfänglichkeit der Pflanzen für Krankheiten. *Phytopath. Z.,* **3,** 1–88 (1931)
102. Waite, B. H. *Fusarium* stalk rot of bananas in Central America. *Plant Disease Reptr.,* **40,** 309–11 (1956)
103. Walker, J. C., Patel, P. N. Splash dispersal and wind as factors in epidemiology of halo blight of bean. *Phytopathology,* **54,** 140–1 (1964)
104. Williams, P. H., Keen, N. T. Relation of cell permeability alternations to water congestion in cucumber angular leaf spot. *Phytopathology,* **57,** 1378–85 (1967)
105. Wills, W. H. Exploratory investigation of the ecology of black shank of tobacco. *Virginia Agr. Expt. Sta. Tech. Bull.,* **181,** 20 pp. (1965)
106. Wilson, A. R. The chocolate spot disease of beans (*Vicia faba* L.) caused by *Botrytis cinerea* Pers. *Ann. Appl. Biol.,* **24,** 258–88 (1937)
107. Yarwood, C. E. Control of powdery mildews with a water spray. *Phytopathology,* **29,** 288–90 (1939)
108. Yarwood, C. E. Effect of soil moisture and nutrient concentration on the development of bean powdery mildew. *Phytopathology,* **39,** 780–88 (1949)
109. Yarwood, C. E. Powdery mildews. *Botany Rev.,* **23,** 235–301 (1957)
110. Yarwood, C. E. Predisposition. In *Plant Pathology,* **1,** 521–62 (Horsfall, J. G., Dimond, A. E., Eds., Academic Press, New York, 674 pp., 1959)
111. Zachos, D. G. Recherches sur la biologie et l'epidemiologie du mildiou de la vigne en Grece. *Ann. Inst. Phytopath. Benaki,* N. S. **2,** 193–335 (1959)
112. Zentmyer, G. A. (Personal communication, 1965)
113. Zentmyer, G. A., Sterling, J. R. Pathogenicity of *Phytophthora cinnamoni* to avocado trees and the effect of irrigation on disease development. *Phytopathology,* **42,** 35–37 (1952)
114. Zimmer, D. E., Urie, A. L. Influence of irrigation and soil infestation with strains of *Phytopthora drechsleri* on root rot resistance of safflower. *Phytopathology,* **57,** 1056–59 (1967)

Copyright 1969. All rights reserved

SOIL WATER IN THE ECOLOGY OF FUNGI[1]

By D. M. Griffin

Department of Agricultural Botany, University of Sydney N.S.W., Australia

Introduction

Few aspects of plant pathology and of soil microbiology have been investigated less satisfactorily than the influence of soil water and the associated soil physical factors. To give but one example, the evidence concerning the influence of soil type and soil water on the root and foot diseases of wheat is remarkably confused and often contradictory (18). This confusion can probably be attributed to the ambiguity of the experimental results, for the relevant factors have rarely, until recent years, been analysed satisfactorily. Even now it is difficult, and sometimes impossible, to devise experiments to elucidate unambiguously some problems in this area.

The physical regime of the soil has a number of component factors of biological interest. The most important of these are temperature, texture, structure, water, aeration, light, and a complex of physico-chemical factors (88). Some of these factors can be further subdivided. This fragmentation of the physical regime suggests a degree of independence amongst the factors which is nonexistent. Thus a change in one of the factors inevitably leads to a change in others. At the moment, to avoid ambiguity, it is usually necessary to simplify the model by attempting to deal with the factors independently. In this article, attention is centered on soil water but other associated factors are briefly considered. [This article is a sequel to, rather than a replacement for, a previous review (52) which should be consulted for much of the earlier literature: somewhat different emphases can be found in other recent reviews (88, 111, 120).]

Some specification of the water content of a soil is the commonest way in which the moisture regime is characterized in plant pathology. Because of this well-established tradition, it is perhaps worthwhile to attempt to emphasize its deficiencies by an analogy. Suppose that the growth of a plant in a heated glasshouse at a certain season of the year is found to be closely correlated with the amount of oil burnt per unit time. For that glasshouse in that season, the correlation may be of value in setting the controls of the oil burner, but it will be of no value in another house of different size and construction unless the size and other, possibly unpredictable, factors are

[1] This review was prepared, in part, whilst I was on study-leave at the East Malling Research Station, Kent, U.K., as a Bursar under the Royal Society and Nuffield Foundation Commonwealth Bursary Scheme.

taken into account. Also, the information on optimum rates of oil consumption will be nearly valueless even in the original glasshouse if the mode of heating is changed from oil to gas or coal. Let us assume that the relevant factor in the analogy is temperature, which will be affected not only by energy input but also by many internal and external factors associated with each house. The important relationship between growth and temperature will at best be revealed poorly, and perhaps not at all, by the relationship between growth and fuel consumption. Similarly in a given soil there may be a correlation between disease and water content but this correlation is of little predictive or interpretive value if the directly operative factor is not water content but some other associated factor, say aeration or solute diffusion. The deficiency in the sole use of measurements of water content thus lies in the variation from soil to soil of the relationship between water content and the specific operative factor.

With few exceptions, only articles in which the specific factors are known with reasonable certainty have been chosen for mention in this review. Thus, the bulk of the literature on the effects of soil water on fungi and plant disease is not considered because of the difficulty or impossibility of interpreting the data.

Basic Concepts and Terminology

There are a number of recent treatments of the physical aspects of water in soils and plants (63, 89, 104, 114) and reference should be made to these for a fuller treatment of the subject. The terms Rose (104) used, and for which he gave the definitions recommended by the International Society of Soil Science, have been used in this article. A useful comparison of the various terminologies used in the soil-plant-atmosphere system has also been published recently (29). The present treatment will be limited to concepts and terms particularly relevant in the present context.

Water content of soil.—I have briefly discussed on another occasion (56) the ways in which the water content of soil is expressed in the literature of plant pathology. The commonest method of expression is as a percentage of the value when the soil is saturated, but such a relative value gives no indication of the amount of water held per unit weight or volume of soil. Furthermore, the value is misleading if workers believe that a soil which is 40 per cent saturated has necessarily twice the air-filled pore volume of a soil which is 70 per cent saturated. Loss of water from many soils, especially from those of heavier texture, is because of compaction and shrinkage as well as from the drainage of stable pores. Only in soils consisting very largely of rigid particles, e.g. sands, will the loss of a given volume of water from a soil be accompanied by the acquisition of an equal volume of gas.

In soil science, water content is normally expressed as grams of water per gram of soil dried at 105° C or as cubic centimetres of water per cubic centimetre of soil. The former measurement is the more easily made but in some cases the latter is the more useful in analysis. Soil moisture content on a percentage weight basis can be converted to a volume basis by multiplying the former by the bulk density of the soil.

Bulk density.—Bulk density is equal to the mass of solids divided by the total volume of all constituents (solid, liquid, and gas). It therefore varies with compaction and with the degree of aggregation of the individual soil particles into structural units. Bulk density also varies with the density of the solid phase, being less in soils with a high content of organic matter and greater in artificial soils made from aluminium oxide grits (53, 54, 61).

Potential; suction.—To absorb water, an organism must be able to overcome the forces tending to retain the water in the soil, i.e. the organism must have the necessary capacity for doing work relative to the moisture potential of the soil water. Moisture potential is usually measured as energy per unit mass of water relative to a stated or implied datum (erg g^{-1}). Thus, the greater the work that must be done by an organism to obtain water, the more negative is the potential of that water. Potential can, however, be expressed as energy per unit volume of water (erg cm^{-3}), which is dimensionally the same as pressure (dyne cm^{-2}). If suction is taken to imply a negative pressure, then moisture potential is equal in magnitude but opposite in sign to the total suction. As suction is probably an easier concept than potential for many biologists, the term suction will be used here.

The total suction to which a soil organism is exposed is the sum of the matric and osmotic suctions. Matric suction is associated with water-solid or water-air interfaces (35, 100, 114) and thus with the size and arrangement of the solid particles. Matric suction is therefore associated primarily with the adhesion of water to solid surfaces and the surface tension at the water-air interfaces in the soil pores. As both are physical phenomena depending upon the solid phase, they are combined under the term "matric" and will not be distinguished subsequently. The suction (or negative pressure) of the water can be expressed in a variety of units such that 1 bar = 10^6 dyne cm^{-2} = 0.987 atm = 75.0 cm mercury suction = 1022 cm water suction. Matric suction is frequently expressed in a logarithmic form where pF x is equivalent to 10^x cm water suction (28). In this review, most data are discussed in terms of the "bar" notation. The relationship between total suction and relative humidity is given by

$$\psi = \frac{RT\,\rho \ln p/p_0}{10^b M}$$

where ψ = total suction (bar), R = gas constant, T = temperature (°A), ρ = absolute density of water at $T°$, p/p_0 = relative humidity (expressed with base of unity, not 100), M = molecular weight of water.

Moisture characteristic curve.—The relationship between soil moisture content and the corresponding matric suction is best expressed graphically as a moisture characteristic curve. The great importance of this relationship has been discussed previously (52) and its utility in soil microbiology is exemplified in recent work (53, 54, 61, 72). The characteristic exhibits hysteresis, a topic fully treated by Youngs (135). The practical importance of hysteresis is that the wetting (absorption) and drying (desorption) boundary curves of the moisture characteristic are quite different. It is therefore necessary to present the particular characteristic relevant to the circumstances of the experiment.

Methods for the determination of water content and suction, and therefore of the moisture characteristic, have been recently summarized (67). Although the pressure membrane technique is satisfactory for bringing soils to a given moisture status for subsequent use, it is invalid to conduct biological experiments within such apparatus if air is used as the compressing gas, because of the toxicity of the enhanced partial pressure of oxygen. Permanent wilting point is attained by the application of 15 atm air pressure, equivalent to 3 atm partial pressure of oxygen; such gaseous environments have been shown to greatly restrict fungal growth (19). This difficulty might be overcome if the composition of the gas was so adjusted that the partial pressure of oxygen was 0.21 atm when the apparatus was at its operating pressure.

Solute diffusion.—It is difficult to state with precision the effect of soil water content on the rate of solute diffusion (20, 21, 126). The difficulty is particularly great in the case of ions or polar molecules, the diffusion of which is greatly modified by the presence of charged surfaces such as those of clay particles. For an unionized, nonpolar molecule, however, simplifying assumptions can be made and the rate of transfer per unit area of cross-section of soil (dq/dt) will be given as a first approximation by $dq/dt = -kD\Theta dc/dx$ where D = coefficient of diffusion of the solute in pure water, Θ = volumetric water content of soil, dc/dx = concentration gradient and k is a tortuosity factor (by analogy with gaseous diffusion). The rate of transfer in a given soil will thus vary directly with its volumetric water content.

Suction and Microbial Activity

Matric suction.—In an earlier paper (52), I stated that

> Although there are few published experiments in which suction has been measured, none suggests a limitation of fungal growth directly due to suction factors in the soils between saturation and the permanent wilting point of mesophytic higher plants. Evidence derived from controlled humidity experiments reveals that most fungi can exert sufficient force to absorb water from atmospheres at a lower relative humidity than those in equilibrium with soil at the permanent wilting point.

Today, this view of the importance of matric suction per se is supported by a considerable mass of data (23, 24, 37, 53–55, 57, 74) which indicates that there is no cessation of fungal growth and little retardation in soils within the suction range 0 to 15 bar (saturation to permanent wilting point).

On the contrary, many species, including plant pathogens, are active at 15 bar and some species grow at suctions up to 400 bar. An ecological differentiation based upon soil water suctions in excess of 15 bar seems well established (23, 55), and at suctions between 145 and 400 bar [relative humidity (RH) 90 to 75 per cent], the active flora consists almost exclusively of species of *Aspergillus* and *Penicillium.* A study of the interaction between temperature and suction showed that *Aspergillus* spp. were of equal or greater prevalence relative to *Penicillium* spp. at all suctions at 30 to 35° C and at 220 to 400 bar at 15 to 20° C (24).

The continued activity of many soil fungi within the suction range 15 to 400 bar is probably of great consequence in arid and semi-arid soils. In these soils, there is a marked increase in ammoniacal nitrogen during the dry season which may partially explain the rapid growth of plants when rain occurs. Although the mechanism is undoubtedly complex (50), Dommergues (37) produced evidence that this accumulation of ammoniacal nitrogen results partly from the activity of fungi in soils too dry to permit the growth of higher plants or the activities of the nitrifying microflora. Nitrifying bacteria are known to be particularly sensitive to increase in either matric or osmotic suctions (90, 113).

The response of bacteria to matric suction has been little investigated (5, 26, 88), but bacteria grow in liquid culture over much the same range of osmotic suctions as fungi (108, 132–134). It would appear, however, that the general activity of the bacterial component of the microflora, compared to that of the fungal, becomes attenuated at smaller suctions (37). Although there is a widespread impression that actinomycetes are favoured by high suctions (74), Chen & Griffin (23) observed actinomycetes to be active only at suctions less than 55 bar. The apparent dominance of actinomycetes in dry soils is likely to be attributable to the resistance of their spores to desiccation rather than to the activity of their hyphae at these suctions (2).

The effect of matric suction on the survival of fungal resting structures in soil is poorly known. It is to be expected that high suctions will have a number of opposing effects. Thus, survival may be enhanced by a reduction in both the level of endogenous respiration and the tendency to spontaneous germination. Such suctions may, however, be directly deleterious if the propagules lose water rapidly. Furthermore, high suctions may still permit the activity of xerophytic fungal antagonists, especially species of *Aspergillus* and *Penicillium,* but prevent the growth of the fungus in question. The effect of suction on the persistence of resting structures is therefore likely to vary greatly from species to species.

In some studies, the number of viable fungal propagules was shown to decline in summer in soils which experienced high suctions, but a similar decline has been difficult to detect in temperate soils (52). The loss of viability of sclerotia of *Phymatotrichum omnivorum* when exposed to air is well documented (44) and provides an example of susceptibility to desiccation in a structure which is generally resistant.

The recovery of *Gibberella zeae* from infested wheat straws lying on the soil surface at 35° C declined with each increase in relative humidity between 32.5 per cent and 100 per cent (17). *G. zeae* grows poorly at 35° C and its decline with increasing soil moisture was attributed to increasing antagonistic activity. In contrast, there was little effect of moisture on recovery at 10° C. At 25° C, recovery was high at 32.5 per cent and 100 per cent RH but very low at 75.5 per cent and 87.0 per cent RH (190 to 390 bar). *G. zeae* is inactive at these intermediate humidities and its poor persistence was attributed to the antagonism of *Penicillium* spp. which were then abundant on the straws. Bruehl & Lai (14) obtained similar results in studies on the survival of *Cephalosporium gramineum* in infested straw at 15° C. Survival was best at the lowest humidity tested (82 per cent RH, 270 bar) and worst between 86 and 90 per cent RH (145 to 200 bar) where *Penicillium* spp. were dominant. The different humidity optima for survival in the two experiments can probably be attributed to the interaction between soil moisture and temperature in controlling fungal activity (24).

The attack by fungi on dormant seeds in soil too dry to permit seed germination provides a simplified example of disease, for here the effect of suction on the host is minimal. Such attack is also of interest for it provides the sole example known to me of a disease in which suction is known to be important through its direct effect on the fungus. In a study of the deterioration of seed in soil, the pattern of fungal colonization of wheat seeds in soil at suctions between about permanent wilting point and 300 bar was similar to that occurring on seed stored at similar suctions away from soil (57). The breakdown of seeds in soil was due to such seed-borne fungi as members of the *Aspergillus glaucus* group. *Penicillium chrysogenum,* and the *P. citrinum* series, and to various soil-borne penicillia and aspergilli, of which *P. expansum* was most important. Only in the 300 bar series did buried seeds survive more than 3 weeks, and the period was less than 1 week in soils near permanent wilting point at 25° C. In this study, shaking the seeds with an organomercurial fungicide failed to enhance longevity but in another (128), such treatment was partially successful. In the latter study, it was shown that a sound seed-coat provided a better protection against seed-decaying organisms than any seed treatment tested. The apparent inactivity of fungicides should probably be attributed to the absence of the necessary free water in these situations (25, 46).

If higher plants are growing in a soil, the system looses water rapidly through transpiration as well as evaporation. It is then impossible to maintain the whole soil mass at a uniform constant suction at any moisture content below that of field capacity. The problems this poses for plant pathologists have been recently discussed (32). An attempted solution is to substitute a cyclical water suction programme for a uniform constant suction. In such a programme, the soil is initially at, say, field capacity but looses water by evaporation and transpiration. When the water content of the soil falls to some selected value (say equivalent to 5 bar), water is added to restore the system to field capacity. The cycle can be repeated for the rest of the experiment, various minimal water contents being selected as the treatments. Such cyclical programmes have been used in the study of diseases of turf grass (30, 31, 91, 92), pine trees (124), and in a modified form, wheat (16). Clearly, however, the data will need very careful analysis if they are to yield information on the effects of suction per se, for all other associated factors change simultaneously. Furthermore, there will be marked gradients in suction and water content within the soil mass so that the values for water content are mean ones only. In general, the soil immediately adjacent to the roots will be at a higher suction than that remote from roots, although this need not always be so (107). Bearing in mind these difficulties in interpretation of the data, it appears that diseases of *Poa pratensis*, *Agrostis tenuis*, and *Pinus taeda* caused by *Sclerotinia homeocarpa* (31), *Pythium ultimum* (91), and *Fomes annosus* (124), respectively, all increase in incidence with increasing matric suction. Gibbs (45) has assembled evidence strongly suggesting that the resistance of conifers to *Fomes annosus* depends largely on the ability of a tree to mobilize resin at the point of infection. In turn, resin mobilization is known to depend on sapwood water content and so, *inter alia,* on the soil water suction.

R. J. Cook, in a paper presented at the First International Congress of Plant Pathology, noted that in many natural situations the surface soil—and thus, the soil adjacent to hypocotyls and surface roots—is drier than the permanent wilting point. The water supply to the plant is then maintained by the deeper roots which penetrate moist soil. The study of the activity of pathogenic fungi at high matric suctions is thus of interest and of direct importance to plant pathologists, for such activity may determine the incidence of diseases of the crown and hypocotyl. Cook stated that *Fusarium culmorum* chlamydospores germinate at suctions of 50 to 60 bar and that antagonism is then less than at 8 to 10 bar. He therefore suggested that the prevalence of 'culmorum' infections of wheat in sites with dry surface soils is because the pathogen escapes antagonism there, but is still able to extract and utilize soil water.

Shear strength.—Trees infected by root-rotting fungi may die whilst standing, the first obvious sign being death of the crown. Alternatively, the tree may topple over whilst the crown is still green, a root examination revealing the presence of the pathogen. Sometimes these different field symptoms appear to be associated with differences in site, sometimes with differences in environment. Thus, trees infected by *Armillariella elegans* in New South Wales usually die whilst standing but occasionally, when in wet soil after heavy rain, they topple over whilst green (60). Although it is beyond the scope of this review to elaborate the hypothesis fully, it is possible that the characteristics are associated with different shear strengths of the soils. Thus, if many of the roots are rotted, the effective volume of soil holding the tree upright is reduced and if the shear strength of the soil is low, a slip-circle failure will occur and the tree will fall even though some of the roots and the crown are alive. Shear strength depends in part on the cohesion of the soil particles brought about by water films, and it increases with matric suction in the range between saturation and permanent wilting point.

Osmotic suction.—That many fungi are tolerant of increased osmotic suction is illustrated by the marine fungi growing in water with a salinity of about thirty-four parts per thousand and an osmotic suction of 22 atm (71). Some terrestrial fungi, too, grow better in agar media made with sea water than in media of lower salinity, although some are adversely affected (102). The soil fungi growing on agar media containing 20 per cent or 30 per cent sodium chloride (water activity 0.87 and 0.80 respectively) were mainly the same species of *Aspergillus* and *Penicillium* as are active in soil or on grain or other substrates at comparable relative humidities (22).

The tolerance of many fungi to reduced RH is known to depend upon both temperature and nutrition (9, 24, 116, 122) and Te Strake (121) has shown a similar dependence by saprolegniaceous fungi grown in saline waters. Thus *Dictyuchus monosporus* was intolerant of salinities in excess of 2 to 8 parts per thousand in estuarine waters but grew on nutrient media of 8.94 per cent salinity.

The clearest example of the importance of salinity in soil mycology is provided by the study of *Coccidioides immitis* (1, 38, 39). The presence of *C. immitis* is associated with high levels of salinity in surface soil and high temperatures at some stage of the year, a distribution probably explained by the growth of the fungus at 40° C on media containing 6 to 8 per cent of sodium chloride or calcium chloride. Under these conditions, two bacteria and a fungus which were inhibitory to the growth of *C. immitis* scarcely grow, and it is likely that there is a great general reduction in microbial activity and a consequent lessening of competition.

In general, the response of fungi to increased osmotic suction is very similar to that produced by an equivalent reduction in RH. Thus, 83.7 atm osmotic suction and 93 to 95 per cent RH are limiting for the growth of *Fomes annosus* (93, 101), the water potentials being similar. The common factor of reduction in total water suction is, therefore, the operative one, and I know of no evidence suggesting that any fungi depend on high ionic strengths for their functioning. Osmophilic fungi thus differ fundamentally from those halophilic bacteria that depend on high ionic strengths of certain salts for their membrane stability (12). Terms such as halophilic, osmophilic, and xerophytic are therefore inadequate for none encompasses the concept of total water suction.

Water Content and Microbial Activity

General.—It seems inherently unlikely that water content as such will have any important direct effect on fungal growth. Fungi, because of their small size, need to absorb little water for growth and, furthermore, they loose little water to the soil atmosphere in contrast to the large transpirational losses of higher plants (52, 100). Fungi have been shown to be active in a number of soils with a water content less than 0.02 g per g soil. (23) and under these conditions it is likely that the fungi absorb much of the water from the vapour phase.

When comparing the number of microorganisms in different soils, numbers (as revealed by dilution plate counts) are correlated with the organic matter content of the soil but not with water content (10, 70). In a given soil at different samplings, however, numbers of bacteria were strongly correlated with water content and fungi showed a similar but less marked trend (70). This latter result is probably attributable not so much to water content itself but to the correlated factors of suction, mobility, growth of higher plants, etc. The number of bacteria appearing on modified Cholodny slides increased with increasing soil moisture whereas the converse was true for fungi (106).

Although absolute water content of a soil is unimportant, it is of importance indirectly through its effect on solute diffusion and the continuity of water pathways.

Solute diffusion.—There are very few published experiments from which the effects of solute diffusion on microbial activity can be deduced. Kerr (72) studied the infection of peas by *Pythium ultimum* and showed that the loss of weight of pea seeds in water was directly related to the amount of sugars diffusing from them. Furthermore, the percentage of infection of peas by the fungus in each of three soils used was directly related to their loss of dry weight, and he has therefore postulated that infection is largely controlled by the amount of substances stimulatory to *P. ultimum* which diffuse away from the seeds. Taking the three soils individually, infection and loss of dry weight were correlated with soil water content and suction. No completely unifying theory connecting the three moisture characteristics, diffusion of exudates and infection could be produced, but this is not surprising because of the complexity of the possible interactions. A further experiment, involving changes in bulk density of the soil at constant gravimetric water content, supported Kerr's contention that the main effect on infection of changes in the moisture regime was mediated by changes in the rate of diffusion from the peas.

Studies on the infection of peas by *Fusarium solani* f. *pisi* (28) suggest that the availability of nutrients by exudation and diffusion from the host are important. Loss of weight by pea seeds during germination was shown to be correlated with chlamydospore germination, both increasing with increasing soil water content. Water contents below 8.7 per cent reduced germination, but when sucrose and ammonium sulphate were mixed into the soil, germination was independent of soil water content over a wide range. The predominating importance of nutrient supply, and therefore, solute diffusion was thus demonstrated. In soil with more than 8.7 per cent water content, germ tubes were extensively lysed 20 to 48 hr after the commencement of the experiment. Such lysis was presumably of microbial origin as it did not occur in sterile soil. The extent of lysis was correlated with loss of dry weight in the period 20 to 48 hr and this observation suggests that the amount of lysis is correlated with the nutrient supply to antagonistic microorganisms, probably bacteria, during the critical period. In this respect, it was shown that the supply of nitrogen, rather than sucrose, was of importance. Infection of the cotyledonary region of peas in pot experiments was greatest in those moisture regimes which favoured both germination and germling survival. The extent of infection thus appears to be determined in large part by factors associated with solute diffusion.

When the sclerotia of *Sclerotium cepivorum* were buried in soil in which the roots of *Allium* spp. were growing, germination increased with increasing soil water content between 20 per cent and 90 per cent saturation. Although the relationship was not linear, change in the diffusion of a soluble stimulant from the root to the sclerotia would appear to be the most likely basis for this effect (27).

The solutes diffusing in the soil water may not always be stimulatory to microorganisms. The significance of soluble inhibitors has often been recognised and much of the literature has been reviewed (68, 86, 94). Seidel (110) produced strong evidence that fungistasis increased in intensity with soil water content, as did the inhibition of spore germination when copper sulphate was incorporated into sand. He interpreted his results in terms of the effect of water content on the rate of diffusion of inhibitors from soil to the spore. His data on fungistasis, however, would appear to support equally the concept of diffusion of nutrients away from the spore as suggested by Ko & Lockwood (73). Diffusion of solutes, and thus water content, would be important in either case. Data on the effect of soil water on the growth of *Coniophora puteana* indicated the likelihood of the diffusion of inhibitors being an important controlling factor (97), although there is an anomalous feature which is worthy of consideration. When actidione was incorporated into sterile soil at a uniform concentration (per unit weight of soil), growth of *C. puteana* decreased with increasing soil moisture content between 25 and 100 per cent saturation. It seems unlikely that either concentration or diffusion of actidione would have been the differential factor causing the changed response at different moisture contents since the concentration of actidione per unit volume of water would have decreased with increasing soil water content and because of the initial uniform distribution of the antibiotic throughout the soil. It is perhaps legitimate to speculate whether frequency of contact between hyphae and water films (and thus actidione) was the significant factor, for these would have been reduced with decreasing water content and with the consequent growth of hyphae in air-filled pores. Such an hypothesis would be tenable only if the hyphae failed to translocate actidone and I am not aware of any evidence on this point. Hyphae are known, however, not to translocate certain other antibiotics (griseofulvin, radicicolin) (41). Whatever the explanation, Pentland's statement (97) that "microhabitats may be more variable in a drier soil than in a wet one because there is less movement of water-soluble products in the drier soil" is valid and important.

Bulk density.—An increase in the bulk density of the soil at constant gravimetric water content causes an increase in the volumetric water content and thus in the rate of solute diffusion. Kerr (72) has shown the importance of this effect in promoting the infection of pea seeds by *Pythium ultimum.*

Fulton, Mortimore & Hildebrand (43), investigating the infection of soybean by *Phytophthora megasperma* var. *sojae,* and Braun & Wilcke (11), investigating collar-rot of trees caused by *P. cactorum,* found that increasing soil bulk density favoured the disease, whereas Winter (129–131) found that increasing soil bulk density decreased the rate of growth of *Ophiobolus graminis* along wheat roots. Change in bulk density, however, causes change in many physical characteristics and in these cases no adequate evaluation of the precise mechanism can be made from the published data.

One situation in which the bulk density of the soil is changed is of particular interest. As a probe, simulating a root, penetrates soil, the soil is to some degree compressed, both in front of, and in an annular zone around the probe. The probe is thus surrounded by a zone of plastic compression which can be subdivided (3, 42). Adjacent to the probe is a minimum voids system, of a thickness comparable to the radius of the root itself, in which the soil particles are close-packed and the bulk density is maximal. This minimum voids system is surrounded by a voids ratio failure zone of a thickness about 2.5 times the root radius. The bulk density in this zone is increased, but not to the extent which occurs in the innermost zone. The outermost zone of plastic compression extends to a radius of about seven times the radius of the probe, but there is relatively little alteration in bulk density within it. In the soils investigated, the thickness of the zones varies from soil to soil and with change in initial bulk density by a factor of two. The implications for microorganisms of the existence of these zones around roots has not been investigated, but the potential significance of the minimum voids and voids ratio failure zones is considerable. Thus an increase in bulk density, particularly that occurring in the minimal voids zone, will affect the rate of solute diffusion and may even delimit the radius of the rhizosphere, in which the microflora is markedly affected by host excretions. Simultaneously, the diffusion paths for oxygen and carbon dioxide will be modified and the matric suction will probably change. The rhizosphere soil is therefore likely to differ physically in many ways from the soil more remote from the root.

Movement of Microorganisms in Soil Water

Active movement through continuous water pathways.—Filamentous fungi can grow through air-filled pores and presumably are able to translocate water to the apices from the sites where their hyphae are in contact with water. Bacteria, however, are dependent upon continuous water pathways for their movement and this limitation has far-reaching consequences (50). Their ability to move appreciable distances in soil has been shown to depend upon an interaction between volumetric water content and matric suction best expressed in terms of the moisture characteristic (61). It is unlikely that bacteria move rapidly or in large numbers through pore-necks of radius less than *ca.* 1.5 μ. Pore-necks of this or greater radius drain off water at 1 bar, a suction which therefore approximately defines one limit for appreciable movement. Another limit, however, is established by the necessity for maintaining continuous water pathways through necks of the requisite radius. This limit is reflected in the necessity for a critical volume of water to be present in pores draining at suctions less than 1 bar. It has been shown experimentally for a small number of particulate systems, that appreciable movement of bacteria occurs at a matric suction of h cm if at least 0.11 cm^3 of water per cm^3 soil is held by the matrix in pores which would drain between h and 1000 cm water.

Lapwood (80) has shown that infection of potato tubers by *Streptomyces scabies* is dependent upon the soil being drier than field capacity during a certain critical stage in tuber development. He has further suggested that successful infection of the lenticels by the actinomycete is dependent upon the absence of a well-developed antagonistic bacterial flora in the lenticels. Such a hypothesis is certainly consistent with the general evidence that the mobility of eubacteria, and thus their ability to colonize lenticels, is reduced as the water content of the soil falls below field capacity.

Greenwood (50) has discussed, and Henis, Keller & Keynan (64) have demonstrated, the importance of the configuration of a substrate and its associated matrix in determining microbial colonization. Thus, cellulose filter paper consisting of intermeshed fibrils was colonized mainly by the bacterium *Cellvibrio*. Cellophane was colonized mainly by the fungus *Stachybotrys*. The difference was attributed to the presence of water films amongst the fibrils which permitted the movement of the bacterium whereas such films were not present, or were not extensive and continuous, on cellophane. Limonard (85), in another context, has shown the importance of matric water content to the outcome of antagonism between bacteria and fungi on seeds.

Bumbieris & Lloyd (15) placed hyphae on the surface of a silt-loam at 25, 19, and 13 per cent water content by weight (0, 0.1, and 3.1 bar, respectively) and found that the rate of lysis was greater in the two wetter soils. They further found that: (*a*) at 25 per cent water content, bacteria and protozoa had increased; (*b*) at 19 per cent water content, the increase of bacteria predominated over that of actinomycetes; and (*c*) at 13 per cent water content, actinomyctes only had markedly increased. The limit for marked bacterial reproduction in the presence of fungal substrates was therefore at about 1 bar suction. Rybalkina & Kononenko (106) had earlier shown that considerable numbers of bacteria occurred on modified Cholodny slides only when the soil was at least as moist as field capacity.

There is thus circumstantial evidence that lysis of mycelium by bacteria becomes much reduced at suctions exceeding 1 bar and this may well be associated with limitations on the movement of bacteria onto and along hyphae. Only in very moist soils will "hyphae serve as roadways along which bacteria follow, dining upon them en route" (26).

Passive movement in water.—Limits derived in a similar way to those for bacteria are presumably applicable to active movement of fungal zoospores. Motility of zoospores, however, is markedly reduced by frequent contact with solid surfaces [which promotes encystment (66)], and by reduced aeration (a concomitant of high soil water content). In practice it seems likely that zoospore movement over appreciable distances in soil is brought about by movement of the water itself rather than by flagella (65).

I have reviewed (52) earlier evidence concerning the effect of particle and pore size on the movement of spores in moving soil water, but additional evidence is now available (78, 79, 136). The depth to which sporangia and zoospores of *Phytophthora infestans* were carried increased with the volume of water applied. Water, equivalent to at least 0.75 in. rainfall, was required to move propagules into soil deeper than 2 in. from the surface. The data suggest that zoospores were carried to greater depths than sporangia, but the difference was not great.

Movement of spores in water appears to be implicated in the disease of carnations caused by *Verticillium cinerescens* (81). In plants grown in sloping troughs with artificial watering, spread of disease from inoculated plants was greater from those placed at the top of the slope than from those placed at the bottom. The number of infected plants was twice as great in troughs watered by hose as in those watered by trickle irrigation, but it is not known whether the difference was due to greater spread of spores by the fluctuating water status of the soil, or to other factors.

Effect of Microorganisms on Soil Water

The distribution of water in soil, and thus aeration, are closely associated with soil structure, that is, the aggregation of particles into compound units. The importance of microorganisms in the genesis and maintenance of soil structure has recently been reviewed (62). In contrast to the beneficial activity of aggregation, microorganisms are also responsible for the coating of soil particles with deleterious organic films. Such films alter the advancing contact angle of water with the particles and in extreme cases make the surfaces water repellent (7, 8, 40). Water-repellent sites were characterized macroscopically by a patchy distribution of plants and a marked loss of productivity. Basidiomycete hyphae, often water-repellent themselves, were particularly associated with the sites. A similar phenomenon appears in the formation of fairy rings (112) and, probably, of barren rings (123). Shantz & Piemeisel have clearly indicated that the death of plants in the former areas is most likely due to drought imposed by the presence of the basidiomycete hyphae. They attributed this to the reduced porosity of the soil brought about by the abundant hyphae, but it seems more likely that the hydrophobic nature of the hyphae, and adjacent soil once dry, are mainly responsible. Fairy rings are characterised by the presence of basidiomycete fruiting bodies, dense mycelium, and zones of higher plant stimulation, but these are absent from barren rings (123). In Australia, in rather dry sheep-grazing country, however, there are many intermediate forms and I am of the opinion that water-repellency of the soil in the barren, or bare, zone is a common factor in all these conditions. The phenomenon is, therefore, one of wide potential importance.

Effect of Shape and Size of Particles and Pores on Microbial Activity

The size and shape of soil particles and aggregates have a profound effect on the moisture characteristic of the soil but there are additional important effects. Thus the disruption of soil aggregates increases the activity of microorganisms by exposing a greater surface area to microbial attack (50, 105).

The activity of bacteria and fungi in aerated glass microbead systems has been studied (95, 96). Growth and respiration of *Bacillus subtilis* were greater in systems of small (37 μ diam) than of large (149 μ diam) beads. The data suggest the existence of a critical competition for surface area in the large beads which had only one quarter the surface area per unit weight of the small beads. *Aspergillus terreus* and *Trichoderma viride,* however, respired more actively and the latter produced more hyphae and spores in the large bead system. The authors suggested that "fungi due to their growth characteristics, have a more critical requirement for inter-bead space than for linear surface" (96). A more precise understanding of this requirement can be obtained. If close-packing of the beads is assumed, it can be shown (34) that the diameter of the pore-necks connecting the major voids in the small bead system is 5.74 μ compared with 23.1 μ for the larger beads. In fact, the beads were not close-packed but a correction factor can be calculated from the data supplied (95). The corrected values for the diameters of the pore-necks in the small and large bead systems are 6.57 μ and 27.1 μ respectively. It is likely that this narrow neck is a limiting factor for fungal growth amongst the small beads, for the presence of even a few hyphae would greatly reduce the rate of movement of oxygen through the neck, even in aerated systems.

In these glass bead systems, the corrected diameters of the inscribed voids in a system with tetrahedral packing (which are measures of the size of the voids) are 9.55 μ and 39.3 μ for small and large beads respectively. It is therefore possible that the reduced sporulation of *V. viride* amongst the small beads (96) can be attributed to the fact that only depauperate conidiophores could be formed. The critical size of void is probably associated with the size of the reproductive structure, for the intensity of sporulation of *Curvularia* sp. (53) and *Pythium ultimum* (54) was reduced in voids of less than *ca.* 45 μ diam. Conidia of the former fungus are about 27 μ long, oogonia of the latter about 20 μ diam, and both are thus much larger than the conidia of *T. viride.*

The effect of the moisture regime in glass bead systems on the sporulation of *Curvularia* sp., *Cochliobolus sativus,* and *Fusarium culmorum* has also been studied (58). Conidial production in the first two species was limited to air-filled pores whereas in the last, conidia were formed principally just beneath the air-liquid menisci.

Soil Aeration

A change in the water content of a soil is normally accompanied by changes in soil aeration caused principally by alteration of the length of the pathways for diffusion through the liquid phase. The relevant literature on oxygen and carbon dioxide is voluminous, and I shall do no more than mention a few facets of the problem.

Oxygen.—The importance of soil aeration in affecting the activity of roots and microorganisms in soil has been widely appreciated and there are a number of general treatments (47, 59, 76, 82–84, 103, 118, 119, 127). Workers have differed on the relative importance of oxygen concentration and oxygen diffusion rate as parameters for evaluating soil aeration. In an attempt to illuminate this issue, I have recently published (59) a theoretical study relating the concentration and diffusion of oxygen to the biology of organisms in soil and have emphasized points of interest to plant pathologists.

The majority of the studies on roots have assumed that diffusion has radial coordinates both outside and inside the root. It is now clear, however, that longitudinal diffusion in the gas phase through intercellular spaces in roots cannot be neglected, particularly in the case of seedlings (47–49, 69). Other workers, however, have shown that oxygen uptake by the roots of some intact plants can be explained in terms of radial diffusion (109), so the relative importance of the radial and longitudinal pathways probably changes with age and species. This is a factor of considerable potential importance in plant pathology.

Most general treatments have assumed, either explicitly or implicitly, that the predominant effect on the plant of oxygen in soil is through the cytochrome system. Bergman (4) has shown that this is not necessarily so in his review of the relationship between oxygen deficiency and plant disease. Recent examples illustrate the diversity of the possible mechanisms. Brown & Kennedy (13) have associated increased seed rot of germinating seeds of *Glycine max* at lowered partial pressures of oxygen with increased exudation by the seeds of sugars stimulatory to *Pythium ultimum.* The physiological basis for the increased exudation is not known with certainty. A quite different basis for increased susceptibility of plants to infection is suggested by the sensitivity of the biosynthesis of some phytoalexins to reduction in the partial pressure of oxygen (33). In the black-pod disease of cocoa, caused by *Phytophthora palmivora,* the polyphenol oxidase system of the host appears to act as a resistance mechanism by inactivating the pectolytic enzyme system of the pathogen (117). The mechanism is oxygen-sensitive and is possibly of significance for subterranean structures in wet soils.

A study of *Armillariella elegans* (115) furnishes an illustration of the role of water films in regard to oxygen supply. Continued growth of rhizomorphs depends not only upon the absence of water films between the atmosphere and the point of origin of rhizomorphs but also upon the presence of a water film over the elongating rhizomorph tip. Such conditions ensure a rapid diffusion of oxygen down the gas-filled central canal of the rhizomorph to its apex simultaneously with a reduced partial pressure of oxygen at the surface of the tip. It appears that the first condition is necessary to ensure an adequate rate of oxygen supply to the dense and rapidly respiring apical mersistem. The second condition is necessary to reduce the rate of reaction of a phenol oxidase and the consequent deposition of melanins in the growing region of the rhizomorph. The melanins, although probably protecting the mature regions of the rhizomorph from bacterial lysis (6, 77, 99), appear to inhibit growth of the apex. Fungal activity can thus be limited by too great a partial pressure of oxygen if the rate of reaction of a phenol oxidase system is critical. The presence of water films may then be beneficial rather than deleterious to the organism.

Carbon dioxide.—Earlier work on the effect of carbon dioxide on fungal growth was briefly noted in my previous review (52) and there have been many subsequent articles relevant to plant pathology. As yet, however, the investigation of the role of carbon dioxide in soil ecology can scarcely be claimed to have passed the introductory stages. Thus the pathways for diffusion cannot be specified with accuracy: indeed, diffusion may be of little importance compared with movement of carbon dioxide in the soil water as it drains. The concentration of carbon dioxide at the source and the sink are generally unknown and the critical concentrations for metabolism and activity are known for few organisms. The situation is further complicated by the reaction of carbon dioxide with water to yield carbonic acid, and the bicarbonate and carbonate ions. The quantitative relationship between these various components of the carbon dioxide-water and carbon dioxide-water-calcium carbonate systems is dependent upon pH (98, 125). The apparent effect of a given partial pressure of carbon dioxide would therefore vary with pH if, in fact, the bicarbonate or carbonate ions (rather than carbon dioxide itself) were the active chemicals. Data supporting such a concept, derived from a number of fungi, have been reported (51, 87).

Soil Water in Plant Pathology

In this article, main attention is given to the relationship between soil microorganisms, especially fungi, and soil water. To the plant pathologist, however, this is but a part of the total situation. For him, the interaction between host, pathogen, and water is of prime importance. Unfortunately, the effect of total water suction on the physiological processes of disease resistance and disease escape is very poorly known, although the existence of such a relationship has been long suggested (36, 44). It is, however, instructive to read a recent review of water absorption, conduction, and transpiration in plants (75), and to realize how profoundly different are the host and the soil-borne pathogen in their relations to water. Absorption of water by roots is largely passive and movement is by mass flow: in fungi absorption is primarily direct from soil into the hyphae and is a molecular, osmotic process. In plants, conduction is through a long and complex pathway of many tissues, extending from the root surface to the stomata: in soil microfungi, the path is short and simple through relatively uniform living cells. Plants lose water rapidly by transpiration into an atmosphere whose relative humidity rarely exceeds 90 per cent for prolonged periods but may often fall below 50 per cent: most fungi of agricultural soils exist in an environment where the relative humidity rarely falls below 90 per cent so that their rate of water loss is correspondingly small. In consequence, the dynamics of water movement in soil are of vital importance to higher plants whereas this is not so for fungi. The effects of a given soil-water regime on the level of activity of the host and the pathogen are therefore rarely comparable. An understanding of the various conditions which favour the two organisms is thus a precursor of one method for the biological control of disease.

LITERATURE CITED

1. Ajello, L. Comparative ecology of respiratory mycotic disease agents. *Bacteriol. Rev.*, **31**, 6–24 (1967)
2. Alexander, M. *Introduction to Soil Microbiology* (Wiley, New York, 472 pp., 1961)
3. Barley, K. P., Greacen, E. L. Mechanical resistance as a soil factor influencing the growth of roots and underground shoots. *Advan. Agron.* **19**, 1–44 (1967)
4. Bergman, H. F. Oxygen deficiency as a cause of disease in plants. *Botan. Rev.*, **25**, 418–585 (1959)
5. Bhaumik, H. D., Clark, F. E. Soil moisture tension and microbial activity. *Soil Sci. Soc. Am. Proc.*, **12**, 234–38 (1947)
6. Bloomfield, B. J., Alexander, M. Melanins and resistance of fungi to lysis. *J. Bacteriol.*, **93**, 1276–80 (1967)
7. Bond, R. D. The influence of the microflora on the physical properties of soils. II. Field studies on water repellent sands. *Australian J. Soil Res.*, **2**, 123–31 (1964)
8. Bond, R. D., Harris, J. R. The influence of the microflora on the physical properties of soils. I. The occurrence and significance of microbial filaments and slimes in soils. *Australian J. Soil Res.*, **2**, 111–22 (1964)
9. Bonner, J. T. A study of the temperature and humidity requirements of *Aspergillus niger*. *Mycologia*, **40**, 728–38 (1948)
10. Borut, S. An ecological and physiological study on soil fungi of the Northern Negev (Israel). *Bull. Res. Council Israel, Sect. D.*, **8**, 65–80 (1960)
11. Braun, H., Wilcke, D. E. Untersuchungen über Bodenverdichtungen und ihre Beziehungen zum Auf-

treten der Kragenfäule *(Phytophthora cactorum). Phytopathol. Z.,* **46,** 71–86 (1963)

12. Brown, A. D., Turner, H. P. Membrane stability and salt tolerance in gram-negative bacteria. *Nature,* **199,** 301–2 (1963)
13. Brown, G. E., Kennedy, B. W. Effect of oxygen concentration on *Pythium* seed rot of soybean. *Phytopathology,* **56,** 407–11 (1966)
14. Bruehl, G. W., Lai, P. Influence of soil pH and humidity on survival of *Cephalosporium gramineum* in infested wheat straw. *Can. J. Plant Sci.,* **48,** 245–52 (1968)
15. Bumbieris, M., Lloyd, A. B. Influence of soil fertility and moisture on lysis of fungal hyphae. *Australian J. Biol. Sci.,* **20,** 103–12 (1967)
16. Burgess, L. W., *Ecology of some Fungi causing Root and Crown Rots of Wheat* (Doctoral thesis, Univ. of Sydney, N.S.W., Australia, 122 pp., 1967)
17. Burgess, L. W., Griffin, D. M. The recovery of *Gibberella zeae* from straw. *Australian J. Exptl. Agr. Animal Husbandry,* **8,** 364–70 (1968)
18. Butler, F. C. Root and foot rot diseases of wheat. *N.S.Wales, Dept. Agr., Sci. Bull.* **77,** 98 pp. (1961)
19. Caldwell, J. Effects of high partial pressures of oxygen on fungi. *Nature,* **197,** 772–74 (1963)
20. Calvet, R. La diffusion dans les systèmes argile-eau. I. Rappels théoretique concernant les phenomenes de diffusion. *Ann. Agron.,* **18,** 217–36 (1967)
21. Calvet, R. La diffusion dans les systèmes argile-eau. II. Diffusion des cations. *Ann. Agron.,* **18,** 429–44 (1967)
22. Chen, A. W.-C. Soil fungi with high salt tolerance. *Trans. Kansas Acad. Sci.,* **67,** 36–40 (1964)
23. Chen, A. W.-C., Griffin, D. M. Soil physical factors and the ecology of fungi. V. Further studies in relatively dry soils. *Trans. Brit. Mycol. Soc.,* **49,** 419–26 (1966)
24. Chen, A. W.-C., Griffin, D. M. Soil physical factors and the ecology of fungi. VI. Interaction between temperature and soil moisture. *Trans. Brit. Mycol. Soc.,* **49,** 551–61 (1966)
25. Christensen, C. M. Deterioration of stored grains by fungi. *Botan. Rev.,* **23,** 108–34 (1957)
26. Clark, F. E. Bacteria in soil, in *Soil Biology,* 15–49 (Burges, A., Raw, F., Eds., Academic Press, London, 532 pp., 1967)
27. Coley-Smith, J. R. Studies of the biology of *Sclerotium cepivorum* Berk. IV. Germination of sclerotia. *Ann. Appl. Biol.,* **48,** 8–18 (1960)
28. Cook, R. J., Flentje, N. T. Chlamydospore germination and germling survival of *Fusarium solani* f. *pisi* in soil as affected by soil water and pea seed exudation. *Phytopathology,* **57,** 178–82 (1967)
29. Corey, A. T., Slatyer, R. O., Kemper, W. D. Comparative terminologies for water in the soil-plant-atmosphere system, in *Irrigation of Agricultural Lands,* Agron. Monograph, **11,** 427–45 (Hagan, R. M., Haise, H. W., Edminster, T. W., Eds., Academic Press, New York, 1180 pp., 1967)
30. Couch, H. B., Bedford, E. R. *Fusarium* blight of turfgrasses. *Phytopathology,* **56,** 781–86 (1966)
31. Couch, H. B., Bloom, J. R. Influence of environment on diseases of turfgrasses. II. Effect of nutrition, pH and soil moisture on *Sclerotinia* dollar spot. *Phytopathology,* **50,** 761–63 (1960)
32. Couch, H. B., Purdy, L. H., Henderson, D. W. Application of soil moisture principles to the study of plant disease. *Virginia Polytechnic Inst., Dept. Plant Pathol. Bull.* **4,** 23 pp. (1967)
33. Cruickshank, I. A. M., Perrin, D. R. Studies on phytoalexins. X. Effect of oxygen tension on the biosynthesis of pisatin and phaseollin. *Phytopathol. Z.,* **60,** 335–42 (1967)
34. Dalla Valle, J. M., *Micromeritics* (Pitman, New York, 428 pp., 1943)
35. Day, P. R., Bolt, G. H., Anderson, D. M. Nature of soil water. In *Irrigation of Agricultural Lands,* Agron. Monograph, **11,** 193–208 (See ref. 29)
36. Dickson, J. G. Influence of soil temperature and moisture on the development of the seedling blight of wheat and corn caused by *Gibberella zeae. J. Agr. Res.,* **23,** 837–70 (1923)
37. Dommergues, Y. Contribution à l'etude de la dynamique microbienne des sols en zone semi-aride et en

zone tropicale seche. *Ann. Agron.*, **13**, 265–324, 379–469 (1962)

38. Egeberg, R. O., Elconin, A. F., Egeberg, M. C. Effect of salinity and temperature on *Coccidioides immitis* and three antagonistic soil saprophytes. *J. Bacteriol.*, **88**,, 473–76 (1964)
39. Elconin, A. F., Egeberg, R. O., Egeberg, M. C. Significance of soil salinity on the ecology of *Coccidioides immitis. J. Bacteriol.*, **87**, 500–3 (1964)
40. Emerson, W. W., Bond, R. D. The rate of water entry into dry sand and calculation of the advancing contact angle. *Australian J. Soil Res.*, **1**, 9–16 (1963)
41. Evans, G., White, N. H. Effect of the antibiotics radicicolin and griseofulvin on the fine structure of fungi. *J. Exptl. Botany*, **18**, 465–70 (1967)
42. Farrell, D. A., Greacen, E. L. Resistance to penetration of fine probes in compressible soil. *Australian J. Soil Res.*, **4**, 1–17 (1966)
43. Fulton, J. M., Mortimore, C. G., Hildebrand, A. A. Note on the relation of soil bulk density to the incidence of *Phytophthora* root and stalk rot of soybeans. *Can. J. Soil Sci.*, **41**, 247 (1961)
44. Garrett, S. D., *Root Disease Fungi* (Chronica Botanica, Waltham, Mass., 177 pp., 1944)
45. Gibbs, J. N. Resin and the resistance of conifers to *Fomes annosus. Ann. Botany (London)*, **32**, 649–66 (1968)
46. Goring, C.A.I. Physical aspects of soil in relation to the action of soil fungicides. *Ann. Rev. Phytopathol.*, **5**, 285–318 (1967)
47. Grable, A. R. Soil aeration and plant growth. *Advan. Agron.*, **18**, 58–106 (1966)
48. Greenwood, D. J. Studies on the transport of oxygen through the stems and roots of vegetable seedlings. *New Phytologist*, **66**, 337–48 (1967)
49. Greenwood, D. J. Studies on oxygen transport through mustard seedlings (*Sinapis alba* L.). *New Phytologist*, **66**, 597–607 (1967)
50. Greenwood, D. J. Measurement of microbial metabolism in soil. In *The Ecology of Soil Bacteria*, 138–57 (Gray, T. R. G., Parkinson, D., Eds., Liverpool Univ. Press, 681 pp., 1967)
51. Griffin, D. H. The interaction of hydrogen ion, carbon dioxide and potassium ion in controlling the formation of resistant sporangia in *Blastocladiella emersonii. J. Gen. Microbiol.*, **40**, 13–28 (1965)
52. Griffin, D. M. Soil moisture and the ecology of soil fungi. *Biol. Rev. Cambridge Phil. Soc.*, **38**, 141–66 (1963)
53. Griffin, D. M. Soil physical factors and the ecology of fungi. I. Behaviour of *Curvularia ramosa* at small soil water suctions. *Trans. Brit. Mycol. Soc.*, **46**, 273–80 (1963)
54. Griffin, D. M., Soil physical factors and the ecology of fungi. II. Behaviour of *Pythium ultimum* at small soil water suctions. *Trans. Brit. Mycol. Soc.*, **46**, 368–72 (1963)
55. Griffin, D. M. Soil physical factors and the ecology of fungi. III. Activity of fungi in relatively dry soil. *Trans. Brit. Mycol. Soc.*, **46**, 373–77 (1963)
56. Griffin, D. M. Soil water terminology in mycology and plant pathology. *Trans. Brit. Mycol. Soc.*, **49**, 367–68 (1966)
57. Griffin, D. M. Fungi attacking seeds in dry seed-beds. *Proc. Linnean Soc. N.S. Wales*, **91**, 84–89 (1966)
58. Griffin, D. M. Observations on fungi growing in a translucent particulate matrix. *Trans. Brit. Mycol. Soc.*, **51**, 319–22 (1968)
59. Griffin, D. M. A theoretical study relating the concentration and diffusion of oxygen to the biology of organisms in soil. *New Phytologist*, **67**, 561–77 (1968)
60. Griffin, D. M. (Unpublished data)
61. Griffin, D. M., Quail, E. G. Movement of bacteria in moist particulate systems. *Australian J. Biol. Sci.*, **21**, 579–82 (1968)
62. Griffiths, E. Microorganisms and soil structure. *Biol. Rev. Cambridge Phil. Soc.*, **40**, 129–42 (1965)
63. Hagan, R. M., Haise, H. W., Edminster, T. W. (Eds.). *Irrigation of Agricultural Lands*, Agron. Monograph, **11** (See ref. 29)
64. Henis, Y., Keller, P., Keynan, A. Inhibition of fungal growth by bacteria during cellulose decomposi-

tion. *Can. J. Microbiol.*, **7**, 857–63 (1961)

65. Hickman, C. J., Ho, H. H. Behaviour of zoospores in plant-pathogenic phycomycetes. *Ann. Rev. Phytopathol.*, **4**, 195–220 (1966)
66. Ho, H. H., Hickman, C. J. Asexual reproduction and behaviour of zoospores of *Phythopthora megasperma* var. *sojae*. *Can. J. Botany*, **45**, 1963–81 (1967)
67. Holmes, J. W., Taylor, S. A., Richards, S. J. Measurement of soil water. In *Irrigation of Agricultural Lands*, Agron. Monograph, **11**, 275–303 (See ref. 29)
68. Jackson, R. M. Antibiosis and fungistasis of soil microorganisms. In *Ecology of Soil-borne Plant Pathogens*, 363–69 (Baker, K. F., Snyder, W. C., Eds., University of California Press, Berkeley, 571 pp., 1965)
69. Jensen, C. R., Stolzy, L. H., Letey, J. Tracer studies of oxygen diffusion through roots of barley, corn and rice. *Soil Sci.*, **103**, 23–29 (1967)
70. Jensen, H. L. Contributions to the microbiology of Australian soils. I. Numbers of microorganisms in soil, and their relation to certain external factors. *Proc. Linnean Soc. N.S.Wales*, **59**, 101–17 (1934)
71. Johnson, T. W., Sparrow, F. K., *Fungi in Oceans and Estuaries* (Cramer, Weinheim, 668 pp., 1961)
72. Kerr, A. The influence of soil moisture on infection of peas by *Pythium ultimum*. *Australian J. Biol. Sci.*, **17**, 676–85 (1964)
73. Ko, W.-H., Lockwood, J. L. Soil fungistasis: relation to fungal spore nutrition. *Phytopathology*, **57**, 894–901 (1967)
74. Kouyeas, V. An approach to the study of moisture relations of soil fungi. *Plant Soil*, **20**, 351–63 (1964)
75. Kramer, P. J. Biddulph, O., Nakayamo, F. S. Water absorption, conduction and transpiration. In *Irrigation of Agricultural Lands*, Agron. Monograph, **11**, 320–36 (See ref. 29)
76. Kristensen, K. J., Lemon, E. R. Soil aeration and plant root relations. III. Physical aspects of oxygen diffusion in the liquid phase of soil. *Agron. J.*, **56**, 295–301 (1964)
77. Kuo, M.-J., Alexander, M. Inhibition of the lysis of fungi by melanins. *J. Bacteriol.*, **94**, 624–29 (1967)
78. Lacey, J. The infectivity of soils containing *Phytophthora infestans*. *Ann. Appl. Biol.*, **56**, 363–80 (1965)
79. Lacey, J. The role of water in the spread of *Phytophthora infestans* in the potato crop. *Ann. Appl. Biol.*, **59**, 245–55 (1967)
80. Lapwood, D. H. The effects of soil moisture at the time potato tubers are forming on the incidence of common scab (*Streptomyces scabies*). *Ann. Appl. Biol.*, **58**, 447–56 (1966)
81. Last, F. T., Ebben, M. H. A preliminary study of factors affecting the spread of *Verticillium cinerescens* Wr. causing carnation wilt. *Rept. Glasshouse Crops Res. Inst. (1963)*, pp. 118–21 (1964)
82. Lemon, E. R. Soil aeration and plant root relations. I. Theory. *Agron. J.*, **54**, 167–70 (1962)
83. Lemon, E. R., Wiegand, C. L. Soil aeration and plant root relations. II. Root respiration. *Agron. J.*, **54**, 171–75 (1962)
84. Letey, J., Stolzy, L. H. Measurement of oxygen diffusion rates with the platinum micro-electrode. I. Theory and equipment, *Hilgardia*, **35**, 545–54 (1964)
85. Limonard, T. Bacterial antagonism in seed health tests. *Netherlands J. Plant Pathol.*, **73**, 1–14 (1967)
86. Lockwood, J. L. Soil fungistasis. *Ann. Rev. Phytopathol.*, **2**, 341–62 (1964)
87. Macauley, B. J., *A Study of the Effects of Carbon Dioxide and the Bicarbonate Ion on Soil Fungi* (Doctoral thesis, Univ. of Sydney, N.S.W., Australia, 105 pp., 1968)
88. McLaren, A. D., Skujins, J. The physical environment of microorganisms in soil, in *The Ecology of Soil Bacteria*, 3–24 (Gray, T.R.G., Parkinson, D., Eds., Liverpool Univ. Press, 681 pp., 1967)
89. Marshall, T. J. Relations between water and soil. *Commonwealth Bur. Soil Sci. (Gt. Brit.), Tech. Commun.*, **5**, 91 pp., 1959
90. Miller, R. D., Johnson, D. D. The effect of soil moisture on carbon dioxide evolution, nitrification and nitrogen mineralization. *Soil Sci. Soc. Am. Proc.*, **28**, 644–47 (1964)
91. Moore, L. D., Couch, H. B., Bloom, J. R. Influence of environment on diseases of turfgrasses. III. Effect

of nutrition, pH, soil temperature, air temperature, and soil moisture on *Pythium* blight of Highland bentgrass. *Phytopathology,* **53,** 53–57 (1963)

92. Muse, R. R., Couch, H. B. Influence of environment on diseases of turfgrasses. IV. Effect of nutrition and soil moisture on *Corticium* red thread of creeping red fescue. *Phytopathology,* **55,** 507–10 (1965)
93. Negrutskii, S. F. O sosushchei sile griba *Fomitopsis annosa* (Fr.) Karst. *Botan. Zh. U.S.S.R.,* **49,** 1480–81 (1964)
94. Park, D. The importance of antibiotics and inhibiting substances, in *Soil Biology,* 435–47 (See ref. 26)
95. Parr, J. F., Norman, A. G. Growth and activity of soil microorganisms in glass micro-beads. I. Carbon dioxide evolution. *Soil Sci.,* **97,** 361–66 (1964)
96. Parr, J. F., Parkinson, D., Norman, A. G. Growth and activity of soil microorganisms in glass microbeads. II. Oxygen uptake and direct observations. *Soil Sci.,* **103,** 303–10 (1967)
97. Pentland, G. D. The effect of soil moisture on the growth and spread of *Coniophora puteana* under laboratory conditions. *Can. J. Botany,* **45,** 1899–1906 (1967)
98. Ponnamperuma, F. N. A theoretical study of aqueous carbonate equilibria. *Soil Sci.,* **103,** 90–100 (1967)
99. Potgieter, H. J., Alexander, M. Susceptibility and resistance of several fungi to microbial lysis. *J. Bacteriol.,* **91,** 1526–33 (1964)
100. Raney, W. A. Physical factors of the soil as they affect soil microorganisms. In *Ecology of Soilborne Plant Pathogens,* 115–18 (See ref. 68)
101. Rishbeth, J. Observations on the biology of *Fomes annosus* with particular reference to East Anglian pine plantations. II. *Ann. Botany (London),* **15,** 1–21 (1951)
102. Ritchie, D. The effect of salinity and temperature on marine and other fungi from various climates. *Bull. Torrey Botan. Club,* **86,** 367–73 (1959)
103. Rixon, A. J., Bridge, B. J. Respiratory quotient arising from microbial activity in relation to matric suction and air-filled pore space of soil. *Nature,* **218,** 961–62 (1968)
104. Rose, C. W., *Agricultural Physics* (Pergamon Press, Oxford, 226 pp., 1966)
105. Rovira, A. D., Greacen, E. L. The effect of aggregate disruption on the activity of microorganisms in the soil. *Australian J. Agr. Res.,* **6,** 659–73 (1957)
106. Rybalkina, A. V., Kononenko, E. V. Méthode d'étude de la microflore active des sols. *Pédologie,* **7** (n. spéc.), 190–96 (1957)
107. Schippers, B., Schroth, M. N., Hildebrand, D. C. Emanation of water from underground plant parts. *Plant Soil,* **27,** 81–91 (1967)
108. Scott, W. J. Water relations of food spoilage microorganisms. *Advan. Food Res.,* **7,** 83–127 (1956)
109. Scotter, D. R., Thurtell, G. W., Tanner, C. B. Measuring oxygen uptake by the roots of intact plants under controlled conditions. *Soil Sci.,* **104,** 374–78 (1967)
110. Seidel, D. Untersuchungen über die Keimhemmung von Pilzsporen im Boden mit Hilfe des Agarschreibentests. *Zentr. Bakteriol., Abt. II,* **119,** 74–87 (1965)
111. Sewell, G. W. F. The effect of altered physical conditions of soil on biological control. In *Ecology of Soilborne Plant Pathogens,* 479–93 (See ref. 68)
112. Shantz, H. L., Piemeisel, R. L. Fungus fairy rings in eastern Colorado and their effect on vegetation. *J. Agr. Res.,* **11,** 191–246 (1917)
113. Sindhu, M. A., Cornfield, A. H. Effect of sodium chloride and moisture content on ammonification and nitrification in incubated soil. *J. Sci. Food Agr.,* **18,** 505–6 (1967)
114. Slatyer, R. O., *Plant-Water Relationships* (Academic Press, London, 366 pp., 1967)
115. Smith, A. M., Griffin, D. M. (Unpublished data)
116. Snow, D. The germination of mold spores at controlled humidities. *Ann. Appl. Biol.,* **36,** 1–13 (1949)
117. Spence, J. A. Black-pod disease of cocoa. II. A study of host-parasite relations. *Ann. Appl. Biol.,* **49,** 723–34 (1961)
118. Stolzy, L. H., Letey, J. Characterizing soil oxygen conditions with a platinum microelectrode. *Advan. Agron.,* **16,** 249–79 (1964)
119. Stolzy, L. H., Letey, J. Measurement

of oxygen diffusion rates with the platinum microelectrode. III. Correlation of plant response to soil oxygen diffusion rates. *Hilgardia,* **35,** 567–76 (1964)

120. Stolzy, L. H., Van Gundy, S. D. The soil as an environment for microflora and microfauna. *Phytopathology,* **58,** 889–99 (1968)
121. Te Strake, D. Estuarine distribution and saline tolerance of some Saprolegniaceae. *Phyton, Buenos Aires,* **12,** 147–52 (1959)
122. Tomkins, R. G. Studies of the growth of molds. I. *Proc. Roy. Soc. (London), Ser. B,* **105,** 375–401 (1929)
123. Toohey, J. I., Nelson, C. D., Krotkov, G. Barren ring, a description and study of causal relationships. *Can. J. Botany,* **43,** 1043–54 (1965)
124. Towers, B., Stambaugh, W. J. The influence of induced soil moisture stress upon *Fomes annosus* root rot of loblolly pine. *Phytopathology,* **58,** 269–72 (1968)
125. Umbreit, W. W., Burris, R. H., Stauffer, J. F., *Manometric Techniques* (Burgess Publishing Co., Minneapolis, 338 pp., 3rd ed., 1957)
126. Viets, F. G. Nutrient availability in relation to soil water. In *Irrigation of Agricultural Lands,* Agron. Monograph, **11,** 458–71 (See ref. 29)
127. Vilain, M. L'aeration du sol. *Ann. Agron.,* **14,** 967–98 (1963)
128. Wallace, H. A. H. Factors affecting subsequent germination of cereal seeds sown in soils of subgermination moisture content. *Can. J. Botany,* **38,** 287–306 (1960)
129. Winter, A. G. Der Einfluss der physikalischen Bodenstruktur auf den Infektionsverlauf **bei der** Ophiobolose des Weizens. *Z. Pflanzenkrankh. Pflanzenschutz,* **49,** 513–59 (1939)
130. Winter, A. G. Untersuchungen über den Einfluss biotischer Faktoren auf die Infektion des Weizens durch *Ophiobolus graminis. Z. Pflanzenkrankh. Pflanzenschutz,* **50,** 113–34 (1940)
131. Winter, A. G. Weitere Untersuchungen über den Einfluss der Bodenstruktur auf die Infektion des Weizens durch *Ophiobolus graminis. Zentr. Bakteriol, Abt. II,* **101,** 364–88 (1940)
132. Wodzinski, R. J., Frazier, W. C. Moisture requirements of bacteria. I. Influence of temperature and pH on requirements of *Pseudomonas fluorescens. J. Bacteriol.,* **79,** 572–78 (1960)
133. Wodzinski, R. J., Frazier, W. C. Moisture requirements of bacteria. II. Influence of temperature, pH, and malate concentration on requirements of *Aerobacter aerogenes. J. Bacteriol.,* **81,** 353–58 (1961)
134. Wodzinski, R. J., Frazier, W. C. Moisture requirements of bacteria. III. Influence of temperature, pH, and malate and thiamine concentration on requirement of *Lactobacillus viridescens. J. Bacteriol.,* **81,** 359–65 (1961)
135. Youngs, E. G. Water movement in soils, in *The State and Movement of Water in Living Organisms,* 89–112 (Fogg, C. E., Ed., 19th Symp. Soc. Exptl. Biology, Cambridge, U.K., 432 pp., 1965)
136. Zan. K. Activity of *Phytophthora infestans* in soil in relation to tuber infection. *Trans. Brit. Mycol. Soc.,* **45,** 205–21 (1962)

Copyright 1969. All rights reserved

EFFECT OF FUNGICIDES ON PROTEIN AND NUCLEIC ACID SYNTHESIS[1,2]

By H. D. Sisler

Department of Botany, University of Maryland
College Park, Maryland

INTRODUCTION

Selective inhibition of one organism in intimate association with another is simplest when the problem is that of controlling a complex species in the presence of a simple one. Numerous examples could be cited to show that selective inhibition of a complex species frequently involves interference with some process that has evolved since the organism separated in evolutionary development from simpler forms. The toxicities of carbon monoxide and botulinum toxin in mammals, carbamates in insects, and 2,4-dichlorophenoxyacetic acid in higher plants are examples of inhibitor action aimed at a level of development higher than that found in simple organisms like bacteria and fungi. On the other hand, a complex organism may have certain advantages over a simpler species in excluding or destroying toxicants which attack processes common to both. These, in fact, seem to be the primary mechanisms by which higher plants are protected from most fungicides presently used to prevent fungal infections. The fact that protective fungicides do not control established fungal infections implies that plant protoplasts are not exposed to fungicidal levels of these toxicants. Therefore, it is unnecessary to propose that higher plant cells are not injured because their metabolism is less sensitive than that of the fungal pathogen.

Chemical control of plant diseases now stands at the threshold of the systemic toxicant. The major defenses of the host against injury from protective fungicides can no longer be utilized to protect against systemic fungicides. Either exclusion or destruction of the chemical by the host would tend to reduce chemotherapeutic effectiveness. Protoplasts of the host may retain some advantage by not accumulating the toxicant to the same extent as those of the pathogen, but for the most part it seems selectivity would need to be based on relatively subtle biochemical differences in processes common to both pathogen and host. Thus, we may have arrived at a point where it is necessary to take advantage of small deviations that have taken place in basic processes since the host and pathogen have separated in the

[1] The author is grateful to Dr. L. R. Krusberg for a critical review of the manuscript.

[2] Preparation of this review was supported in part by PHS Research Grant No. AI00225 from the Institute of Allergy and Infectious Diseases.

evolutionary scheme. This does not appear to be an attractive choice, but it may be a necessary one because fungi and bacteria are simple organisms possessing few processes differing fundamentally from those found in host plants. Nature has achieved success at this level of specificity in developing a number of highly selective antibiotics that inhibit protein synthesis in bacteria but not in higher organisms. Some antifungal antibiotics also display selective action based on minor differences in a fundamental process common to all organisms. Few, if any, organic fungicides are so discriminating in their action as these antibiotics, but such compounds should emerge as research in plant chemotherapy progresses.

Two major areas of metabolism in which variations among organisms offer opportunities for selective action of chemicals are protein and nucleic acid synthesis. Consideration will be given in this article to the mode of action and to factors regulating selectivity of some fungitoxic compounds inhibiting these processes. Compounds interfering with the conversion of free amino acids into proteins and with the synthesis and polymerization of nucleotides into nucleic acids will be regarded as inhibitors of protein and nucleic acid synthesis, respectively. Detailed examination will be limited to toxicants used to control fungal pathogens of plants, although frequent reference will be made to pertinent information derived from studies of other compounds.

INHIBITION OF PROTEIN SYNTHESIS

According to present concepts, synthesis of protein from free amino acids involves three major steps: (*a*) an activation step in which amino acids react with ATP to form enzyme-bound aminoacyl adenylate derivatives; (*b*) reaction of the aminoacyl adenylate derivatives with specific transfer RNA's (tRNA's) to form aminoacyl-tRNA intermediates; and (*c*) transfer of the aminoacyl t-RNA intermediates to ribosome-messenger-RNA complexes where the amino acids are joined by peptide linkages to growing peptides.

Some amino acid analogues may undergo activation and react with appropriate transfer RNA's. These analogues are then incorporated into polypeptides in place of the corresponding normal amino acids. Protein synthesis may continue for a period at a linear rather than at an exponential rate, yielding modified products with varying degrees of activity and changes in physical properties (79). Growth may likewise continue for a limited period at a linear rate. *p*-Fluorophenylalanine is representative of toxic amino acid analogues which produce such effects (67). An inhibitory analogue which is not incorporated into protein prevents incorporation of the corresponding normal amino acid, and as a consequence, the incorporation of all other amino acids as well (79). Reduction, amidation, or esterification of the carboxyl group of tyrosine yields analogues which are incapable of activation, and, of course, these analogues cannot be incorporated into proteins. Nevertheless, they are bound to the tyrosine activating enzyme and are good competitive inhibitors of tyrosine activation (10). Presumably such analogues inhibit protein synthesis more rapidly than those which are incorporated into protein.

Most effective antifungal or antibacterial compounds known to inhibit protein synthesis interfere with the last of the three steps mentioned above. This step is rather complicated and may be affected by toxicants in a variety of ways. At this level a toxicant may interfere with binding of aminoacyl-tRNA's to ribosomes, with the attachment of the anticodon region of tRNA to the coding site of the messenger RNA, or with the polymerization of the incoming amino acid to the growing peptide. An inhibitor may act as an analogue of aminoacyl-tRNA like puromycin and prematurely terminate the growth of peptide chains (68, 107), or, like streptomycin in bacterial systems, it may cause misreading of the messenger RNA code (47).

Cytoplasmic Versus Mitochondrial Protein Synthesis

Before proceeding to detailed discussion of specific toxicants, some mention should be made of the two systems of protein synthesis found in fungi and other eukaryotic organisms (43, 74, 87). One system is associated with 80S ribosomes in the cytoplasm, and the other is located in the mitochondria. The latter system resembles the one found in prokaryotic organisms (bacteria and blue-green algae) and differs from the cytoplasmic system particularly in regard to its sensitivity to certain antibiotics. The cytoplasmic system in fungi is insensitive to chloramphenicol (6, 82, 88, 95) and to various other antibiotics that inhibit protein synthesis in bacteria (12, 53), whereas the mitochondrial system is relatively sensitive to these antibiotics. Chloramphenicol, erythromycin, tetracycline, carbomycin, spiramycin, oleandomycin, and lincomycin partially or completely inhibit formation of cytochromes *a*, a_3, *b*, and *c*, in growing cells of *Saccharomyces cerevisiae* (12, 13, 43, 58). Failure of the cells to form these cytochromes and the inner cytoplasmic membrane (13) is attributed to specific inhibition of mitochondrial protein synthesis by the antibiotics. The disappearance of the cytochromes is paralleled by a drop in respiratory capacity to less than 5 per cent of that in normal cells (43). The defective cells appear to be phenocopies of a respiratory-deficient cytoplasmic petite mutant (13). Growth is supported by energy derived from glycolysis, and cell yields are normal when an abundant supply of carbohydrate is available (5 per cent glucose) but are reduced when the substrate is limited (1 per cent glucose).

Erythromycin resistance in yeast is extra-chromosomally inherited (57, 84), suggesting that mitochondrial DNA may be the genetic determinant of resistance. Mutants resistant to chloramphenicol and tetracycline have a reduced permeability to the antibiotics which is chromosomally inherited (57).

Chloramphenicol inhibits synthesis of cytochromes in *Pythium ultimum* in much the same manner as in *S. cerevisiae,* but at lower concentrations (62). In contrast to most fungi, *P. ultimum* is relatively sensitive to chloramphenicol. A concentration of 100 μg/ml strongly inhibits dry weight increase, but a slow growth persists even when the inhibitor concentration is increased to 1000 μg/ml. Surprisingly, the antibiotic has little effect on linear growth of the fungus. Mycelium grown in the presence of chloramphenicol is devoid of cytochromes *a*, a_3, and *b*, but contains increased levels of cytochrome *c*. Electron micrographs reveal the absence of stalked particles normally present on cristae of mitochondria of *P. ultimum.* Oxygen uptake of chloramphenicol-grown cells of both *P. ultimum* and *Rhodotorula glutinis* is nearly identical with that of normally grown cells on a dry weight basis (87). However, antimycin A, an inhibitor of electron transport between cytochromes *b* and *c* (73) does not inhibit oxygen uptake by chloramphenicol-grown cells of either species, but does inhibit oxygen uptake of normally grown cells of both species. These results suggest that a mechanism of terminal oxidation differing from the normal cytochrome pathway must operate in the chloramphenicol-grown cells. The system is apparently inefficient in regard to energy production and is capable of maintaining only a slow growth rate in these two obligate aerobes, particularly in *P. ultimum* (87).

Chloramphenicol, at levels which do not inhibit protein synthesis by cytoplasmic ribosomes, inhibits protein synthesis in isolated mitochondria of fungi (33, 53, 74, 104) and higher organisms (1, 74). On the other hand, cycloheximide at high levels does not inhibit protein synthesis in isolated mitochondria of yeast (53, 74) and mammals (1, 3, 59) at concentrations which strongly inhibit protein synthesis by cytoplasmic ribosomes. Insensitivity of the mitochondrial system to cycloheximide does not appear to be related to mitochondrial impermeability (1, 3). In contrast to cycloheximide and chloramphenicol, puromycin inhibits protein synthesis in both mitochondrial (3) and cytoplasmic systems (68). Most mitochondrial proteins are apparently synthesized in the cytoplasm and are then incorporated into the mitochondria. Synthesis of these proteins is inhibited by cycloheximide and, therefore, in experiments with intact cells, their appearance in the mitochondria is suppressed by the antibiotic (1, 61).

Growth of many bacterial species is inhibited by 0.5 to 5 μg/ml of chloramphenicol (64), a concentration range well below that required to inhibit growth of the most sensitive fungi. Even though mitochondrial protein synthesis may resemble bacterial protein synthesis in sensitivity to chloramphenicol, it is obvious that sensitivity of the bacterial system in intact cells is much higher than that of the mitochondrial system in intact fungal cells. The difference could result from the necessity of the antibiotic to transverse two membrane systems in the fungal cells and only one in the bacterial cells. The possibility that permeability plays a role in this differential sensitivity is suggested by the observation that mitochondrial protein synthesis is strongly inhibited in isolated mitochondria of *Saccharomyces carlsbergensis* by 10 μg/ml of chloramphenicol (33), a concentration comparable to those preventing growth of sensitive bacteria. Similar results are also reported for isolated mitochondria of *S. cerevisiae* (53). Chloramphenicol at 10 μg/ml inhibits mitochondrial protein synthesis 73 per cent, but much higher concentrations are required to suppress cytochrome formation in intact cells of this yeast (12). These results are consistent with the observation that increased resistance to chloramphenicol in mutants of *S. cerevisiae* is due to decreased cell permeability (57).

The minute amount of protein synthesis carried out by isolated mitochondrial preparations could result from the presence of a relatively small number of contaminating bacteria. The systems are insensitive to ribonuclease and behave in practically all respects like bacterial cells. However, considerable evidence indicating that the protein synthesis is not due to contaminating bacteria has been presented by Roodyn & Wilkie (74).

Cycloheximide

Although the value of cycloheximide for chemical control of plant diseases has been known for more than 20 years (29, 31), only in recent years has the compound achieved prominence in the fields of physiology and biochemistry. Ready availability, together with high potency and specificity, have made it the compound of choice as an inhibitor of protein synthesis in a variety of fungal, algal, higher plant, and animal cells. Cycloheximide is without doubt now more widely known as a selective inhibitor of protein synthesis than as an antibiotic which controls certain plant diseases.

Literature through 1965 relating to the mechanism of action of cycloheximide and related glutarimide antibiotics has been reviewed elsewhere (85), but much additional literature has accumulated during the past 3 years. Space permits consideration of only a limited number of these articles here.

Site of action.—Substantial evidence indicates that fungitoxicity of cycloheximide results specifically from inhibition of protein synthesis (36, 52, 78, 81) at the stage where amino acids are transfered from tRNA into polypeptides (5, 27, 71, 80, 82, 101). The transfer process affected involves reactions associated with a complicated polyribosomal system consisting of messenger RNA, aminoacyl-tRNA and ribosomes. Although the antibiotic action at the polyribosomal level has been the subject of several investigations, the precise mechanism of inhibition remains obscure. The solution is difficult primarily because of problems encountered in attempting to separate the process into its component parts so that the effects of cycloheximide on each can be determined.

There is general agreement that the antibiotic does not cause breakdown of polyribosomal aggregates while inhibiting protein synthesis (15, 32, 90, 94, 100, 103). It does not cause premature detachment of peptides from a polyribosomal complex (15, 27) as does puromycin (68). Cycloheximide inhibits polyribosome assembly and peptide chain initiation (2, 15, 56, 63, 90, 94). Some investigators (15, 30, 90, 94) have suggested that these effects result indirectly from a slowing of peptide chain elongation and movement of ribosomes along the messenger. Others suggest that the antibiotic primarily affects the ribosome activation process dependent on ATP (63) or the peptide chain initiation process (2). The results of Godchaux et al. (32) indicate that assembly of polyribosomes from single ribosomes is not the process mainly affected by low toxicant concentrations which inhibit protein synthesis. It has been suggested that two sites of quite different sensitivities to cycloheximide occur in polyribosomal systems (2, 32, 56). Therefore, some of the conflicting results reported in the literature might be attributed to the differences in levels of inhibitor used by various investigators. Since resistance or susceptibility to cycloheximide in yeast cells is determined by the 60S ribosomal subunit (71), the antibiotic must interfere with some factor associated with this particle.

New evidence continues to lend support to earlier conclusions (27, 82) that cycloheximide is a specific inhibitor of protein synthesis, although the possibility that other sites of action may be involved is not completely eliminated at this time. A striking effect of cycloheximide on conversion of exogenously supplied uridine-^{14}C to RNA cytidylic acid was observed in cocklebur leaf discs, and it was suggested that the antibiotic may interfere with interconversion of uridine triphosphate and cytidine triphosphate (75). This interesting case deserves further investigation, but, in view of data cited below and until inhibition can be demonstrated in a cell-free system, the author reserves the opinion that the observed effect is a consequence of inhibited protein synthesis. In some fungal (40, 52), algal (66), and mammalian cells (4) inhibition of protein synthesis by cycloheximide is accompanied by strong inhibition of DNA synthesis. However, this inhibition as well as that of RNA synthesis (23, 26) is probably a reflection of an initial interference with protein synthesis. The following evidence supports this conclusion.

(*a*) Tolerance of cell-free protein synthesizing systems to cycloheximide is correlated with the tolerance of whole cells to the antibiotic (11, 16, 29, 71, 83). In one strain of *S. cerevisiae,* a marked increase in tolerance is controlled by a single gene. The ribosomes of this strain reflect the resistance seen in the intact organism (16).

(*b*) In cells exposed to cycloheximide, RNA synthesis often continues with little or no inhibition after protein synthesis is almost completely inhibited (18, 27, 36, 81, 89, 96, 108). In the slime mold, *Dictyostelium discoideum,* protein synthesis is inhibited almost completely within one-half hour, whereas RNA synthesis remains unaffected even after 4 hr (93).

(*c*) Actinomycins indirectly prevent protein synthesis by inhibiting DNA-dependent RNA synthesis (72). These compounds inhibit synthesis of certain enzymes in intact cells as effectively as cycloheximide. However, different mechanisms of action for the actinomycins and cycloheximide are indicated by the fact that the former must be added earlier than the latter in order to produce comparable inhibition. For example, actinomycin D prevents synthesis of UDP-GAL polysaccharide transferase in *D. discoïdeum* providing it is added 4 hr before the enzyme normally appears (93). Cycloheximide, added at any time before the enzyme normally appears, prevents its appearance, and, if added after appearance begins, immediately prevents any further rise in enzyme level. Actinomycin is ineffective in preventing the rise in enzyme level when added at the onset of enzyme appearance.

A similar distinction between the action of actinomycin C and cycloheximide was observed in the myxomycete, *Physarum polycephalum* (77). Thymidine kinase begins to increase 50 min before mitosis and reaches a maximum at the end of mitosis. Actinomycin C prevents this increase only when added 1 hr before the expected onset of the increase. Addition of cycloheximide at any time prior to or during the period of activity increase, prevents or immediately interrupts the increase. Effects of cycloheximide and actinomycin D on the action of auxin in artichoke tubers are similar to those described above. Both antibiotics inhibit auxin induced growth when added simultaneously with indolacetic acid (IAA), but only cycloheximide is effective when added 24 hr after IAA (69).

The preceding experiments indicate that inhibitors (actinomycins) which act at the level of transcription of RNA from DNA must be added earlier than one (cycloheximide) which acts at the translational level of protein synthesis in order to produce similar effects. Although these experiments do not eliminate the possibility that cycloheximide may affect RNA synthesis directly, they indicate that inhibition of protein synthesis is a primary effect of the antibiotic.

(*d*) Acetoxycycloheximide (27, 84, 108) and puromycin (68, 108) are both inhibitors of protein synthesis but differ markedly in structure and specific mode of action. However, their effects on protein, DNA, and RNA synthesis in HeLa cells are nearly identical (108). Both severely inhibit protein synthesis and cause an intermediate degree of inhibition of DNA synthesis, but have no immediate effect on RNA synthesis.

(*e*) Cycloheximide inhibits protein and DNA synthesis to about the same extent in one type of mammalian cells, but has no effect on DNA synthesis in cell-free preparations of the same cell type (4). On the other hand, inhibition of protein synthesis has invariably been demonstrated in a variety of cell-free systems derived from cells which are sensitive to the antibiotic. In some cell types, onset of inhibition of protein synthesis precedes that of DNA synthesis, and the degree of inhibition of the former process exceeds that of the latter (27, 81, 108).

(*f*) Cycloheximide permits a burst of DNA synthesis to take place in the absence of protein synthesis when it is added to cultures of *P. polycephalum* during late prophase (19).

Mechanisms of resistance.—The specific mode of action of cycloheximide and the striking differences in sensitivity of closely related fungi to the antibiotic indicate that development of resistant strains should be anticipated. Such strains have been described in several species of fungi (85).

A detailed study of cycloheximide resistance in *S. cerevisiae* was made by Wilkie & Lee (102). Eight genes for resistance were detected in various strains of the organism. Four of these were semi-dominant, one was dominant, and three were recessive. In addition to these eight genes, recessive modifier genes were found which do not themselves confer resistance, but rather act to increase the effectiveness of the resistance genes. Positive interaction of genes in recombinant strains yielded cells which were highly resistant to the antibiotic. A strain carrying one semi-dominant gene, one recessive gene, and a modifier gene was resistant to 1000 μg/ml compared to less than 0.5 μg/ml for the wild type.

A multigenic system controlling resistance to a toxicant which apparently acts at a specific site poses some interesting questions regarding the mechanisms of gene action. Conceivably a decrease in permeability, an alteration of the target site, or the loss or acquisition of the ability to metabolize the toxicant are factors which could operate synergistically in such a system of resistance.

S. pastorianus is highly sensitive to cycloheximide, whereas *S. fragilis* is extremely resistant. Resistance of the latter organism does not appear to result from impermeability (98) or ability to detoxify the antibiotic (85). Cell-free protein synthesizing systems derived from cells of *S. fragilis* are far more resistant to cycloheximide than are those derived from cells of *S. pastorianus* (83). The difference in sensitivity at the ribosomal level probably accounts for the difference in sensitivity seen in the intact organisms. Resistance or susceptibility in the cell-free systems is determined by the ribosomes and not by soluble enzymes or other factors present in the supernatant fraction. More recently, resistance of the *S. fragilis* ribosomes was shown to be a property of the 60S subunit (71).

Cooper et al. (16) studied cycloheximide resistance in *S. cerevisiae* using the strains described by Wilkie & Lee (102). Strains carrying the recessive resistance gene $ac_8{}^r$ (resistant to 20 μg/ml) yielded cell-free protein synthesizing systems resistant to cycloheximide. As with *S. fragilis*, resistance was clearly associated with the ribosomes. The heterozygous diploid carrying $ac_8{}^r$ and its wild type allele, while phenotypically sensitive, yielded ribosomes with intermediate resistance *in vitro*. The diploid cells apparently contain a mixed complement of both sensitive and resistant ribosomes which gives rise to mixed polysomes. With both types of ribosomes utilizing the same messenger RNA's simultaneously, the nonsensitive ribosomes may be blocked from movement along the messenger RNA by the sensitive ribosomes whose movement is impeded by the antibiotic (16). The semi-dominant genes, $AC_5{}^r$ and $AC_7{}^r$, do not influence resistance of the ribosomes or permeability of the cells to the antibiotic (16). However, these genes individually confer resistance to cycloheximide ranging from 0.5 to 1 μg/ml, which is not a marked increase in tolerance over that of the wild type. Assuming that permeability in the wild type is only slightly above the threshold limiting toxicity, it still seems possible that a minor decrease in permeability could account for the observed increase in resistance seen in strains carrying genes AC_5^r and AC_7^r.

Permeability seems to be the factor regulating fungitoxicity of the two cycloheximide analogues, acetoxycycloheximide, and streptovitacin A (84). In *S. pastorianus,* both analogues are far more effective as inhibitors of protein synthesis in cell-free systems than as inhibitors of cell growth. The reverse is true for cycloheximide.

Cunninghamella blakesleeana converts cycloheximide in good yield to the acetate derivative (41) which is nontoxic (54, 86). However, no comment was made (41) concerning the possibility that acetylation of the antibiotic may be a resistance mechanism in *C. blakesleeana.*

Hsu (42) isolated two strains of *Neurospora crassa* with intermediate resistance to cycloheximide controlled in each case by a single gene. In one strain the resistance gene was located in linkage group I, but in the other it was located in linkage group V. Progeny from crosses between these two strains containing both resistance genes grew at a slow rate. However, heterokaryons containing both genes were highly resistant and grew normally. Apparently no attempt has been made to analyze the mechanism of action of these two genes in conferring resistance to cycloheximide.

Rothschild & Suskind (76) made the interesting observation that spontaneous or 2-amino purine-induced mutants of *N. crassa* which are resistant to cycloheximide are susceptible to streptomycin. Resistance to cycloheximide ranged from 0.01 to 5 μg/ml and was controlled by several genes. Ribosomal proteins of three mutants investigated showed electrophoretic or antigenic differences from the wild type.

Finally, strains of amoeba of *P. polycephalum* have been isolated in which resistance to cycloheximide is controlled by a single pair of alleles. However, plasmodia, homozygous or heterozygous for the allele determining resistance, are no more tolerant to cycloheximide than plasmodia homozygous for the allele determining sensitivity (20).

Application in host-parasite interactions.—Specific inhibitors of protein synthesis can aid in elucidating the role of protein synthesis in biochemical processes such as those involved in resistance and host-parasite interactions. Zucker & El-Zayat (110) describe an interesting phenomenon in potato tuber discs which is an appropriate example of the application of such an inhibitor. Freshly cut discs, or those incubated in water for 30 hr, lose their natural resistance to a *Pseudomonas* sp. when treated with cycloheximide. These results indicate that maintenance of resistance to this bacterium requires a continuous synthesis of specific proteins. A similar requirement appears to exist for tissues of carrot, beet, and eggplant because these also become badly infected with the same *Pseudomonas* sp. when treated with cycloheximide.

Increased susceptibility to bacteria is not always the consequence of treating plant tissues with cycloheximide. Crown gall disease of cherry trees can be controlled by applications of the acetate or thiosemicarbazone derivatives of cycloheximide (37). Activity of these derivatives most likely results from cycloheximide released enzymatically or by a H^+ ion-catalyzed hydrolysis (86). Control of the disease is somewhat surprising in light of the fact that cycloheximide is not generally toxic to bacteria and presumably is not toxic to *Agrobacterium tumefaciens,* the pathogen in this case. A likely possibility for the mechanism of disease control is the suppression of gall formation by action of the antibiotic on host protein synthesis. Cycloheximide might prove useful in studying other bacterial diseases where auxins are implicated in the disease process. Phytotoxicity of the antibiotic may present some difficulties, but the concentration might be adjusted to suppress significantly the host response to auxin without appreciable damage to normal processes.

Anisomycin and Emetine

Anisomycin is an antibiotic isolated from *Streptomyces* sp., and emetine (an amebicidal agent) is the principal alkaloid of ipecac, the ground roots of *Uragoga ipecacuanha.* These compounds are mentioned here because they resemble cycloheximide in configuration and conformation as well as in mode of action. Both compounds inhibit protein synthesis at a stage subsequent to the formation of aminoacyl-tRNA in cell-free extracts of yeast and animal cells, but not in those of *Escherichia coli* (34, 35). Anisomycin is toxic to yeast cells, but emetine is not, probably because penetration is restricted by cell permeability. In intact HeLa cells a strong inhibition of DNA synthesis and a normal or stimulated RNA synthesis accompany inhibition of protein synthesis by anisomycin. This pattern of effects is characteristic of those produced by other inhibitors of protein synthesis in intact cells. Since neither emetine nor anisomycin inhibits protein synthesis in extracts from *E. coli* cells, these compounds are unlikely to inhibit mitochondrial protein synthesis. In one respect anisomycin differs from cycloheximide. The former compound effectively inhibits protein synthesis in cell-free systems from *S. fragilis,* but the latter does not.

Blasticidin S

Blasticidin S is used as a protective and therapeutic agent to control rice blast disease caused by *Pyricularia oryzae.* Many species of fungi are not affected by low concentrations of the antibiotic, but *P. oryzae* is highly sensitive. The antibiotic is toxic to bacteria and in this respect differs from cycloheximide (65).

Protein synthesis in whole cells or in cell-free systems of *P. oryzae* is highly sensitive to blasticidin S, and toxicity of the antibiotic to this fungus is attributed specifically to inhibitory effects on protein synthesis (44, 65). Further elucidation of the mechanism of action has been attempted using cell-free systems from *E. coli.* The antibiotic effectively inhibits puromycin-induced release of polypeptides from polyribosomal complexes in the *E. coli* system (17, 106). It interferes with the transfer of peptides from peptidyl-tRNA's to incoming aminoacyl-tRNA's, but not with the preceding steps in protein synthesis (44, 106).

Factors affecting selectivity of blastocidin S have been examined and two mechanisms of resistance have been found (45). *Pellicularia sasakii,* a species naturally resistant to the blasticidin S, possesses a protein synthesizing system which in either whole cells or in cell-free systems is highly tolerant to the antibiotic. A strain of *P. oryzae* highly resistant to blastocidin S was selected. Tolerance in this organism is probably due to impermeability because protein synthesis in intact cells is unaffected by 1000 μg/ml of antibiotic, whereas in cell-free systems it is inhibited by 1 μg/ml. It would be interesting to know whether or not these blasticidin S-resistant organisms exhibit parallel resistance to cycloheximide and conversely, whether or not cycloheximide-resistant strains of *S. cerevisiae* (102) and *N. crassa* (42, 76) exhibit parallel resistance to blasticidin S.

Low concentrations of blasticidin S effectively inhibit tobacco mosaic virus multiplication in leaves of bean and tobacco plants when applied at the time of inoculation (38). Even though normal plant protein synthesis is inhibited, detrimental effects on the virus are much greater than on the host. Apparently the antibiotic permits synthesis of the small amount of protein and nucleic acid required for cell maintenance but largely prevents the increases necessary for rapid virus synthesis. Blastocidin S is also highly effective in preventing the increase in activities of certain enzymes which normally follows wounding of sweet potato tissue (46, 49).

BOTRAN (2,6-DICHLORO-4-NITROANILINE)

Among the organic fungicides, inhibition of protein synthesis has been implicated in the toxic action of 2,6-dichloro-4-nitroaniline (DCNA). The fungicide at 1 μg/ml prevents an increase in protein in growing hyphae of *Rhizopus arrhizus* but permits nearly normal increases in DNA and RNA (97). DCNA inhibits incorporation of leucine into protein by 36 per cent, an effect which seems rather moderate to account for complete inhibition of growth. Definite assignment of the toxic action of the compound to a site directly in the pathway to protein synthesis will require further studies in cell-free systems. An isolate of *R. arrhizus* tolerated 1000 μg/ml of DCNA (97); therefore, resistance to the toxicant must not be too difficult to achieve. These findings suggest a specific site of action unless the toxicant is extremely prone to metabolic destruction within the resistant strain or is readily excluded from it. Although DCNA is not toxic to mature tobacco plants it is toxic to young seedlings and to cells in tissue cultures (55). Inhibition of protein synthesis was suggested as the basis of toxicity. The bases of resistance of mature tobacco plants and of certain strains of *R. arrhizus* to DCNA seem worthy of further investigation.

NUCLEIC ACID SYNTHESIS

Variations among organisms in the multitude of enzymes involved in nucleic acid biosynthesis offer some promising possibilities for exploitation in selective toxicity. Certain cases of these variations have been revealed in studies with purine and pyrimidine analogues (7, 25, 39). Although the synthesis of nucleotides from simple precursors involves several enzymes and intermediates, it is not uncommon for organisms to convert an exogenously supplied purine or pyrimidine to a nucleotide in a single reaction that is not part of the main pathway. By way of example, uracil can be converted to uridine-5-phosphate by the enzyme uridylic acid pyrophosphorylase (7). The same enzymes that upgrade the natural purines and pyrimidines to nucleosides or nucleotides also upgrade purine or pyrimidine analogues to nucleoside or nucleotide analogues. An organism possessing the enzyme necessary to upgrade a particular analogue in this manner is often adversely affected by the analogue, whereas one which lacks this enzyme is not (25, 39). This particular type of variation in enzyme capacity constitutes one of the important mechanisms responsible for the selective toxicity of purine and pyrimidine analogues. In a general way, the analogues also tend to inhibit selectively cells undergoing the most rapid multiplication (39). Thus, the rapidly dividing cells of an invading fungal pathogen may be much more adversely affected by such compounds than the relatively quiescent cells of the host.

Once conversion of a purine or pyrimidine analogue to a nucleotide has taken place, metabolism may be affected in a variety of ways. The nucleotide analogue may inhibit some step in the synthesis of nucleotides directly, or it may exert feedback control of the synthesis much in the same manner as an excess of a normal intermediate. The fraudulent nucleotides may be incorporated into nucleic acids yielding defective products. Conceivably, the nucleotide analogues may also be incorporated into coenzymes in place of the normal nucleotides, or interfere otherwise with the utilization of normal metabolites for coenzyme synthesis (25, 39).

6-Azauracil

The pyrimidine analogue, 6-azauracil (AZU) has proven experimentally effective in systemic control of powdery mildew and scab of cucumber (21, 22). A study of the mechanism of action (21) revealed that the scab fungus, *Cladosporium cucumerinum,* converts AZU to 6-azauridine (AZUR) which is then converted to 6-azauridine-5-monophosphate (AZUMP). Metabolism of orotic acid by the fungus is strongly inhibited when cells are treated with either AZU or AZUR, presumably because the toxicant inhibits orotidine-5-monophosphate (OMP) decarboxylase. However, when studies were extended to cell-free extracts, neither AZU nor AZUR inhibited OMP decarboxylase activity; only AZUMP effectively inhibited the enzyme. It seems evident, therefore, that AZU and AZUR are only precursors to the toxicant responsible for inhibition of fungal growth. Upgrading of AZU, at least to the level of AZUMP, appears to be the minimum conversion necessary for toxicity.

The decisive importance of metabolic conversion of AZU for toxicity was demonstrated in studies utilizing strains of *C. cucumerinum* resistant to the inhibitor. Among eight strains selected for high tolerance to AZU, four were tolerant to AZUR and four were sensitive. The AZUR-sensitive strains were unable to convert AZU to AZUR. Inability to accomplish this conversion is obviously the basis of resistance of these strains to AZU. Enzymes necessary to convert AZUR to AZUMP were present in the wild type and in all eight AZU resistant strains. It seems unlikely, therefore, that tolerance of four of the strains to AZUR is based on the inability to convert AZUR to AZUMP. Resistance in these is due to some mechanism not yet defined (21).

Ideally, selectivity of compounds like AZU is best when only the pathogen is able to convert the compound to a biologically active form. However, cucumber plants are relatively insensitive to AZU, even though they readily convert the analogue to AZUR (21). Two factors were suggested to operate in favor of the host over the pathogen in this case. A more rapid rate of RNA synthesis in the fungal cells renders them more prone to injury by the inhibitor than host cells, and high OMP decarboxylase activity in host cells make them less vulnerable to the toxicants than cells of the pathogen.

Benzimidazoles

Much interest has developed in benzimidazoles following the recent introduction of the two new systemic fungicides, methyl-1-(butylcarbamoyl)-2-benzimidazolecarbamate (Benlate) (24, 28) and 2-(4′-thiazolyl)-benzimidazole (Thiabendazole) (91). Little is known at present regarding the mode of action of either of these compounds. One group reports that thiabendazole affects transamination reactions and protein synthesis in fungi and that toxicity of the compound is antagonized by pyridoxine, biotin, and guanine (91).

Benzimidazole has long been recognized as a purine analogue because its toxicity to *S. cerevisiae* is largely reversed by adenine and guanine (105). Therefore, conversion of benzimidazole to a nucleotide in biological systems would not be unexpected in view of the results obtained with other purine analogues (39). In wheat leaves, benzimidazole nucleoside has been identified as a product of benzimidazole metabolism (50). The nucleotide may be incorporated into nucleic acids because labeled benzimidazole is released by acid hydrolysis from the nucleic acid fraction of leaves fed benzimidazole-2-^{14}C. Benzimidazole nucleotide may also be incorporated into a coenzyme in wheat plants. Enzyme preparations made from wheat embryos catalyze the substitution of benzimidazole for the nicotinamide moiety of NAD (51) to form benzimidazole adenine dinucleotide. Benzimidazole delays senescence in detached wheat leaves and thus appears to have a beneficial effect.

However, the appearance of conversion products of the compound in vital cellular components indicate that it may be detrimental, at least in growing cells.

Metabolism of benzimidazole to the level of a nucleotide in wheat plants suggests the possibility that Benlate and Thiabendazole may be similarly metabolized in fungal cells. Assuming that a nucleotide derivative is the active fungitoxicant, the situation would be analogous to the one described for 6-azauracil (21). Although there is no direct evidence to indicate that these compounds undergo such conversions in fungal cells, the lag period associated with Benlate toxicity is consistent with the idea that is upgraded to a toxic derivative. Benlate breaks down rapidly in aqueous solution yielding 2-benzimidazole carbamic acid methyl ester (BCM)—a compound that is as toxic to *N. crassa* and *Rhizoctonia solani* as Benlate (14). It seems likely that Benlate acts fungicidally via a BCM intermediate. A period of exposure of 6 to 8 hr to either Benlate or BCM is required before growth (dry weight increase) of *N. crassa* is inhibited. This lag period may be the time required for the formation of a toxic conversion product such as a BCM nucleotide. A derivative of this type would not necessarily interfere with nucleic acid biosynthesis, but might affect any of a variety of reactions in which the corresponding normal nucleotides are involved.

Benzimidazole as such is a poor fungicide, but simple addition at position two of a carbamate ester group produces the highly effective fungicide, BCM. Possibly the carbamate ester group permits binding of the benzimidazole moiety to a protein via a carbamylation reaction comparable to that which occurs between carbamate insecticides and cholinesterase (70). The delayed effect of BCM on fungal growth also suggests the possibility that this carbamate derivative may be acting as an inhibitor of cell division. Future research may well reveal a mode of action for BCM and Benlate differing from that of any fungicide presently used for control of plant diseases.

One benzimidazole derivative is known to affect nucleic acid synthesis. The compound, 2-mercapto-1-(β-4-pyridethyl) benzimidazole (MPB) strongly inhibits RNA synthesis in a number of mammalian cell types (9, 92). Toxicity of the compound is readily reversed by changing the incubation medium (92) but is not reversed by adenosine, guanosine, or dimethylbenzimidazole, a component of vitamin B_{12} (9). Onset of inhibition of RNA synthesis following addition of MPB is rapid, and therefore it is unlikely that any metabolic conversion of the compound is necessary for toxicity. Conversion of the benzimidazole to a nucleotide seems unlikely because the nitrogen atom to which ribose phosphate would normally be attached is already substituted with another group.

There is general agreement that acidic benzimidazole herbicides and insecticides, such as the 2-trifluoromethyl derivatives, uncouple respiratory chain phosphorylation (8, 48). Only the anionic forms resulting from dissociation of the hydrogen from the ring nitrogen atom are effective uncoupling agents. Substitutions which produce compounds with a pK low enough for high dissociation in the biological pH range yield effective toxicants. Structure-activity relationships suggest that Benlate and Thiabendazole are unlikely to be uncoupling agents, although this possibility should not be overlooked in a search for their mode of action.

Glyodin

Imidazole derivatives are likely candidates as inhibitors of nucleic acid synthesis because of structural similarity to purine nucleotides and their precursors. West & Wolf (99) demonstrated that toxicity of the imidazole fungicide glyodin (2-heptadecyl-2-imidazoline) is annulled by guanine and xanthine and suggested that the compound is a competitive inhibitor of purine biosynthesis. Although this may be the mechanism of action of low concentrations of the toxicant in some organisms, other effects may also be produced by such surface-active compounds. Kerridge (52) studied the action of glyodin in *S. carlsbergensis* and found it to inhibit both nucleic acid and protein synthesis. Effects of low concentrations of toxicant were partially relieved by guanine. These results were somewhat complicated, however, because glyodin also induced membrane damage. In view of the fact that cationic surface-active agents generally are toxic, it seems doubtful whether the imidazole group of glyodin is sufficiently specific to permit a mechanism of action distinctly different from that of other cationic surface-active agents. These compounds probably interfere with various cell structures and functions by competing with natural cations for anionic sites.

Phytoactin

Phytoactin is a polypeptide antibiotic (109) that has been used to a limited extent in plant disease control. Studies of the effect of the antibiotic on several aspects of metabolism in *S. pastorianus* revealed that RNA synthesis is most strongly affected (60). Incorporation of ^{14}C from glucose into nucleotide precursors of RNA is not appreciably affected by toxic doses of the antibiotic, but incorporation into RNA is inhibited 80 to 90 per cent. Whether this inhibition results from a direct action of the antibiotic in the pathway to RNA synthesis, or from an indirect effect of action at another site remains to be determined. The pattern of inhibition suggests a mode of action resembling that of the actinomycins which bind to DNA and inhibit DNA-dependent RNA polymerase.

Chloroneb

The fungicide, 1,4-dichloro-2,5-dimethoxybenzene (chloroneb), is highly toxic to only a few fungi including *R. solani* and *Sclerotium rolfsii.* The mode of action of the compound has been investigated in *R. solani* (40). Respiration of the fungus is not affected even after exposure to the toxicant for 24 hr. Synthesis of protein and RNA is only moderately suppressed but DNA synthesis is strongly inhibited. Whether this inhibition results from a direct action of the inhibitor in the pathway to DNA synthesis or from indirect effects of inhibition at other sites has not been determined. Differences in sensitivity of fungi to chloroneb do not appear to be due to differences in amount of toxicant taken up. Neither *N. crassa* nor *S. pastorianus* is very sensitive to chloroneb, but these organisms take up as much of the compound as the sensitive fungus, *R. solani*. Furthermore, ability to metabolize the toxicant is apparently not an important factor in determining sensitivity to the toxicant. The basis of the highly selective action of the relatively simple chloroneb molecule remains an interesting question.

LITERATURE CITED

1. Ashwell, M. A., Work, T. S. Contrasting effects of cycloheximide on mitochondrial protein synthesis *'in vivo'* and *'in vitro.' Biochem. Biophys. Res. Commun.,* **32,** 1006–12 (1968)
2. Baliga, B. S., Pronczuk, A. W., Munro, H. N. Site of action of cycloheximide on protein synthesis by liver polysomes. *Federation Proc.,* **27,** 766 (1968)
3. Beattie, D. S., Basford, R. E., Koritz, S. B. The inner membrane as the site of *in vitro* incorporation of l-[^{14}C]-leucine into mitochondrial protein. *Biochemistry,* **6,** 3099–106 (1967)
4. Bennett, L. L., Jr., Smithers, D., Ward, C. T. Inhibition of DNA synthesis in mammalian cells by Actidone. *Biochim. Biophys. Acta,* **87,** 60–69 (1964)
5. Bennett, L. L., Jr., Ward, V. L., Brockman, R. W. Inhibition of protein synthesis *in vitro* by cycloheximide and related glutarimide antibiotics. *Biochim. Biophys. Acta,* **103,** 478–85 (1965)
6. Bretthauer, R. K., Marcus, L., Chaloupka, J., Halvorson, H. O., Bock, R. M. Amino acid incorporation into protein by cell-free extracts of yeast. *Biochemistry,* **2,** 1079–84 (1963)
7. Brockman, R. W., Anderson, E. P. Pyrimidine analogues. In *Metabolic Inhibitors,* **I,** 239–85 (Hochster, R. M., Quastel, J. H., Eds., Academic Press, New York, 669 pp., 1963)
8. Büchel, K. H., Korte, F., Uncoupling of the oxidative phosphorylation in mitochondria by NH-acidic benzimidazoles. *Angew. Chem. (Intern. Ed. Engl.),* **4,** 788–89 (1965)
9. Bucknall, R. A., Carter, S. B. A reversible inhibitor of nucleic acid synthesis. *Nature,* **213,** 1099–101 (1967)
10. Calendar, R., Berg, P., The catalytic properties of tyrosyl ribonucleic acid synthetases from *Escherichia coli* and *Bacillus subtilis. Biochemistry,* **5,** 1690–95 (1966)
11. Clark, J. M., Jr., Chang, A. Y. Inhibitors of the transfer of amino acids from amino acyl soluble ribonucleic acid to protein. *J. Biol. Chem.,* **240,** 4734–39 (1965)
12. Clark-Walker, G. D., Linnane, A. W. *In vivo* differentiation of yeast cytoplasmic and mitochondrial protein synthesis with antibiotics. *Biochem. Biophys. Res. Commun.,* **25,** 8–13 (1966)
13. Clark-Walker, G. D., Linnane, A. W. The biogenesis of mitochondria in *Saccharomyces cerevisiae.* A comparison between cytoplasmic respiratory-deficient mutant yeast and chloramphenicol inhibited wild type cells. *J. Cell Biol.,* **34,** 1–14 (1967)
14. Clemons, G. P., Sisler, H. D. Formation of a fungitoxic derivative from Benlate. *Phytopathology,* **59,** 705–6 (1969)
15. Colombo, B., Felicetti, L., Baglioni, C. Inhibition of protein synthesis by cycloheximide in rabbit reticulocytes. *Biochem. Biophys. Res. Commun.,* **18,** 389–95 (1965)
16. Cooper, D., Banthorpe, D. U., Wilkie, D. Modified ribosomes conferring resistance to cycloheximide in mutants of *Saccharomyces cerevisiae. J. Mol. Biol.,* **26,** 347–50 (1967)
17. Coutsogeorgopoulos, C. Inhibitors of the reaction between puromycin and polylysyl-RNA in presence of ribosomes. *Biochem. Biophys. Res. Commun.,* **27,** 46–52 (1967)
18. Cummins, J. E., Brewer, E. N.,

Rusch, H. P. The effect of Actidione on mitosis in the slime mold *Physarum polycephalum*. *J. Cell Biol.*, **27**, 337–41 (1965)

19. Cummins, J. E., Rusch, H. P. Limited DNA synthesis in the absence of protein synthesis in *Physarum polycephalum*. *J. Cell Biol.*, **31**, 577–83 (1966)
20. Dee, J. Genetic analysis of Actidione — resistant mutants in the Myxomycete, *Physarum polycephalum*, Schw. *Genet. Res.*, **8**, 101–10 (1966)
21. Dekker, J. The development of resistance in *Cladosporium cucumerinum* against 6-azauracil, a chemotherapeutant of cucumber scab, and its relation to biosynthesis of RNA-precursors. *Neth. J. Plant Path.*, **74**, Suppl. 1, 127–36 (1968)
22. Dekker, J., Oort, A. J. P. Mode of action of 6-azauracil against powdery mildew. *Phytopathology*. **54**, 815–18 (1964)
23. deKloet, S. R. Ribonucleic acid synthesis in yeast. The effect of cycloheximide on the synthesis of ribonucleic acid in *Saccharomyces carlsbergensis*. *Biochem. J.*, **99**, 566–81 (1966)
24. Delp, C. J., Klopping, H. L. Disease control with Du Pont fungicide 1991. *Abstr. of Papers, First Intern. Congr. Plant Pathol.*, p. 44, London (1968)
25. Dods, R. F. Drug design in chemotherapy: The azanucleotides and related compounds. *Bull. Inst. Cellular Biol., Univ. of Conn.*, **9**, 1–10 (1967)
26. Ennis, H. L. Synthesis of ribonucleic acid in L cells during inhibition of protein synthesis by cycloheximide. *Mol. Pharmacol.*, **2**, 501–619 (1966)
27. Ennis, H. L., Lubin, M. Cycloheximide: Aspects of inhibition of protein synthesis in mammalian cells. *Science*, **146**, 1474–76 (1964)
28. Erwin, D. C., Mee, H., Sims, J. J. The systemic effect of 1-(butylcarbomyl)-2-benzimidazole carbamic acid, methyl ester, on *Verticillium* wilt of cotton. *Phytopathology*, **58**, 528–29 (1968)
29. Felber, I. M., Hamner, C. L. Control of mildew on bean plants by means of an antibiotic. *Botan. Gaz.*, **110**, 324–25 (1948)
30. Felicetti, L., Colombo, B., Baglioni, C. Inhibition of protein synthesis in reticulocytes by antibiotics. II. The site of action of cycloheximide, streptovitacin A, and pactamycin. *Biochim. Biophys. Acta*, **119**, 120–29 (1966)
31. Ford, J. H., Klomparens, W., Hamner, C. L. Cycloheximide (Actidione) and its agricultural uses. *Plant Disease Reptr.*, **42**, 680–95 (1958)
32. Godchaux, W., Adamson, S. D., Herbert, E. Effect of cycloheximide on polyribosome function in reticulocytes. *J. Mol. Biol.*, **27**, 57–72 (1967)
33. Grivell, L. A. Amino acid incorporation by mitochondria isolated, essentially free of micro-organisms, from *Saccharomyces carlsbergensis*. *Biochem. J.*, **105**, 44c–46c (1967)
34. Grollman, A. P. Structural basis for inhibition of protein synthesis by emetine and cycloheximide based on an analogy between ipecac alkaloids and glutarimide antibiotics. *Proc. Natl. Acad. Sci. U.S.*, **56**, 1867–74 (1966)
35. Grollman, A. P. Inhibitors of protein biosynthesis. II. Mode of action of anisomycin. *J. Biol. Chem.*, **242**, 3226–33 (1967)
33. Haidle, C. W., Storck, R. Inhibition by cycloheximide of protein and RNA synthesis in *Mucor rouxii*. *Biochem. Biophys. Res. Commun.*, **22**, 175–80 (1966)
37. Helton, A. W., Williams, R. E. Control of aerial crown-gall disease in cherry trees with spray-applied systemic fungicides. *Phytopathology*, **58**, 782–87 (1968)
38. Hirai, T., Hirashima, A., Itoh, T., Takahashi, T., Shimomura, T., Hayashi, Y. Inhibitory effect of blasticidin S on tobacco mosaic virus multiplication. *Phytopathology*, **56**, 1236–40 (1966)
39. Hitchings, G. H., Elion, G. Purine analogues. In *Metabolic Inhibitors*, **I**, 215–37 (See ref. 7)
40. Hock, W. K., Sisler, H. D. Specificity and mechanism of antifungal action of chloroneb. *Phytopathology*, **59**, 627–32 (1969)
41. Howe, R., Moore, R. H. Acetylation of cycloheximide by *Cunninghamella blakesleeana*. *Experientia*, **24**, 904 (1968)
42. Hsu, K. S. The genetic basis of Actidione resistance in *Neurospora*.

J. Gen. Microbiol., **32**, 341–47 (1963)

43. Huang, M., Biggs, D. R., Clark-Walker, G. D., Linnane, A. W. Chloramphenicol inhibition of the formation of particulate mitochondrial enzymes of *Saccharomyces cerevisiae*. *Biochim. Biophys. Acta*, **114**, 434–36 (1966)
44. Huang, K. T., Misato, T., Asuyama, H. Effect of blasticidin S on protein synthesis of *Piricularia oryzae*. *J. Antibiotics (Tokyo), Ser. A*, **17**, 65–70 (1964)
45. Huang, K. T., Misato, T., Asuyama, H. Selective toxicity of blasticidin S to *Piricularia oryzae* and *Pellicularia sasakii*. *J. Antibiotics (Tokyo), Ser. A*, **17**, 71–74 (1964)
46. Hyodo, H., Uritani, I. The inhibitory effect of some antibiotics on increase in *o*-diphenol oxidase activity during incubation of sliced sweet potato tissue. *Agr. Biol. Chem.(Tokyo)*,**30**, 1083–86 (1966)
47. Jacoby, G. A., Gorini, L. The effect of streptomycin and other aminoglycoside antibiotics on protein synthesis. In *Antibiotics I. Mechanism of Action*, 726–47 (Gottlieb, D., Shaw, P. D., Eds., Springer Verlag, New York, 785 pp., 1967)
48. Jones, O. T. G., Watson, W. A. Properties of substituted 2-trifluoromethylbenzimidazoles as uncouplers of oxidative phosphorylation. *Biochem. J.*, **102**, 564–73 (1967)
49. Kanazawa, Y., Shichi, H., Uritani, I. Biosynthesis of peroxidases in sliced or black rot-infected sweet potato roots. *Agr. Biol. Chem. (Tokyo)*, **29**, 840–47 (1965)
50. Kapoor, M., Waygood, E. R. Metabolism of benzimidazole in wheat. I. Formation of benzimidazole nucleotide. *Can. J. Biochem.*, **43**, 153–64 (1965)
51. Kapoor, M., Waygood, E. R. Metabolism of benzimidazole in wheat. II. Formation of benzimidazole adenine dinucleotide and its products. *Can. J. Biochem.*, **43**, 165–71 (1965)
52. Kerridge, D. The effect of Actidione and other antifungal agents on nucleic acid and protein synthesis in *Saccharomyces carlsbergensis*. *J. Gen. Microbiol.*, **19**, 497–506 (1958)
53. Lamb, A. J., Clark-Walker, G. D., Linnane, A. W. The biogenesis of mitochondria 4. The differentiation of mitochondrial and cytoplasmic protein synthesizing systems *in vitro* by antibiotics. *Biochim. Biophys. Acta*, **161**, 415–27 (1968)
54. Lemin, A. J., Magee, W. E. Degradation of cycloheximide derivatives in plants. *Plant Disease Reptr.*, **41**, 447–48 (1957)
55. Lewis, J., Weber, D. J., Venketeswaren, S. Mode of action of 2,6-dichloro-4-nitroaniline in plant tissue culture. *Phytopathology*, **59**, 93–97 (1969)
56. Lin, S., Mosteller, R. D., Hardesty, B. The mechanism of sodium fluoride and cycloheximide inhibition of hemoglobin biosynthesis in the cell-free reticulocyte system. *J. Mol. Biol.*, **21**, 51–69 (1966)
57. Linnane, A. W., Lamb, A. J., Christodoulou, C., Lukins, H. B. The biogenesis of mitochondria. VI. Biochemical basis of the resistance of *Saccharomyces cerevisiae* toward antibiotics which specifically inhibit mitochondrial protein synthesis. *Proc. Natl. Acad. Sci. U.S.*, **59**, 1288–93 (1968)
58. Linnane, A. W., Saunders, G. W., Gingold, E. B., Lukins, H. B. The biogenesis of mitochondria, V. Cytoplasmic inheritance of erythromycin resistance in *Saccharomyces cereviseae*. *Proc. Natl. Acad. Sci. U.S.*, **59**, 903–10 (1968)
59. Loeb, J. N. Hubby, B. G. Amino acid incorporation by isolated mitochondria in the presence of cycloheximide. *Biochim. Biophys. Acta*, **166**, 745–48 (1968)
60. Lynch, J. P., Sisler, H.D. Mechanism of action of phytoactin in *Saccharomyces pastorianus*. *Phytopathology*, **57**, 367–73 (1967)
61. Mahler, H. R., Perlman, P., Henson, C., Weber, C. Selective effects of chloramphenicol, cycloheximide, and nalidixic acid on the biosynthesis of respiratory enzymes in yeast. *Biochem. Biophys. Res. Commun.*, **31**, 474–80 (1968)
62. Marchant, R., Smith, D. G. The effect of chloramphenicol on growth and mitochondrial structure of *Pythium ultimum*. *J. Gen. Microbiol.*, **50**, 391–97 (1968)
63. Marcus, A., Feeley, J. Ribosome activation and polysome formation *In vitro*: Requirements for ATP.

Proc. Natl. Acad. Sci. U.S., **56**, 1770–77 (1966)

64. McLean, I. W., Schwab, J. L., Hillegas, A. B., Schlingman, A. S. Susceptibility of micro-organisms to chloramphenicol (chloromycetin). *J. Clin. Invest.*, **28**, 953–63 (1949)
65. Misato, T. Blasticidin S. In *Antibiotics I. Mechanism of Action*, 434–39 (See ref. 47)
66. Morris, I. Inhibition of protein synthesis by cycloheximide (Actidione) in *Chlorella*. *Nature*, **211**, 1190–92 (1966)
67. Munier, R., Cohen, G. N., Incorporation d' analogues structuraux d' aminoacides dans les protéines bactériennes au cours de leur synthèse *in vivo*. *Biochim. Biophys. Acta*, **31**, 378–91 (1959)
68. Nathans, D. Puromycin. In *Antibiotics I. Mechanism of Action*, 259–77 (See ref. 47)
69. Nooden, L. D. Studies on the role of RNA synthesis in auxin induction of cell enlargement. *Plant Physiol.*, **43**, 140–50 (1968)
70. O'Brien, R. D., *Insecticides, Action And Metabolism* (Academic Press, New York, 332 pp., 1967)
71. Rao, S. S., Grollman, A .P. Cycloheximide resistance in yeast: A property of the 60 S ribosomal subunit. *Biochem. Biophys. Res. Commun.*, **29**, 696–704 (1967)
72. Reich, E., Cerami, A., Ward, D. C. Actinomycin. In *Antibiotics I. Mechanism of Action*, 714–25 (See ref. 47)
73. Rieske, J. S. Antimycin A. In *Antibiotics I. Mechanism of Action*, 542–84 (See ref. 47)
74. Roodyn, D. B., Wilkie, D., *The Biogenesis of Mitochondria* (Methuen, London, 123 pp., 1968)
75. Ross, C. Influence of cycloheximide (Actidione) upon pyrimidine nucleotide metabolism and RNA synthesis in cocklebur leaf discs. *Biochim. Biophys. Acta*, **166**, 40–47 (1968)
76. Rothschild, H., Suskind, S. R. Ribosomal proteins of cycloheximide — resistant mutants of *Neurospora crassa*. *Bacteriol. Proc. (Abstr.)*, 129 (1968)
77. Sachsenmaier, W., Fournier, D. V., Gürtler, K. F. Periodic thymidine kinase production in synchronous plasmodia of *Physarum polycephalum*: Inhibition by actinomycin and Actidione. *Biochem. Biophys. Res. Commun.*, **27**, 655–60 (1967)
78. Shepherd, C. J. Inhibition of protein and nucleic acid synthesis in *Aspergillus nidulans*. *J. Gen. Microbiol.*, **18**, IV (1958)
79. Shive, W., Skinner, C. G. Amino acid analogues. In *Metabolic Inhibitors I*, 1–73 (See ref. 7)
80. Siegel, M. R., Sisler, H. D. Inhibition of protein synthesis *in vitro* by cycloheximide. *Nature*, **200**, 675–76 (1963)
81. Siegel, M. R., Sisler, H. D. Site of action of cycloheximide in cells of *Saccharomyces pastorianus*. I. Effect of the antibiotic on cellular metabolism. *Biochim. Biophys. Acta*, **87**, 70–82 (1964)
82. Siegel, M. R., Sisler, H. D. Site of action of cycloheximide in cells of *Saccharomyces pastorianus*. II. The nature of inhibition of protein synthesis in a cell-free system. *Biochim. Biophys. Acta*, **87**, 83–89 (1964)
83. Siegel, M. R., Sisler, H. D. Site of action of cycloheximide in cells of *Saccharomyces pastorianus*. III. Further studies on the mechanism of action and the mechanism of resistance in *Saccharomyces* species. *Biochim. Biophys. Acta*, **103**, 558–67 (1965)
84. Siegel, M. R., Sisler, H. D., Johnson, F. Relation of structure to fungitoxicity of cycloheximide and related glutarimide derivatives. *Biochem. Pharmacol.*, **15**, 1213–23 (1966)
85. Sisler, H. D., Siegel, M. R. Cycloheximide and other glutarimide antibiotics. In *Antibiotics I. Mechanism of Action*, 283–307 (See ref. 47)
86. Sisler, H. D., Siegel, M. R., Ragsdale, N. N. Factors regulating toxicity of cycloheximide derivatives. *Phytopathology*, **57**, 1191–96 (1967)
87. Smith, D. G., Marchant, R. Chloramphenicol inhibition of *Pythium ultimum* and *Rhodotorula glutinis*. *Arch. Mikrobiol.*, **60**, 262–74 (1968)
88. So, A. G., Davie, E. W. The incorporation of amino acids into protein in a cell-free system from yeast. *Biochemistry*, **2**, 132–36 (1963)
89. Soeiro, R., Vaughan, M. H., Darnell,

J. E. The effect of puromycin on intranuclear steps in ribosome biosynthesis. *J. Cell Biol.*, **36,** 91–101 (1968)
90. Stanners, C. P. The effect of cycloheximide on polyribosomes from hamster cells. *Biochem. Biophys. Res. Commun.*, **24,** 758–64 (1966)
91. Staron, T., Allard, C., Darpoux, H., Grabowski, H., Kollman, A. Persistance du thiabendazole dans les plantes. Propriétes systemiques des ses sels et quelques données nouvelles sur son mode d' action. *Phytiat. Phytopharm.*, **15,** 129–33 (1966)
92. Summers, W. P., Mueller, G. C. A study of factors regulating RNA synthesis In Hela cells using MPB, a reversible inhibitor of RNA synthesis. *Biochem. Biophys. Res. Commun.*, **30,** 350–55 (1968)
93. Sussman, M., Inhibition by Actidione of protein synthesis and UDP-GAL polysaccharide transferase accumulation in *Dictyostelium discoideum. Biochem. Biophys. Res. Commun.*, **18,** 763–67 (1965)
94. Trakatellis, A. C., Montjar, M., Axelrod, A. E. Effect of cycloheximide on polysomes and protein synthesis in the mouse liver. *Biochemistry,* **4,** 2065–71 (1965)
95. Van Etten, J. L. Protein synthesis during fungal spore germination 1. Characteristics of an *in vitro* phenylalanine incorporating system prepared from germinated spores of *Botryodiplodia theobromae. Arch. Biochem.*, **125,** 13–21 (1968)
96. Warner, J. R., Girard, M., Latham, H., Darnell, J. E., Ribosome formation in Hela cells in the absence of protein synthesis. *J. Mol. Biol.*, **19,** 373–82 (1966)
97. Weber, D. J., Ogawa, J. M. The mode of action of 2, 6-dichloro-4-nitro-aniline in *Rhizopus arrhizus. Phytopathology,* **55,** 159–65 (1955)
98. Wescott, E. W., Sisler, H. D. Uptake of cycloheximide by a sensitive and a resistant yeast. *Phytopathology,* **54,** 1261–64 (1964)
99. West, B., Wolf, F. T. The mechanism of action of the fungicide, 2-heptadecyl-2-imidazoline. *J. Gen. Microbiol.*, **12,** 396–401 (1955)
100. Wettstein, F. O., Noll, H., Penman, S. Effect of cycloheximide on ribosomal aggregates engaged in protein synthesis *in vitro. Biochim. Biophys. Acta,* **87,** 525–28 (1964)
101. Widuczynski, I., Stoppani, A. O. M. Action of cycloheximide on amino acid metabolism in *Saccharomyces ellipsoideus. Biochim. Biophys. Acta,* **104,** 413–26 (1965)
102. Wilkie, D., Lee, B. K. Genetic analysis of Actidione resistance in *Saccharomyces cerevisiae. Genet. Res.*, **6,** 130–38 (1965)
103. Williamson, A. R., Schweet, R. Role of the genetic message in polyribosome function. *J. Mol. Biol.*, **11,** 358–72 (1965)
104. Wintersberger, E. Proteinsynthese in isolierten Hefe-Mitochondrien. *Biochem. Z.*, **341,** 409–19 (1965)
105. Woolley, D. W. Some biological effects produced by benzimidazole and their reversal by purines. *J. Biol. Chem.*, **152,** 225–32 (1944)
106. Yamaguchi, H., Tanaka, N. Inhibition of protein synthesis by blasticidin S. II. Studies on the site of action in *E. coli* polypeptide synthesizing enzymes. *J. Biochem. (Tokyo),* **60,** 632–42 (1966)
107. Yarmolinski, M. B., de la Haba, G. L. Inhibition by puromycin of amino acid incorporation into protein. *Proc. Natl. Acad. Sci. U.S.*, **45,** 1721–29 (1959)
108. Young, C. W. Inhibitory effects of acetoxycycloheximide, puromycin, and pactamycin upon synthesis of protein and DNA in asynchronous populations of Hela cells. *Mol. Pharmacol.*, **2,** 50–55 (1966)
109. Ziffer, J. S., Ishihara, S. J., Cairney, T. J., Chow, A. W. Phytoactin and phytostreptin, two new antibiotics for disease control. *Phytopathology,* **47,** 539 (1957)
110. Zucker, M., El-Zayat, M. M. The effect of cycloheximide on the resistance of potato tuber discs to invasion by a fluorescent *Pseudomonas* sp. *Phytopathology,* **58,** 339–44 (1968)

Copyright 1969. All rights reserved

HEAT THERAPY OF VIRUS DISEASES OF PERENNIAL PLANTS

By George Nyland, Department of Plant Pathology
AND
A. C. Goheen, Agricultural Research Service,
U. S. Department of Agriculture
University of California, Davis, California

Introduction

Clean stocks of perennial crop plants are important for optimum production, and are even more essential for physiological studies since diseased plants differ greatly from healthy plants in their physiology. For example, infection with the grapevine leafroll virus causes a deficiency of cation uptake in leaves. As a result, early investigators of leafroll, considering it a disorder in mineral nutrition, were baffled by the lack of response to fertilizers. Cook & Goheen (38) showed that the mineral-deficiency symptoms were a response to a virus disease. Investigators need no longer work with perennial plants of unknown virus content since clean stocks can be obtained by heat therapy, alone or combined with meristem-tip culture.

Heat is an important therapeutic agent for treating diseases in plants. Over 100 years ago, Scots gardeners immersed bulbs in hot water before planting, thus being the first known to use heat for therapy of plants (209). The therapeutic benefits of heat used against a specific disease became widely known with work of Jensen (110). Wilbrink (204) suggested that Jensen's work may have prompted Sayer to try hot-water treatment of sugarcane infected with sereh disease. Although that treatment was not used extensively in treating sereh-diseased cane setts, it came into widespread use when sugarcane chlorotic streak (127) and, later, ratoon stunt (181) appeared.

Kunkel found that peach yellows was cured by either dry heat or hot-water treatment (112), and used both methods in curing 11 diseases of the yellows group, including aster yellows (115–117). No virus diseases other than those of the yellows group were successfully cured by heat until Kassanis (100) reported curing potato leafroll. If Blodgett (14) had chosen potato leafroll instead of potato mosaic for heat treatment, he would have antedated Kunkel by 13 years, and Kassanis by 26. His maximum treatment of 35° C for 4 months would most certainly have cured potato tubers of leafroll although it did not inactivate potato mosaic virus, probably potato virus X (PVX). Recent reports (42) of the association of mycoplasma organisms with diseases of the yellows group could mean that diseases spread by leafhoppers, and perhaps some others that also are easily hot-water-labile *in vivo,* have similar etiologic agents. This would mean that no viruses *sensu* Bawden (7) are hot-water-labile *in vivo* at 50 and 52° C for 20 to 30 min. We are not aware of any unequivocal evidence of the virus nature of any disease that is easily cured by hot-water treatments.

Several earlier reviews treat the subject of heat therapy of virus diseases or closely related subjects. Some cover heat therapy of plant virus diseases directly (13, 90, 102–105, 110, 125, 126, 139); others cover heat therapy of the virus diseases of a specific group of crop plants (79, 145). Still others are concerned with virus-disease control, and consider heat therapy an important tool for curing plants (72, 82, 91, 92, 170, 187). Finally, some reviews cover the general therapy of plant diseases, providing sections on heat therapy of plant viruses (57, 182). Related reviews treat the subject of the effect of temperature on plant disease and disease control (3, 4, 6, 206, 207), and the physiological responses of biological systems to temperature (59, 61, 96). Heat therapy is used frequently along with meristem culture (135, 159, 160). Heat therapy has also become a subject for sections in books on plant viruses (7, 24).

Heat therapy is in progress in several locations, and new issues of many journals contain references to successful treatment of additional viruses. According to Kristensen (110), approximately 1000 papers on heat therapy or related subjects appeared before 1966. The number of viruses successfully inactivated *in vivo* has steadily increased. Fifteen were reported by 1950, 75 by 1960, and 100 by 1966 (90). We report about 120 to the present. The space available does not permit listing all of the papers that have appeared on this subject or a review of heat treatment of viruses in seed.

We review the methods and results of heat treatment and factors affecting them and point out the important uses of heat therapy. We attempt to group viruses with similar heat reactions on the basis of the incomplete data available in the literature. We discuss thermal constants and the mechanism of inactivation, and suggest that the heat stability of plant viruses is indicated more meaningfully by half-life *in vivo* at 38° C (when it can be obtained) than by the thermal inactivation point. Such precise data might reveal more clearly the relationships and mechanisms involved in inactivation.

Methods Used in Heat Treatment

For asexually propagated plants that are completely infected with virus there are five methods reported to obtain virus-free explants: (*a*) nucellar embryony, which is useful in citrus (201); (*b*) chemotherapy (72); (*c*) meristem (135, 160) or tip culture (88); (*d*) leaf removal, as can be practiced only in the case of Abutilon mosaic virus, so far as is known (90, 109); and (*e*) heat therapy. The most important of the five methods listed is heat therapy. Heat is applied either by hot water in controlled-temperature tanks or by hot air in controlled-temperature chambers. To obtain virus-free explants from plants infected with viruses not completely labile with heat treatments alone, heat therapy is frequently combined with meristem and tip culture.

Hot water and moist air.—Hot water was used earlier than hot air in treatments, and is used extensively with sugarcane diseases. Early workers established the desirability of rapid circulation of water around the individual pieces of plant tissue (68). They also revealed the greater tolerance of host tissue to hot moist air treatments, in which heat is applied in water-saturated warm air (119, 181). Baker (3) suggests that some of the causes of damage to plants or plant parts treated in water at temperatures necessary for therapy are leaching, water-soaking, and asphyxiation of host tissues.

Hot-water baths and chambers for applying moist heat vary in size from laboratory models to equipment large enough to treat a ton or more of sugarcane at one time. Control of temperature and time is very important because of the small spread between the heat inactivation point of the virus and the maximum tolerance of the host.

Hot-air treatments.—Chambers used to treat infected plants in containers for days or weeks at 35 to 40° C have varied from boxes slightly larger than a dog kennel (79) to walk-in rooms capable of treating large numbers of plants (143). Chambers described in earlier reports (17, 51, 102, 112, 152) used natural light. Natural light is ideal if supplemented by artificial light during short days. Artificial light alone is also satisfactory if of sufficient intensity. Either glass or plastic is used for the tops and sides of heat chambers exposed to natural or artificial light. A sophisticated chamber permits indefinite treatment of some plant species at 38° C (65, 178, 186). The improvement in both equipment and techniques has permitted treatment of strawberry plants and chrysanthemums for up to 8 months (21, 132), and citrus for 3 years (188). Some plants grow even better in the heat chamber than in the greenhouse (76). Satisfactory heat-treatment chambers can be constructed at a modest cost (49).

Preconditioning increased the survival time of plants in the heat room and of plants or plant parts in hot water. Thus, exposing plants to 27 to 35° C for a week or two prior to exposure at 38° C increased survival time in the heat room (51, 145, 203). Also beneficial is a gradual increase in the temperature (16, 79). Factors favoring survival of plants or plant parts under heat treatment are high carbohydrate reserve, maturity of tissues, partial dehydration that accompanies storage, previous growth at high temperature, and low humidity (3, 65).

The temperature in the root zone and in the plant during treatment is usually lower than the air temperature. Lower root temperature favors survival of the plants and does not seem to interfere with therapy (47, 107, 131, 144, 152). For this reason, some prefer clay pots to metal or plastic pots (131, 153). The temperature may be 5.5° C lower in the root zone of strawberry plants in clay pots than in the air (131, 132) but in plastic pots there is less difference. The temperature in the tissue of the tops of the plants may average 6° C lower than the air temperature; the difference in temperature among individual plants may be 5° C (49). In hot-air treatments for short periods, the temperature at the center of scions placed in an oven at 75° C reached 50° C in 10 min and did not reach 55° C after 20 min (106).

Plant survival is increased by intermittent application of heat. Heating plants on alternate days did not inactivate peach yellows (112). Hamid & Locke (71), however, successfully treated eye-pieces of potato tubers infected with leafroll virus at 40 or 45° C for 2 hr alternating with 22 hr at room temperature for 8 weeks, or 2 weeks, respectively. The thermal-death-time of potato tuber eye-pieces at a constant temperature of 34, 37, or 40° C was less than that of the virus. Larson (118) found that chrysanthemum plants infected with virus B survived 4 weeks at a constant temperature of 38° C, but survived for 12 weeks at a daily temperature cycle of 38 or 40° C for up to 16 hr, and 20° C for 8 hr. He could not propagate tips from plants held constantly for 4 weeks at 38° C. The percentage of tips free of virus was high only when the high temperature was held for at least 16 hr daily. Citrus stubborn virus was not inactivated when exposed *in vivo* for 37 days at alternating temperatures of 44.5° C for 9 hr and 30° C for 15 hr (148).

Temperatures fluctuating between 35 and 43° C have been reported as more favorable for plant survival than constant temperature at 38° C (131, 132), but there is little information on the effect of intermittent treatments or fluctuating temperature on the efficiency of virus inactivation. Weil (202) found that the total time required to inactivate TMV *in vitro* was less with separate intermittent exposures to 90° C than with continuous exposure. This may also be true with *in vivo* treatments of some viruses. Hamid & Locke (71) with intermittent exposure inactivated potato leafroll in 4 days at 40° C, whereas with continuous exposure Rozendaal et al. (167) needed 12 days at the same temperature. Comparison of continuous and intermittent treatment of chrysanthemums may show a similar difference if this is expressed as the size of tips that can be taken (70, 118).

The survival of infected plants in the heat chamber is also affected by plant age, length of time since transplanting into containers, and seasonal effects. A few specific examples can be mentioned. Kunkel (112) noted that older trees survived better than younger trees. Plants well established in the containers survived better than those recently transplanted (16, 102, 144, 152, 198). Late summer or autumn treatment is preferred for grapes (64), potato tubers (175), and strawberries (17). Grapes treated for 60 days in autumn produced 3 to 5 times as many tips as grapes treated in spring. Fruit-tree species are best treated in late summer (145) if lateral buds are removed during or after treatment (144), or if dormant buds are placed in seedlings prior to treatment (93). If shoot tips produced during treatment are to be removed, however, Campbell (27) preferred to treat in the spring. April or May is the time preferred by some to treat chrysanthemums (111), but summer is also recommended (83).

Plants in the heat chamber usually show temporary damage or abnormalities. Some plants develop abnormal color and shape of leaves (186, 187) or fail to develop normal storage roots (74). Other temporary changes were devernalization and fasciation in chrysanthemums (21).

Special techniques in heat therapy.—Special techniques for therapy have been developed for some species, to be used with or without heat treatment (87, 93). Galzy (52, 53) heat-treated explants of grape growing on agar at 35° C, studied the effect of virus on rooting and established plants free of fanleaf by removing tip cuttings from the treated explants. The method should be applicable to other species and lend itself to various physiological studies. Rich (164) removed eyes, both before and after heat treatment, from potato tubers infected with PVX, and implanted them in tubers of a variety immune to PVX. In both cases he obtained plants free of virus. Mellor & Fitzpatrick (131) excised and rooted axillary buds that developed on the older parts of the crowns of strawberry plants after or during heat treatment.

Virus therapy is assisted by combining chemical treatments with heat. One treatment aimed at reducing the respiration rate with KCN (39), and another aimed at inhibiting the virus with malachite green (89) in the water in which plant parts were treated. Thiouracil, 2,4-D, and indoleacetic acid applied to carnation plants or to the medium used for meristem culture, gave beneficial results (159). Bolton (16) applied 150 ml of one per cent potassium permanganate solution each week to strawberry plants during treatment at 39° C or above to control root-disease organisms.

The success of heat therapy in air depends in most cases upon removal of a small to large portion of the treated plant after the prescribed exposure to high temperature. Combining heat treatment with meristem culture or tip removal increased the efficiency of therapy (10, 27, 45, 70, 149, 159). Among the easiest methods to apply, where they will work, are removing lateral buds after treatment and placing them in clean seedlings in a nursery, or simply dividing the stems of treated plants into cuttings and rooting them. Explants may or may not be free of virus, so they require indexing.

Heat Stability and Virus Grouping

Kassanis suggested (101) that heat treatment might be useful in grouping viruses when more is known about the mechanism of action. Our attempt to do so might be considered a beginning, though still premature. We have tabulated (Tables I, II) the known facts published on heat therapy and listed the last two of the four groups of Gibbs' (58) cryptograms to relate these facts to other properties of the viruses. We used the Kew list (128) for virus names. For each virus we list only a single reference to the minimum temperature-time treatment that was effective.

Viruses of the yellows group are placed together in the first section in Table I. A few viruses are included that resemble yellows but their vectors are not known to be leafhoppers. Most within this group are inactivated at 10 to 20 min at about 50° C. Some require longer times. Times reported are

TABLE I

PLANT VIRUSES INACTIVATED *in vivo* BY HEAT

Group and virus name	Cryptogram[a]	Host treated	Treatment			Authority
			Temp. °C	Time	Type	
YELLOWS TYPE						
Aster yellows	*/*:S, I/Au	Vinca	45	2.5 hr	W[b]	114
			42	2 wk	A	114
Bayberry yellows	*/*:S/*	Vinca	42	6 da	A	162
Cherry little cherry	*/*:S/Au	Prunus	37.5	3 wk	A	147
Chrysanthemum flower distortion	*/*:S/Au	Chrysanthemum	35	2 mo	A	22
Citrus yellow shoot		Citrus	50	40 min	A	119
Clover dwarf	*/*:S/Au	Vinca	40	10 da	A	199
Clover phyllody	*/*:S/Au	strawberry	41	14 da	A	155
		Vinca	40	10 da	A	199
Clover wound tumor	S/S:S, I/Au	sweet clover	40	12 da	A	171
			14	8 wk	A	171
Crimean yellows	*/*:S/Au	Vinca	40	10 da	A	199
Delphinium yellows	*/*:S/Au	Vinca	41	3 wk	A	155
Grapevine flavescence doree	*/*:S/Au	grape	30	3 da	W	32
Guatemala grass spikiness	*/*:S/*	Guatemala grass	52	20 min	W	137
Lucerene witches' broom		Vinca	40	7 da	A	117
Mulberry dwarf	*/*:S/Au	mulberry	55	40 min	W	189
			50	1 da	A	189
Opuntia witches' broom		Opuntia	45	5 hr	W	130
			45	15 da	A	142
Parastolbur	*/*:S/Au	Vinca	40	10 da	A	199
Peach phony	*/*:S/Au	peach	48	40 min	W	94
Peach rosette	*/*:S/Au	peach	50	10 min	W	112
			35	2 wk	A	112
Peach x-disease	*/*:S/Au	peach	50	6 min	W	73
Peach yellow leafroll	*/*:S/Au	peach	50	10 min	W	141
Peach yellows	*/*:S/Au	peach	50	10 min	W	112
			35	2 wk	A	112
Pennisetum streak	*/*:S/*	Pennisetum	52	20 min	W	26
Potato witches' broom	*/*:S/Au	potato tubers	36	6 da	A	115
		Vinca	42	13 da	A	115
Rubus stunt	*/*:S/Au	Rubus	45	1 hr	W	196
			37	21 da	A	95
Stolbur	*/*:S/Au	Vinca	40	10 da	A	199
Sugarcane chlorotic streak	*/*:S/Au	sugarcane	52	20 min	W	127
			54	8 hr	A	1
Sugarcane grassy shoot	*/*:S/(Ap)	sugarcane	50	30 min	W	174
			54	8 hr	A	174
Sugarcane sereh	*/*:S/*	sugarcane	45	30 min		
			+52	30 min	W	204
Sugarcane whiteleaf	*/*:S/Au	sugarcane	54	50 min	W	120
			54	8 hr	A	120
Vaccinium (cranberry) false-blossom	*/*:S/Au	cranberry	42	8 da	A	116
Vaccinium stunt	*/*:S/Au	blueberry	52	2 hr	A	185

[a] Third and fourth pairs of Gibbs (58), third pair=outline of particle/outline of "nucleocapsid"; S=spherical, E=elongated with parallel sides, ends not rounded, U=elongated with parallel sides, ends rounded. Fourth pair=kinds of hosts infected/kinds of vector; F=fungus, I=invertebrate, S=seed plant/Ac=mite, Al=white fly, Ap=aphid, Au=leaf plant, or tree hopper, Cc=mealy bug, Fu=fungus; *=unknown and ( )=doubtful.

[b] W=water; A=air.

TABLE I—(*Continued*)

Group and virus name	Cryptogram[a]	Host treated	Treatment			Authority
			Temp. °C	Time	Type	
HOT WATER LABILE—*Not sap transmissible*						
Cherry necrotic rusty mottle	*/*:S/*	cherry	50	10 min	W	143
Hibiscus leafcurl	*/*:S/AL	Hibiscus	40	25 min	W	136
			35	25 da	A	136
Potato leafroll	S/S:S, I/Ap	potato tubers	50	17 min	W	140
			36	20 da	A	100
Strawberry complex		strawberry	43	30 min	W	134
HOT WATER LABILE—*Sap transmissible*						
Hop nettlehead complex		hop	45	5 min	W	62
Prune dwarf	S/S:S/*	plum	35	30 hr	W	36
		peach	38	17 da	A	144
Prunus necrotic ringspot	S/S:S/*	cherry	35	36 hr	W	44
			38	17 da	A	144
Cherry line pattern complex		cherry	45.5	3 hr	W	69
Sugarcane ratoon stunting		sugarcane	50	2 hr	W	181
			54	8 hr	A	180
ONLY HOT AIR LABILE—*Easily inactivated*						
Abutilon mosaic	(S)/*:S/Al	Abutilon	37	21 da		102
Alfalfa mosaic	U/U:S/Ap	alfalfa	36	8 da		50
Apple chlorotic leafspot	E/E:S/*	apple	38	7 da		203
Apple leaf pucker	*/*:S/*	apple	38	7 da		203
Apple (Malus) platycarpa dwarf	*/*:S/*	Malus	37	20 da		27
Apple (Malus) platycarpa scaly bark	*/*:S/*	Malus	37	20 da		27
Apple rubbery wood	*/*:S/*	apple	38	7 da		203
Apple Spy 227 epinasty and decline	*/*:S/*	apple	38	7 da		203
Apple stem pitting	E/E:S/*	apple	38	7 da		203
Arabis mosaic	S/S:S/Ne	*Nicotiana clevelandii*	38	21 da		81
Bean yellow mosaic	E/E:S/Ap	sweet clover	10	6 wk		9
Black currant reversion	*/*:S/Ac	Ribes	34	20 da		28
Carnation ringspot	S/S:S/*	carnation	37	21 da		149
Carnation streak	*/*:S/*	carnation	37	28 da		192
Cassava mosaic	*/*:S/Al	carnation	39	28 da		35
Chrysanthemum green flower	*/*:S/*	Chrysanthemum	38	28 da		183
Chrysanthemum ringspot	*/*:S/*	Chrysanthemum	36	21 da		83
Chrysanthemum stunt (English strain)	*/*:S/*	Chrysanthemum	36	21 da		83
Citrus greening	*/*:S/Ps	citrus	40	28 da		121
Citrus infectious variegation	S/S:S/Ap	citrus	38	30 da		123
Citrus stubborn	*/*:S/*	citrus	51	1.5 hr		148
Citrus tristeza	E/E:S/Ap	citrus	39	28 da		41
Clover (white) new		*N. clevelandii*	38	28 da		86
Clover yellow vein	E/E:S/Ap	*N. clevelandii*	38	4 wk		84
Cucumber mosaic	S/S:S/Ap	Passiflora	37	14 da		190
Fig mosaic	*/*:S/Ac	fig	37	24 da		31
Gooseberry veinbanding	*/*:S/Ap	Ribes	35	14 da		98
Grapevine fanleaf (strain of Arabis mosaic)	S/S:S/Ne	grape	35	21 da		53
Mushroom, 1 & 2	S/S:Fu/*	mushroom	33	?		54
Mushroom, 3	U/U:Fu/*	mushroom	33	14 da		55
Peach (Muir) dwarf	*/*:S/*	peach	38	17 da		144
Pear bark necrosis	*/*:S/*	pear	37	28 da		154
Pear stony pit	*/*:S/*	pear	70	10 min		36

TABLE I—(*Continued*)

Group and virus name	Cryptogram[a]	Host treated	Treatment			Authority
			Temp. °C	Time	Type	
Pear vein yellows	*/*:S/*	pear	37	28 da		154
Potato A	E/E:S/Ap	potato	35	14 da		195
Potato Y	E/E:S/Ap	potato	38	7 da		194
Prunus necrotic ringspot	S/S:S/*	Prunus	38	17 da		144
Prunus necrotic ringspot (apple mosaic strain)	S/S:S/*	Malus	37	20 da		193
Raspberry mosaic heat-labile components	*/*:S/Ap	Rubus	38	4 da		37
Rose mosaic	*/*:S/*	rose	35	28 da		88
Rose yellow mosaic	*/*:S/*	rose	38	14 da		198
Strawberry crinkle	*/*:S/Ap	strawberry	38	3 da		10
Strawberry latent C	*/*:S/Ap	strawberry	35	22 da		122
Strawberry mild yellow edge	*/*:S/Ap	strawberry	37	26 da		153
Strawberry mottle	*/*:S/Ap	strawberry	37	7 da		17
Strawberry veinbanding	*/*:S/Ap	strawberry	42	10 da		15
Strawberry, unspecified		strawberry	37.5	10 da		18
Tobacco ringspot	S/S:S/Ne	cowpea, tobacco	36	28 da		77
Tomato aspermy	S/S:S/Ap	Chrysanthemum	38	14 da		111
Tomato ringspot (Peach yellow bud mosaic strain)	S/S:S/Ne	peach	38	21 da		146
Tomato spotted wilt	S/S:S/Th	tomato	38	28 da		79
Turnip mosaic	E/E:S/Ap	horseradish	37	21 da		89
ONLY HOT AIR LABILE—*Difficult to inactivate*						
Broad bean mottle	S/S:S/*	*N. clevelandii*	38	6 wk		86
Carnation etched ring	S/S:S/*	carnation	38	6 wk		20
Carnation I.R.	S/S:S/*	*N. clevelandii*	38	6 wk		86
Carnation latent	E/E:S/Ap	carnation	38	8 wk		20
Carnation mottle	S/S:S/I	carnation	40	6 wk		159
Carnation vein mottle	E/E:S/Ap	carnation	40	8 wk		159
Cherry (sour) green ring mottle	*/*:S/*	sour cherry	38	6 wk		146
Chrysanthemum B	E/E:S/Ap	Chrysanthemum	37	2 mo		70
Chrysanthemum rosette	*/*:S/*	Chrysanthemum	35	2 mo		23
Citrus exocortis	*/*:S/*	lemon	38	33 wk		188
Citrus psorosis	*/*:S/I	citrus	35–43	12 wk		66
Citrus yellowing		lemon	38	22 wk		188
Grapevine asteroid mosaic	*/*:S/*	grape	38	6 wk		64
Grapevine corky bark	*/*:S/*	grape	38	14 wk		64
Grapevine leafroll	*/*:S/*	grape	38	8 wk		64
Grapevine yellow vein (strain of ringspot)		grape	38	6 wk		63
Hydrangea ringspot	E/E:S/*	hydrangea	35	12 wk		19
Peach stubby twig	*/*:S/*	peach	38	5 wk		146
Pelargonium leafcurl	S/S:S/*	*N. clevelandii*	37	30 da		80
Plum pox	E/E:S/Ap	Prunus	38	39 da		107
Poplar mosaic	E/E:S/*	poplar	38	6 wk		11
Potato S	E/E:S/Ap	potato plants	35	6 wk		178
Potato X	E/E:S/(Fu)	potato plants	35	15 wk		178
Strawberry witches' broom	*/*:S/*	strawberry	35–46	8 wk		131
Sweet potato internal cork	E/E:S/Ap	sweet potato plants	38	3 mo		74
Sweet potato leafspot	*/*:S/Ap	sweet potato plants	38	1 mo		76
Sweet potato yellow dwarf	*/*:S/Al	sweet potato plants	38	1 mo		76
Turnip crinkle	S/S:S/Cl	Brassica	38	6 wk		86

TABLE II

VIRUSES NOT INACTIVATED BY HEAT TREATMENT *in vivo*

Virus name	Cryptogram	Host treated	Treatment			Authority
			Temp. °C	Time	Type	
Apple (Virginia crab) decline		apple	38	4.5 wk	A[a]	203
Cacao swollen shoot	U/*:S/Cc	Cacao budwood	50	25 min	W	138
Cherry mottle leaf	*/*:S/Ac[b]	cherry	38	5 wk	A	146
Chrysanthemum D	*/*:S/*	Chrysanthemum	37	4 wk	A	83
Chrysanthemum E	*/*:S/*	Chrysanthemum	36	4 wk	A	83
Chrysanthemum stunt (American strain)	*/*:S/*	Chrysanthemum	38	12 wk	A	118
Chrysanthemum vein mottle	E/E:S/Ap	Chrysanthemum	36	4 wk	A	83
Citrus xyloporosis	*/*:S/*	Citrus	35	75 da	A	67
Dioscorea green-banding	*/*:S/Ap	Dioscorea tubers	50	1 hr	W	168
			37	2 wk	A	168
Mulberry mosaic	*/*:S/Ap	mulberry	40	21 da	A	163
Narcissus yellow stripe	E/E:S/Ap	Narcissus	37	21 da	A	25
Peach calico	*/*:S/*	peach	37	40 da	A	33
Peach mosaic	*/*:S/Ac	peach	50	32 min	W	113
			35	30 da	A	113
Plum bark split	*/*:S/*	plum	37	39 da	A	45
Potato aucuba mosaic	E/E:S/Ap	potato tubers	38	5 wk	A	166
Potato mop top virus	E/E:S/Fu	potato	37	7 wk	A	99
Potato spindle tuber	*/*:S/*	potato tubers	35	39 da	A	46
Quince sooty ringspot	*/*:S/*	quince	50	8 min	W	154
			37	28 da	A	154
Raspberry leaf curl	*/*:S/Ap	raspberry	37	28 da	A	177
Raspberry mosaic heat stable components	*/*:S/Ap	raspberry	37	90 da	A	176
Raspberry vein chlorosis	*/*:S/Ap	raspberry	50	?	W	34
Raspberry yellows	*/*:S/*	raspberry	50	?	W	34
Strawberry necrotic shock	*/*:S/*	strawberry	20–44	6 mo[c]	A	48
Sweet potato russet crack	*/*:S/*	sweet potato	38	4 mo	A	75
Tobacco rattle virus	E/E:S/Ne	potato tubers	38	5 wk	A	166
Vaccinium (blueberry) ringspot (=tobacco ringspot?)	*/*:S/*	Vaccinium	60	1 hr	W	184
			38	?	A	184

[a] W=water; A=air.
[b] Personal communication, L. S. Jones, USDA Entomology, Riverside, California.
[c] Treated in the field at Meloland, California.

often minimum times tested and not necessarily the thermal death-point. This is true for cherry little cherry, chrysanthemum flower distortion, Crimean yellows, delphinium yellows, parastolbur, and stolbur. In some cases water treatments were not tried, only air. Mycoplasma organisms have been associated with several (42, 120) of the diseases listed in this part of Table I. Mycoplasmas associated with animal diseases are also easily heat-inactivated (169).

The second section of Table I includes four diseases easily inactivated that are not sap- or leafhopper-transmitted, and have not yet been proved to be viruses. On the basis of heat sensitivity the etiologic agent of these diseases could be similar to that of yellows.

The third section of Table I includes five diseases also easily inactivated for which definite virus particles have been described, although some question still remains for ratoon stunting. All are sap-transmissible.

The fourth section of Table I includes viruses designated as easily inactivated on the arbitrary basis that inactivation in less than 29 days was sufficient to permit propagation of explants free of infection (half-life less than 29 days at *ca* 38° C). Insect vectors and virus particles have been described for some in this group but not for others.

The last section of Table I lists viruses that we classify as difficult to inactivate (half-life greater than 28 days at *ca* 38° C). Treatment periods shown were the minimum required to obtain at least some explants free of infection. Almost all treatments in this group were combined with either meristem culture or removal of tips or axillary buds.

The basis for successful heat inactivation *in vivo* is the difference between the host and virus in tolerance to high temperature. There might be an equal difference in tolerance to low temperature between the host and virus that could be utilized in therapy. Two quite different viruses—clover wound tumor and bean yellow mosaic in clover—were inactivated when infected plants were grown at temperatures near minimum for plant growth. Plants infected with potato virus Y were also reported to be cured at low temperature (90). Additional diseases, both yellows and nonyellows, might respond to low-temperature therapy, but we are not aware of any others that were tested.

Table II includes some viruses that have been tested many times but have resisted all attempts to inactivate them, as well as others that have not been adequately tested. The exceptional tolerance to heat of some suspected viruses such as chrysanthemum stunt, sweet potato russet crack, strawberry necrotic shock, and the heat-stable components of raspberry mosaic suggests that these viruses may exist in the host as nucleic acid, as has been shown for citrus exocortis (172). This virus is extremely heat stable but was inactivated after treatment at 38° C for 33 weeks (188).

Virus Shape and Stability

As apparent in the tables, the morphology of virus particles is not associated consistently with ease or difficulty of inactivation by heat *in vivo*. Up to 1964, heat therapy was successful only with spherical viruses or with viruses of unknown particle morphology (102, 104). Among the viruses grouped under "easy to inactivate" for which particle morphology is known, 10 are rod-shaped and 13 are spherical. In the group designated "difficult to inactivate," 9 are rod-shaped and 6 are spherical. Of the viruses not yet inactivated (Table II), 6 are rod-shaped and 19 are of unknown morphology.

Probably all viruses can be inactivated *in vivo* with the right combination of temperature and time and attention to all factors favoring plant survival (8).

Indexing for Proof of Virus Inactivation

Virus inactivation is assessed either by observation of treated plants or explants, or by indexing procedures. Indexing by graft inoculation is more reliable than assay by sap transmission, insects, electron microscopy, or serology (37, 152, 153, 178). Heat-tolerant strains or the original virus may remain undetected during one or more index tests after treatment (34, 79, 153, 156). Inactivation of a virus in different hosts may vary within an experiment and between experiments (152, 153, 203). In most cases, observation alone cannot be considered proof of inactivation (40, 79, 150, 159, 178, 179).

Viruses like PVS in potatoes are particularly difficult to detect since they can be demonstrated only by serology or electron microscopy (178). In indexing for PVX and PVS (178), the virus was assumed to be PVS when flexuous rods were present and the PVX test on *Gomphrena* was negative. Half of the plants that were negative for PVS in electron microscopy still indexed positive for PVX serologically.

Cross-protection tests and reinoculation of treated plants are also used for proof of virus inactivation (114, 166).

Thermal Constants and the Mechanism of Heat Inactivation

The usual thermal constant given for plant viruses is the thermal inactivation point [the temperature required to inactivate virus in an aliquot of infected pressed sap in 10 min *in vitro* (97)]. With proper care, this point can be determined for viruses that are mechanically transmissible (12, 133), but accuracy is affected by the virus source as well as the concentration and extraction methods (133). Christoff (36) and Kegler (106) attempted to determine thermal inactivation points for certain tree fruit viruses by heating budwood in water and air for the prescribed 10-min periods and then budding to healthy seedlings. Data were obtained by indexing the seedling or the explant if they grew. The difficulty with *in vivo* studies is that viruses may have thermal inactivation points higher than the thermal-death-point of the host plant.

Thermal inactivation of plant viruses *in vitro* follows the course of a first-order chemical reaction (2, 12, 157, 202). Noting this fact, Price (157) suggested that the velocity constant of inactivation at a specified temperature would better express temperature relationships of a virus than its thermal inactivation point, since this constant could be determined with greater accuracy from an equivalent amount of data. He further suggested that a true picture of the heat stability of the virus could be drawn if the velocity constants of inactivation were determined for a virus through a range of temperatures under specified conditions. Strangely, only a few kinetic studies have been made of the inactivation *in vitro* of plant viruses (2, 12, 157,

202) or plant virus nucleic acid (60) in line with Price's suggestion, and none has been made for the warm-air inactivation of plant viruses *in vivo*. On the other hand, kinetic studies with animal viruses (151, 205) have enabled the formulation of sophisticated theories on the mechanisms of thermal inactivation at the molecular level.

Although no specific study has been made of the inactivation of a plant virus *in vivo* through a temperature range, Fenrow et al. (46) plotted the thermal-death-curve of potato leafroll virus between 32 and 40° C from their own data and data from the literature on inactivation of this virus in tubers. The curve was a straight line when temperature was plotted against the logarithm of inactivation time. This indicates that the inactivation of potato leafroll virus *in vivo* proceeds as a first-order reaction even when inactivation times are determined under a wide array of environmental conditions.

An advantage is readily apparent from using velocity constants of inactivation for *in vivo* studies of viruses in perennial crops. One can show that half-life is an alternative way of expressing this constant; and with sufficient populations of infected test plants in the heat chamber the half-life at a specified temperature for any virus, whether or not mechanically transmissible to a local-lesion assay host, can be estimated by statistical methods (37). Thus the kinetics of the inactivation of any plant virus that is inactivated before its host plant dies from the treatments can be studied through a range of temperatures. For viruses that are cured in infected perennial plants, we suggest that half-life of the virus *in vivo* at 38° C is a more meaningful statistic than the *in vitro* thermal inactivation point. The half-life *in vivo* at 38° C expresses the thermal stability of the virus, complementing the half-life *in vitro,* which expresses longevity as suggested by Yarwood & Sylvester (208). The concept of Yarwood & Sylvester should be defined more precisely as half-life *in vitro* at 20° C since temperature affects any biological reaction rate (96).

Most published studies give only fragmentary consideration to the kinetics of heat tolerance of the host plant, possibly on account of the expense of more than a single heat chamber but more probably because the purpose in most studies has been the production of clean mother stocks rather than understanding of the principles of heat therapy. Straight-line survival curves, where determined (14, 158), indicate the host killing is a rate process (96) much like inactivation of the virus. Whether the temperature coefficient of inactivation of the virus is the same as that of its host is unknown for most viruses. Data of KenKnight (108) suggest that the thermal-death-curve of peach rosette virus and dormant *Prunus* buds in hot water might intersect at high temperatures. Kunkel (112) found that the difference between the minimum inactivation period for peach yellows virus and the maximum survival period of peach buds was greater at lower than at higher temperatures. This probably indicates the reason that most successful virus inactivations are carried out near 38° C, a temperature that is lethal to viruses but that can be withstood for long periods with adequate light by a conditioned host, particularly a perennial.

There is no support for an early hypothesis that virus is not completely inactivated during heat treatments (101). Within the plant there may be centers where virus still remains even with relatively prolonged heat treatment, though the danger of such persistent infections is eliminated by propagating heated buds or tips. Clones of grapevines and fruit trees are still free of virus that were freed of virus by heat treatments at Davis 10 years ago and protected against reinfection in the field. In early work at Davis we were unable to free entire grapevine plants of fanleaf by heating at 38° C for 64 days, but when the tolerance of the host to heat was increased and the exposure extended to 102 days (64), the virus was eliminated completely. Many workers have had similar experience.

The hypothesis that heat therapy results from a shift in balance between virus synthesis and degradation (101, 104, 105) in which synthesis is reduced is probably true at temperatures sublethal for both the virus and the host. Here, titer may be the expression of balance between synthesis and degradation of virus. At higher temperatures, however, synthesis does not occur, and inactivation of the virus results from heat. The hypothesis does not account for the ultimate extinction of virus in the surviving host tissues after prolonged treatments at the critical temperature-time.

The most plausible hypothesis for the mechanism of heat therapy is that high temperatures cause the destruction of essential chemical activities in both virus and host but that the host is better able to recover from the damage. In other words, the temperature coefficient of thermal inactivation for the host exceeds that of the virus at certain temperatures, as suggested by Geard (57). This mechanism is probably not different from that of heat treatments used for killing other pathogens intimately associated with a host plant.

The inactivation process *in vivo* may be purely physical, as it seems to be *in vitro,* or it may be aided by biological changes induced in the host metabolism by high temperatures (56).

Campbell (27) speculated that therapy might result from the immobilization of virus, so that new tips on rapidly growing tissues in the chamber remain virus-free. Welsh & Nyland (203) found, however, that buds already formed and infected before heating were freed of virus by the heating. Mellor & Stace-Smith (132) and McGrew (122) found that the only obviously significant variables for heat-therapy success were temperature, length of treatment, and size of tips taken for starting explants; and that position or rate of growth in the chamber had no bearing on it. These findings suggest that heat therapy results from inactivation of virus, not immobilization.

Whether temperature-time treatments or the size of tips taken for starting explants is more important in plant-virus therapy will depend on the specific virus and its invasiveness within the host. Holmes (88) found that tips from dahlia plants infected with spotted-wilt virus propagated free from the disease at normal growing temperatures. On the other hand, Stace-Smith & Mellor (178) found that axillary bud tips 1 to 3 mm long from potato plants infected with PVX were generally free of the virus only after 26 weeks at 35° C, whereas a few tips less than 500 μ long were free without heating. Whether any virus is completely systemic within its host plant is not established. Sheffield (173) detected tobacco mosaic virus in the apices of roots and stems of tomato plants less than 100 μ long, and Walkey & Webb (200), in electron-microscope studies, observed virus particles and tubules enclosing particles in the meristem dome of *Nicotiana rustica L.* infected with cherry leafroll virus. Many workers (10, 135, 160), however, have obtained clean plants through cultures of meristems from virus-infected mother plants.

Mathews & Lyttleton (129) found that turnip yellow mosaic virus in plants held at 35° C lost infectivity without significant changes in other virus properties. Babos & Kassanis (2), in *in vitro* studies, observed that inactivation of tobacco necrosis virus (TNV) proceeded exponentially at different temperatures, as if it were a two-component system with rapidly inactivating and slowly inactivating fractions. Nucleic acid extracts of TNV inactivated similarly to the intact virus at 51° C, indicating that inactivation resulted from changes in virus nucleic acid. Those workers also found inactivation to occur without significant changes in other virus properties or virus morphology. Additional kinetic studies of viruses *in vivo* as well as *in vitro* should shed further light on the mechanisms of successful heat therapy of plant viruses.

Uses of Heat Therapy

Heat therapy has both basic and practical applications. It is a tool both in identifying viruses and in separating them from disease complexes. Practically, it is indispensable to producing clean nuclear stocks where clonally propagated crop plants have become completely infected with one or more viruses. It has been used both deliberately and unintentionally as a direct treatment for controlling different plant virus diseases under field conditions.

The identification of viruses or virus relationships is impossible for many viruses of perennial plants because the viruses cannot be isolated from the diseased host. Thus, properties such as morphology and serology which might demonstrate relationships or lack of relationship of the etiologic agents cannot be studied directly. A very useful tool for indicating relationships among such viruses could be the half-life *in vivo* at 38° C. Heat treatment might succeed where serological methods fail (155).

Heat inactivation *in vivo* can help identify individual viruses where complexes exist (152, 203), or it might even establish that a specific disease is caused by a virus (53).

Heat therapy has aided understanding of some problems of rootstock-scion incompatibilities and might be useful in studying others. Stubbs (187) found that Josephine pear budded onto an apple rootstock grew vigorously when held in a heat chamber for 2 to 3 months at *ca* 38° C, whereas without heat the graft combination failed.

Pure cultures of heat-stable viruses from virus complexes are possible through differential heat inactivation in cases where virus vectors or differential host plants are unknown (45, 146). Heat treatments are particularly useful for separating labile from stable components in raspberry virus studies (34, 37, 176). Hildebrand (75) found that the symptoms of sweet potato russet crack were masked until heat therapy eliminated the more labile internal cork virus.

The chief practical achievements of heat therapy are clean planting stock in a few cases, and mother plants free of virus disease. Sugarcane viruses (chlorotic streak and ratoon stunting) are controlled by hot-water treatments of the cane cuttings (setts). Annually, several thousand tons are treated in large tanks in hot water or in hot-air ovens, where accurate control of temperature permits virus eradication without appreciable damage to the setts themselves (68, 104, 181).

After Kassanis (100) eradicated leafroll from potatoes it seemed that heat treatments of seed tubers might be used effectively in potato production. Such treatments, however, caused too much damage to the tubers for direct planting (71, 165); and Dutch workers (166, 197) reported that the treatments caused mutations to develop in the new plants from the heated tubers. The symptoms that they observed, however, appear to have been temporary growth responses resulting from partial inactivation or reinvasion by other potato viruses in the young plants. No further reports of plant mutations resulting from heat treatments are known. Perhaps even the effect of tuber damage is overestimated as recent work from Czechoslovakia (175) indicates that heat treatments, especially when applied to tubers in the fall, reduced the incidence of leafroll and increased potato yields. In any event heat treatments are useful for developing sources of nuclear potato stocks (165).

Carnation growers are using heat therapy directly. In south France, carnation stocks are produced in greenhouses where temperatures are held at 40° C for 1 month before tip cuttings are taken for propagation (161). The production system also involves the indexing of cuttings to *Chenopodium amaranticolor* Coste & Reyn. and should prove very effective for the suppression of certain carnation virus diseases if standards for indexing are sufficiently high.

Heat treatments have been employed most widely for ensuring virus freedom in mother plants used for the production of foundation stocks. Here, heat therapy is followed with thorough indexing over a period of 1 to 3 years before the plants are considered free of known viruses. They are then grown in isolated plantings or screenhouses as source plants for clean planting stocks. This procedure has been very successful in California for the production of clean stocks of grapevines and stone fruits. Disease-free stock produced for spring planting in 1969 included over 2,000,000 grapevine cuttings and almost 100,000 fruit trees. Advantages from stocks originating from heat-treated mother plants can be cited for several crops: apples (30), carnations (85, 149), chrysanthemums (78), peaches (124), pears (29), and strawberries (5), to list a few.

One's point of view sometimes determines whether freedom from disease is beneficial or not. Certain cactus fanciers measure their success by the number of species or forms of cacti that they have in their collections. Nozeran & Neville (142) noted that such fanciers inoculated their plants with *Opuntia* witches' broom virus to create additional monstrosities. This virus is very easily inactivated by heat, and one might suppose that in this isolated case heat therapy would be deleterious if it reduced the fancier's collection to half its original number. Some nurserymen look with trepidation on our current investigations on heat inactivation of camellia variegation virus, since it might lessen the demand for all variegated forms. In some cases, inactivation of mild strains of some viruses might not be desirable.

Heat need not be applied in sophisticated heat chambers for the successful eradication of plant viruses. Sometimes, in fact, reduction of natural heat can favor disease. Thus, potato leafroll was unknown on the plains of the river valleys of India when seed tubers were stored in thatched huts for 6 months during the hot summer season, but it became a very serious problem when cold-storage facilities were built or when seed tubers were brought in from the cooler mountain districts (191). Frazier et al. (48) found that natural heat eliminated certain strawberry viruses if plants were grown in an area with high summer temperatures, and observations on the effects of high summer temperatures on peach yellows started Kunkel's scientific investigations of the possibility of heat therapy for controlling virus disease.

One can take advantage of natural heating to control certain virus diseases. Thus, Fulton (51) set the minimum temperature in a plastic chamber in his greenhouse at 31° C, and natural heating eliminated a virus complex from strawberry plants growing in the chamber. In South Africa, McClean (121) built polyethylene tents over citrus trees infected with greening virus and found that on hot clear days the maximum temperature within the tent could go to 45 to 54° C. When the temperature exceeded 43° C under these tents, trees exposed for periods of 3 days to a week survived and showed considerable remission of greening symptoms in comparison with uncovered adjacent checks.

Heat treatment is being utilized in California to assist in moving plant material from areas quarantined for certain virus diseases. Heat treatment assures freedom from tomato ringspot in stone fruits and is substituted for a 3 year index on indicator plants. Several accessions of plants introduced into the U.S.A. from other countries have been made available through heat treatment when they would otherwise have been excluded. Heat treatment should be adopted as standard procedure for the movement of plants through quarantines.

In the 1890's, careful clean-stock programs were put into effect in Java to control the sereh disease of sugarcane (204). Only clean stocks grown in certain mountain districts were used for planting new fields. The success of these programs can be deduced from the fact that the sereh disease no longer exists (43). Kobus and his associates and Wilbrink (204) were able to inactivate the etiologic agent with hot-water treatments, and it is on this basis that we assume that it was caused by a virus. Peach yellows in the eastern U.S.A. has reached the near-vanishing point because of tree removal and other control measures. Complete eradication may not be necessary; the disease may disappear when the incidence falls below a survival threshold. Clean-stock programs based on heat therapy and indexing should relegate many other plant virus diseases to the limbo of sugarcane sereh and peach yellows.

LITERATURE CITED

1. Abbot, E. V. Diseases in relation to variety planting in 1959. *Sugar Bull.*, **37,** 272–73, 283–84 (1959)
2. Babos, P., Kassanis, B. Thermal inactivation of tobacco necrosis virus. *Virology,* **20,** 490–97 (1963)
3. Baker, K. F. Thermotherapy of planting material. *Phytopathology,* **52,** 1244–55 (1962)
4. Baker, K. F. Principles of heat treatment of soil and planting material. *J. Australian Inst. Agr. Sci.,* **28,** 118–26 (1962)
5. Baldini, E., Goidanich, G. Aspetti agronomici e sanitari della degenerazione della fragola. *Riv. Ortoflorofrutticolt. Ital.,* **48,** 381–409 (1964)
6. Baumann, Giselle. Möglichkeiten und Methoden der Heilung von Pflanzenkrankheiten durch Wärmeeinwirkung. *Deut. landwirtsch., Berlin,* **5,** 76–80 (1954)
7. Bawden, F. C. *Plant Viruses and Virus Diseases* (Ronald Press, New York, 361 pp., 1964)
8. Bawden, F. C. Some reflections on thirty years of research on plant viruses. *Ann. Appl. Biol.,* **58,** 1–11 (1966)
9. Baxter, L. W., Jr., McGlohon, N. E. A method of freeing white clover plants of bean yellow mosaic virus. *Phytopathology,* **49,** 810–11 (1959)
10. Belkengren, R. O., Miller, P. W. Culture of apical meristems of *Fragaria vesca* strawberry plants as a method of excluding latent A virus. *Plant Disease Reptr.,* **46,** 119–21 (1962)
11. Berg, T. M. Studies on poplar mosaic virus and its relation to the host. *Mededel. Landbouwhogeschool Wageningen,* **64**(11), 1–72 (1964)
12. Best, R. J. Thermal inactivation of tomato spotted wilt virus. *Australian J. Exptl. Biol. Med. Sci.,* **24,** 21–25 (1946)
13. Bhargava, K. S. Treatments for the control of virus diseases in plants. *Sci. and Culture (Calcutta),* **23,** 389–93 (1958)
14. Blodgett, F. M. Time-temperature curves for killing potato tubers by heat treatments. *Phytopathology,* **13,** 465–75 (1923)
15. Bolton, A. T. Virus-free strawberries and raspberries in eastern Canada. *Res. for Farmers, Canada,* **10**(2), 14–16 (1965)
16. Bolton, A. T. The inactivation of veinbanding and latent C viruses in strawberries by heat treatment. *Can. J. Plant Sci.,* **47,** 375–80 (1967)
17. Bovey, R. Guérison de fraisiers, atteints de virus, par traitement thermique. *Landwirtsch. Jahrb. Schweiz.,* **68,** 1041–47 (1954)
18. Bovey, R. Les bases scientifiques de la production de plants de fraisiers

sains. *Revue romande agr. viticult. et arboricult.*, **12,** 5–8 (1956)

19. Brierley, P. Virus-free Hydrangeas from tip cuttings of heat-treated ringspot-affected stock plants. *Plant Disease Reptr.*, **41,** 1005 (1957)
20. Brierley, P. Heat cure of carnation viruses. *Plant Disease Reptr.*, **48,** 143 (1964)
21. Brierley, P., Lorentz, P. Healthy tip cuttings from some mosaic-diseased Asiatic chrysanthemums; some benefits and other effects of heat treatment. *Phytopathology,* **50,** 404–08 (1960)
22. Brierley, P., Smith, F. F. Symptoms of chrysanthemum flower distortion, dodder transmission of the virus, and heat cure of infected plants. *Phytopathology,* **47,** 448–50 (1957)
23. Brierley, P., Smith, F. F. Some characteristics of eight mosaic and two rosette viruses of chrysanthemum. *Plant Disease Reptr.*, **42,** 752–63 (1958)
24. Broadbent, L. Control of plant virus diseases. In *Plant Virology,* 330–64 (Corbett, M. K., Sisler, H. D., Eds., University of Florida Press, Gainesville, Fla., 527 pp., 1964)
25. Broadbent, L., Green, D. E., Walker, P. Narcissus virus diseases. *Roy. Hort. Soc. Daffodil-Tulip Yearbook,* **28,** 154–60 (1962)
26. Bruehl, G. W. Chlorotic streak disease of *Pennisetum purpureum. Plant Disease Reptr.*, **37,** 34–35 (1953)
27. Campbell, A. I. Apple virus inactivation by heat therapy and tip propagation. *Nature,* **195,** 520 (1962)
28. Campbell, A. I. The inactivation of black currant reversion virus by heat therapy. *Ann. Rept. Long Ashton Agr. Hort. Res. Sta., 1964,* pp., 89–92 (1965)
29. Campbell, A. I. The growth of young pear trees after elimination of some viruses by heat treatment. *Ann. Rept. Long Ashton Agr. Hort. Res. Sta., 1965,* 111–16 (1966)
30. Campbell, A. I., Coles, J. S. The growth of heat-treated apple trees on several rootstocks. *Ann. Rept. Long Ashton Agr. Hort. Research Sta., 1965,* 117–22 (1966)
31. Casalicchio, G. Applicazione della termoterapia nel risanamento della piante di fico affette da mosaico. *Phytopathol. Medit.*, **3,** 184–85 (1964)
32. Caudwell, A. l'inhibition *in vivo* du virus de la flavescence dorée par la chaleur. *Ann. Epiphyties. Suppl.*, **17,** 61–66 (1966)
33. Chamberlain, E. E., Atkinson, J. D., Hunter, J. A. Occurrence of peach calico in New Zealand. *New Zealand J. Sci. Technol.*, **38,** 813–19 (1957)
34. Chambers, J. The production and maintenance of virus-free raspberry plants. *J. Hort. Sci.*, **36,** 48–54 (1961)
35. Chant, S. R. A note on the activation of mosaic virus in cassava (*Manihot utilissima* Pohl.) by heat treatment. *J. Exptl. Agr.*, **27,** 55–58 (1959)
36. Christoff, A. Die Obstvirosen in Bulgarien. *Phytopathol. Z.*, **31,** 381–436 (1958)
37. Converse, R. H. Effect of heat treatment on the raspberry mosaic virus complex in Latham red raspberry. *Phytopathology,* **56,** 556–59 (1966)
38. Cook, J. A., Goheen, A. C. The effect of a virus disease, leafroll, on the mineral composition of grape tissue and a comparison of leafroll and potassium deficiency symptoms. *Am. Inst. Biol. Sci., Publ. 8,* 338–54 (1961)
39. Cornuet, P., Marrou, J. Essais d'amelioration du rendement de la thermotherapie du fraisier. *Advan. Hort. Sci.*, **1,** 164–67 (1961)
40. Cropley, R. Comparison of some apple latent viruses. *Ann. Appl. Biol.*, **61,** 361–72 (1968)
41. Desjardins, P. R., Wallace, J. M., Wollman, E. S. H., Drake, R. J. A separation of virus strains from a tristeza-seedling-yellows complex by heat treatment of infected lime seedlings. In *Citrus Virus Diseases,* 91–95 (Wallace, J. M., Ed., Agricultural Publications, University of California, Berkeley, 243 pp., 1959)
42. Doi, Y., Teranaka, M., Yora, K., Asuyama, H. Mycoplasma—or PLT group-like microorganisms found in the phloem elements of plants infected with mulberry dwarf, potato witches' broom, aster yellows, or Paulownia witches' broom (in Japanese with English abstract).

Ann. Phytopathol. Soc. Japan, **33,** 259–66 (1967)

43. Edgerton, C. W. *Sugarcane and Its Diseases* (Louisiana State University Press, Baton Rouge, La., 290 pp., 1955)
44. Ehlers, C. G. Separation and identification of viruses that incite diseases of stone fruits. *Dissertation Abstr.,* **17,** 1879–80 (1957)
45. Ellenberger, C. E. Heat inactivation of some viruses in plum varieties and rootstocks. *Ann. Rept. East Malling Res. Sta., Kent, 1959,* 99–101 (1960)
46. Fenrow, K. H., Peterson, L. C., Plaisted, R. L. Thermotherapy of potato leafroll. *Am. Potato J.,* **39,** 445–51 (1962)
47. Fitzpatrick, R. E., Stace-Smith, R., Mellor, F. C. Heat inactivation of some strawberry and raspberry viruses. *Proc. Can. Phytopathol. Soc.,* **22,** 13 (1954)
48. Frazier, N. W., Voth, V., Bringhurst, R. S. Inactivation of two strawberry viruses in plants grown in a natural high-temperature environment. *Phytopathology,* **55,** 1203–5 (1965)
49. Fridlund, P. R. An inexpensive heat chamber for curing virus diseased plants. *Plant Disease Reptr.,* **46,** 703–5 (1962)
50. Frosheiser, F. L. Freeing alfalfa clones from alfalfa mosaic virus by heat treatment. *Phytopathology,* **59,** 391–92 (1969)
51. Fulton, J. P. Heat treatment of virus-infected strawberry plants. *Plant Disease Reptr.,* **38,** 147–49 (1954)
52. Galzy, R. Confirmation de la nature virale du court-noué de la vigne par des essais de thermothérapie sur des cultures *in vitro. Compt Rend. Acad. Sci., Paris,* **253,** 706–8 (1961)
53. Galzy, R. Action de la température 35°C sur *Vitis rupestris* atteint de court-noue. *Bull. Soc. franc. Physiol. Vegetale,* **12,** 391–99 (1966)
54. Gandy, D. G. A transmissible disease of cultivated mushrooms ("watery stipe"). *Ann. Appl. Biol.,* **48,** 427–30 (1960)
55. Gandy, D. G., Hollings, M. Die-back of mushrooms: a disease associated with a virus. *Glasshouse Crops Res. Rept., 1961,* 103–8 (1962)
56. Gay, J. D., Kuhn, C. W. Specific infectivity of cowpea chlorotic mottle virus from five hosts. *Phytopathology,* **58,** 1609–15 (1968)
57. Geard, I. D. The role of therapy in the control of plant diseases. *J. Australian Inst. Agr. Sci.,* **24,** 312–18 (1958)
58. Gibbs, A. J. Cryptograms. In *Plant Virus Names,* 135–49 (Martyn, E. B., Ed., Commonwealth Mycological Inst., Phytopathol. Papers, **9,** Kew, Surrey, England, 204 pp., 1968)
59. Giese, A. C. Temperature as a factor in the cell environment. In *Cell Physiology,* 193–207 (W. B. Saunders Co., Philadelphia, Pa., 592 pp., 1962)
60. Ginoza, W. Kinetics of heat inactivation of ribonucleic acid of tobacco mosaic virus. *Nature,* **181,** 958–61 (1958)
61. Ginoza, W. Inactivation of viruses by ionizing radiation and by heat. In *Methods in Virology,* 139–209 (Maramorosch, K., Koprowski, H., Eds., Academic Press, New York, Vol. 4, 764 pp., 1968)
62. Glazewska, Z. Proby zwalczania lisciozweju Chmielu metoda termoterapii. *Biul. Inst. Ochr. Rosl., Poznan,* **24,** 131–34 (1963); *Rev. Appl. Mycol.,* **43,** 1370 (1964)
63. Goheen, A. C. (Unpublished data, 1969)
64. Goheen, A. C., Luhn, C. F., Hewitt, W. B. Inactivation of grapevine viruses *in vivo. Proc. Intern. Conf. Virus Vector Perennial Hosts, Davis, Calif.,* 255–65 (1965)
65. Goheen, A. C., McGrew, J. R., Smith, J. B. Tolerance of strawberry plants to hot-water therapy. *Plant Disease Reptr.,* **40,** 446–51 (1956)
66. Grant, T. J. Effect of heat treatments on tristeza and psorosis viruses in citrus. *Plant Disease Reptr.,* **41,** 232–34 (1957)
67. Grant, T. J., Jones, J. W., Norman, G. G. Present status of heat treatment of citrus viruses. *Proc. Florida State Hort. Soc., 1959,* **72,** 45–48 (1960)
68. Greenaway, S. Notes on the hot-water treatment of cane setts in the Mackay District, 1953. *Proc. Queensland Soc. Sugarcane Technologists,* **21,** 201–5 (1954)

69. Gualaccini, F. Prove di inattivazione termica dell'agente virosico della maculatura lineare del ciliegio. *Boll. Staz. Patol. Vegetale, III,* **18,** 31–47 (1960)
70. Hakkaart, R. A., Quak, F. Effect of heat treatment of young plants on freeing chrysanthemums from virus B by means of meristem culture. *Neth. J. Plant Pathol.,* **70,** 154–57 (1964)
71. Hamid, A., Locke, S. B. Heat inactivation of leafroll virus in potato tuber tissues. *Am. Potato J.,* **38,** 304–10 (1961)
72. Härdtl, H. Die Übertragung und Bekämpfung pflanzlicher Viruskrankheiten. *Anz. Schaedlingskunde,* **35,** 86–91 (1962)
73. Hildebrand, E. M. Rapid transmission of yellow-red virosis in peach. *Contrib. Boyce Thompson Inst.,* **11,** 485–96 (1941)
74. Hildebrand, E. M. Heat treatment for eliminating internal cork virus from sweet potato plants. *Plant Disease Reptr.,* **48,** 356–58 (1964)
75. Hildebrand, E. M. Russet crack: A menace to the sweet potato industry. I. Heat therapy and symptomatology. *Phytopathology,* **57,** 183–87 (1967)
76. Hildebrand, E. M., Brierley, P. Heat treatment eliminates yellow dwarf virus from sweet potato. *Plant Disease Reptr.,* **44,** 707–9 (1960)
77. Hitchborn, J. H. The effect of high temperature on the multiplication of two strains of tobacco ring spot virus. *Virology,* **3,** 243–44 (1957)
78. Hoffmann, G. M., Lemper, J. Möglichkeiten der Gesunderhaltung von Chrysanthemen-Kulturen. *Gartenwelt,* **61,** 417–20 (1961)
79. Hollings, M. Heat treatment in the production of virus-free ornamental plants. *G. Brit. Natl. Agr. Advis. Serv. Quart. Rev.,* **57,** 31–34 (1962)
80. Hollings, M. Studies of pelargonium leaf curl virus. I. Host-range transmission and properties *in vitro. Ann. Appl. Biol.,* **50,** 189–202 (1962)
81. Hollings, M. Cucumber stunt mottle, a disease caused by a strain of arabis mosaic virus. *J. Hort. Sci.,* **38,** 138–49 (1963)
82. Hollings, M. Disease control through virus-free stock. *Ann. Rev. Phytopathol.,* **3,** 367–96 (1965)
83. Hollings, M., Kassanis, B. The cure of chrysanthemums from some virus diseases by heat. *J. Roy. Hort. Soc.,* **82,** 339–42 (1957)
84. Hollings, M., Nariani, T. K. Some properties of clover yellow vein, a virus from *Trifolium repens* L. *Ann. Appl. Biol.,* **56,** 99–109 (1965)
85. Hollings, M., Stone, O. M. Investigations of carnation viruses I. Carnation mottle. *Ann. Appl. Biol.,* **53,** 103–18 (1964)
86. Hollings, M., Stone, O. M. Studies of pelagronium leaf curl virus. II. Relationships to tomato bushy stunt and other viruses. *Ann. Appl. Biol.,* **56,** 87–98 (1965)
87. Holmes, F. O. Elimination of spotted wilt from dahlias by propagation of tip cuttings. *Phytopathology,* **45,** 224–26 (1955)
88. Holmes, F. O. Rose mosaic cured by heat treatment. *Plant Disease Reptr.,* **44,** 46–47 (1960)
89. Holmes, F. O. Elimination of turnip mosaic virus from a stock of horseradish. *Phytopathology,* **55,** 530–32 (1965)
90. Holmes, F. O. Cures for virus diseases. *Plants and Gardens,* **22**(4), 28–31, 54 (1967)
91. ten Houten, J. G., Quak, F., van der Meer, F. A. Heat treatment and meristem culture for the production of virus-free plant material. *Neth. J. Plant Pathol.,* **74,** 17–24 (1968)
92. ten Houten, J. G., Quak, F., van der Meer, F. A. Heat treatment and meristem culture for the production of virus-free plant material. *World Rev. Pest Control,* **7,** 115–20 (1968)
93. Hunter, J. A., Chamberlain, E. E., Atkinson, J. D. Note on a modification in technique for inactivating apple mosaic virus in apple wood by heat treatment. *New Zealand J. Agr. Res.,* **2,** 945–46 (1959)
94. Hutchins, L. M., Rue, J. L. Promising results of heat treatments for inactivation of phony disease virus in dormant peach nursery trees. (Abstr.) *Phytopathology,* **29,** 12 (1939)
95. Jaroslavcev, E., Pomazkov, J. Disinfection of raspberry stock (in Russian) *Sadovodstvo,* **3,** 26 (1967)
96. Johnson, F. H., Eyring, H., Polissar,

M. J. Temperature. In *The Kinetic Basis of Molecular Biology,* 187–285 (John Wiley and Sons, New York, 874 pp., 1954)

97. Johnson, J. The classification of plant viruses. *Wisconsin Univ. Agr. Expt. Sta. Res. Bull.,* **76,** 16 (1927)
98. Jones, O. P., Vine, S. J. The culture of gooseberry shoot tips for eliminating virus. *J. Hort. Sci.,* **43,** 289–92 (1968)
99. Jones, R. A. C., Harrison, B. D. Potato mop-top virus. *Ann. Rept. Scot. Hort. Res. Inst., 1967,* 59–60 (1968)
100. Kassanis, B. Heat inactivation of leaf-roll virus in potato tubers. *Ann. Appl. Biol.,* **37,** 339–41 (1950)
101. Kassanis, B. The control of plant viruses by therapeutic methods. *Proc. Conf. Potato Virus Diseases, Lisse-Wageningen, 1951,* 48–50 (1952)
102. Kassanis, B. Heat-therapy of virus-infected plants. *Ann. Appl. Biol.,* **41,** 470–74 (1954)
103. Kassanis, B. Effects of changing temperature on plant virus diseases. *Advan. Virus Res.,* **4,** 221–41 (1957)
104. Kassanis, B. Therapy of virus-infected plants. *J. Roy. Agr. Soc. Engl.,* **126,** 105–14 (1965)
105. Kassanis, B., Posnette, A. F. Thermotherapy of virus infected plants. *Recent Advan. Botany,* **1,** 557–63 (1961)
106. Kegler, H. Untersuchungen über Virosen des Kernobstes. I. Das Apfelmosaikvirus. *Phytopathol. Z.,* **37,** 170–86 (1960)
107. Kegler, H. Eeine einfache Apparatur zur Wärmebehandlung viruskranker Obstpflanzen. *Arch. Gartenbau,* **15,** 69–74 (1967)
108. KenKnight, G. Thermal stability of peach rosette virus. *Phytopathology,* **48,** 331–35 (1958)
109. Keur, J. Y. Studies of the occurrence and transmission of virus diseases in the genus Abutilon. *Bull. Torrey Botan. Club,* **61,** 53–70 (1934)
110. Kristensen, H. R. Virussygndomme hos planter og termoterapi. *Horticultura,* **20,** 171–80 (1966)
111. Kristensen, H. R., Thomsen, A. Chrysanthemum-viroser. *Tidsskr. Planteavl.,* **62,** 627–69 (1958)
112. Kunkel, L. O. Heat treatments for the cure of yellows and other virus diseases of peach. *Phytopathology,* **26,** 809–30 (1936)
113. Kunkel, L. O. Peach mosaic not cured by heat treatments. *Am. J. Botany,* **23,** 683–86 (1936)
114. Kunkel, L. O. Heat cure of aster yellows in periwinkles. *Am. J. Botany,* **28,** 761–69 (1941)
115. Kunkel, L. O. Potato witches' broom transmission by dodder and cure by heat. *Proc. Am. Phil. Soc.,* **86,** 470–75 (1943)
116. Kunkel, L. O. Studies on cranberry false blossom. *Phytopathology,* **35,** 805–21 (1945)
117. Kunkel, L. O. Transmission of alfalfa witches' broom to nonleguminous plants by dodder, and cure in periwinkle by heat. *Phytopathology,* **42,** 27–31 (1952)
118. Larsen, E. C. Daily temperature-cycles in heat inactivation of viruses in chrysanthemum and apple. *Proc. Intern. Hort. Congr., 17th, College Park, Md.,* **1,** 104 (1966)
119. Lin, K. H., Lo, H. H. A preliminary study on thermotherapy of yellow shoot disease of citrus. *Acta Phytopathol. Sinica,* **4,** 169–75 (1965); *Rev. Appl. Mycol.,* **45,** 1067 (1966)
120. Liu, H. P. The nature of the casual agent of white leaf disease of sugarcane. *Virology,* **21,** 593–600 (1963)
121. McClean, A. P. D. (Personal communication, 1969)
122. McGrew, J. R. Eradication of latent C virus in the Suwanee variety of strawberry by heat plus excised runner-tip culture. *Phytopathology,* **55,** 480–81 (1965)
123. Majorana, G. Infezioni naturali di "variegatura infettiva" su mandarino (*Citrus reticulata* Blanco). Nota preventiva. *Phytopathol. Mediter.,* **5,** 185–87 (1966)
124. Marenaud, C., Saunier, R. Thermotherapie sur pecher. *Ann. Amelioration Plantes,* **17**(1), 13–21 (1967)
125. Martelli, G. P. Termoterapia delle virosi. *L'Ital. Agr.,* **103,** 513–28 (1966)
126. Martelli, G. P., Ciccarone, A. La termoterapia della virosi. *Frutticoltura,* **26,** 613–20 (1964)
127. Martin, J. P. Pathology. *Hawaiian*

Sugar Planters' Assoc. Exptl. Sta. Ann. Rept., 1932, 23–42 (1933)

128. Martyn, E. B., Ed. *Plant Virus Names* (Commonwealth Mycological Inst., Phytopathol. Papers, **9,** Kew, Surrey, England, 204 pp., 1968)
129. Mathews, R. E. F., Lyttleton, J. W. Heat inactivation of turnip yellow mosaic virus *in vivo. Virology,* **9,** 332–342 (1959)
130. van der Meer, F. A. The effect of hot water treatment on a virus of *Opuntia exaltata. Neth. J. Plant Pathol.,* **73,** 58–59 (1967)
131. Mellor, F. C., Fitzpatrick, R. E. Strawberry viruses. *Can. Plant Disease Surv.,* **41,** 218–55 (1961)
132. Mellor, F. C., Stace-Smith, R. Eradication of potato virus X by thermotherapy. *Phytopathology,* **57,** 674–78 (1967)
133. Milbrath, J. A. An investigation of thermal inactivation of alfalfa mosaic virus. *Phytopathology,* **53,** 1036–40 (1963)
134. Miller, P. W. Effect of hot-water treatment on virus-infected strawberry plants and viruses. *Plant Disease Reptr.,* **37,** 609–11 (1953)
135. Morel, G. Regeneration des variétiés virosées par la culture des méristémes apicaux. *Rev. Hort., Paris,* **136,** 733–40 (1964)
136. Mukherjee, A. K., Raychaudhuri, S. P. Therapeutic treatment against leaf curl of some malvaceous plants. *Plant Disease Reptr.,* **50,** 88–90 (1966)
137. Mulder, D. Spikiness disease of Guatemala grass (*Tripsacum laxum* Nash) : a virus disease. *Tea Quart.,* **34,** 16–18 (1963)
138. Murray, D. B., Swarbrick, J. T. Heat treatment of cacao budwood with reference to cacao virus. *St. Augustine, Trinidad, Imp. Coll. Trop. Agr. Rept., 1957–1968,* 65–66 (1959)
139. Nagaich, B. B. Role of thermotherapy in the control of virus diseases of fruit trees and other vegetatively propagated crops. *Indian Phytopathology. Soc. Bull.,* **I,** 43–48 (1963)
140. Nagaich, B. B., Upreti, G. C. Heat inactivation of potato leaf roll virus. *Indian Potato J.,* **6,** 96–102 (1964)
141. Nichols, C. W., Nyland, G. Hot water treatment of some stone fruit viruses. (Abstr). *Phytopathology,* **42,** 517 (1952)
142. Nozeran, R., Neville, P. Nature 'virale' de certaines transformations observées chez des cactacées. *Naturalia monspeliensia Série Botanique, Montpellier,* **15,** 109–14 (1963)
143. Nyland, G. Hot-water treatment of Lambert cherry budsticks infected with necrotic rusty mottle virus. *Phytopathology,* **49,** 157–58 (1959)
144. Nyland, G. Heat inactivation of stone fruit ringspot virus. *Phytopathology,* **50,** 380–82 (1960)
145. Nyland, G. Thermotherapy of virus-infected fruit trees. *Proc. Europ. Symp. Fruit Tree Virus Disease, 5th, Bologna, 1962,* 156–58 (1964)
146. Nyland, G. (Unpublished data, 1969)
147. Nyland, G., Reeves, E. L. Heat inactivation of the little cherry virus in trees of Shiro-fugen flowering cherry. *Intern. Symp. Fyto Farmacie en Fytiatrie,* **14,** 1060–61 (1962)
148. Olson, E. O., Rogers, B. Effects of temperature on expression and transmission of stubborn disease of citrus. *Plant Disease Reptr.,* **53,** 45–49 (1969)
149. van Os, H. Production of virus-free carnations by means of meristem culture, *Neth. J. Plant Pathol.,* **70,** 18–26 (1964)
150. Paludan, N. Undersøegelser vedrørende nellike-viroser. *Tidsskr. Planteavl.,* **69,** 38–46 (1965)
151. Pollard, Ernest C. Theory of the physical means of the inactivation of viruses. *Ann. N. Y. Acad. Sci.,* **83,** 654–60 (1960)
152. Posnette, A. F. Heat inactivation of strawberry viruses. *Nature,* **171,** 312 (1953)
153. Posnette, A. F., Cropley, R. Heat treatment for the inactivation of strawberry viruses. *J. Hort. Sci.,* **33,** 282-88 (1958)
154. Posnette, A. F., Cropley, R., Wolfswinkel, L. D. Heat inactivation of some apple and pear viruses. *Ann. Rept. East Malling Res. Sta., Kent, 1961,* 94–96 (1962)
155. Posnette, A. F., Ellenberger, C. E. Further studies on green petal and other leafhopper transmitted viruses infecting strawberry and clover. *Ann. Appl. Biol.,* **51,** 69–83 (1963)

156. Posnette, A. F., Jha, A. The use of cuttings and heat treatment to obtain virus-free strawberry plants. *Ann. Rept. East Malling Res. Sta., Kent, 1959,* 98 (1960)
157. Price, W. C. Thermal inactivation rates of four plant viruses. *Arch. Ges. Virusforsch.,* **1,** 373–86 (1940)
158. Price, W. C., Knorr, L. C. Kinetics of thermal destruction of citrus tissues in relation to the virus disease problem. *Phytopathology,* **46,** 657–61 (1956)
159. Quak, Frederika. Heat treatment and substances inhibiting virus multiplication, in meristem culture to obtain virus-free plants. *Advan. Hort. Sci.,* **1,** 144–48 (1961)
160. Quak, F. Meristeemcultuur als middel ter verkrijging van virusvrije planten. *Landbouwk. Tijdschr.,* **78,** 301–5 (1966)
161. Ravel d'Esclapon, C. de. La regeneration des varietes d'oeillots Americains par thermotherapie. *Rev. Hort.,* **139,** 1227–29 (1967)
162. Raychaudhuri, S. P. Studies on bayberry yellows. *Phytopathology,* **43,** 15–20 (1953)
163. Raychaudhuri, S. P., Ganguly, B., Basu, A. N. Further studies on the mosaic disease of mulberry. *Plant Disease Reptr.,* **49,** 981 (1965)
164. Rich, A. E. Inactivation of potato virus X in Green Mountain potatoes. (Abstr.) *Phytopathology,* **58,** 402 (1968)
165. Roland, G. Quelques reserches sur l'enroulement de la pomme de terre (Solanum virus 14, Appel & Quanjer). *Parasitica Gembloux,* **8,** 150–58 (1952)
166. Rozendaal, A. Demonstration of experiments with potato viruses. *Proc. Conf. Potato Virus Disease, Lisse-Wageningen, 1951,* 63–65 (1952)
167. Rozendaal, A., Thung, T. H., van der Want, J. P. H. Curing virus diseases by heat. *Proc. Intern. Botan. Congr. 7th, Stockholm, 1950,* **7,** 710–12 (1953)
168. Ruppel, E. G., Delpin, H., Martin, F. W. Preliminary studies on a virus disease of a sapogenin-producing *Dioscorea* species in Puerto Rico. *Univ. of Puerto Rico, J. Agr.,* **50,** 151–57 (1966)
169. Sabin, A. B. The filterable microorganisms of the pleuropneumonia group. *Bacteriol. Rev.,* **5,** 1–68 (1941)
170. Schmelzer, K. Sind viruskranke Pflanzen heilbar? *Wiss. Fortschingsber,* **3,** 105–9 (1967)
171. Selsky, M. I., Black, L. M. Effect of high and low temperature on the survival of wound-tumor virus in sweet clover. *Virology,* **16,** 190–98 (1962)
172. Semancik, J. S., Weathers, L. G. Exocortis virus of citrus: Association of infectivity with nucleic acid preparations. *Virology,* **36,** 326–28 (1968)
173. Sheffield, F. M. L. Presence of virus in the primordial meristem. *Ann. Appl. Biol.,* **29,** 16–17 (1942)
174. Singh, K. Heat therapy of sugarcane. *Indian Sugar,* **17,** 181–86 (1967); *Rev. Appl. Mycol.,* **46,** 3555 (1967)
175. Šip, V. Pokusy s termoterapii u viru svinutky brambor. Czechoslovak republic. *Ministerstvo Zemedlstvi. Lesniho a vodniho hospordarstvi. Ustav vedeckotechickych Informaci Sbornik UVTI: Ochrana Rostlin,* **1**(3), 35–42 (1965)
176. Stace-Smith, R. Current status of bramble viruses. *Can. Plant Disease Surv.,* **40**(1), 24–42 (1960)
177. Stace-Smith, R. Studies on Rubus virus diseases in British Columbia VIII. Raspberry leaf curl. *Can. J. Botany,* **40,** 651–57 (1962)
178. Stace-Smith, R., Mellor, F. C. Eradication of potato viruses X and S by thermotherapy and axillary bud culture. *Phytopathology,* **58,** 199–203 (1968)
179. Stadler, L., Schütz, F. Virusefreie Erdbeeren. *Schweiz. Z. Obst-Weinbau,* **68,** 30–35, 53–61 (1959)
180. Steib, R. J., Thaung, M. M., Wang, L. Effect of heat treatment tests on the germination and yields of sugarcane. (Abstr.) *Phytopathology,* **44,** 507 (1954)
181. Steindl, D. R. L., Hughes, C. G. Ratoon stunting diseases. *Cane Growers Quart. Bull.,* **16,** 79–95 (1953)
182. Stevens, N. E. Departures from ordinary methods in controlling plant diseases. *Botan. Rev.,* **4,** 429–45, 677–78 (1938)
183. Stone, Olwen M., Hollings, M. Tip culture and the production of virus-free clones, *Glasshouse*

Crops Res. Inst. Rept., 1693, 88 (1964)

184. Stretch, A. W. Red ringspot virus of cultivated highbush blueberry. *Proc. Ann. Blueberry Open House, 33rd, Pemberton, New Jersey,* 3–4 (1965)

185. Stretch, A. W. (Personal communication, 1969)

186. Stubbs, L. L. A phytotron cabinet for heat-therapy studies with virus-infected plants. *Commonwealth Phytopathol. News,* **9,** 49–52 (1963)

187. Stubbs, L. L. Production of virus-free propagating material. *J. Agr. (Victoria),* **61,** 421–28 (1963)

188. Stubbs, L. L. Apparent elimination of exocortis and yellowing viruses in lemon by heat therapy and shoot-tip propagation. *Proc. Conf. Intern. Organ. Citrus Virologists, 4th, Italy, 1966,* 96–99 (1969)

189. Tahama, Y. Studies on the dwarf disease of mulberry tree. VIII. Recovery by heat treatment. *Ann. Phytopathol. Soc. Japan,* **29,** 39–42 (1964)

190. Taylor, R. H. An investigation of the viruses which cause Woodiness of passion fruit. (Abstr.) *J. Australian Inst. Agr. Sci.,* **25,** 71 (1959)

191. Thirumalachar, M. J. Inactivation of potato leafroll by high temperature storage of seed tubers in Indian plains. *Phytopathol Z.,* **22,** 429–436 (1954)

192. Thomsen, A. Termoterapeutiske behandlinger af nellike. *Horticultura,* **15,** 136–39 (1961)

193. Thomsen, A. Frugttrae-vira inaktiveret ved termotarapi. *Tidsskr. Planteavl.,* **72,** 141–52 (1968)

194. Thomson, A. D. Heat treatment and tissue culture as a means of freeing potatoes from virus Y. *Nature,* **177,** 709 (1956)

195. Thomson, A. D. The elimination of viruses from potato tissue. *Proc. Conf. Potato Virus Diseases, 3rd, Lisse-Wageningen, 1957,* 156–59 (1958)

196. Thung, T. H. Waarnemingen omtrent de dwergziekte bij framboos en wilde braam. II. *Tidschr. Plantenziekten,* **58,** 255–59 (1952)

197. Thung, T. H. Herkenning en genezing van enige virusziekten. *Mededel. Directeur Tuinbouw.,* **15,** 714–21 (1952)

198. Traylor, J. A., Williams, H. E., Nyland, G. Heat therapy of rose mosaic. (Abstr.) *Phytopathology,* **57,** 1010 (1967)

199. Valenta, V. Thermal inactivation of yellows-type viruses *in vivo. Acta Virol.,* **6,** 94 (1962)

200. Walkey, D. G. A., Webb, M. J. W. Virus in plant apical meristems. *J. Gen. Virol.,* **3,** 311–13 (1968)

201. Weathers, L. G., Calavan, E. C. Nucellar embryony—a means of freeing citrus clones of viruses. *Proc. Conf. Intern. Organ. Citrus Virologists, 1st, Univ. California, Riverside, Calif., 1957,* 197–202 (1959)

202. Weil, B. Thermale inaktivierung von zwei Pflanzenviren, ein Beitrag zur Wärmetherapie pflanzlicher Viruskrankheiten. *Phytopathol. Z.,* **31,** 45–78 (1957)

203. Welsh, M. F., Nyland, G. Elimination and separation of viruses in apple clones by exposure to dry heat. *Can. J. Plant Sci.,* **45,** 443–54 (1965)

204. Wilbrink, Gerarda. Warmwaterbehandeling van stekken als geneesmiddel tegen de serehziekte van het suikerriet. *Arch. Suikerind. Ned. Ind. Surabaya,* **31,** 1–15 (1923)

205. Woese, C. Thermal inactivation of animal viruses. *Ann. N. Y. Acad. Sci.,* **83,** 741–51 (1960)

206. Yarwood, C. E. Temperature and plant disease. *World Rev. Pest Control,* **4,** 53–63 (1965)

207. Yarwood, C. E. Heat activation and inactivation of plant viruses. *Deut. Akad. Landwirtsch-Wiss., Berlin, Tagungsber.,* **74,** 119–24 (1965)

208. Yarwood, C. E., Sylvester, E. S. The half-life concept of longevity of plant pathogens. *Plant Disease Reptr.,* **43,** 125–28 (1959)

209. Zandbergen, M. Hot water treatment for bulbs. *Roy. Hort. Soc. Daffodil-Tulip Yearbook, 1965,* **30,** 187 (1964)

Copyright 1969. All rights reserved

MULTILINE CULTIVARS AS A MEANS OF DISEASE CONTROL[1,2]

By J. Artie Browning and K. J. Frey
Iowa State University, Ames, Iowa

Introduction

"The multiline theory for the production of composite varieties is one of the truly new concepts of the century in breeding self-pollinated crops" (59). In spite of such strong opinions, and the fact [known for mid-America at least since 1898 (4)] that crop mixtures buffer against disease loss, use of multiline cultivars has been shrouded in controversy to the point that only three research groups are actively developing multiline cultivars and only three series have been released. Literature on multiline cultivars is sparse, and much of it is found in symposia papers, abstracts, and annual reports. Nevertheless, this paper will attempt to synthesize a consistent body of theory underlying the need for the use of genetic diversity in small grains, to analyze critically the available pertinent data on multiline cultivars, and to suggest promising experiments.

Mutiline or mutilineal cultivars?—One school uses the term multilineal cultivar (7); most others use multiline cultivar (e.g. 16, 19, 36, 38). Used adjectivally, multilineal is the correct form, but we prefer multiline since it parallels its counterpart, pure line, long accepted in plant breeding literature. There is precedent too, for use of line as an adjective. Consider line chief, line drive and line officer, for instance.

Resistance types.—Resistance types have been referred to by divers names (36). Use of the term hypersensitivity as a general term for immunity and high resistance should be avoided. It assumes host cells die in response to the pathogen which dies subsequently. Actually, the mycelium is not dead but quiescent; and it can resume development under proper conditions after periods of at least 9 days after inoculation (by which time the characteristic resistant response is fully formed) for *Puccinia graminis avenae* (14), *P. coronata avenae* (14, 88), and 20 days for *P. graminis tritici* (23).

Van der Plank (86) introduced the concepts of vertical (VR) and horizontal (HR) resistance. Vertical resistance is effective against some, but

[1] Published with the approval of the Iowa Agricultural and Home Economics Experiment Station, Ames, Iowa as Journal Paper No. J-6199, Project 1176.

[2] The following abbreviations are used: resgene *for* resistance gene (87); Virugene *for* virulence gene; HR *for* horizontal resistance; and VR *for* vertical resistance.

not all, races of a pathogen; HR is effective against all races. Horizontal resistance usually is polygenic in inheritance while VR usually is conditioned by oligogenes (66). Plant pathologists have been preoccupied with individual plants and individual races but it is population of plants that feed increasing populations of people. We like the concept of "population resistance" to describe the situation where the fungus population cannot increase and damage the well-buffered (2) host population which may possess VR, HR, or combinations of them.

The Risk in Growing Small Grains

Crowd diseases, especially those caused by the continental rust fungi, are a primary risk in small grain culture in many areas of the world. Theoretically, the rusts should be held in check by several mechanisms: (*a*) interspecific diversification; (*b*) intraspecific diversification; (*c*) physiologic resistance; and (*d*) stabilizing tendencies in the rust fungi. These are interrelated, with (*c*) and (*d*) basic to the operaton of the others. When cultivation began, man decreased interspecific diversification by controlling weeds, and during the past seven decades plant breeders have eliminated intraspecific diversity by producing pure line cultivars. Early pure line selection resulted in homogeneous genotypes for a given agricultural area. There were several genotypes within an area, and they differed in contiguous areas. The situation became acute when plant breeders began using hybridization extensively. They incorporated a single resistance gene (resgene) into cultivars sown from border to border. For example, the Bond and Hope genes for crown and stem rust resistance in oats and wheat, respectively, were distributed in cultivars grown extensively in the United States and Canada. Widespread use of pure line cultivars also largely eliminated the competition among rust clones, thus interfering with stabilizing tendencies of the rust fungi and giving new virulent biotypes a selective advantage. These factors have been of paramount importance in such biotypes becoming widespread.

Several decades ago, rust control appeared to have an easy solution: Produce cultivars with homogeneous genetic resistance to all known races. Stanton et al. (72), reflecting the optimism of the 1930's, said: "The development of strains of oats highly resistant to smut, crown rust, and stem rust, and with desirable grain characters . . . has been achieved." Stakman & Christensen (71), who earlier had been just as optimistic, said 36 years later: "The hopes (which periodically soared very high) of controlling . . . rust have not been completely realized. Results have not matched expectations."

Variability of rust fungi.—Why? What went wrong? The rusts are highly variable. Leeuwenhoek (28) may have been the first author to note pathogenic specialization in the rust when, in 1678, he observed that some grasses were contaminated with rust whereas others were not. He did not realize the capital significance of his observation and left it to Eriksson to discover pathogenic specialization on different crop species, and to Stakman and colleagues, on different crop varieties (43, 71). The hosts, too, are variable. Biffen (5) discovered that rust reaction was subject to Mendelian inheritance.

Since these landmark discoveries, man has unwittingly guided the evolution of the rust fungi over vast areas. In general, plant breeders searched for sources of immunity from or high resistance to rust pathogens, studied their inheritance, and incorporated resgenes into commercial cultivars. Whenever possible, they sought the pure line (the unit) of "greatest value" (2). Plant pathologists, in turn, collected, identified, and mapped the prevalence of pathogenic races [the so-called units of plant pathology (53)]. They have even been accused of making pathogenic races the end rather than a means to better understanding.

The vicious circle.—This resulted in man's homogenizing to one genotype where there had been many. This one, selected for resistance to prevalent races, excluded those races. However, deviants in the rust population, virulent on the new resistant varieties, increased without competition wherever resistant cultivars were released—the USA, Canada, Australia, Kenya (43). Breeders responded by incorporating another resgene into the host population, and the pathogen countered with another virugene in its population. And so a vicious circle was maintained. Several reviews have discussed this problem relative to wheat and wheat stem rust in North America (43, 70, 71), but possibly it is best illustrated with the racial dynamics of oat crown and stem rust (Fig. 1). With oats, the appearance of a new pathogen, *Helminthosporium victoriae,* forced an abrupt change in oat cultivars which became polar for crown and stem rust response (52, 71).

The cultivar change for Iowa is representative of the North Central region. In 1941, some 99 per cent of the Iowa oat acreage was sown to Richland and similar rust-susceptible pure line cultivars. By 1945, Victoria derivatives, cultivars of hybrid origin, occupied 92 percent of that acreage. *H. victoriae* (virulent on the Victoria derivatives) caused losses to the Iowa oat crop estimated at 5, 25, 32, and 1 per cent in 1945, 1946, 1947 and 1948, respectively, so Victoria derivatives were replaced rapidly by blight-resistant Bond derivatives (from 0.05 per cent in 1945 to 98 per cent in 1949). A single Bond derivative, Clinton, occupied 75 per cent of the total oat acreage of the USA in 1950.

The sudden shift from Victoria-Richland cultivars inevitably caused parallel shifts in clones of the pathogen populations (Fig. 1). The Victoria derivatives were resistant to *P. graminis avenae* races 1, 2, and 5, but susceptible to rare races 8 and 10. Expectedly, races 8 and 10 increased, reaching 45 per cent of the stem rust cultures identified in 1946. As the Victoria derivatives were forced from production by *H. victoriae,* itself a product of an extensive homogeneous host population, the Bond derivatives increased. These were resistant to races 8 and 10, but susceptible to race 7. After a two-year lag, the race-7 increase paralleled that of Bond derivatives peak-ing at over 80 per cent of the total isolates in 1953. The Bond derivatives also selected for crown rust clones virulent on them. Identification of cultures virulent on Bond derivatives, which earlier had been considered highly resistant to crown rust, began in the mid 1930's. Although race 45 was a super race on cultivars grown at that time, it did not predominate

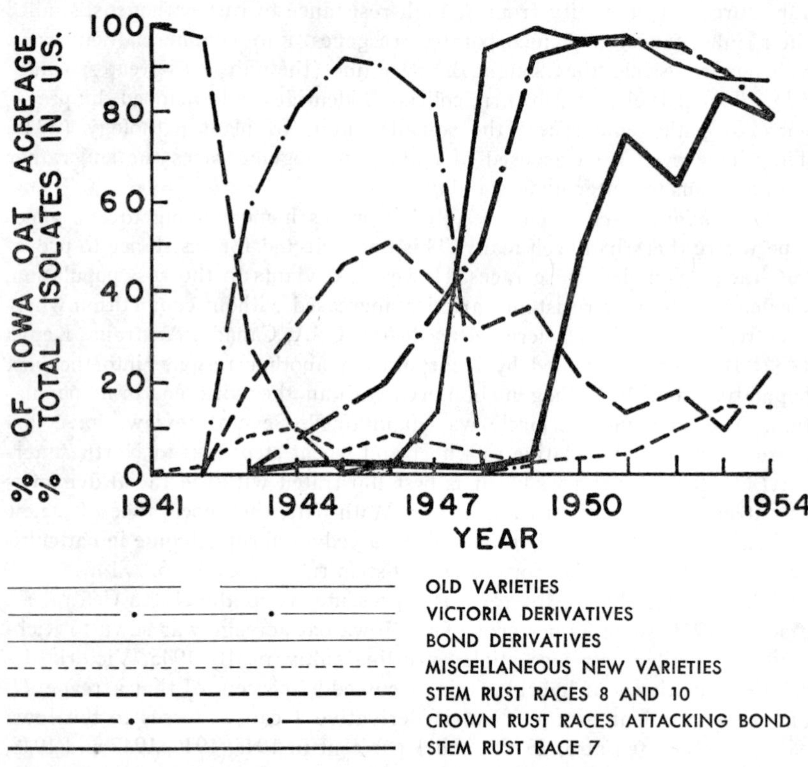

FIG. 1. Percentage of Iowa oat acreage planted to cultivars of different origin, and percentage of total United States isolates of crown and stem rust races identified during 1941–1954, inclusive. After (52).

until the commercial production of Bond derivatives began. Then Bond races jumped to over 95 per cent of all cultures of crown rust identified in the USA. The Bond resistance has been supplemented several times, and after each there has been a subsequent change in rust race prevalence. The result is that oats in the USA continue to be grown, but with no protection from certain races. This illustrates the point that the pathogenic response of the rust can be "causally related to the man-made modification of their hosts" (43).

That oligogenes giving VR were widely used by plant scientists was understandable. They were easy to manipulate in greenhouse and field, superior in yield trials, and there was always the possibility that the next resgene might not be ephemeral. Plant pathologists, especially, were unaccustomed to working with populations, and breeders pursued the boom yields that a well adapted cultivar in a monoculture (weed free and rust free) could give. They apparently were unaware that, in selecting for maximum adaptation in a self-pollinated crop, they simultaneously selected for minimum adaptibility (66), so that in the face of adversity the crop had no biological means of containing the fungus or adjusting to its presence.

Warnings.—There were numerous warnings that all was not well (21, 37, 38, 57, 75, 77, 79, 87). Stevens (73, 74) showed that disease importance increased with the degree of inbreeding in several crop species, and stated (76) that no control measure yet devised could equal the efficiency of the inherent capacity of a cross-pollinated crop to protect itself. Likewise, Rosen (63) stated: "The epidemics experienced on oats during the past decade were in part caused by the extensive use of varieties possessing the same germ plasm." A solution suggested by Suneson (81) was the use of nonuniform crop varieties.

Breaking the Vicious Circle

The vicious circle of new oat and wheat cultivars selecting new, virulent forms of their respective pathogens—a phenomenon which necessitates still newer cultivars—probably can be broken with a continental control program utilizing: (*a*) fungicides [reviewed by Rowell (64)]; (*b*) tolerant cultivars; (*c*) pure line cultivars with polygenic horizontal resistance; or (*d*) multiline cultivars with heterogenic horizontal resistance.

Tolerance.—A tolerant cultivar is one which supports susceptible-type uredia but which shows significantly smaller yield reductions from rust infection than a susceptible one (67). Although susceptible, the tolerant cultivar endures the pathogen. When compared with their respective isogenic lines, Benton oat cultivar was tolerant to a massive crown rust epiphytotic whereas Clinton was severely damaged although both appeared susceptible (21). Cherokee and several other "susceptible" oat cultivars were more tolerant to three races of crown rust than were Benton and Clinton (67).

Van der Plank (85) examined data from Hayden (34) showing that Lee and Sentry wheats were tolerant to *P. graminis tritici* race 15B when compared to Marquis, Carleton, and other varieties. Heavy versus light inoculations made relatively little difference on Marquis and Carleton, but did on Lee and Sentry; furthermore, the rate of rust spread was slower on Lee and Sentry. Van der Plank interpreted the tolerance of these cultivars as partial resistance manifested not by lower infection type, but by a slower rate of increase. He suggested that, if such cultivars were grown in Central North America, they would cumulatively delay a rust epiphytotic out of existence.

Tolerance also should be more stable than high resistance, since it should provide no mechanism for new races to prevail over established ones (21), but this has not been tested. Tolerance to oat crown rust is inherited as a quantitative character, with heritability (calculated from components of variance) ranging up to 55, 75, and 52 per cent, respectively, when measured as yield, kernel weight, and kernel density response (68). Tolerance is poorly understood and difficult to manipulate, but promising.

Pure line cultivars with horizontal resistance.—Unlike VR, HR is effective against all races of a pathogen (86). It is manifested by a reduced rate of pathogen increase but not necessarily by a reduced infection type (86). No sharp line separates pure line tolerance from pure line HR, yet tolerance probably is an entity "separable from resistance" (36). In general, a tolerant cultivar looks susceptible and yields resistant; a cultivar with HR should both look and yield resistant. There is strong evidence that generalized (horizontal) resistance has prevented corn rust from being of economic importance in the USA (36). As defined (86), HR is the indicated control measure for the cereal rusts. However, it is impossible to prove a given resistance type is effective against all clones of a pathogen, present and future; therefore, HR must be regarded as a very useful concept which merits diligent study. Pure line HR and tolerance will be compared later with population resistance of multiline cultivars.

Multiline cultivars with synthetic horizontal resistance.—Multiline cultivars were suggested by Jensen (38) for oats, and by Borlaug & Gibler (6, 10) for wheat. The following three programs relate to the introduction of multiline cultivars:

(*a*) The New York program. Jensen (38) proposed the development of multiline cultivars by blending several compatible pure lines of different genotypes. The pure lines thus blended would be nonuniform for disease resistance (38) and agronomic phenotype (40). Such cultivars require judicious selection of components (32, 39). Jensen & Kent (41) are developing multiline cultivars in three maturity classes for New York state.

(*b*) The Rockefeller Foundation program. Borlaug & Gibler (10) used a modified backcross approach described by Borlaug (7) to develop wheat lines for multiline cultivars. The best available recurrent parents were chosen and a large group of "donor parents" was selected from International Wheat Rust Nursery data. These were crossed and backcrossed. Bc_1 plants were tested with a tester race and resistant plants most like the recurrent parent agronomically were backcrossed to the recurrent parent, and the process repeated for three backcrosses. Eight to 16 of the best isogenic lines were selected to be blended mechanically to form the multiline cultivar. A large number of lines was considered desirable to minimize the visual impact of one slightly off-type line. The composition of the multiline cultivar could be varied from year to year in response to race changes.

While enthusiasm for multilines continues in CIMMYT (Centro Internacional de Mejoramiento de Maiz y Trigo), a multiline wheat cultivar has not been released from the Mexican program. With new semi-dwarf wheats, progress in yield and lodging resistance was so rapid through conventional breeding that the release of multiline cultivars built on creole recurrent parents was precluded (59). This points out that the backcross method of developing multiline cultivars is agronomically conservative (19) and its use must be preceded by a breeding program which has developed agronomically superior recurrent parents. Doubtless multiline cultivars would have been released in Mexico if rust had become a limiting factor in production.

The first multiline cultivar for commercial production was Miramar 63, released from the Colombian program (59) for use in the high, cool savannahs of the Andes where stripe rust is serious. To develop Miramar 63, the Brazilian wheat Frocor was crossed with some 600 varieties and lines. Over 1,200 lines, similar phenotypically to Frocor but with resistance from the 600 non-recurrent parents, were produced. Miramar 63 was a mechanical mixture of equal parts of ten of the best lines giving resistance to stripe and stem rust. Miramar 63 was widely accepted, and its yield more than doubled those of older varieties in some areas (60). Within 2 years stem rust was parasitizing two component lines; but total losses were always less than the theoretical maximum of 20 per cent. The two stem rust susceptible lines, plus two others, were dropped from the composite variety, and four new ones from the reserve of over 600 lines were added to form a new multiline cultivar, Miramar 65. This illustrates the plasticity inherent in multiline cultivars (61).

(*c*) The Iowa program. The third group developing multiline cultivars is in Iowa where two series of multiline cultivars have been released (15–17). Two recurrent parents have been used: a Clintland type (C.I. 7555) and an early experimental line (C.I. 7970). Over 25 apparently different resgenes have been incorporated into the isogenic lines of the recurrent parents via conventional backcrossing.

For seedling resistance, we make crosses and first backcrosses in the greenhouse, and resistant Bc_1 plants are identified by rust test and backcrossed again to the recurrent parent. Three or four generations are obtained in the greenhouse per year by using an 18-hour photoperiod. All lines are backcrossed six times, after which F_2 plants from Bc_6 are grown in space-planted rust nurseries. The F_3 survivors are grown in 4- or 8-foot rows and selected rigorously for rust reaction and agronomic type. The F_4 survivors are further selected in 4×8-foot yield plots, and lines phenotypically indistinguishable from the recurrent parent are bulked.

The program is slower when adult plant resistance is involved, for rust testing must be done in the mature plant stage. With some resgenes, resistance is apparent only in the field. This allows fewer generations per year but permits better opportunity for selecting for agronomic type in early generations.

The number of backcrosses required in multiline development is a moot question. The backcross method is "well-suited for effecting the small number of gene substitutions to increase the usefulness of successful varieties, without the risk of breaking up the existing combinations of desirable genes which have made them outstanding in many respects" (12). Having started with well adapted recurrent parents, we use six backcrosses to recover their types. Six backcrosses should eliminate at least 97 per cent of the unlinked genes and, therefore, "should reconstitute quite precisely the genotype of the recurrent parent, but three backcrosses would be inadequate to accomplish this purpose" (12). CIMMYT workers have used three backcrosses, but their crosses have been made in the fields where rigid selection could be practiced in early generations. This may have added the equivalent of one or two more backcrosses.

Two series of multiline cultivars have come from the Iowa program. Multiline E68 (early) and Multiline M68 (midseason) have been released; Multiline E69 and Multiline M69 are being increased; increase of Multiline E70 and Multiline M70 will begin in the summer of 1969. At present it takes five years from the time Breeder's Seed of individual isogenic lines is increased to the time farmers sow Certified Seed and harvest grain which cannot be resold as Certified Seed.

Much work is needed to place a sound scientific foundation under the system of determining the composition of a multiline cultivar initially and altering it subsequently. Under "limited generation" seed certification rules, growers of Certified Seed of all cultivars (pure line, multiline, etc.) must return to Registered Seed each year. Limited generation rules permit altering the composition of the multiline cultivars whenever desirable. In Iowa, the component isolines, candidate component isolines, multilines, candidate recurrent parents, and recommended pure line cultivars are tested each year at several locations in yield trials. They also are tested in 10 to 12 single-race rust nurseries. Data from these trials are used to determine the end product, the multiline cultivar, and to update the cultivar before each release. We select a series of candidate lines whose production is equivalent to that of the recurrent parent and array their reactions to rust races. From this array, lines are selected and composited in proportions to give adequate population resistance to contemporary crown rust races. Admittedly, some of the methodology we use is subjective. For instance, immune lines and those that give an intermediate reaction are hypothesized to have equal value in a multiline. We mix isolines in unequal proportions based on the relative resistance of each line to each race and the relative, actual, and anticipated prevalence of the different races. Our Multiline E68 contains ten component lines; Multiline E69, eight; Multiline E70, eleven; Multiline M68, eight; Multiline M69, nine; and Multiline M70, seven components.

Action of Diverse Stands in Disease Situations

Many experiments have been reported involving mixtures of different genotypes, but relatively few considered the disease factor. In some cases disease influence was not reported (e.g. 56) even though it was known to be a factor (54, 55).

Most literature on diseases in mixtures is concerned with small grains, but one interesting exception is with cotton. Mixed cropping of cotton is common in the Malwa plateau of Central India. The common mixture contains native diploid *Gossypium arboreum* 'Desi' (itself a mixture) and American tetraploid *G. hirsutum* 'Upland.' The Upland component when in the mixture is less damaged by red leaf and leafroll than when in pure stands, and Desi is protected from Fusarium wilt. The result is a higher yielding, higher quality, more stable product which farmers continue to grow in spite of a penalty imposed under the Cotton Transport Act and the Cotton Ginning and Pressing Act for not growing a pure crop (1, 66).

Blends of unrelated pure lines.—Jensen (38) reported yields from oats-barley, wheat-oats-barley, and oats-oats field blends, were consistently in favor of the blend over the component means. While the main thrust of his paper was directed toward a future disease risk situation, he did not mention the impact of disease on his experiments.

Wheat-wheat mixtures have been grown to control leaf rust (11). Wheat-oat mixtures under rust attack yielded more than either component (4) and prevented rust penetration through the plot (81). In the severe stem rust epiphytotic of 1935, mixtures of resistant oats and susceptible barley yielded more than the mean of the pure stands (46). Oat-barley mixtures have been suggested as an agro-technical method for controlling the oat sterile-dwarf virus disease (84).

Pfahler worked with mixture of cultivated oat species for grain (54) and grain and forage yield (55). The species had different crown rust responses, but he did not separate the effect of disease from the remainder of the environment. The fitness of the composites exceeded that of the components for both grain and forage yield. Mixtures of Clintland and Mo. 0–205 oat cultivars yielded 13 per cent more than the mean of the pure stands when under stem rust attack (13), but this blend produced a 5.5 per cent advantage when no rust was present (15). Apparently this superiority of blends of cultivars over the means of the components is nearly universal (2, 66).

Mixtures of near-isogenic lines.— Yield response of mixtures of near-isogenic lines of wheat or oats to rust are shown in Table I. In each case the mixture yield showed a nonlinear response. The superior yield resulted because less rust, presumably causing less damage, was present on susceptible plants in blends than in pure stands (4, 13, 81).

Evidence also comes from direct estimates of pathogen development. Blends and pure stands of Clintland *A* (susceptible to stem rust, race 8), Clintland *D* (susceptible to race 7), and Clintland *BD* (resistant to both 7 and 8) were grown in 100 × 100-foot plots. Susceptible-type uredia per ten susceptible culms were counted at frequent intervals as rust emanated from foci. However, workers handling plants supplemented natural air currents in disseminating spores. Resistant plants served as effective barriers to air dissemination of inoculum but not to worker dissemination. This was most obvious near the foci and along entry paths where worker activity was greater, and data from there were discarded. Twelve feet from the foci, however, away from most early-season worker activity, rust had increased by rust climax at the rate of $r = 0.42$ per day for a 1:1 mixture and $r = 0.50$ per day for the susceptible pure stands (16). With the lower rate of increase, susceptible plants in the mixture had fewer rust pustules than did those in a pure stand (Table II) (15).

Leonard & Kent (47, 49) studied increase of oat stem rust in the same Clintland lines. Assuming that rust increase on resistant plants was negligible and that the probability of a urediospore reaching a susceptible plant was proportional to the percentage of susceptible plants in the mixture, they calculated that the rate of stem rust increase was proportional to the logarithm of the proportion of susceptible plants in the mixture. Increase of *P. graminis avenae* races 6F and 7A in the Clintland lines was described by the equation $r_m = r_s + c \log_e m$, where r_m was the rate of increase in the mixture, r_s was the rate of increase in a susceptible plot, m was the proportion of susceptible plants in the mixture, and c is a constant with respect to m. The reciprocal of the generation time of the fungus i.e., l/g, can be substituted for c. The generation time of the fungus, defined as the time that elapsed between inoculation and the production of the average spore in a pustule, was about 21 days.

TABLE I

OBSERVED AND EXPECTED[a] RELATIVE YIELDS OF NEAR-ISOGENIC WHEAT AND OAT LINES IN PURE STANDS AND MIXTURES UNDER RUST EPIPHYTOTIC CONDITIONS

Crop	Pathogen	Cultivars Susceptible:Resistant	Yield of designated Susceptible: Resistant blends in percentage of the resistant							
			1:0	0:1	3:1		1:1		1:3	
					Obs.	Exp.	Obs.	Exp.	Obs.	Exp.
Wheat[b]	Stem Rust	Onas 41:Onas 53	43	100	94	57	—	—	—	—
Oats[c]	Crown Rust Race 203	Clinton:Clintland	56	100	75	67	86	80	94	89
Oats[d]	Stem Rust Race 7	Clintland *D*:Clintland *BD*	72	100	—	—	92	86	—	—
Oats[e]	Fungicide sprayed	Clintland *D*:Clintland *BD*	99	100	—	—	98	99.5	—	—

[a] Assuming a simple linear response of the components. [b] 1957 data from (81). [c] 1962 data from (15). [d] Averages for 1957–60 from (15). [e] Averages for 1958–60 from (15).

Clifford (25) studied crown rust dispersal in a 1:1 susceptible:immune oat mixture and found that, in comparison to the pure stand, the spread of rust to susceptible plants in the plot periphery was reduced, reflecting the buffering effect of the resistant component.

Since a multiline cultivar contains several individual lines each containing a different VR gene, the cultivated unit (i.e. the heterogeneous population of plants) has quantitative population resistance to the rust population similar to that possessed by a pure line cultivar with tolerance or HR. Vertical resistance appears best, relative to other types of protection, in small plots where abundant extraneous inoculum inundates susceptible lines. Cultivars suspected of having tolerance or HR should be tested in large plots so that rust which causes damage must increase upon the variety being evaluated (86). Probably Suneson (81) found such a great advantage (37 per cent) for the 3:1 susceptible:resistant wheat blend over the mean of the pure stand (Table I) because he used large plots (5 × 36 feet with a single rust focus at one end). With oats, our yield advantage for blends was never this high but we used standard 4 × 8-foot yield plots. If no rust entered from outside the plot (which is progressively less likely, the smaller the plot) rust spores had to travel 4.5 times as far in Suneson's wheat plots as in our plots. Since rust dispersal is a function of time and distance, large plots would provide a more precise measure of the yield advantage of a multiline cultivar. Current standard plot size at the Iowa Station for multi-

TABLE II

MEAN NUMBER OF OAT STEM RUST UREDIA AT RUST CLIMAX ON TEN SUSCEPTIBLE PLANTS AT EIGHT POINTS OF THE COMPASS TWELVE FEET FROM THE FOCUS IN MIXTURES OF CLINTLAND ISOGENIC LINES[a]

Mixture	Virulent Race	No. Uredia Per Cent
Clintland *A*	8	103
Clintland *D*	7	97
Mean of above		100[b]
Clintland *BD*	None	0
Clintland *A*+*B*	8	59
Clintland *D*+*B*	7	63
Clintland *A*+*D*	7+8	56
Clintland *A*+*D*+*BD*	7+8	36

[a] 1961 data from (15). [b] 100% = mean of 380 susceptible-type uredia/ten culms.

line cultivar experiments under rust epiphytotic conditions is 50 × 50 feet. Plots are separated by 50 feet of a resistant variety of oats and are arranged in a E-W line to minimize interplot spread of inoculum by prevailing southerly winds. People may not enter the plots during periods of rust increase. Such plots should more closely approximate farm conditions and minimize the representational or "cryptic" error (86).

Yield of spores as a measure of disease severity.—Grain yield or rust prevalence can be used as measures of the effectiveness of multiline cultivars. Measuring the spore yield of the pathogen enables one to study the developing epiphytotic from outside the plot. Cournoyer, Browning & Jowett (26, 27) did this using a blend of six isogenic lines and its pure line susceptible and resistant counterparts. They inoculated 50 × 50-foot plots by transplanting infected seedlings into the center of each plot. Each transplant bore 12 uredia of one crown rust race. Half the plots were inoculated only with race 264 which was virulent on all components of the blend, and the other half was inoculated with six races including race 264. Spores were trapped with Rotorod spore samplers (3) at the periphery of each plot downwind from the focus for 2 hr (13:00 to 15:00) daily. The results (Fig. 2) showed very different spore yields from susceptible pure lines and the multilines. The epiphytotic developed to completion in the susceptible pure line as indicated by the plateau of spore production apparent on July 13 and 14. The development of the pathogen was terminated by the lack of uninfected tissue. The epiphytotic did not develop to completion on the multiline, however, but was still increasing when it was terminated by host maturity. Further statistical treatment of these data showed that the logistic growth model adequately describes the development of an artificially induced crown rust epiphytotic (35). Consequently, the spore yield data support the theory that multiline cultivars buffer against loss from a crown rust epiphytotic by delaying the rate of increase and degree of ultimate development of the pathogen.

Mechanism of Action of Multilines

Reduction of x_0 *and* r.—The mechanism by which multiline cultivars probably buffer against disease loss has been described in part by several authors (7, 16, 38, 47, 66, 86). According to van der Plank (86), the pathogen increases from initial inoculum x_0 at the rate r in time t, and results in x proportion of susceptible tissue becoming infected. A variety with VR, being selectively resistant to the race population, reduces x_0, but since r remains unchanged and is usually so high as not to be limiting, x may be very large by the end of the disease season. Vertical resistance, therefore, is valuable only as long as it gives resistance to all prevalent races and keeps x_0 very small for all. This is why "vertical resistance is well able to control focal outbreaks and mild epidemics of stem rust .... But vertical resistance on its own has been unsatisfactory against great epidemics" (86).

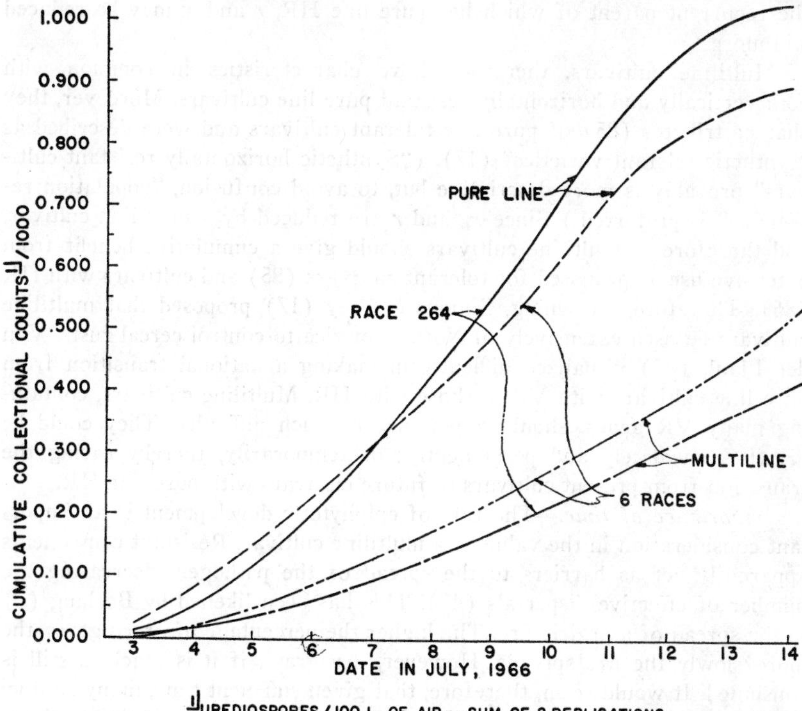

FIG. 2. Cumulative daily collectional counts from spore samplers outside 50 × 50-foot plots (summed over two replications), of oat crown rust unrediospores/100 liters of air, expressed as proportions of 1000 and plotted against time for the four treatment combinations. From (26).

A variety with HR, supposedly resistant to all races of the pathogen, does not reduce x_0; it reduces r (86). With r small, the rate of epiphytotic development is reduced to the point where the host matures with small x and little measurable damage.

Since a multiline cultivar possesses many genes for VR, x_o is reduced. Assuming a viable spore of a race virulent on two of ten components of a miltiline, the probability is only 0.2 that the plant on which the spore lands will be susceptible. If the spore lands on a susceptible plant, infects the tissue and forms a pustule, the progeny have a probability of 0.2 that surrounding plants will be compatible hosts. Thus, not only is x_o reduced, as is characteristic of a cultivar with VR, but r also is reduced as is characteristic of HR. This explains the reduced number of uredia shown in Table II, the reduced slope of the multiline curve in Fig. 2, and the data of Clifford (25) and Leonard (47). If VR genes are incorporated into isogenic lines the recurrent parent of which has pure line HR, r and x may be reduced still more.

Multiline cultivars, therefore, have characteristics in common with both vertically and horizontally resistant pure line cultivars. Moreover, they share attributes (85) of pure line tolerant cultivars and were described as "synthetic tolerant varieties" (17). ("Synthetic horizontally resistant cultivars" probably is more descriptive but, to avoid confusion, "population resistance" is preferred.) Since x_0 and r are reduced by a multiline cultivar, and therefore x, multiline cultivars should give a cumulative benefit from extensive use as proposed for tolerant cultivars (85) and cultivars with HR (86). Therefore, Browning, Simons & Frey (17) proposed that multiline cultivars be used extensively in North America to control cereal rusts. Van der Plank (86) visualized difficulty in making a national transition from pure line cultivars with VR to those with HR. Multiline cultivars, containing many VR genes, should experience no such difficulty. They could be grown immediately and permanently; or temporarily, thereby easing the transition from present cultivars to future cultivars with pure line HR.

Importance of time.—The rate of epiphytotic development is an important consideration in the value of a multiline cultivar. Resistant components apparently act as barriers to the spread of the pathogen, decreasing the number of effective dispersals (47). This has been likened by Borlaug (7) to the spread of a prairie fire. The higher the percentage of green grass, the more slowly the fire spreads. However, dry grass, if it is reached, still is consumed. It would seem, therefore, that given sufficient time, many subfoci would develop, the barrier effect would be minimized, and susceptible plants would be damanged by rust building up upon themselves and other susceptible plants near each subfocus—unless, as in Iowa, time runs out in the short disease season and rust increase on the multiline is stopped by host maturity (Fig. 2). That the situation is more complex is indicated by several cases in which diversity protects the host over a much longer growing season. Rust is present throughout the growing season in Colombia, but Miramar 63 buffered effectively against losses from stripe and stem rust (61). Corn rust is a serious disease on individual inbred lines in Central America but not on open pollinated varieties which are very diverse for rust response (7, 36). Also, less rust occurs on synthetic varieties of sunflower in Manitoba than on inbred lines (58). Finally, *P. graminis* on orchard grass in Iowa (22) and *P. coronata* on perennial ryegrass in New Zealand (31) have been effectively controlled with synthetic varieties which develop over long growing seasons.

Time of epiphytotic development is a doubly important factor on grain crops because the grain (the economic portion) is produced at the end of the growing season. In Iowa, oat groats fill at the rate of 5 per cent per day; hence, delaying a rust epiphytotic one rust cycle (about 8 days) could be reflected in a 40 per cent increase in grain yield. With only a 1:1 field blend (Table II), we calculated $r = 0.42$ per day for the blend as compared to $r = 0.50$ for the pure line susceptible. This was enough to delay the epiphytotic 4 days and increase yield of a midseason cultivar some 20 per cent (16). Time also is of critical importance for wheat and wheat rusts (64).

Advantages of Multilines

Advantages of multiline cultivars may be summarized as follows: (*a*) They provide a mechanism to synthesize instant, well-buffered, horizontally resistant cultivars which, unlike pure line cultivars, can utilize without difficulty several resgenes at the same locus or resgenes which happen to be linked (7) in the repulsion phase. (*b*) They should extend indefinitely the useful life of a given resgene and enable a resistance breeding program eventually to be reduced in size (7) while the breeder carries on parallel improvement in the recurrent parent (7). (*c*) Removing the rust hazard should stabilize the cultivars used and enable farmers to optimize production for a given multiline cultivar on a given farm (7). (*d*) They offer a means whereby a center of variety development can distribute host cultivars far and wide without risk of homogenizing the pathogen population on a global scale.

Multiline varieties as instant synthetic horizontally resistant cultivars.— While there are many compelling reasons for use of genetic diversity in small grains (66), the *raison d'être* for developing and using multiline cultivars is for meeting future risk situations in high rust-hazard areas. The average life of a new, rust-resistant oat cultivar in the Corn Belt was only 4 to 5 years (77). For wheat, the average life of a new type of resistance was 15 years in the USA and Canada, but only 5 years in Mexico (9, 62).

Cultivars with HR or generalized resistance are the preferred means of effecting rust control and obtaining cultivar stability (19, 36, 86). If such cultivars were resistant to all races of the pathogen, they should effectively break the vicious circle. When tested in a rust nursery where abundant inoculum was supplied from a susceptible spreader, Iowa oat line C237-89-IV apparently was susceptible to all crown rust races. In 1968 (15), it was used as a spreader for 11 races of *P. coronata avenae,* and it failed to support adequate development of ten of the races—they were unable to increase and cause an epiphytotic. To the eleventh race, 326, however, C237-89-IV was highly susceptible. C237-89-IV apparently had HR, i.e., it was rated susceptible under heavy rust conditions (designed to test for VR) but failed to support the intensification of rust upon itself—except for race 326. Thus, HR is a useful concept; but only with time and spactial distribution can a cultivar be proven to have HR, in the strict sense of the definition that requires a cultivar with HR to resist all races.

Multiline cultivars, on the other hand, have built-in instant and predictable population HR. Of course, considerable diversity of resistance sources is required to develop a multiline cultivar. For wheat stem rust and oat crown rust, there are adequate numbers of genes for VR, but for oat stem rust a multiplicity of new sources of resistance will be needed.

Linkage and allelism.—In developing a multiline cultivar, one can incorporate more than one allele at the same locus without difficulty. Also, strong linkage in the repulsion phase would not preclude the use of genes at two linked loci.

Extending the useful life of resgenes.—Multilines should greatly extend the useful life of a resgene. In the past, when a resgene, e.g. the Bond gene, lost its effectiveness, it was kept in the population although it was masked by genes subsequently incorporated. With multiline cultivars, if a gene lost its effectiveness, as in Miramar 63 (61), the line containing that gene could be withdrawn and held in reserve until the corresponding virugene became insignificant in the pathogen population.

Necessity of genetic diversity for centers of cultivar development.—Mexico and Colombia have become centers for development of broadly-adapted wheat cultivars. A major potential advantage of multiline cultivars is that they offer a way of utilizing such centers safely. Wheat cultivars from Mexico and Colombia, being well-adapted to diverse conditions in those countries and insensitive to photoperiod, have excelled wherever they have been grown, from 36° S lat to 50° N lat, and from sealevel to 10,000 feet elevation (60, 62). Colombian cultivars are grown much more extensively outside than within that country (60), and the same probably is true of Mexican wheats. They are truly cosmopolitan cultivars that have revolutionized wheat production in India and Pakistan. One cannot help being thrilled at their prospects for helping feed a hungry world, nor alarmed at their potential for guiding the evolution of major pathogens on a global scale.

In this situation, adequate genetic diversity is indicated. The resultant "genetic homeostasis" might well prove to be "the gyroscope that holds the ship steady in a surging sea" (44). Borlaug (8, 9, 61) compared the degree of variability in pure line, multiline, and multiline hybrid wheat cultivars, and developed sophisticated plans for multiline hybrid wheat cultivars, "essentially making every plant in such a hybrid population different for rust resistance, while maintaining a single phenotype for agronomic and quality characteristics" (61). Hybrid multiline cultivars must be preceded by the development of an outstanding single hybrid with proven commercial value. Ultimately, the superior rust control which promises to come from multiline hybrids may be one of the greatest benefits to be realized from hybrid wheats (61, 62).

To cut costs and simplify seed production and distribution (62) commercial companies developing commercial pure line or, especially, hybrid cultivars, also should work from a center where cultivars with broad adaptation are obtained. Even with genetically diverse material emanating from the center of development, it still might be necessary for workers in recipient countries to make adjustments to meet local conditions (24).

CRITICISM OF THE MULTILINE APPROACH

The multiline approach for controlling pathogens with great epiphytotic potential has been criticized as expensive (81), agronomically conservative (19, 36), and a breeding ground for new races (36, 66, 86) possibly even a super race (19).

Cost.—The development of multiline cultivars may be more costly than the development of varieties using evolutionary plant breeding (81). However, it does provide an opportunity to heterogenize only the genetics associated with rust resistance. The end product of evolutionary plant breeding, e.g. Harland barley (82), may possess unwanted diversity for some traits. Harland could be quite satisfactory as a feed barley, but for malting, the genetic diversity likely would be objectionable. Further, the genetic diversity of Harland could not be adjusted readily by plant scientists, but would be adjusted only through the slow natural evolutionary process. For example, Harland is the product of a 40-year evolutionary breeding program while, in contrast, Multiline E68 was developed in only 7 years.

Agronomically conservative.—Caldwell (19) said multiline breeding programs are "laborious and would require most of the resources of most breeding programs, thus limiting much that might otherwise be done in creating inherently better crop varieties. The method is highly conservative, restricting the breeder to older crop genotypes of proven performance."

If multiline cultivars are agronomically conservative, they are pathologically progressive. As presently conceived, they are recommended only for areas with a high rust hazard where pure line cultivars with VR have fed the vicious circle. In such areas, the indicated way of controlling rust appears to be with HR, yet HR is usually of "multigenic inheritance, making it of rare occurrence and costly to recognize and recover from hybrid progeny" (19). Multiline cultivars, on the other hand, give instant HR where a rust hazard justifies the development of such well-buffered populations.

Some cultivars which represented major breakthroughs in agronomic type (e.g. Clinton oats which set new standards in quality and straw strength) were detected only because they fortuitously had resistance to then-prevalent pathogens. Possibly greater breakthroughs in agronomic potential could be obtained if breeders could select for agronomic traits in nurseries protected with fungicides, so they need not be so concerned whether superior agronomic types carry concomitant disease resistance. (More often than not, disease resistance has been the object of selection with agronomic type concomitant.) A fungicide-protected nursery would be nearly ideal for breeding cultivars with superior agronomic characteristics for use as recurrent parents, since the latter must be rust susceptible to avoid masking resgenes during backcrossing. Of course, parallel testing would be necessary to insure adequate resistance to diseases not buffered against by multilineness.

Also, one can raise the valid question whether, in view of the "cumulatively very powerful" (66) evidence that there are benefits derived from mixing small grains, the truly agronomically conservative program is not one which concentrates only on pure lines.

Multilines as a breeding ground for new races.—Since the reasons advanced for developing multiline cultivars are that they minimize rust epiphytotics and stabilize host cultivars, a very pertinent question involves the effect of multiline cultivars on the rust population.

Speculation has been advanced that multilines would provide maximum chance for individual rust races to increase stepwise in complexity of virugenes (86) by providing the "the ideal situation for breeding new races" (19) through mutation, heterokaryosis, or somatic hybridization (43). Another viewpoint holds that extensive use of multiline cultivars would result in a highly heterogeneous rust population with biotypes collectively virulent on all components of the multiline, but that a super race would not develop (7, 36, 66).

Basic to this view is van der Plank's axiom (86) that simple races of the pathogen are fittest to survive on simple cultivars. A simple race is one with no or few virugenes, and a simple host cultivar is one with no or few VR genes. Natural selection for simple races is the basis for VR. Races of *Tilletia caries, T. foetida, Melampsora lini, Phytophthora infestans,* and *Puccinia graminis tritici* in Australia support this concept (86). This is impressive support, especially since the development of the first three pathogens is concomitant with a sexual-like process and possible hybridization, whereas the last two can exist indefinitely as vegetatively propagated clones. To these can be added recent evidence from *P. coronata* (51), *P. graminis avenae* (48), and *Trichometasphaeria turcica* (65).

P. coronata avenae races identified in the USA during 1961–1965 (51) are given in Table III. Race group 202 is pathogenic on Bond but not on Victoria, Landhafer, or Trispernia. Race group 216 is similar but has added virulence for Victoria. Race group 290 attacks Bond and Landhafer but not Trispernia. Race group 264 parasitizes all hexaploid sources of resistance, and is equivalent to a super race of crown rust. The Puccinia Path constitutes a single interdependent north-south spore distribution system for Central North America (33). Victoria derivatives are commonly grown in the South and Landhafer derivatives in the North. A smaller acreage of other derivatives persists in both regions. Race group 202 predominated in the mid-1950's, but it then declined, nationwide, except in the Northeast where buckthorn frequently is rusted and which is outside the Puccinia Path. Race groups 216 and 290 next became the most common races. We interpret race group 216 to be fittest to survive in the South, and race group 290 fittest to survive in the North. Each, however, to predominate in the Puccinia Path, had also to be reasonably fit to survive in the area where the other was fittest. They had the fewest unnecessary genes for virulence on varieties in the Puccinia Path but possessed the necessary genes for epiphytotic potential. Race group 264, on the other hand, with virulence for all oats in the

Puccinia Path and with several unnecessary genes for virulence, not only failed to increase in prevalence, but actually decreased. Apparently it lacked those fitness genes necessary for epiphytotic potential.

The above data represent natural selection by a commercial crop over a large area. The following experiments were designed specifically to test the hypothesis that simple races are fittest to survive on simple hosts. Scheifele, Nelson & Wernham (65) studied stabilizing selection in *T. turcica.* They

TABLE III

MAJOR RACE GROUPS OF *Puccinia coronata avenae* COLLECTED IN THE USA, 1961–1965[a]

Year	Race group (percentage distribution)				
	202	216	264	290	Misc.
1961	26.4	38.3	6.7	28.6	0.4
1962	10.6	44.4	8.9	35.7	0.3
1963	18.3	19.0	8.3	52.2	2.2
1964	8.5	28.3	6.9	54.7	1.6
1965	6.0	40.6	2.9	48.9	1.6

[a] From (51).

selected a simple and a complex isolate which differed in mating type, and used the constant association of pathogenicity and mating type in the conidial stage to identify isolates. When equal quantities of inoculum of the simple and the complex isolate were inoculated onto a simple host, the simple isolate comprised 58, 76, and 96 per cent, after the first, second, and third generations, respectively, thus tending to support van der Plank's axiom (86).

In an ingenious experiment with *P. graminis avenae,* Leonard (48) inoculated barberry with sporidia from telia-bearing *Dactylis glomerata* stems. He then inoculated Craig oats (no major resgenes) with aeciospores, and for eight generations allowed the rust to recycle on Craig and Clintland *A* (resgene *A*). At each generation the stem rust differential varieties were inoculated and the resultant resistant- and susceptible-type pustules counted. Assuming a random distribution of modifier genes, pustule counts would be unbiased measures of survival values associated with different alleles for virulence. Virulence occurred in the rust population for oat stem rust resgenes *A, B, D, E,* and *F.* Differential varieties containing these genes gave an increasing number of resistant-type uredia with increasing numbers of pathogen generations. Isolates with virulence for particular differential varieties had survival values 14 to 46 per cent lower than those of isolates avirulent on the same differentials.

Thus evidence accumulates in favor of survival advantages for strains of fungi with the least numbers of unnecessary virugenes, but few suggestions have been made as to why. Since virugenes are generally recessive in the rusts, it may be that for the dicaryotic rust fungi there is survival advantage in the heterozygotic condition (30). In animals and plants, "good buffering is a feature of heterozygosity" (2), and in higher plants which are vegetatively propagated, selection of a superior clone implies selection for high heterozygosity (66). Rust races are clones, or mixtures of related clones. Possibly, fittest pathogenic clones and clones of higher plants are alike in being heterozygous. Of course, this reasoning assumes that heterozygosity at virulence loci is associated with some vigor trait.

Before assuming that the automatic selection of simple races will protect components of a multiline, we should consider evidence to the contrary. In experiments with *T. turcica* paralleling the one cited above, strain T-8 completely predominated over strain R58 although R58 had fewer unnecessary genes for virulence (65), possibly indicating that factors other than virulence genes per se dictate fitness. Temperature and other ecological factors may influence variations in the relative prevalence of what stem rust races (45, 50).

Unnecessary virugenes presumably would have to be homozygous recessive to increase and be detected in a population. Although a mechanism that selects for simple races of fungi also should select against unnecessary virugenes, there are fairly frequent reports of unnecessary virugenes. For example, virulence for the diploid oat crown rust differential Saia originated apparently by mutation from race 216 and became sufficiently prevalent to be collected in four provinces of Canada even though it had no known selective advantage on commercial oat cultivars (29). Similarly, virulence for the oat crown rust differential Ascencao occurred commonly for several years even though the Ascencao resistance was not used in commercial cultivars. The Ascencao-attacking race (213A in group 216) subsequently dropped from the rust population as mysteriously as it entered (51, 69). Finally, virulence for oat cultivars with resgene *F* and *H* have predominated in the stem rust population (78) without apparent selection advantage among commercial plantings. Race 6AF, even with fewer virugenes than race 6AFH, is a super race on contemporary oat cultivars and has predominated with unnecessary virugenes. Clearly, in some cases at least, factors beside host resistance markedly influence selection of rust biotypes.

That a super race, if it developed, might predominate in a multiline was not substantiated experimentally for crown rust race 264 (Fig. 2) (26, 27). Race 264 yielded significantly more spores on a pure line than did a mixture of six races which included race 264, but the mixture of races significantly outyielded race 264 on the multiline.

To summarize this section, only further research and experience with multiline cultivars planted over an extensive area can determine the rust race situation therein. Almost certainly the race population will be quite heterogeneous (43). Probably *r* will be reduced enough that the amount of inoculum will be small except, probably, in years of first class epiphytotics on pure line cultivars. We agree with Borlaug (7) and Hooker (36) that a super race should not arise. A multiline cultivar is a step toward the complete diversity of nature (17) in which a super race has not arisen and predominated in a wild population.

Flax rust, being autoecious and having both sexual and heterokaryotic systems, is well suited to supply data on this intriguing problem. The necessary host lines and parasite strains probably are available and, unlike the cereal rusts, they can be grown in isolation. To do so, a minimum of two large isolated fields of flax is suggested. To test whether a super race would develop in a multiline, field one would be planted to a mixture of, say, ten isogenic lines containing different major resgenes. The field would be inoculated with several clones of *Melampsora lini,* preferably ten, each containing a single virugene for a component of the multiline. The field should be replanted to the same lines for several years, leaving infested debris of the previous crop to propagate the rust fungus. The fungus population should be sampled periodically to map changes in virulence patterns. Field two would be planted to a universal suscept, probably Bison, and inoculated with two or more complex races of flax rust. Field two would be replanted and sampled as with field one to see if the races lost complexity on the simple variety. The results should be highly suggestive of those to be expected for heterogeneous populations of the cereals and their rust fungi.

How Resistant Should a Multiline Be?

Related is the question: How much resistance should be incorporated into a multiline? With 25 per cent of the host population susceptible, no significant reduction in yield was caused by a stem rust epiphytotic in California (81). Others have suggested that the susceptible plants in a multiline would not suffer from rust damage if as many as 6 to 12.5 per cent (7) or 40 per cent (41) were susceptible.

Considering not the level of resistance necessary to protect susceptible plants, but the level of susceptibility necessary to stabilize the rust population, Leonard (48) calculated that when 35 to 40 per cent of the plants are susceptible to all races of the pathogen, the simple races will be maintained and the race structure will stabilize. He considered that the more complex race would be virulent on all components of the multiline and would multiply at a normal rate. Resistant plants in the population would reduce the reproduction rate of the fitter simple races to that of the complex race and so an equilibrium would be established.

Although diseases may not cause marked genetic shifts in some composite bulk populations (80), there is some evidence that a heterogeneous host population may stabilize well above the level of susceptibility calculated by Leonard (48) to be necessary to stabilize the pathogen population. A composite made by mixing F_2 seed from 250 oat crosses was planted in drill strips and subjected to crown rust epiphytotics for several consecutive years. Remnant seed from each generation was stored for subsequent analysis. Seed from one line of descent was winnowed to reject lighter seed, presumably from rust-infected plants. The frequency of resgenes in the winnowed population shifted from 0.19 in the F_2 to 0.39 in the F_5 after which it tended to stabilize (83).

Since adequate diversity is not available, an experiment on controlling oat stem rust with multiline cultivars which incorporated 35 to 40 per cent susceptible plants is worth trying. The remainder of the multiline cultivar should consist of isogenic lines containing single resgenes. Since the current prevalent race 6AF (78) is a super race on contemporary cultivars, planting the entire oat acreage of the Puccinia Path to such a multiline should not worsen the present stem rust threat. To minimize extraneous inoculum and make interpretation easier, however, the experiment should be run in an isolated area, possibly a valley in the West. Several acres should be planted and inoculated initially with races 6AF and 6AFH, along with simple races 1, 2, 7, and 8. Inoculum would have to be collected at random near the end of each season to supply race prevalence data and to store in liquid nitrogen for reinitiating the epiphytotic the following year. Seed should be replanted, with some stored for subsequent genotype analysis. Paired rusted and fungicide-sprayed plots in the field would give a measure of the level of protection afforded by the multiline. Genotype analysis and race prevalence data should reveal the nature and extent of shifts in the host and pathogen populations.

Gene Deployment

Leonard (48) suggested that, unless the increase rate of the complex race is reduced and the races stabilized, "the effect of the multiline variety on the overall rate of increase of rust will be the same as that of a pure line variety which contained all of the genes for resistance." If this is so, it probably is true only for small epidemiological units such as isolated mountain valleys in the Andes where rust might recycle primarily upon the compatible, cultivated crop. It need not be true for the Puccinia Path where rust can be made to cycle between incompatible hosts.

"The South and the North are mutually supplementary in the annual development of wheat stem rust . . . in a vast area of North America extending from Mexico through the Mississippi basin of the USA and onward to the prairie provinces of Canada, a distance of some 2,500 miles" (33). The uredial stage seldom survives the summer in southern Texas and northern Mexico, nor the winter north of Texas. Hence, each area, North and South, is dependent on the other for inoculum in season. Barberry eradication unified this area, the Puccinia Path, for stem rust by requiring the fungus to survive on gramineous hosts in the North and the South in turn. This was an important contribution of that vast program; yet we have not taken advantage of the unity that resulted.

Van der Plank (86) suggested that, where the route of an epiphytotic is known, VR genes should be added to the cultivars where the epiphytotic ends but not where it starts. This would give maximum opportunity for selecting simple races against which the VR genes would be effective. In a single, mutually-interdependent epidemiological unit like the Puccinia Path of mid-America, it is a moot point whether the epiphytotic begins in the North or the South, but since the disease covers a smaller area for a longer time during the winter months, we may take that for the beginning. The overwintering area then becomes the primary epiphytotic area, and the North the secondary epiphytotic area. Even though planting simple cultivars in the primary epiphytotic area might result in no more rust loss than is experienced with current practices, it is unlikely that simple cultivars would be acceptable as a sole protection from rust in that area. For this reason, planned deployment of oat crown rust resgenes has been suggested (18). Resgenes would be deployed by agreement; resgenes used in the North would not be used in the South, and vice versa. Scientists in each zone could use them as desired in simple or complex cultivars, in pure lines or in multilines. Inoculum coming into each area would come from host genotypes other than those being grown within the area and should be avirulent on them; i.e., inoculum would "come from outside the crop" (36) and the unity of the Puccinia Path would be broken.

Interregional diversity of resgenes was suggested earlier (18, 20, 38, 42, 66, 86). It was subjected to a limited test in Indiana where wheat leaf rust seldom overwinters. Vigo wheat possessed leaf rust resistance different from that in other cultivars in Indiana or to the south. Races virulent on the Fultz type of resistance in Vigo were not selected in the primary epiphytotic area to the south, and Vigo remained free of rust damage in Indiana even in severe leaf rust years (20).

Resgene deployment is indicated for the Puccinia Path against those continental fungi for which an adequate diversity of resistance is available.

Concluding Remarks

Despite notable successes in breeding pure line cultivars (19), pathogens, especially the rusts, remain threats to small grain production in many areas of the world. In North America, a coordinated continental program utilizing fungicides, pure line tolerant cultivars, or pure line cultivars with horizontal or generalized resistance, or both types of cultivars holds promise for controlling the shifty, continental cereal rust fungi. But economical, FDA-cleared fungicides are not yet available, and pure line cultivars with tolerance or horizontal resistance are conceptual, difficult and expensive to work with, and for the future. Multiline cultivars, however, have characteristics in common with both pure line tolerant and horizontally resistant cultivars. Being easy to develop, they hold great promise as a dynamic, natural, biological system of effectively buffering the host population against the rust population. Planned gene deployment should undergird multiline cultivars, as it would pure line cultivars and fungicides. There are compelling arguments for using increased heterogeneity in small grains, not the least of which is additional adaptability toward a future disease-risk situation.

> It can . . . be argued that homogenous populations can be produced which will cope with unpredictable fluctuations in environment (e.g., diseases) as well as heterogeneous populations. This may be the case, but in the meantime populations in which there is an appropriate compromise between the demands for uniformity and the advantages of diversity appear to have much to offer in terms of improving and stabilizing performance (2).

LITERATURE CITED

1. Aiyer, A. K. Y. N. Mixed cropping in India. *Indian J. Agr. Sci.*, **19**, 439–543 (1949)
2. Allard, R. W., Hansche, P. E. Some parameters of population variability and their implications in plant breeding. *Advan. Agron.*, **16**, 281–325 (1964)
3. Asai, G. N. Intra- and inter-regional movement of uredospores of black stem rust in the upper Mississippi River Valley. *Phytopathology*, **50**, 535–41 (1960)
4. Atkinson, J. Field experiments. *Iowa Agr. Expt. Sta. Bull.*, **45**, 216–29 (1900)
5. Biffen, R. H. Mendel's laws of inheritance and wheat breeding. *J. Agr. Sci.*, **1**, 4–48 (1905)
6. Borlaug, N. E. New approach to the breeding of wheat varieties resistant to *Puccinia griminis tritici*. *Phytopathology*, **43**, 467 (1953) (Abstr.)
7. Borlaug, N. E. The use of multilineal or composite varieties to control airborne epidemic diseases of self-pollinated crop plants. *Proc. Intern. Wheat Genet. Symp., 1st., Winnipeg, 1958*, 12–26 (1959)
8. Borlaug, N. E. Wheat, rust, and people. *Phytopathology*, **55**, 1088–98 (1965)
9. Borlaug, N. E. Basic concepts which influence the choice of methods for use in breeding for disease resistance in cross-pollinated and self-pollinated crop plants. In *Breeding Pest-Resistant Trees*, 327–44 (Pergamon Press, Oxford, England, 505 pp., 1966)
10. Borlaug, N. E., Gibler, J. W. The use of flexible composite wheat varieties to control the constantly changing stem rust pathogen. *Agron. Abstr.*, **81** (1953)
11. Borojevic, S., Misic, T. The value of growing wheat varieties in mixtures. *Savr. Polj.*, **1**, 3–15 (1962). *Field Crop Abstr.*, **17**, 663 (1964)
12. Briggs, F. N., Allard, R. W. The current status of the backcross method of plant breeding. *Agron. J.*, **45**, 131–38 (1954)
13. Browning, J. A. Studies of the effect of field blends of oat varieties on stem losses. *Phytopathology*, **47**, 4–5 (1957) (Abstr.)
14. Browning, J. A. Altering the effect of oat rust resistance genes by certain physical means. *Phytopathology*, **50**, 630 (1960) (Abstr.)
15. Browning, J. A., Frey, K. J. (Unpublished data)
16. Browning, J. A., Frey, K. J., Grindeland, R. L. Breeding multiline oat varieties for Iowa. *Iowa Farm Sci.*, **18**, No. 8, 5–8 (1964)
17. Browning, J. A., Simons, M. D., Frey, K. J. The potential value of synthetic tolerant or multiline varieties for control of cereal rusts in North America. *Phytopathology*, **52**, 726 (1962) (Abstr.)
18. Browning, J. A., Simons, M. D., Frey, K. J., Murphy, H. C. Regional deployment for conservation of oat crown rust resistance genes. *Iowa Agr. Expt. Sta. Spl. Rept.* (In press)
19. Caldwell, R. M. Advances and challenges in the control of plant diseases through breeding. In Pest Control by Chemical, Biological, Genetic, and Physical Means. *AAAS Symp.* (Sect. O on Agr.), Dec. 1964, Knipling, E. F., Chmn., ARS, Montreal, Canada, *U. S. Dept. Agr. ARS*, 33–110, 117–26 (1966)
20. Caldwell, R. M., Schafer, J. F., Compton, L. E., Patterson, F. L. A mature-plant type of wheat leaf rust resistance of composite origin. *Phytopathology*, **47**, 690–92 (1957)
21. Caldwell, R. M., Schafer, J. F., Compton, L. E., Patterson, F. L. Tolerance to cereal leaf rusts. *Science*, **128**, 714–15 (1958)
22. Carlson, I. T. (Unpublished data)
23. Chakravarti, B. P. Attempts to alter infection processes and aggressiveness of *Puccinia graminis* var. *tritici*. *Phytopathology*, **56**, 223–29 (1966)
24. Chandler, R. F., Jr. The case for research. In Strategy for the Conquest of Hunger, 92–97. *Rockefeller Foundation Symp.* August 1968 (The Rockefeller Foundation, New York, 131 pp., 1968)
25. Clifford, B. C. Relations of disease resistance mechanisms to pathogen dynamics in oat crown rust epidemiology. *Dissertation Abstr.*, **29**, 835–B (1968)
26. Cournoyer, Blanche M. *Crown Rust Intensification Within and Dissemination from Pure Line and*

Multiline Varieties of Oats (Master's thesis, Iowa State Univ., Ames, Iowa, 70 pp., 1967)

27. Cournoyer, B. M., Browning, J. A., Jowett, D. Crown rust intensification within and dissemination from pure-line and multiline varieties of oats. *Phytopathology,* **58,** 1047 (1968) (Abstr.)
28. Dobell, C. Antony van Leeuwenhoek and his "Little Animals" (Dover Publications, Inc., New York, 435 pp., 1960)
29. Fleischmann, G. The origin of a new physiologic race of crown rust virulent on the oat varieties Victoria and Saia. *Can. J. Botany,* **41,** 1613–15 (1963)
30. Flor, H. H. Epidemiology of flax rust in the North Central States. *Phytopathology,* **43,** 624–28 (1953)
31. Gibbs, J. G. Field resistance in *Lolium* sp. to leaf rust (*Puccinia coronata*). *Nature,* **209,** 420 (1966)
32. Grafius, J. E. Rate of change of lodging resistance, yield, and test weight in varietal mixtures of oats, *Avena sativa* L. *Crop Sci.,* **6,** 369–70 (1966)
33. Hamilton, Laura M., Stakman, E. C. Time of stem rust appearance on wheat in the western Mississippi basin in relation to the development of epidemics from 1921 to 1962. *Phytopathology,* **57,** 609–14 (1967)
34. Hayden, E. B. Differences in infectibility among spring wheat varieties exposed to spore showers of race 15B of *Puccinia graminus* var. *tritici. Phytopathology,* **46,** 14 (1956) (Abstr.)
35. Henriquez, O. M. *Logistic Models for Spore Count Data from Artificially Induced Epiphytotics* (Master's thesis, Iowa State Univ., Ames, Iowa, 67 pp., 1968)
36. Hooker, A. L. The genetics and expression of resistance in plants to rusts of the genus Puccinia. *Ann. Rev. Phytopathol.,* **5,** 163–82 (1967)
37. Hutchinson, J. B. The application of genetics to plant breeding. I. The genetic interpretation of plant breeding problems. *J. Genet.,* **40,** 271–82 (1940)
38. Jensen, N. F. Intra-varietal diversification in oat breeding. *Agron. J.,* **44,** 30–34 (1952)
39. Jensen, N. F. Population variability in small grains. *Agron. J.,* **57,** 153–62 (1965)
40. Jensen, N. F. Broadbase hybrid wheats. *Crop. Sci.,* **6,** 376–77 (1966)
41. Jensen, N. F., Kent, G. C. New approach to an old problem in oat production. *Farm Res.,* **29,** No. 2. 4–5 (1963)
42. Johnson, T. Regional distribution of genes for rust resistance. *Robigo,* **6,** 16–17 (1958)
43. Johnson, T. Man-guided evolution in plant rusts. *Science,* **133,** 357–62 (1961)
44. Jones, D. F. Heterosis and homeostasis in evolution and in applied genetics. *Am. Naturalist,* **92,** 321–28 (1958)
45. Katsuya, K., Green, G. J. Reproductive potentials of races 15B and 56 of wheat stem rust. *Can. J. Botany,* **45,** 1077–91 (1967)
46. Klages, K. H. W. Changes in the proportions of the components of seeded and harvested cereal mixtures in abnormal seasons. *J. Am. Soc. Agron.,* **28,** 935–40 (1936)
47. Leonard, K. J. Factors affecting rates of stem rust increase in mixed plantings of susceptible and resistant oat varieties. *Phytopathology* (In press)
48. Leonard, K. J. Selection in heterogeneous populations of *Puccinia graminis* f. sp. *avenae. Phytopathology* (In press)
49. Leonard, K. J., and Kent, G. C. Increase of stem rust in mixed plantings of susceptible and resistant oat varieties. *Phytopathology,* **58,** 400–1 (1968) (Abstr.)
50. Loegering, W. Q. Survival of races of wheat rust in mixtures. *Phytopathology,* **41,** 56–65 (1951)
51. Michel, L. M., Simons, M. D. Pathogenicity of isolates of oat crown rust collected in the USA, 1961–1965. *Plant Disease Reptr.,* **50,** 935–38 (1966)
52. Murphy, H. C. Protection of oats and other cereal crops during production. In *Food Quality. Effects of production practices and processing,* 99–113 (Irving, G. W., Jr., Hoover, S. R. Eds., Am. Assoc. Adv. Sci. Publ. 77, 298 pp., 1965)
53. Person, C. Development of concepts in rust genetics. *Proc. Genet. Soc. Canada,* **3,** 25–29 (1958)
54. Pfahler, P. L. Fitness and variability

in fitness in the cultivated species of Avena. *Crop Sci.*, **4**, 29–31 (1964)

55. Pfahler, P. L. Genetic diversity for environmental variability within the cultivated species of Avena. *Crop Sci.*, **5**, 47–50 (1965)
56. Pfahler, P. L. Environmental variability and genetic diversity within populations of oats (cultivated species of Avena) and rye (*Secale cereale L.*). *Crop Sci.*, **5**, 271–75 (1965)
57. Poehlman, J. M. Are our varieties of oats too resistant to disease? *Agron. Abstr.*, 15–16 (1952)
58. Putt, E. D. Breeding for rust resistance in sunflowers. *Forage Notes, Ottawa*, **3**, No. 2, 1–7 (1957)
59. Rockefeller Foundation, Program in Agr. Sci., Ann. Rpt. 1962–63, 310 pp., (1963)
60. Rockefeller Foundation, Program in Agr. Sci., Ann. Rpt. 1963–64, 285 pp. (1964)
61. Rockefeller Foundation, Program in Agr. Sci., Ann. Rpt. 1964–65, 262 pp. (1965)
62. Rodriquez, R., Quinones, M. A., Borlaug, N. E., Narvaez, I. Hybrid wheats: Their development and food potential. *Centro Internacional de Mejoramiento de Maiz y Trigo Res. Bull.* **3**, 37 pp. (1967)
63. Rosen, H. R. New germ plasm for combined resistance to Helminthosporium blight and crown rust of oats. *Phytopathology*, **45**, 219–21 (1955)
64. Rowell, J. B. Chemical control of the cereal rusts. *Ann. Rev. Phytopathol.*, **6**, 243–62 (1968)
65. Scheifele, G. L., Nelson, R. R., Wernham, C. C. Studies on stabilizing selection in *Trichometasphaeria turcica* (*Helminthosporium turcicum*). *Plant Disease Reptr.*, **52**, 427–30 (1968)
66. Simmonds, N. W. Variability in crop plants, its use and conservation. *Biol. Rev. Cambridge Phil. Soc.*, **37**, 422–65 (1962)
67. Simons, M. D. Relative tolerance of oat varieties to the crown rust fungus. *Phytopathology*, **56**, 36–40 (1966)
68. Simons, M. D. Heritability of crown rust tolerance in oats. *Phytopathology* (In press)
69. Simons, M. D. (Unpublished data)
70. Stakman, E. C. Problems in preventing plant disease epidemics. *Am. J. Botany*, **44**, 259–67 (1957)
71. Stakman, E. C., Christensen, J. J. The problem of breeding resistant varieties. In *Plant Pathology: An Advanced Treatise*, **3**, 567–624 (Horsfall, J. G., Dimond, A. E., Eds., Academic Press, New York, 675 pp., 1960)
72. Stanton, T. R., Murphy, H. C., Coffman, F. A., Humphrey, H. B. Development of oats resistant to smuts and rusts. *Phytopathology*, **24**, 165–67 (1934)
73. Stevens, N. E. Disease, damage and pollination types in "grains". *Science*, **89**, 339–40 (1939)
74. Stevens, N. E. Botanical research by unfashionable technics. *Science*, **93**, 172–76 (1941)
75. Stevens, N. E. How plant breeding programs complicate plant disease problems. *Science*, **95**, 313–16 (1942)
76. Stevens, N. E. Disease damage in clonal and self-pollinated crops. *J. Am. Soc. Agron.*, **40**, 841–44 (1948)
77. Stevens, N. E., Scott, W. O. How long will present spring oat varieties last in the Central Corn Belt? *Agron. J.*, **42**, 307–9 (1950)
78. Stewart, D. M., Romig, R. W. Prevalence of physiologic races of *Puccinia graminis* f. sp. *avenae* and *tritici* in the USA in 1967. *Plant Disease Reptr.*, **52**, 642–46 (1968)
79. Suneson, C. A. Effect of stem rust on the yield of wheat. *Agron. J.*, **46**, 112–14 (1954)
80. Suneson, C. A. An evolutionary plant breeding method. *Agron. J.*, **48**, 188–91 (1956)
81. Suneson, C. A. Genetic diversity—a protection against plant diseases and insects. *Agron. J.*, **52**, 319–21 (1960)
82. Suneson, C. A. Harland barley. *Calif. Agr.*, August, p. 9 (1968)
83. Tiyawalee, D. *Mass Selection for Crown Rust Resistance in Oat Populations* (Master's thesis, Iowa State Univ., Ames, Iowa, 48 pp., 1967)
84. Vacke, J. Investigation of agro-technical methods for the control of oat sterile-dwarf virus disease. *Ved. Pr. Ustr. Uyzk. Ust. rostl. Vyroby Praze-Ruzyni*, **11**, 213–25

(1967). *Field Crop Abstr.*, **21,** 743 (1968)

85. van der Plank, J. E. Analysis of epidemics. In *Plant Pathology: An Advanced Treatise,* **3,** 229–89 (Horsfall, J. G., Dimond, A. E., Eds., Academic Press, New York, 675 pp., 1960)
86. van der Plank, J. E. *Plant diseases: Epidemics and Control* (Academic Press, New York, 349 pp., 1963)
87. Watson, I. A., Singh, D. The future for rust resistant wheat in Australia. *J. Australian Inst. Agr. Sci.*, **18,** 190–97 (1952)
88. Zimmer, D. E., Schafer, J. F. Relation of temperature to reaction type of *Puccinia coronata* on certain oat varieties. *Phytopathology,* **51,** 202–3 (1961)

Copyright 1969. All rights reserved

PROGRESS IN THE DEVELOPMENT OF DISEASE-RESISTANT RICE

By S. H. Ou

The International Rice Research Institute, Los Banos, Laguna, Philippines

AND

P. R. Jennings

Inter-American Rice Program, Centro Internacional de Agricultura Tropical
Cali, Colombia

Diseases of the rice plant take a large toll from production. Many improved varieties developed in the tropics in recent years have been discarded after a brief period because of their susceptibility to the blast disease. Bacterial blight recently has become very destructive in tropical Asia. The hoja blanca virus severely damages the rice crop in several Latin American countries, and the tungro virus causes heavy losses in Southeast Asia. Many other diseases reduce rice production in varying degrees. It is expected that rice diseases will become more important with the trend toward greater use of fertilizers, closer plant spacing, and year-round culture of the new photoperiod-insensitive dwarf varieties. These new varieties, such as IR8, now grown extensively from the Philippines to Pakistan and in several Latin American countries, intensify the danger of future epidemics through the replacement of hundreds of local varieties by one genotype.

Work on the control of rice diseases has been more limited than that on diseases of other major food crops. Japanese rice workers have been major contributors to our knowledge of rice diseases. Since World War II, rice disease investigations in Japan have emphasized chemical control, although some outstanding work has been done in breeding for resistance to a few diseases.

Only recently have disease control measures for rice received attention in the tropics. The tropical climate permits pathogens as well as host plants to thrive throughout the year. Chemical protection over the vast rice belt of Southeast Asia against high populations of pathogens for prolonged periods under the monsoon climate is more difficult than in Japan. The social and economic conditions in the tropics present other obstacles. On the other hand, the warm climate in the tropics permits the conducting of breeding programs and disease testing at the rate of two or three generations throughout the year, without need for expensive facilities.

The Blast Disease

Blast is the most important disease of the rice plant. It occurs in all rice-growing areas of the world, causing heavy, and occasionally total, losses in yield. The airborne conidia of the fungus, *Pyricularia oryzae* Cav., infect the leaves, nodes, panicles, and other above-ground parts of the rice plant. Leaf blast, node blast, panicle blast (neck rot or rotten neck) are the terms commonly used to refer to these infected plant parts.

Sources of resistance.—Varietal differences in resistance to blast have been observed for many decades. For example, from 1900 to 1910 the Japanese varieties Kameji and Aikoku were considered highly resistant while Shinriki was thought to be very susceptible (55).

To assess disease reaction more accurately and to handle a large number of varieties in a short time, various methods for testing resistance to blast have been developed in different countries, and a uniform testing method was adopted for the International Blast Nursery Program (109). Numerous field tests in many countries have identified several resistant varieties, some of which have been used in breeding programs (see section on breeding for Blast Resistance). Some reports have been summarized (110).

Varietal reaction varies from country to country, locality to locality, and season to season in the same location. In general, the work involved only a relatively small number of varieties, a few seasons, or limited geographic areas. The resistant varieties selected were not exposed to many pathogenic races and, consequently, the majority did not have a very broad base of resistance. This deficiency partly accounts for the limited success of past breeding programs. To illustrate the changes in varietal reaction, some of the work done in the Philippines in recent years is briefly described below.

From 1962 to 1964, 8214 varieties were tested at the blast nursery of the International Rice Research Institute (IRRI). Of these, 1457 were found highly resistant (53a) and were tested seven times in the same nursery in 2 years. The 450 selected from these were then tested at seven stations in different regions of the Philippines. After repeated tests, 75 varieties proved resistant at all stations (53b). In future tests, some of these varieties will probably become susceptible. These results illustrate the change in varietal reaction with season and locality as well as the rapid changes in the pathogenic races of the fungus. Some varieties, however, remained resistant at all times and places in the Philippines. It is these varieties with a broad spectrum of resistance that are believed to be most useful in breeding for blast resistance.

To identify material as having a broad spectrum of resistance, it is imperative that it be tested over a wide range of geographic regions in the rice growing countries. Thus, an international program is necessary.

During the 1963 symposium on the rice blast disease at IRRI, the International Uniform Blast Nurseries, initiated by FAO-IRC, were modified and strengthened (109). A total of 258 selected varieties were tested in 50 stations of 26 countries, and the results are summarized in three reports [*Intern. Rice Comm. Newsletter,* **13**(3), 22–30 (1964); **15**(3), 1–10 (1966); **17**(3), 1–23 (1968)]. This work suggests the following: (*a*) Some varieties have a broad spectrum of resistance, although none seems to be resistant in all tests. For instance, Tetep was recorded as having a susceptible reaction in only two of the 56 tests in 1964–65 and none in the 58 tests in 1966–67. It was resistant at all stations in the Philippines and against hundreds of artificial inoculations conducted in cooperative studies in Japan and the United States (13, 41). Since 1966, an additional 321 varieties selected from the IRRI nursery have been included in the international tests to identify more varieties with broad spectrum resistance. (*b*) Varietal reactions usually show regional patterns, i.e., resistance or susceptibility throughout the test stations of a given geographic area. Where broad spectrum resistance is not available, regionally resistant varieties could be used for breeding purposes in specific areas. (*c*) Many of the introduced varieties are resistant, at least initially, i.e., many japonica varieties are resistant in tropical Asia (India, Pakistan) while many indica varieties are resistant in temperate Asia (Japan, Korea).

Several workers have searched for sources of resistance in the wild species of *Oryza* (66, 67, 69, 88, 91, 99) using different strains of wild rice and different races of the fungus. Some species reported resistant by one author were found susceptible by another. In general, the reactions to blast of these wild species were similar to those of cultivated varieties.

Field resistance, or horizontal resistance, to blast has great potential value, but has not been studied in sufficient detail. Sakurai & Toriyama (137) claimed that varieties St. 1 and Chugoku 31, developed for stripe virus resistance, also showed field resistance to blast. They produced small numbers of lesions when several isolates were inoculated. Recently, however, St. 1 was severely infected in other localities in Japan and in Korea (verbal communication). The search for field resistance to blast is under way at IRRI and in Colombia.

Evaluation of leaf blast resistance is generally based upon the reaction of seedlings or young plants. In the field, certain varieties resistant to leaf blast were subsequently susceptible to neck rot, thus suggesting that resistance in one plant part was not necessarily correlated with resistance in another. Ou & Nuque (114) injected spore suspensions of many isolates of different fungus races into leaf sheaths of partially emerged panicles of varieties with known reactions to the isolates at the seedling stage and found that varieties resistant to the isolates at the seedling stage showed no neck rot while those susceptible to the isolates developed a high percentage of neck rot, often reaching 100 per cent. It was concluded that resistance to leaf blast is closely correlated with resistance to neck rot, at least in the majority of cases. The field observation that plants resistant at the seedling stage become susceptible to neck rot is explained by shifts in races during the growing period. Testing for resistance only at the leaf stage is advantageous since field tests for neck rot resistance are difficult to conduct.

The pathogenic races.—The existence of strains of *P. oryzae* differing in pathogenicity was first noticed by Sasaki (138, 139), who showed that rice varieties resistant to one strain were severely infected by another. Intensive studies on pathogenic races were not conducted in Japan until 1950, when some varieties, such as Futaba, known for about 10 years to be resistant suddenly became very susceptible. In about 1960, 12 varieties were finally selected as differentials and 13 pathogenic races were identified (38). In the United States, pathogenic races were first reported by Latterell et al. (84). In further studies using additional isolates from the United States, Asia, and Latin America, 15 races were identified (83). Race 16 was added later (8). In Taiwan, 19 races have been identified from 16 differential varieties (25); in Korea, 5 and 10 races (4, 85); in the Philippines, 25 races from 12 differentials (17); in India, 11 races (118); and in Colombia, 14 races have been reported (33). The number of identified races continues to increase. By 1968, about 100 races were reported in the Philippines and 60 were identified in the IRRI blast nursery.

Since most race studies were started independently in each country, varying sets of differential varieties have been used so that races identified in one country cannot be compared with those known in another. A cooperative study was started in 1963 between Japan and the United States to develop an international set of differential varieties. During the 3-year study, hundreds of isolates collected in Japan and the United States were tested extensively on the 39 differential varieties which have been used in Japan, Taiwan, and the United States. Eight varieties were selected and 32 race groups were characterized. These races were called international races and were designated as IA, IB, etc to IH, followed by numbers (13, 41). The standardization of international race numbers has been suggested by IRRI in a paper submitted for publication in *Phytopathology*.

The international and national differential varieties used in varying countries are listed in Table I.

Conidia from a single lesion comprise many pathogenic races and those produced from monoconidial cultures are also differentiated into many pathogenic races (36, 112). Single conidia seem to be unstable and great variability in the organism is apparent. Suzuki (144–146) reported that cells of the conidia, appresoria, and mycelium are heterocaryotic and that anastomosis is common. He concluded that heterocaryosis is the basis of variation. Chu & Lee (28) also reported that the cells are multinucleate. Yamasaki & Niizeki (162), however, reported that most cells are uninucleate. They found a portion of the cells to be multinucleate only in certain strains. Giatong & Frederiksen (36) also found that most cells are uninucleate. Using an electron microscope, Wu & Tsao (159) recently reported a uninucleate condition. Since no perfect stage of the organism is known, the other possible mechanisms of the genetic variability are parasexualism and heterocytosome. Parasexualism was indicated by Yamasaki & Niizeki (162) who noted hyphal anastomosis and migration and apparent fusion of nuclei. The haploidization of these heterozygous diploids could cause further genetic change. Detailed studies on the mechanism of variation are needed.

TABLE I

DIFFERENTIAL VARIETIES FOR *P. oryzae* USED IN DIFFERENT COUNTRIES

International[a]	Japan[b]	U.S.A.[c]	Taiwan, China[d]	Philippines[e]	India[f]	Korea[g]	Colombia[h]
Raminad Str. 3	Tetep	Zenith	Kung-shan-wu-shan-ken	Kataktara DA-2	AC. 1613	Zenith	Raminad Str. 3
Zenith	Tadukan	Lacrosse	Taichung 65	CI 5309	CR. 906	Ishikari-shiroke	Zenith
NP-125	Usen	Caloro	Pai-kan-tao	Chokotou	Bangawan	Pi 1	NP-125
Usen	Chokotou	Sha-tiao-tsao(P) (CI 8970-P)	Taichung 171	Co 25	S. M. 6	Sensho	Usen
Dular	Yakeiko	Sha-tiao-tsao(s)	Chianung 242	Wagwag	Mas	Kanto 51	Dular
Kanto 51	Kanto 51	CI 5309	Kwangfu 1	Pai-kan-tao	Intan	Ayanishiki	Kanto 51
Sha-tiao-tsao(s) (CI 8970-S)	Ishikari-shiroke	PI 180061	Chianung 280	Peta	CR 907	Norin 17	Sha-tiao-tsao(s)
Caloro	Homare-nishiki	PI 201902	Taichung line 33	Raminad Str. 3	BJ-1	Norin 22	Caloro
	Ginga	Wagwag	Kanto 51	Taichung t-c-w-c	S. 67	Norin 1	Aichi-asahi
	Norin 22	Raminad Str. 3	Norin 21	Lacrosse		Tonewase	Ishikari-shiroke
	Aichi-asahi	(Rexoro)	Sensho	Sha-tiao-tsao(s)			(Napal)
	Norin 20	(Taichung 65)	Cutsugulcul	Khao-teh-haeng 17			(Bluebonnet 50)
			Natala				
			Kao-chio-lin-chou				
			Kaohsiung-ta-li-chen-yu				
			Taichung-ti-chio-wu-chien				

[a] See literature (13)
[b] (38)
[c] (11) and Latterell, F. M. & Marchetti, M. A., Relationships among races of *Pyricularia oryzae*. *Phytopathology*, **58**, 1057 (1968)
[d] (25)
[e] (17)
[f] (118)
[g] (4)
[h] (33)

Breeding for blast resistance.—Breeding for blast resistance in Japan, the United States, India, Taiwan, and Thailand has been reviewed in detail by Ito (55), Atkins et al. (11), Padmanabhan (117), Chang et al. (22), and Dasananda (30), respectively.

In Japan, systematic breeding for blast resistance was begun in 1927. Variety Norin 6 (Joshu × Senichi) was developed in 1935 and Norin 8 (Ginbozu × Asahi) in 1936. Norin 6 was then found resistant to neck blast but susceptible to leaf blast, while Norin 8 was just the reverse. Hybridization of the two produced Norin 22 and Norin 23 which are resistant to both leaf and neck blast. Later, the resistance factors of Norin 22 were introduced into other susceptible varieties. Several resistant varieties have since been developed by various stations to further improve the agronomic traits of both Norin 22 and 23.

The highly resistant upland variety Sensho was crossed with lowland varieties by Iwatsuku, resulting in the resistant varieties Shinju and Futaba (55). These two resistant varieties later produced four others: Wakaba and Koganenishiki (Futaba × Norin 6 × Norin 22) and Yutakasenbon and Senbonmasari (Shinju × Senbonasahi × Tokaisenbon).

Matsuo (90) found two Chinese japonica varieties, Toto and Reishiko, to be highly resistant. They were used as sources of resistance and to avoid hybrid sterility which often occurs in japonica-indica crosses. Hybridization with these two introductions resulted in Kanto Nos. 51 to 55 (77). From Kanto 53 × Norin 29, Kusabue was selected which has better quality and yielding ability.

Katamura developed Pi Nos. 1 to 4 through crosses with the indica variety Tadukan (Tadukan × Senbonasahi and Norin 8) (55). These Pi varieties generally yield poorly, but are highly resistant to blast and have been used extensively for breeding purposes in recent years.

In the United States, the first attempt to control blast by breeding for varietal resistance was the introduction of 12 supposedly resistant Italian varieties that were later found to be susceptible (92). Zenith, Nira, and Fortuna were the most resistant varieties known in the early 1940's.

A program of testing and breeding for resistance to blast was started in 1959. The resistance of commercial varieties to races present in the United States was determined (11). The first tests involved inoculations with race 6, the most common in the field, and with race 1 to which Zenith is susceptible. Subsequent tests involved other races and field tests (11). The highly resistant variety Dawn (Century Patna 231 × HO12-1-1) was released in 1966 (19). Dawn is also moderately resistant to brown spot.

In Madras State, India, breeding for blast resistance started early in the 1920's. Variety CO 4 and later TKM 1 were found resistant. Hybridization in 1927 and 1928 between CO 4 and ADT 10 resulted in two new resistant varieties, CO 25 and CO 26, which rapidly replaced the popular ADT 10. The crosses CO 4 × CO 13 and CO 4 × GEB-24 produced CO 29 and CO

30. Variety CO 4 has also been crossed with many other varieties to develop new blast-resistant varieties for Madras.

In other states of India, including Andhra Pradesh, Mysore, Maharashtra, Uttar Pradesh, and Kashmir, breeding for blast resistance was also started, using CO 4 or its progenies and other varieties as sources of resistance (117).

In the Central Rice Research Institute of India at Cuttack, a large number of varieties have been screened for blast resistance with subsequent testing in other states (117).

In Taiwan, more than 55 foreign introductions and many local selections have been used as sources of resistance. Breeding work has been carried out in the District Agricultural Improvement Stations at Taichung, Chiayi, Kaohsiung, Taipei, and Taitung for many years. Numerous hybrid lines were tested for blast resistance in special localities where blast epidemics occur each year. The majority of these varieties resulted from crosses among local ponlai varieties. Most of the programs emphasized yield, quality, and early maturations as well as blast resistance.

In Thailand, serious attention has been given to blast since 1959. Numerous varieties have been tested for resistance and hundreds of crosses made with progenies tested in various regions. Although no new varieties have been released, the susceptible older varieties gradually have been eliminated and more resistant ones are in wider use.

At IRRI, various sources of resistance to blast such as Dawn, Zenith, Ktaktara DA-2, Leuang Yai 34, and H-105 have been used in the breeding program. All hybrid lines at various stages of breeding are tested in blast nurseries of the Institute. Those selected as resistant are further tested in various cooperative stations in and outside the Philippines.

A limited attempt has been made to induce mutations for blast resistance by radiation both in Taiwan and Thailand (30, 51, 86, 87). X-rays, thermal neutrons, and gamma rays have been employed. Satisfactory resistance has not been obtained.

In reviewing the past work on breeding for blast resistance, the authors note considerable temporary success. However, a high degree of lasting resistance has not been achieved in most cases. Varieties usually become susceptible after a brief period of culture or when introduced into other areas.

Genetics of resistance.—Sasaki (140) in Japan studied the inheritance of resistance to blast in 1917 shortly after Biffen & Nilsson-Ehle demonstrated that disease resistance in crop plants is controlled by genes. Takahashi (148) has summarized the continuing Japanese studies through 1959.

From these studies, it appears that genes controlling resistance vary from one to three pairs and that resistance is generally dominant. The differences reported might have resulted from genetically different materials, methods used for inoculation, and criteria used for classifying reaction. All these experiments were made prior to the identification of the physiologic races and it was not known what race or races were involved. Sasaki (139) determined resistance by presence or absence of lesions, Nakatomi (98) by percentage of dead plants, Hashioka (42) by lesion color, and Okada & Maeda (106) and Abumiya (1) by lesion types. Obviously there is no uniform basis for comparing results and little information can be derived from these early experiments.

Following the recognition of physiologic races, Goto (37) using the sheath inoculation method reported that resistance was dominant in about 50 per cent of the crosses, incompletely dominant in 25 per cent, and recessive in the remaining 25 per cent. He concluded that four pairs of genes are responsible for resistance and that gene effects are cumulative.

More recently, studies of the inheritance of blast resistance have been made with known races of the disease. Atkins & Johnston (12) studied three crosses inoculated with races 1 and 6. When only one race was inoculated, the F_2 population segregated three resistant to one susceptible and a 1:2:1 ratio was observed in the F_3. When both races were inoculated, the F_2 population showed 9:3:3:1 ratio. Their results suggested two independent dominant genes for resistance to races 1 and 6, designated as *Pi-1* and *Pi-6.*

Yamasaki, Kiyosawa and others (70–75, 161) grouped a number of Japanese and Chinese varieties into five types based upon the reaction of these varieties to seven Japanese races. The varieties of the Aichi Asahi type carry one completely dominant gene, *Pi-a,* conferring resistance to the fungus isolates Ina 72 and Ina 168. Kanto 51 type has one gene, *Pi-k,* controlling medium resistance to fungus isolate P-2b and high resistance to isolates Hoku 1, Ken 54–20, Ken 54–04, and Ina 168. Ishikari Shiroke type has one gene, *Pi-i,* for medium resistance to isolates P-2b, Ina 72, Ken 54–20, Ken 54–04, and Ina 168. Genes *Pi-k* and *Pi-i* varied from complete to incomplete dominance according to environmental conditions. The three genes behaved independently.

Kiyosawa (70–73) further identified two genes for resistance in Tadukan, the *Pi-a* gene and a new gene *Pi-ta* and two genes *Pi-a* and *Pi-z,* in Zenith. The same genes were found in other varieties. In Norin 22, a medium-resistant variety, Kiyosawa, Matsumoto & Lee (75) found major and minor genes for resistance to fungus isolates Ken 54–04 and Ken Ph-03. Varieties of Toto type have *Pi-a* and *Pi-k* genes, and variety Minehikari (the same type) has gene *Pi-m* (74) in addition. Hsieh & Chien (48) found four resistance genes *Pi-4, Pi-13, Pi-22,* and *Pi-25,* in several varieties in Taiwan.

One of the most difficult problems in the study of inheritance of resistance to blast is the variable reaction in the populations of hybrid varieties. There often is a full range of reactions from resistance to susceptibility. Classification of the reactions depends on the discretion of the investigator. Since nearly isogenic lines have not been used in these genetic studies, it has been difficult to evaluate gene number accurately. It has also been assumed that conidia of the fungus used in inoculation are uniform in pathogenicity. Recent experiments (112) show that the spores of at least some monoconidial isolates differ in pathogenicity. This difference raises doubts about the validity and value of genetic analyses and, in fact, confuses the long-held concept of the nature of physiological races.

Brown Spot, Stem Rot, Sheath Blight, Cercospora Leaf Spot and Other Fungus Diseases

Brown spot disease.—This disease, caused by *Cochiobolus miyabeanus* (Ito et Kuribayashi) Drechsler ex Dastur, is usually found in its conidial stage, *Helminthosporium oryzae* Breda de Haan. It has been recognized for decades (21). It was considered to be the major factor in the Bengal famine in 1943. Despite its occasional importance, the evaluation and use of resistance has been so limited that it is of little, if any, practical value.

Suematsu (143) and Chiu (26) first reported brown spot resistant varieties. Ganguly (34) found six resistant varieties in Bengal. Yoshii & Matsumoto (164) tested 20 varieties introduced to Japan, and Asada and coworkers (7) found Hainan 217 and Chin Tsu Chin to be resistant. Kawai & Kakisaki (68) classified Hukubozu, Diakokuwase, Tamahaku, Honomaru, and Norin 17 as resistant. Among 490 varieties tested by Ganguly & Padmanabhan (35) and Padmanabhan, Ganguly & Chandwani (119) in India only nine were considered resistant.

Recent studies (15, 141) connected with the physiological disease known in Japan as "akiochi" show that brown spot is associated with soils deficient in nutrients and those containing excessive amounts of H_2S or organic acids. Baba (15) suggested the use of diluted H_2S solution to test resistance to root rot in connection with akiochi and brown spot. This test may prove useful for the evaluation of brown spot resistance pending confirmation of its association with root rot resistance.

Little is known about the pathogenicity of the organism. Nisikado (104) noted that strains of the organism differ in morphology, cultural characters, physiology, and sporulation. Tochinai & Sakamoto (151) studied 132 monosporic strains on four kinds of media and tested them on 15 rice varieties. The separated ten growth types and a wide range of pathogenicity from strongly to weakly virulent. Nawaz & Kausar (100) reported similar findings. Padmanabhan (116) in India, however, considered that there was no specialization in pathogenicity. Misra & Chatterjee (93) found a great difference in sporulation ability and pathogenicity in two isolates.

Resistance to the brown spot disease has been studied in Texas. The CI 9515 variety was found to be highly resistant. Several lines of rice were resistant to brown spot and blast, but generally showed low yield potential (Atkins, personal communication). In 1966 a new variety, Dawn, having high blast resistance and moderate resistance to brown spot was released.

Nagai & Hara (97) working with a Korean strain of *H. oryzae,* found that resistance was a dominant character, while Adair (2) indicated that resistance was recessive and polygenic.

Stem rot.—Stem rot is caused by two similar organisms, *Helminthosporium sigmoideum* Cav. and *H. sigmoideum* var. *irregulare* Cralley & Tullis. The former has a perfect stage, *Leptosphaeria salvinii* Catl. Both are commonly found in their sclerotial stage and have been called *Sclerotium oryzae* Catl. The sclerotia of the organisms are abundant in rice fields and considerable production losses have been reported.

Varietal reaction to the disease has been observed by many workers. In the United States Tisdale (150) found that Early prolific is very susceptible while Japanese varieties are less susceptible. Hector (47) stated that the Dudshar, a Bengalese variety, was apparently immune. Park & Bertus (120) observed that Hmawbi 37 was severely attacked, while Hondurawela was free from the disease. In the Philippines Reyes (123) found that Ramay, Elonelon, Arabon, and Kinaturay showed resistance. He later (124) developed the resistant variety Raminad Str. 3 from a cross between Ramay (R) and Inadhica (S). Cralley (29) reported that short-grain, early-maturing varieties are less susceptible and that the degree of resistance is a varietal character, chiefly in japonica types but only indirectly related to the time of maturity. Goto & Fukatsu (39) have reported earlier varieties to be more susceptible than late ones. Ono (107) found considerable differences among rice varieties in their reactions to stem rot. Adair & Cralley (3) found no variety to be highly resistant. Ono & Suzuki (108) inoculated both *L. salvinii* and *H. sigmoideum irregulare* on a number of varieties and found some varieties to be more resistant or susceptible to one than to the other. Hsieh (49) and IRRI (53b) tested many varieties with similar results.

In the past, varietal resistance was evaluated either in naturally infested fields or through artificial inoculation with sclerotia. An efficient testing method was developed recently (40, 53b) in which cut stems are used. Resistance may be measured within two weeks.

Very little is known about pathogenic strains and the inheritance of stem rot resistance. Apparently no breeding program is concerned at present with the development of stem rot resistant varieties.

Sheath blight.—The fungus causing sheath blight has been referred to as *Corticium sasakii* (Shirai) Matsumoto or as *Rhizoctonia solani* Kuhn. The disease is favored by high temperature and humidity and has often caused serious damage in both temperate and tropical regions.

Hashioka (43) reported from field tests that varieties introduced into Taiwan from India, Thailand, Burma, Europe, and North America were more resistant than local varieties. Hsieh, Wu & Chien (50) in Taiwan observed wide variations in varietal reaction. Relatively few were resistant and none was highly resistant or immune. Indica varieties were more resistant than japonicas. Varieties at IRRI (53c) showed a similar tendency.

Kang, Lee & Kim (64) in Korea reported that it is difficult to combine earliness with resistance. Investigations at IRRI (53c) showed that seedling evaluation is a satisfactory improvement over the conventional procedure of testing adult plants.

In Taiwan, Chien & Chung (23) classified 300 fungal isolates into seven culture types and six physiological races. The susceptibility and resistant reactions used to separate the races, however, were not discontinuous. Akai, Ogura & Sato (5) found that strains with poor mycelial growth are less pathogenic. Tu (154) noted that strains with less aerial mycelium in culture are usually more pathogenic.

Although inoculation and evaluation techniques seem to be satisfactory, no breeding program for resistance to sheath blight is known to the authors.

Cercospora leaf spot.—Spaerulina oryzae Hara which causes this disease is usually found in its conidial stage, *Cercospora oryzae* Miyake. During the 1930's and 1940's, it was considered an important disease in the United States and a successful breeding program was begun by Tullis (155).

Among more than 100 varieties, 58 were resistant. Ryker & Jodon (133) studied numerous rice varieties. A few were highly resistant. Host reaction may vary because of different pathogenic races in the fungus. Some varieties, including Asahi and Kamrose, were resistant to all strains in the United States (3).

Ryker (129) reported five distinct physiologic races of the fungus based upon reaction of the four differential varieties, Blue Rose, Blue Rose 41, Fortuna, and Caloro and later (130) reported races 6, 7, and 8 based upon the reaction of these and other varieties. Several varieties developed since then are resistant.

The inheritance of resistance to *Cercospora* leaf spot has been studied. Ryker & Jodon (133) in five of six crosses, found an F_2 segregation of three resistant to one susceptible. In the sixth cross a single recessive gene controlled resistance. Further work showed that a single dominant factor for resistance does not satisfy all cases. Adair (2) found that the factor for resistance is dominant over susceptibility in several resistant varieties, while Supreme Blue Rose carries at least one dominant factor for susceptibility. Jodon, Ryker & Chilton (61) reported that in 35 out of 48 crosses a single dominant factor for resistance was involved. In the other crosses, two, three or even more genes were concerned.

Ryker (128) and Ryker & Chilton (131) found that the dominant gene for resistance to race 1 is closely linked with a recessive gene for susceptibility to race 2. Cross-over plants resistant to both races rarely are found. Jodon & Chilton (60) found no association between resistance and the morphological characteristics of the rice plants.

Bakanae disease.—The disease is caused by *Gibberella (Fusarium) moniliforme* (Sheld.) Winel and is marked by the elongation (the "Bakanae" phenomenon) or stunting of seedlings and adult plants.

Ito & Kimura (56) in Japan found that Shiroke, Akage 3, and Kairyo-Mochi 1 are resistant. In India, Thomas (149) found a few resistant varieties that have less than 5 per cent infection as did Reyes (125) in the Philippines. Hashioka (44) tested 200 varieties and reported that those from temperate regions are less affected than those from the tropics. Rajagopalan (121) found six resistant varieties among the 20 tested in India. Varieties Lemello and Roverbella are resistant in Italy (6). As the disease is effectively controlled by seed treatment, no attempt to develop resistant varieties seems to have been made.

Nisikado & Matsumoto (105) reported that among the 66 strains of the fungus isolated from rice and the five strains of *G. moniliforme* var. *majus,* there were marked differences in pathogenicity as indicated by the degree of elongation. Chiu (27) studied a number of isolates and found that some caused dwarfing, some caused elongation and others had no effect on the size of rice seedlings, thus confirming an earlier report by Seto (142).

Bacterial Blight and Bacterial Streak

Bacterial blight.—The bacterial blight or bacterial leaf blight disease caused by *Xanthomonas oryzae* (Uyeda & Ishiyama) Dowson has been studied in Japan since the beginning of this century. Although it is one of the major diseases of the rice plant in Japan, its importance in tropical Asia was recognized only recently. The favorable climatic conditions and the more virulent strains of this disease found in the tropics cause severe production losses. The disease is not known in America, Africa, or Europe.

Varietal resistance has been studied by many Japanese workers and resistant varieties have been developed (147). In the earlier studies, resistant varieties were selected from naturally infected fields, and artificial inoculation by spraying bacterial suspensions was used. In 1951, a needle inoculation technique which offers a more precise evaluation was introduced by Muko & Yoshida (95). Recently, to obtain a large number of infected plants, seedlings in flooded nursery beds have been submerged with bacterial suspensions (31, 166). In most of the studies, local strains of the organisms were used and varietal reactions differed from station to station.

Several sources of resistance to bacterial blight have been used in Japan and many resistant varieties developed (147) although none is now considered highly resistant to the more virulent strains recently isolated. Fujii & Okada (32) recently reviewed the breeding work carried out in Japan. A resistant selection, Kono 35, was found in the field of Shinriki in 1926 and used to develop Asakaze and other resistant varieties.

Another resistant variety, Shobei, was the source of resistance in the development of other new varieties.

Resistance to bacterial blight in Shig-sekitori 11 and resistance to blast in Reishiko were combined in the new variety Oyodo. Many other varieties resistant to bacterial blight were developed in several Japanese agricultural experiment stations (147).

At IRRI several thousand varieties were tested against the Philippine strains of bacterial blight. Those having a high degree of resistance are listed in the IRRI Annual Report for 1967. Zenith, CI 9210, and several other varieties have been used as sources of resistance in the Institute's breeding program. Since these varieties may be susceptible to virulent strains from other tropical countries, an international testing program is required to identify those with broad resistance. A project was recently established between IRRI and the University of Hawaii to study the virulence of strains of the bacterium to be collected from ten countries in Asia.

In 1957, the resistant variety, Asakaze, was severely affected by the bacterial blight disease in Japan. As a result, studies to classify strains of the organism on the basis of their virulence were undertaken (147). Recent reports (78, 79) divided strains into two groups based upon lesion size on a number of resistant, intermediate, and susceptible varieties. Group I strains attacked all varieties and Group II strains produced large lesions only on intermediate or susceptible varieties. Each group may be further divided into different types based on the size of lesions or the level of virulence on resistant, moderate, and susceptible variety groups.

Fifty strains were studied in the Philippines (53b). Typical pathogenicity patterns are illustrated in Figure 1. Isolate B6 is very virulent, causing moderate reaction in resistant varieties, severe reaction in intermediate, and death in susceptible varieties. Isolate B23 is least virulent, causing an immune reaction on resistant varieties and only moderate damage on susceptible ones. Isolate B72 causes distinct reactions in resistant, intermediate, and susceptible varieties. In contrast, isolate B59 causes similar reactions among these test varieties corresponding to Group I of the Japanese reports mentioned above. Another reaction pattern shows a minor host reponse by resistant and intermediate varieties and severe reaction in susceptible varieties. A few virulent strains may cause moderately susceptible reactions in a few of the resistant varieties.

Not much detailed information is available on the genetics of bacterial blight resistance. Studies made with the less virulent temperate zone strains indicate that genes for resistance may be dominant or recessive and monogenic or polygenic (65). In Norin 27 × Asahi 1 and Norin 27 × Norin 18, the gene for resistance in Norin 27 is monogenic and dominant. In a cross between Norin 18 and Asahi 1, the moderate resistance of Norin 18 showed no dominance over the susceptibility of Asahi 1. In crosses between the susceptible Murasaki-myozetsu and resistant varieties, Mishimura & Sakaguchi (101) found a phenotypic expression of resistance in the F_2 generation with a ratio of 3:1.

Washio, Kariya & Toriyama (157) recently reported that resistance to bacterial group A in variety, Kidama, was controlled by two complementary dominant genes, X_1 and X_2. Resistance to bacterial group A in Norin 274 and Kanto 60 was also conditioned by a dominant gene. Resistance to bacterial group C in Aikoku-sato-sango-kei was controlled by a dominant gene or

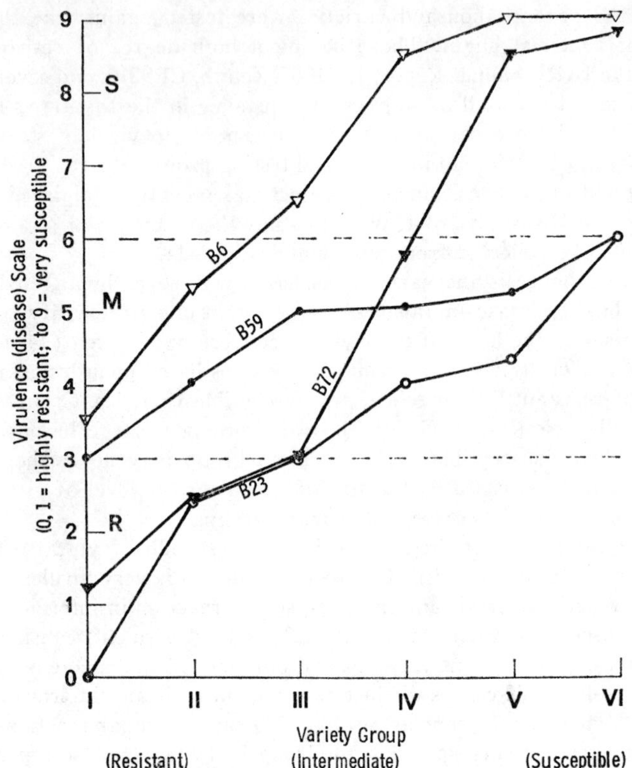

FIG. 1. Pathogenicity patterns of strains of *Xanthomonas oryzae*. (IRRI Annual Report, 1966)

genes. It was also found that resistance factors for bacterial blight, blast, and stripe disease are independent.

Sakaguchi reported two genes for resistance, Xa_1 and Sa_2 (Murata, 96). The Kidama group of varieties has Xa_1 controlling resistance to group I of the bacterium. The Rantaj-emas group carries Xa_1 and Xa_2 of which Xa_2 is the principal factor for resistance to group II of the bacterium.

Many crosses between resistant and susceptible varieties in the Philippines show that resistance may be recessive or dominant depending upon variety combinations and degree of resistance in the resistant parent (53c).

Bacterial streak.—Bacterial streak or bacterial leaf streak, caused by *Xanthomonas translucens* (J.J.R.) Dowson f. sp. *oryzae* [Fang, Lin & Chu (31)] Pordesimo, is found only in tropical Asia. The organism is quickly spread by rain and irrigation water. Under favorable conditions, the disease causes as much damage as bacterial blight.

Studies of the disease in the tropics were started only recently. An artificial inoculation method has been developed (53c), and of 750 varieties screened for varietal resistance against one of the most virulent strains in the Philippines, were highly resistant. The wild species *Oryza sativa* f. *spontanea* and *O. rufipogon* were also resistant.

The strains of the organism differ in virulence. A survey of 150 Philippine isolates showed that weak strains produce lesions less than 0.5 cm long while virulent strains produce lesions 5 to 10 cm long under standard testing conditions. Susceptible varieties, however, were always more susceptible to all isolates and resistant varieties were more resistant, indicating that the pathogen is not differentiated into pathogenic races (53b).

Stripe, Hoja Blanca, Tungro, Dwarf, and Other Viruses

Development of stripe-resistant varieties.—The stripe virus is transmitted by *Leodelphax striatellus* (Fall.), *Unkanodes sapporonus* (Mats.) and *Ribautodelphax albifascia* (Mats). It has become increasingly important in Japan as a result of the recent practice of early planting.

In identifying the resistant varieties in the field, Suzuki et al. (146) observed that early planting is favorable for disease development. Sakurai and co-workers (135, 136) developed a useful seedling test method in the laboratory. All 123 Japanese lowland varieties are susceptible, but resistance was found in Japanese upland varieties and foreign varieties. Forty-six of the 53 Japanese upland varieties are resistant. Various degrees of resistance were found in foreign varieties, many of which were highly resistant. Both upland and lowland sources of resistance have been used in breeding for new resistant varieties although resistance in the Japanese upland varieties is stronger and more stable. All ponlai varieites from Taiwan are susceptible (134).

Genes from the resistant upland variety Kanto 72, were used to produce several commercial varieties that have resistance to blast and bacterial blight as well (152).

Foreign varieties (indica type) were also used as breeding material. Following repeated back-crosses, Chugoki 31 was selected from the cross between Norin 8 and Modan, a resistant Pakistani variety (153).

Toriyama (152) concluded that resistance to the stripe virus in Japanese upland varieties is controlled by two pairs of complementary dominant genes, St_1 and St_2, and resistance in indica varieties is controlled by one pair of dominant genes, St_3. Yamaguchi, Yasuo & Ishii (160) reported that the resistance of the F_1 plant was influenced by the female genotype.

Hoja blanca.—This virus, first identified in 1957, is second to blast in importance in several Latin American countries. *Sogatodes oryzicola* (Muir) and *S. cubanus* (Crawf.) are the vectors. Although all major United States varieties grown commercially in Latin America are susceptible, the active collaboration of the Colombian Institute of Agriculture, the

United States Department of Agriculture, and The Rockefeller Foundation has resulted in the successful development of hoja blanca-resistant varieties. No physiological strains have appeared in the field to date.

An emergency varietal field screening program in 1957 (9) which involved several thousand varieties showed that most of the resistant varieties are japonicas from China, Japan, Italy, and Spain. Most indicas are susceptible. While field tests in Colombia and Cuba were successful, greenhouse tests by artificial inoculation were developed and perfected (80–82) for evaluating small numbers of plants and for special studies.

The United States breeding program featured such sources of resistance as Gulfrose, Lacrosse, and PI 215936, a japonica from Taiwan. Lacrosse has been used in the development of Northrose (Lacrosse × Arkrose), Nova (Lacrosse × Zenith-Nira), and Nova 66, a selection from Nova (20, 62, 63). Progeny of the crosses were tested in several Latin American countries and a uniform hoja blanca nursery and virus × strain nursery were established (80).

Recently in Colombia, several nearly immune lines and varieties were developed for commercial use. These lines are also highly tolerant to the vectors. This material, apparently more resistant than any of the several local and introduced parental sources of resistance, was selected for several successive generations under conditions of severe epiphytotics. There has been no difficulty in combining resistance with other desired agronomic traits.

Resistance to hoja blanca appears to be dominant over susceptibility and is relatively simply inherited (18).

Tungro.—The tungro virus was identified only recently (53, 126). The "yellow-orange leaf" disease reported from Thailand, "penyakit merah" of Malaysia, and perhaps the "mentek" disease of Indonesia and "yellowing" of India are caused by similar or closely related viruses (111, 113, 115, 122, 127, 158). One of the most important virus diseases in Southeast Asia, it is transmitted by *Nephotettix impicticeps* Ishihara and *N. apicalis* (Mats.).

Varietal evaluation was initiated soon after the identification of the virus and a mass screening technique was developed to handle a large number of lines and varieties (89). Varieties differ in percentage of tungro infection as well as in recovery from infection, and many resistant varieties have been found (53b, 89). Several commercial varieties in the Philippines and Indonesia are resistant. All these varieties were selected from a cross between Latisail and Tjina made by Van der Meulen (156) in Indonesia. The Indian variety Pankari 203 is the most resistant known to date and is the only one resistant to both the vector and the virus. Thousands of other varieties were field tested in Thailand and a number were found resistant (158).

Many of these resistant varieties have been used in the breeding program at IRRI. The two highly successful varieties released by IRRI, IR8, and IR5, have acceptable levels of tungro resistance derived from their common parent Peta. IR8 is also resistant to the vector.

Preliminary experiments with several crosses between resistant and susceptible varieties have shown that the F_1 population is generally resistant, particularly when a highly resistant parent is involved, and the F_2 segregates in approximately a 9:7 ratio of resistant to susceptible plants (53b). Recent studies at IRRI indicated that resistance at the seedling stage and in the adult plant may be controlled by separate genes.

Rice dwarf.—Transmitted by *Nephotettix cincticeps* (Uhler), *Inazuma dorsalis* (Mats.), and *N. apicalis* (Mats.), this virus was identified in Japan at the turn of the century and has subsequently caused considerable losses.

Much attention has been given to various aspects of the disease, but little to varietal resistance. Yasuo Yamaguchi & Ishii (163) tested a number of varieties in the field and found that, while Japanese varieties were susceptible, foreign varieties showed marked differences in level of resistance. The varieties Hyahunichi-to, Pe Bi Hun, Tetep, Loktjan, Kaladumai, and Dahrial were considered highly resistant.

Yellow dwarf.—The vectors are *Nephotettix cincticeps* (Uhler), *N. impicticeps* Ishihara, and *N. apicalis* (Mats.). The disease is found in all rice-growing countries in Asia, but the apparent damage is usually minor. The disease is now suspected to be due to a mycoplasma-like organism.

In field tests in Taiwan, Hashioka (45) found considerable differences in resistance among the more than 300 varieties tested. Komori & Takano (76) in Japan found a glutinous variety, Saitama Mochi 10, and several introduced varieties to be resistant.

A seedling test method was developed by Morinaka & Sakurai (94) and several resistant varieties were identified. Except for Tetep, these are all glutinous rice.

Preliminary experiments showed that F_1 plants of a cross between Saitama-Mochi 10 (R) and Manryo (S) were moderately resistant (134).

Black-streaked dwarf.—The virus is transmitted by *Laodelphax striatellus* (Fallen), *Unkanodes sapporonus* (Mats.), and *Ribautodelphax albifascia* (Mats.) and is confined to Japan. It infects many grasses and cereals including wheat and corn but causes relatively little damage to rice.

Using a seedling test method similar to that for stripe, Morinaka & Sakurai (94) found several resistant varieties.

Transitory yellowing.—The transitory yellowing virus in Taiwan, transmitted by *Nephotettix apicalis* (Mats.) and *N. cincticeps* (Uhler), has caused considerable damage the last few years. Field tests have shown that several commercial varieties are resistant.

Straight-Head, "Akiochi," and Other Physiological Diseases

Straight-head.—Straight-head is associated with the flooding of certain types of soil. First identified in the United States, similar symptoms have been observed in several parts of the world. Damage may be greatly reduced or eliminated by temporary drainage midway in the growing season.

Although the nature of the disease is not completely known, rice varieties show different degrees of resistance (10).

Akiochi and akagare.—These Japanese diseases are caused by deficiencies of potassium, nitrogen, silica, and others, and are associated with soils that produce injurious levels of hydrogen sulfide and organic acids which prevent the absorption of nutrients, or in soils where little potassium but much nitrogen is absorbed by rice plants.

No Japanese varieties have been found resistant to akiochi. Varieites Ginnen and Tetep are highly resistant while Bomba, Tadukan, Chyokato, Louisiana Nonbeard, and Jamaica are resistant (16). Varieties resistant to akiochi are also less susceptible than others to akagare and are resistant to brown spot (15).

White Tip and Stem Nematode

White tip.—The white tip disease caused by *Aphelenchoides besseyi* Christie is widely distributed in the rice-growing areas of Asia and North America. Considerable damage has been reported (14, 165).

Nishizawa & Yamamoto (102, 103), Goto & Fukatsu (40) and others in Japan, and Atkins & Todd (14) in the United States have found a wide range of varieties resistant to the white tip disease.

Hung (52) found ponlai varieties to be very susceptible in Taiwan, but the leading indica varieties showed no symptoms.

No special attention has been given to breeding for resistance to the white tip disease.

Stem nematode.—The stem nematode disease, caused by *Ditylenchus angustus* (Butler) Filipjev, is found only in some swampy areas of eastern Bengal and Uttar Pradesh in India, and in Malaysia, Thailand, Burma, Egypt (UAR), and Malagasy. Severe damage occurs in only relatively small areas.

The only information on varietal resistance to the disease is that reported by Hashioka (46) in Thailand, who used the coleoptile inoculation method. The japonica varieties seemed to be more resistant partly because their coleoptiles are smaller at the time of inoculation.

Root-knot nematode.—Caused by *Meloidogyne* spp., the disease is found on rice in rotation with other more susceptible crops.

Israel, Rao & Rao (54) reported that rice varieties differ in susceptibility to *M. incognita* var. *acrita* in India. Variety Ch-47 had an average of only 0.7 galls per plant while HR-19 had 53.5 galls and others were intermediate.

Discussion

This brief review suggests that the development of disease-resistant rice varieties has been successful for a few diseases but is generally inadequate for most. Although much work has been done on the blast disease, progress toward its control has been limited because of the great variability of the causal organism. Breeding for resistance against such viruses as hoja blanca, stripe, and tungro generally has been successful because the causal agents are less variable. Bacterial blight and bacterial streak organisms differ in strain virulence but do not seem as variable as the blast organism although the relative virulence of the bacterial strains in various countries is not sufficiently known. Recent work shows promise in transferring resistance to these diseases into the new high-yielding varieties. Considerable success was recorded toward the control of *Cercospora* leaf spot and the straight-head disease. The efforts expended in the study of the remaining diseases have not been sufficient.

Certain positive measures have been taken recently to ensure more rapid progress. IRRI has collected, maintains, and provides seed of some 10,000 cultivated varieties and wild species. By screening the collection, many sources of resistance to the more important pathogens have been identified. The Institute, in cooperation with many agencies in various countries, is establishing international disease nurseries to evaluate further these sources of resistance in distinct environments.

The worldwide impact of the new, high-yielding, fertilizer-responsive, and photoperiod-insensitive plant types typified by IR8 will have a profound influence on future breeding for resistance. The new widely adapted plant types offer a unique opportunity to breeders for disease resistance. First, workers can concentrate on building resistance into relatively few and rather similar plant types and need not be concerned with the resistance levels of vast numbers of narrowly adapted varieties. Secondly, success in fixing a high level of resistance into an international variety such as IR8 will have a broad and substantial impact on national yield averages.

Several genetic studies of the major diseases have shown that their resistance genes are not linked so that resistance to several diseases may be combined in a single variety. Several new hybrid lines at IRRI show promise of resistance to two or more major diseases.

The most difficult problem in the development of disease-resistant rice varieties is the blast disease. In general, genetic information, inoculation techniques, large-scale testing methods, disease scoring, and the knowledge of plant reactions in successive growth stages are well developed. Deficiencies in these areas cannot be offered to rationalize past failure. Rather, the difficulty is directly concerned with the immense genetic variability of the pathogen. The pathogenic plasticity of *Pyricularia* has been continually underestimated by workers since they first became concerned with the disease problem. The most recent evidence of the tremendous instability of most individual races questions the practicality, if not the concept, of conventional race identification and the search for resistance against specific races.

We suggest that blast-resistance workers might best face the issue directly and consider the following general points as valid, even if unpalatable: (*a*) Programs based on conventional major gene resistance have not achieved the expected success during the past half century. There is little value in the continued use of parental sources resistant to certain races but fully susceptible to others. The continued search for narrow spectrum major genes, either naturally occurring or through induced mutation, seems pointless. (*b*) Studies on the inheritance of blast resistance have been scattered and uncoordinated. Unless isogenic parent lines and stable races are used, results are predestined to be suspect. (*c*) The complexity of the problem forces the conclusion that isolated research efforts are likely to fail. A vigorous international effort is indispensable to success.

The only rational approach available, considering the past decades of experience, is the use of broad-spectrum resistance conferring stable protection for prolonged periods of field culture. Three possible approaches seem worthy of serious attention:

(*a*) An international search for materials showing consistent resistance over several seasons in numerous geographical areas. As described, the International Uniform Blast Nurseries program has had some promising initial results. Presumably, this type of resistance is conferred by major genes with the inability to combat changes in pathogen genotypes. Nevertheless, the approach should not be discarded until all hopeful varieties are fully tested.

(*b*) Multilineal or composite varieties consisting of a mixture of identical varietal phenotypes combining an array of major resistance genes. As one or more components of the variety become susceptible, they would be replaced with others carrying distinct resistance factors. This technique suffers from two serious difficulties, apart from the problems inherent in the breeding process and in the maintenance and blending of isogenic lines for commercial use. First, in the environment of the tropics, where the vast majority of rice is grown, and where the problem is most acute, the fungus is likely to change faster than the most efficient breeder can develop isogenic replacements for susceptible lines. Secondly, repeated backcrossing to a common variety is an integral part of the procedure. The final product, therefore, is limited to the yield potential of the recurrent parent. A minimum of 4 to 5 years with two crops per year is required to produce and stabilize the isogenic lines. There has been recent spectacular progress in increasing field yields two- and threefold with the new dwarf types through conventional breeding. If this rising curve in breeding for yield is to continue, the question is whether the loss in potential yield advance during the 4- to 5-year period of backcrossing would be compensated for by a gain in yield protection through multilineal resistance carried in a variety of lower yielding ability. The loss of one to gain the other would seem to be an unnecessary sacrifice.

(*c*) Horizontal (field, partial, or minor gene) resistance. The brilliant success of horizontal resistance against late blight in potatoes serves as a model. This polygenic type of resistance, having no interaction with pathogen races, has not received much attention. It is not known whether this type of resistance exists and if so what it would look like. The expression of field resistance to blast might involve small-sized lesions and/or a reduced number of lesions. Small-sized lesions are often observed in blast nurseries. They are, however, probably the results of major-gene resistance. To distinguish between major-gene resistance and field resistance, repeated tests with various pathogenic races (genes) would be necessary. Only those constantly producing small lesions may be considered to have field resistance. Varieties showing small numbers of lesions are also often seen in blast nurseries and are due, in many cases, to the low populations of the conidia of a particular race or races. This may be verified by isolating the fungus from the lesions and reinoculating the varieties exhibiting such reactions. Varieties which continue to produce a small number of lesions despite artificial inoculations, may have field resistance. The potential stability of horizontal resistance would far outweigh the difficulties in transferring and fixing it in an autogamous seed-propagated plant such as rice. By definition, a search for such resistance necessarily would involve international testing as the first step. Such a program, built around the existing international programs, might offer the most practicable solution to the old problem of rice blast.

LITERATURE CITED

1. Abumiya, H. Phytopathological studies on the breeding of rice varieties resistant to blast disease. *Bull. Tohoku Natl. Agr. Expt. Sta.*, **17,** 1–101 (1959)
2. Adair, C. R. Inheritance in rice of reaction to *Helminthosporium oryzae* and *Cercospora oryzae. U. S. Dept. Agr. Tech. Bull. 772,* 1–18 (1941)
3. Adair, C. R., Cralley, E. M. 1949 Rice yield and disease control tests. *Arkansas Agr. Expt. Sta. Rept., Ser. 15,* 1–20 (1950)
4. Ahn, C. J., Chung, H. C. Studies on the physiologic races of rice blast fungus, *Piricularia oryzae* in Korea. *Seoul Univ. J. Biol. Agr., Ser. D* (i.e., B), **11,** 77–83 (1962)
5. Akai, S., Ogura, H., Sato, T. Studies on *Pellicularia filamentosa* (Pat.) Rogers. I. On the relation between the pathogenicity and some characters on culture media. *Ann. Phytopathol. Soc. Japan,* **25,** 125–30 (1960)
6. Anonymous. New rice varieties. (Abstr.) *Rev. Appl. Mycol.,* **39,** 469 (1960)
7. Asada, Y., Akai, S., Fukutomi, M. Varietal differences in susceptibility of rice plants to *Helminthosporium blight.* (Prelim. rept.) *Japan. J. Breeding,* **4,** 51–53 (1954)
8. Atkins, J. G. Prevalence and distribution of pathogenic races of *Piricularia oryzae* in the U. S. (Abstr.) *Phytopathology,* **52,** 2 (1962)
9. Atkins, J. G., Adair, C. R. Recent discovery of hoja blanca, a new rice disease in Florida, and varietal resistance tests in Cuba and Venezuela. *Plant Disease Reptr.,* **41,** 911–15 (1957)
10. Atkins, J. G., Beachell, H. M., Crane, L. E. Reaction of rice varieties to straighthead. *Texas Agr. Expt. Sta. Progr. Rept. 1865,* 1–2 (1956)
11. Atkins, J. G., Bollich, C. N., Johnston, T. H., Jodon, N. E., Beachell, H. M., Templeton, G. T. Breeding for blast resistance in the United States. *Proc. Symp. Rice Blast Disease, Intern. Rice Res. Inst., 1963,* 333–41 (1965) (See Ref. **22**)
12. Atkins, J. G., Johnston, T. H. Inheritance in rice of reaction to races 1 and 6 of *Piricularia oryzae. Phytopathology,* **55,** 993–95 (1965)
13. Atkins, J. G., Robert, A. L., Adair, C. R., Goto, K., Kozaka, T., Yanagida, Y., Yamada, M., Matsumoto, S. An international set of rice varieties for differentiating

races of *Piricularia oryzae. Phytopathology,* **57,** 297–301 (1967)

14. Atkins, J. G., Todd, E. H. White tip disease of rice. III. Yield tests and varietal resistance. *Phytopathology,* **49,** 189–91 (1959)
15. Baba, I. Nutritional studies on the occurrence of *Helminthosporium* leaf spot and "akiochi' of the rice plant. *Bull. Natl. Inst. Agr. Sci. (Japan), Ser. D,* **7,** 1–157 (1958)
16. Baba, I., Harada, T. Physiological diseases of rice plant in Japan. *Japan. J. Breeding,* **4,** 101–51 (1954)
17. Bandong, J. M., Ou, S. H. The physiologic races of *Piricularia oryzae* Cav. in the Philippines. *Philippine Agr.,* **49,** 655–67 (1966)
18. Beachell, H. M., Jennings, P. R. Mode of inheritance of hoja blanca resistance in rice. (Abstr.) *Proc. Rice Tech. Working Group, 9th, Lafayette, La., 1960,* 11–12 (1961)
19. Bollich, C. N., Atkins, J. G., Scott, J. E., Webb, B. D. Dawn—a blast resistant, early maturing, low grain rice variety. *Rice J.,* **69**(4), 14, 16, 18, 20 (1966)
20. Bollich, C. N., Scott, J. E., Beachell, H. M. Gulfrose rice. (Reg. No. 28) *Crop Sci.,* **5,** 288 (1965)
21. Breda de Haan, J. Vorlaufige Beschreibung von Pilzen bein Kulturpflanzen beobachtet. *Bull. Inst. Bot. Buitenzorg,* **16,** 11–13 (1900)
22. Chang, T. T., Wang, M. K., Lin, K. M., Cheng, C. P. Breeding for blast resistance in Taiwan. In *The Rice Blast Disease, Proc. Symp. Intern. Rice Res. Inst., 1963,* 371–77 (Johns Hopkins Press, Baltimore, Md., 507 pp., 1965)
23. Chien, C. C., Chung, S. C. Physiologic races of *Pellicularia sasakii* in Taiwan. *Agr. Res. (Taiwan),* **12,** 1–6 (1963)
24. Chiu, R. J. Virus diseases of rice in Taiwan. *FAO Intern. Rice Comm., Working Party Rice Prod. Protect., 10th, Manila, 1964* (Mimeographed)
25. Chiu, R. J., Chien, C. C., Lin, S. Y. Physiologic races of *Piricularia oryzae* in Taiwan. *Proc. Symp. Rice Blast Disease, Intern. Rice Res. Inst., 1963,* 245–55 (1965) (See Ref. 22)
26. Chiu, W. F. Studies on helminthosporiose of rice. III. *Bull. Coll. Agr. Forestry, Nanking, N. S.,* **48,** 1–10 (1936)
27. Chiu, W. F. A preliminary study on the physiological differentiation of *Fusarium fujikuroi* (Saw.) Wr. *Nanking J.,* **9,** 305–321 (1940)
28. Chu, M. Y., Li, H. W. Cytological studies of *Piricularia oryzae* Cav. *Botan. Bull. Acad. Sinica,* **6,** 116–30 (1965)
29. Cralley, E. M. Resistance of rice varieties to stem rot. *Arkansas Agr. Expt. Sta. Bull.,* **329,** 1–31 (1936)
30. Dasananda, S. Breeding for blast resistance in Thailand. *Proc. Symp. Rice Blast Disease, Intern. Rice Res. Inst., 1963,* 379–96 (1965) (See Ref. 22)
31. Fang, C. T., Liu, C. F., Chu, C. L. A preliminary study on the disease cycle of the bacterial leaf blight of rice. *Acta Phytopathol. Sinica,* **2,** 173–85 (1956)
32. Fujii, K., Okada, M. Progress in breeding of rice varieties for resistance of bacterial leaf blight in Japan. *Symp. Rice Diseases and Their Control by Growing Resistant Varieties and Other Measures, Tokyo, 1967,* E1–E18 (Agr. Forestry and Fisheries Res. Council, Ministry of Agriculture and Forestry, Tokyo, 1967) (Mimeographed)
33. Galvez-E., G. E., Lozano-T., J. C. Identification of races of *Piricularia oryzae* in Colombia. *Phytopathology,* **58,** 294–96 (1968)
34. Ganguly, D. *Helminthosporium* disease of paddy in Bengal. *Sci. Cult. (Calcutta),* **12,** 220–23 (1946)
35. Ganguly, D., Padmanabhan, S. Y. *Helminthosporium* disease of rice. III. Breeding resistant varieties—selection of resistant varieties from genetic stock. *Indian Phytopathol.,* **12,** 99–110 (1959)
36. Giatong, P., Frederiksen, R. A. Variation in pathogenicity of *Piricularia oryzae.* (Abstr.) *Phytopathology,* **57,** 460 (1967)
37. Goto, I. On the inheritance of resistance to the blast disease. *Conf. Phytopathol. Soc. Japan, 1959* (Takahashi, 1965)
38. Goto, K. Physiologic races of *Piricularia oryzae* in Japan. *Proc. Symp. Rice Blast Disease, Intern. Rice*

Res. Inst., 1963, 237–42 (1965) (See Ref. 22)

39. Goto, K., Fukatsu, R. Studies on the stem rot of rice plant. I. Varietal resistance and seasonal development. *Bull. Div. Plant Breeding Cultivation, Tokai-Kinki Natl. Agr. Expt. Sta.,* **1,** 27–39 (1954)
40. Goto, K., Fukatsu, R. Studies on the white tip of rice plant III. Analysis of varietal resistance and its nature. *Bull. Natl. Inst. Agr. Sci., (Japan), Ser. C,* **6,** 123–49 (1956)
41. Goto, K., Kozaka, T., Yanagita, K., Takahashi, Y., Suzuki, H., Yamada, M., Matsumoto, S., Shindo, K., Atkins, J. G., Robert, A. L., Adair, C. R. U.S.-Japan cooperative research on the international pathogenic races of the rice blast fungus, *Piricularia oryzae* Cav., and their international differentials. *Ann. Phytopathol. Soc. Japan,* **33** (extra issue), 1–87 (1967)
42. Hashioka, Y. Studies on the mechanism of prevalence of rice blast fungus disease in the tropics. *Taiwan Agr. Res. Inst. Tech. Bull. 8,* 1–237 (1950)
43. Hashioka, Y. Studies on pathological breeding of rice. IV. Varietal resistance of rice to the sclerotial diseases. *Japan. J. Breeding,* **1,** 21–26 (1951)
44. Hashioka, Y. Studies on pathological breeding of rice. V. Varietal resistance of rice to the "Bakanae" disease. *Japan. J. Breeding,* **1,** 167–71 (1952)
45. Hashioka, Y. Studies on pathological breeding of rice. VI. Varietal resistance of rice to the brown spot and yellow dwarf. *Japan. J. Breeding,* **26,** 14–16 (1953)
46. Hashioka, Y. The rice stem nematode *Ditylenchus angustus* in Thailand. *FAO Plant Protect. Bull.,* **11,** 97–102 (1963)
47. Hector, G. P. Annual report of the first Economic Botanist to the Government of Bengal for the year 1930–31. (Abstr.) *Rev. Appl. Mycol.,* **11,** 157–58 (1932)
48. Hsieh, S. C., Chien, C. C. Recent status of rice breeding for blast resistance in Taiwan, with special regard to races of the blast fungus *Symp. on Rice Diseases,* F1–F28 (See ref. 32)
49. Hsieh, S. P. Y. Stem rot of rice in the Philippines (Master's thesis, Univ. of the Philippines, Manila, 1966)
50. Hsieh, Y. T., Wu, Y. L., Shian, K. A. *Screening for Sheath Blight Resistance in Rice Varieties* (Kaohsiung District Agr. Improvement Sta., Pingtung, Taiwan, 92 pp., 1965)
51. Huang, C. S. Induction of mutations for rice improvement in Taiwan. *Chinese-Am. Joint Comm. on Rural Reconstr. (Taiwan), Plant Ind. Ser.,* **22,** 59–76 (1961)
52. Hung, Y. P. White tip disease of rice in Taiwan. *Plant Protect. Bull. (Taiwan),* **1,** 1–4 (1959)
53. *Intern. Rice Res. Inst. Ann. Rept.,* 105–8 (1963) ; (a) 128–56 (1964) ; (b) 82–104 (1966) ; (c) (1967, in press)
54. Israel, P., Rao, Y. S., Rao, V. N. Rice nematodes—host and parasite relationship. *FAO Intern. Rice Comm. Working Party Rice Prod. Protect., 10th, Manilla, 1964* (Mimeographed)
55. Ito, R. Breeding for blast resistance in Japan. *Proc. Symp. Rice Blast Disease, Intern. Rice Res. Inst., 1963,* 361–70 (1965) (See Ref. 22)
56. Ito, S., Kimura, J. Studies on the "Bakanae" disease of the rice plant. *Hokkaido Agr. Expt. Sta. Rept.,* **27,** 1–94 (1931)
57. Jodon, N. E. Louisiana releases Sunbonnet through Seed Growers Assn. *Rice J.,* **56**(3), 40 (1953)
58. Jodon, N. E. Breeding for improved varieties of rice and other cereal grains. *Rice J.,* **57**(5), 32, 34–36 (1954)
59. Jodon, N. E. Toro and Sunbonnet, two new midseason, long grain rice varieties compared. *Rice J.,* **58(12), 8–11 (1955)**
60. Jodon, N. E., Chilton, S. J. P. Some characters inherited independently of reaction of physiologic races of *Cercospora oryzae* in rice. *J. Am. Soc. Agron.,* **38,** 864–72 (1946)
61. Jodon, N. E., Ryker, T. C., Chilton, S. J. P. Inheritance of reaction to physiologic races of *Cercospora oryzae* in rice. *J. Am. Soc. Agron.,* **36,** 497–507 (1964)
62. Johnston, T. H. Registration of rice varieties. *Agron. J.,* **50,** 694–700 (1958)

63. Johnston, T. H., Templeton, G. E., Sims, J. L., Hall, V. L., Evans, K. O. Performance in Arkansas of Nova 66 and other medium grain rice varieties, 1960 to 1965. *Arkansas Agr. Expt. Sta. Rept., Ser. 148,* 1–24 (1966)
64. Kang, I. M., Lee, E. K., Kim, Y. K. Studies on resistance of rice variety of sheath blight *(Corticium sasakii)* in field. *Res. Rept. Korea Office of Rural Develop.,* **8,** 235–41 (1965)
65. Kariya, K., Washio, Y. Effect of the selection during early segregating generations for bacterial leaf blight resistance in rice. *Chugoku Agr. Res.,* **5,** 39–40 (1956)
66. Katsuya, K. Susceptibility of wild and foreign cultivated rice to the blast fungus, *Piricularia oryzae. Ann. Rept. Natl. Inst. Genet., Japan,* **9,** 48–49 (1959)
67. Katsuya, K. Susceptibility of wild rice to the blast fungus, *Piricularia oryzae, Ann. Rept. Natl. Inst. Genet., Japan,* **10,** 76–77 (1960)
68. Kawai, I., Kakisaki, T. Studies on the brown spot of the rice plant. I. Environment to outbreak and its control. *Nogyo Kairyo Gijutso,* **70,** 1–32 (1955)
69. Kawamura, E. Reaction of certain species of the genus *Oryza* to the infection of *Piricularia oryzae. Kjusu Imp. Univ. Sci. Fak. Terkult. Bull.,* **9,** 157–66 (1940)
70. Kiyosawa, S. Studies on inheritance of resistance of rice varieties to blast. II. Genetic relationship between the blast resistance and other characters in rice varieties Reishiko and Sekiyama 2. *Japan. J. Breeding,* **16,** 87–95 (1966)
71. Kiyosawa, S. Studies on inheritance of resistance of rice varieties to blast. III. Inheritance of resistance in rice variety Pi no. 1 to the blast fungus. *Japan. J. Breeding,* **16,** 243–50 (1966)
72. Kiyosawa, S. The inheritance of resistance of the Zenith type varieties of rice to the blast fungus. *Japan. J. Breeding,* **17,** 99–107 (1967)
73. Kiyosawa, S. Inheritance of resistance of the rice variety Pi No. 4 to blast. *Japan. J. Breeding,* **17,** 165–72 (1967)
74. Kiyosawa, S. Inheritance of blast resistance in some Chinese rice varieties and their derivatives. *Japan. J. Breeding,* **18,** 193–204 (1968)
75. Kiyosawa, S., Matsumoto, S., Lee, S. C. Inheritance of resistance of rice variety Norin 22 to two blast fungus strains. *Japan. J. Breeding,* **17,** 1–6 (1967)
76. Komori, N., Takano, S. Varietal resistance of rice plant to rice yellow dwarf in the field. *Proc. Kanto-Tosan Plant Protect. Soc.,* **11,** 22 (1964)
77. Koyama, T. On the breeding of highly resistant varieties to rice blast by the hybridization between Japanese varieties and foreign varieties in Japanese type of rice. *Japan. J. Breeding,* **2,** 25–30 (1952)
78. Kuhara, S., Kurita, T., Tagami, Y., Fujii, H., Sekiya, N. Studies on the strain of *Xanthomonas oryzae* (Uyeda et Ishiyama) Dowson, the pathogen of the bacterial leaf blight of rice with special reference to its pathogenicity and phage-sensitivity. *Bull. Kyushu Agr. Expt. Sta.,* **11,** 263–312 (1965)
79. Kusaba, T., Watanabe, M., Tabei, H. Classification of the strains of *Xanthomonas oryzae* (Uyeda et Ishiyama) Dowson on the basis of their virulence against rice plants. *Bull. Natl. Inst. Agr. Sci. (Japan), Ser. C,* **20,** 67–82 (1966)
80. Lamey, H. A. Varietal resistance to hoja blanca. In *The Virus Diseases of the Rice Plant, Proc. Symp., Intern. Rice Res. Inst., 1967* (In press)
81. Lamey, H. A., Lindberg, G. D., Brister, C. D. A greenhouse testing method to determine hoja blanca reaction of rice selections. *Plant Disease Reptr.,* **48,** 176–79 (1964)
82. Lamey, H. A., McMillian, W. W., McGuire, H. H. Transmission and host range studies on hoja blanca. (Abstr.) *Proc. Rice Tech. Working Group, 9th, Lafayette, La., 1960,* 20–21 (1961)
83. Latterell, F. M., Tullis, E. C., Collier, J. W. Physiologic races of *Piricularia oryzae* Cav. *Plant Disease Reptr.,* **44,** 679–83 (1960)
84. Latterell, F. M., Tullis, E. C., Otten, R. T., Gubernik, A. Physiologic races of *Piricularia oryzae.*

(Abstr.) *Phytopathology,* **44,** 495–96 (1954)

85. Lee, S. C., Matsumoto, S. Studies on the physiologic races of rice blast fungus in Korea during the period of 1962–1963. *Ann. Phytopathol. Soc. Japan,* **32,** 40–45 (1966)
86. Li, H. W., Hu, C. H., Chang, W. T., Wang, T. S. The utilization of X-radiation for rice improvement. *Proc. Symp. Effects Ionizing Radiations on Seeds, Karlsruhe, 1960,* 484–94 (1961)
87. Lin, K. M., Lin, P. C. Radiation-induced variation in blast disease resistance in rice. *Japan. J. Breeding,* **10,** 19–22 (1960)
88. Lin, K. M., Lin, P. C. Survey of intervarietal variation in resistance to blast disease. *Taiwan Agr. Res. Inst. Spec. Bull.,* **3,** 38–44 (1961)
89. Ling, K. C. Testing rice varieties for resistance to tungro disease. *Proc. Symp. Virus Diseases Rice Plant, Intern. Rice Res. Inst., 1967* (In press)
90. Matsuo, T. Genecological studies on cultivated rice. *Bull. Natl. Inst. Agr. Sci. (Japan), Ser. D,* **3,** 1–111 (1952)
91. Mello-Sampayo, T., Vianna e Silva, M. Preliminary experiments on the determination of resistance of some cultivated forms of rice to *Piricularia oryzae* Br. and Cav. *Rev. Appl. Mycol.,* **34,** 481–82 (1955)
92. Metcalf, H. The story of a plant introduction. *J. Wash. Acad. Sci.,* **11,** 474 (1921)
93. Misra, A. P., Chatterjee, A. K. Comparative study of two isolates of *Helminthosporium oryzae* Breda de Haan. *Indian Phytopathol.,* **16,** 275–81 (1963)
94. Morinaka, T., Sakurai, Y. Varietal resistance by seedling test method to black-streaked dwarf of rice plant. (Abstr.) *Ann. Phytopathol. Soc. Japan,* **32,** 89–90 (1966)
95. Muko, H., Yoshida, K. A needle inoculation method for bacterial leaf blight disease of rice. *Ann. Phytopathol. Soc. Japan,* **15,** 179 (1951)
96. Murata, N. Genetic aspects on resistance to bacterial leaf blight in rice and variation of its causal bacterium. *Symp. Rice Diseases,* D1–D10 (See Ref. 32)
97. Nagai, I., Hara, S. On the inheritance of variegation disease in strain of rice plant. *Japan. J. Genet.,* **5,** 140–44 (1930)
98. Nakatomi, S. On the variability and inheritance of the resistance of rice plants to rice blast disease. *Japan. J. Genet.,* **4,** 31–38 (1926)
99. Narise, T. Variation in susceptibility to *Piricularia* disease in wild rice populations. *Ann. Rept. Inst. Genet., Japan,* **10,** 72–73 (1960)
100. Nawaz, M., Kausar, A. G. Cultural and pathogenic variation in *Helminthosporium oryzae. Biologia (Pakistan),* **8,** 35–48 (1962)
101. Nishimura, Y., Sakaguchi, S. Inheritance of resistance in rice to bacterial leaf blight, *Bacterium oryzae* (Uyeda et Ishiyama) Nakata. *Japan. J. Breeding,* **9,** 58 (1959)
102. Nishizawa, T., Yamamoto, S. Studies on the varietal resistance of rice plant to the rice nematode disease "Senchu Singare Byo." II. A test of the leading varieties and part of breeding lines of rice plants in Kyushu. *Kyushu Noji Shikinjo,* **8,** 91–92 (1951)
103. Nishizawa, T., Yamamoto, S., Mizuta, H. Studies on the varietal resistance of rice plant to the rice nematode disease "Senchu Shingare Byo." VII. *Bull. Kyushu Agr. Expt. Sta.,* **2,** 71–80 (1953)
104. Nisikado, Y. Comparative studies on the *Helminthosporium* disease of rice in the Pacific region. *Ann. Phytopathol. Soc. Japan,* **2,** 14–25 (1927)
105. Nisikado, Y., Matsumoto, H. Studies on the physiological specialization of *Gibberella fujikuroi,* the causal fungus of the rice "bakanae" disease. *Trans. Tottori Soc. Agr. Sci.,* **4,** 200–11 (1933)
106. Okada, M., Maeda, H. Inheritance of resistance to leaf blast in crosses between foreign and Japanese varieties of rice. *Bull. Tohoku Natl. Agr. Expt. Sta.,* **10,** 59–68 (1956)
107. Ono, K. Varietal resistance to stem-rot in rice. *Ann. Phytopathol. Soc. Japan,* **13,** 14–18 (1949)
108. Ono, K., Suzuki, H. Studies on mechanism of infection and ecology of blast and stem-rot of rice plant. *Spec. Rept. Plant Disease Insect Pest Forecast Serv., Ministry Agr. and Forestry, Japan,* **4,** 94–152 (1960)

109. Ou, S. H. A proposal for an international program of research on the rice blast disease. *Proc. Symp. Rice Blast Disease, Intern. Rice Res. Inst., 1963,* 441–46 (1965) (See Ref. 22)

110. Ou, S. H. Varietal reactions of rice to blast. *Proc. Symp. Rice Blast Disease, Intern. Rice Res. Inst., 1963,* 223–34 (1965)

111. Ou, S. H. Rice diseases of obscure nature in tropical Asia with special reference to "mentek" disease in Indonesia. *Intern. Rice Comm. Newsletter,* **14**(2), 4–10 (1965)

112. Ou, S. H., Ayad, M. R. Pathogenic races of *Pyricularia oryzae* originating from single lesions and monoconidial cultures. *Phytopathology,* **58,** 179–82 (1968)

113. Ou, S. H., Goh, K. G. Further experiment on "penyakit merah" disease of rice in Malaysia. *Intern. Rice Comm. Newsletter,* **15**(2), 31–32 (1966)

114. Ou, S. H., Nuque, F. The relation between leaf and neck resistance to the rice blast disease. *Intern. Rice Comm. Newsletter,* **12**(4), 30–35 (1963)

115. Ou, S. H., Rivera, C. T., Navaratnam, S. J., Goh, K. G. Virus nature of "penyakit merah" disease of rice in Malaysia. *Plant Disease Reptr.,* **49,** 778–82 (1965)

116. Padmanabhan, S. Y. Specialization in pathogenicity of *Helminthosporium oryzae. Abstr. 18, pt. 4, Proc. Indian Sci. Congr., 40th, New Delhi, 1953*

117. Padmanabhan, S. Y. Breeding for blast resistance in India. *Proc. Symp. Rice Blast Disease, Intern. Rice Res. Inst., 1963,* 343–59 (1965) (See Ref. 22)

118. Padmanabhan, S. Y. Physiologic specialization of *Piricularia oryzae* Cav., the causal organism of blast disease of rice. *Current Sci.,* **34,** 307–8 (1965)

119. Padmanabhan, S. Y., Ganguly, D., Chandwani, G. H. *Helminthosporium* disease of rice. VIII. Breeding resistant varieties: selection of resistant varieties of early duration from genetic stock. *Indian Phytopathol.,* **19,** 72–75 (1966)

120. Park, M., Bertus, L. S. Sclerotial diseases of rice in Ceylon. 2. *Sclerotium oryzae* Catt. *Ann. Roy. Botan. Gard., Peradeniya,* **11,** 342–59 (1932)

121. Rajagopalan, K. Screening rice varieties for resistance to foot-rot disease. *Current Sci.,* **30,** 145–47 (1961)

122. Raychaudhuri, S. P., Misra, M. D., Ghosh, A. Preliminary note on transmission of a virus disease resembling tungro of rice in India and other virus-like symptoms. *Plant Disease Reptr.,* **51,** 300–1 (1967)

123. Reyes, G. M. A preliminary report on the stem-rot of rice. *Philippine Agr. Rev.,* **22,** 313–31 (1929)

124. Reyes, G. M. Rice hybrids versus stem-rot disease. *Philippine J. Agr.,* **7,** 413–18 (1936)

125. Reyes, G. M. Rice diseases and methods of control. *Philippine J. Agr.,* **10,** 419–36 (1939)

126. Rivera, C. T., Ou, S. H. Leafhopper transmission of "tungro" disease of rice. *Plant Disease Reptr.,* **49,** 127–31 (1965)

127. Rivera, C. T., Ou, S. H., Tantere, D. M. Tungro disease of rice in Indonesia. *Plant Disease Reptr.,* **52,** 122–24 (1968)

128. Ryker, T. C. Linkage in rice of two resistant factors to *Cercospora oryzae.* (Abstr.). *Phytopathology,* **31,** 19–29 (1941)

129. Ryker, T. C. Physiologic specialization in *Cercospora oryzae. Phytopathology,* **33,** 70–74 (1943)

130. Ryker, T. C. New pathogenic races of *Cercospora oryzae* affecting rice. (Abstr.). *Phytopathology,* **37,** 19–20 (1947)

131. Ryker, T. C., Chilton, S. J. P. Inheritance and linkage of factors for resistance to two physiological races of *Cercospora oryzae* in rice. *J. Am. Soc. Agron.,* **34,** 836–40 (1942)

132. Ryker, T. C., Cowart, L. E. Development of *Cercospora*-resistant strains of rice. (Abstr.). *Phytopathology,* **38,** 23 (1948)

133. Ryker, T. C., Jodon, N. E. Inheritance of resistance to *Cercospora oryzae* in rice. *Phytopathology,* **30,** 1041–47 (1940)

134. Sakurai, Y. Varietal resistance to stripe, dwarf, yellow dwarf, and black-streaked dwarf. In *The Virus Diseases of the Rice Plant.*

Proc. Symp. Intern. Rice Res. Inst., 1967 (Johns Hopkins Press, Baltimore, Md.) (In press)

135. Sakurai, Y., Ezuka, A. The seedling test method of varietal resistance of rice plant to stripe virus disease. 2. The resistance of various varieties and strains of rice plant by the method of seedling test. *Bull. Chukogu Agr. Expt. Sta.*, **A 10,** 51–70 (1964)
136. Sakurai, Y., Ezuka, A., Okamoto, H. The seedling test method of varietal resistance of rice plants to stripe virus disease (part 1) *Bull. Chugoku Agr. Expt. Sta.*, **A 9,** 113–25 (1963)
137. Sakurai, Y., Toriyama, K. Field resistance of the rice plant to *Piricularia oryzae* and its testing method. *Symp. Rice Diseases,* J1–J20 (See Ref. 32)
138. Sasaki, R. Existence of strains in rice blast fungus, I. *J. Plant Protect. (Japan),* **9,** 631–44 (1922)
139. Sasaki, R. Existence of strains in rice blast fungus, II. *J. Plant Protect. (Japan),* **10,** 1–10 (1923)
140. Sasaki, R. Inheritance of resistance to *Piricularia oryzae* in different varieties of rice. *Japan. J. Genet.*, **1,** 81–85 (1922)
141. Sato, K. Studies on the blight disease of rice plant. *Bull. Inst. Agr. Res. Tohoku Univ.*, **15,** 199–237 (1964) ; **15,** 239–342 (1965) ; **16,** 1–54 (1965)
142. Seto, F. Experimentelle Untersuchungen über die hemmende und die beschleunigende Wirkung des Erregers der sogenannten "Bakanae"-Krankheit, *Lisea fujikuroi* Sawada auf das Wachstum der Reiskeimlinge. *Mem. Coll. Agr. Kyoto Imp. Univ., Agr. Econ. Ser.*, **18,** 1–23 (1932)
143. Suematsu, N. On the resistant varieties. *Ann. Phytopathol. Soc. Japan,* **1,** 53–56 (1921)
144. Suzuki, H. Origin of variation in *Piricularia oryzae. Proc. Symp. Rice Blast Disease, Intern. Rice Res. Inst., 1963,* 111–49 (1965) (See Ref. 22)
145. Suzuki, H. *Studies on Biologic Specialization in* Pyricularia oryzae *Cav.* (Tokyo Univ. Agr. and Tech., Tokyo, 235 pp., 1967)
146. Suzuki, H., Kato, T., Kawaguchi, K., Sasanuma, H. On the testing method of resistance of rice varieties to rice stripe, *Oryza* virus 2 Kuribayashi, in the frequently affected paddy field. *J. Tochigi Agr. Expt. Sta.*, **4,** 1–15 (1960)
147. Tagami, Y., Mizukami, T. Historical review of the researches on bacterial leaf blight of rice. *Xanthomonas oryzae* (Uyeda et Ishiyama) Dowson. *Spec. Rept. Plant Disease Insect Pest Forecast Serv., Ministry Agr. and Forestry, Japan,* **10,** 1–112 (1962)
148. Takahashi, Y. Genetics of resistance to the rice blast disease. *Proc. Symp. Rice Blast Disease, Intern. Rice Res. Inst., 1963,* 303–29 (1965) (See Ref. 22)
149. Thomas, K. M. The foot rot of paddy and its control. *Madras Agr. J.*, **21,** 263–72 (1933)
150. Tisdale, W. H. Two sclerotium diseases of rice. *J. Agr. Res.*, **21,** 649–58 (1921)
151. Tochinai, Y., Sakamoto, M. Studies on the physiologic specialization of *Ophiobolus miyabeanus* Ito and Kuribayashi. *J. Fac. Agr. Hokkaido Imp. Univ.*, **41,** 1–96 (1937)
152. Toriyama, K. Genetics of and breeding for resistance to rice virus diseases. *Proc. Symp. Virus Diseases of the Rice Plant, Intern. Rice Res. Inst., 1967* (In press)
153. Toriyama, K., Sakurai, Y., Washio, O., Ezuka, A. A newly bred rice line, Chugoku No. 31 with stripe disease resistance transferred from an indica variety. *Bull. Chugoku Agr. Expt. Sta., Ser. A,* **13,** 41–54 (1966)
154. Tu, J. C. Strains of *Pellicularia sasakii* isolated from rice in Taiwan. *Plant Disease Reptr.*, **51,** 682–84 (1967)
155. Tullis, E. C. *Cercospora oryzae* on rice in the United States. *Phytopathology,* **27,** 1005–8 (1937)
156. Van der Meulen, J. G. J. Rice improvement by hybridization and results obtained. *Contrib. Gen. Agr. Res. Sta., Bogor, Indonesia,* **116,** 1–38 (1950)
157. Washio, O., Kariya, K., Toriyama, K. Studies on breeding rice varieties for resistance to bacterial leaf blight. *Bull. Chugoku Agr. Expt. Sta., Ser. A,* **13,** 55–86 (1966)
158. Wathanakul, L., Weerapat, P. Virus

diseases of rice in Thailand. *Proc. Symp. Virus Diseases of the Rice Plant, Intern. Rice Res. Inst., 1967* (In press)

159. Wu, H. K., Tsao, T. H. The ultrastructure of *Piricularia oryzae* Cav. *Botan. Bull. Acad. Sinica,* **8** (Spec. no.), 353–63 (1967)

160. Yamaguchi, T., Yasuo, S., Ishii, M. Studies on rice stripe disease. II. Study on the varietal resistance to stripe disease of rice plant. *J. Central Agr. Expt. Sta. (Japan),* **8,** 109–60 (1965)

161. Yamasaki, Y., Kiyosawa, S. Studies on inheritance of resistance of rice varieties to blast. I. Inheritance of resistance of Japanese to several strains of the fungus. *Bull. Natl. Inst. Agr. Sci. (Japan), Ser. D,* **14,** 39–69 (1966)

162. Yamasaki, Y., Niizeki, H. Studies on variation of rice blast fungus *Piricularia oryzae* Cav. I. Karyological and genetical studies on variation. *Bull. Natl. Inst. Agr. Sci. (Japan), Ser. D,* **13,** 231–74 (1965)

163. Yasuo, S. T., Yamaguchi, T., Ishii, M. Experimental results in plant diseases. *Studies on the stripe and dwarf of rice plant* (Central Agr. Expt. Sta., Japan, 1960). (Mimeographed)

164. Yoshii, H., Matsumoto, M. Studies on the resistance to helminthosporiose of the rice varieties introduced to Japan, I. *Sci. Rept. Matsuyama Agr. Coll.,* **6,** 23–60 (1951)

165. Yoshii, H., Yamamoto, S. On some methods for the control of rice nematode disease. *Sci. Bull. Fac. Agr. Kyushu Univ.,* **12,** 123–31 (1951)

166. Yoshimura, S., Iwata, K. Studies on examination method of varietal resistance to bacterial leaf blight disease of rice plant. I. Immersion method of inoculation and its applied method. *Proc. Assoc. Plant Protect. Hokuriku,* **13,** 25–31 (1965)

Copyright 1969. All rights reserved

USE OF PLANT PATHOGENS IN WEED CONTROL[1]

By Charles L. Wilson[2]

University of Arkansas, Fayetteville, Arkansas

Introduction

The idea of using plant pathogens to control weeds is almost as old as the science of plant pathology itself. This idea has occurred to scientists and laymen alike, as is indicated by a letter from a farmer that Dr. Byron D. Halsted received at the New Jersey Agricultural Experiment Station before 1893. The letter reads:

> Two years ago about an acre of our farm was overrun with Canada thistle, but by the time they were in full bloom a rust struck them and hardly any seed of the plant matured. We plowed the land in the fall, and last year scarcely a thistle appeared. If this rust could be widely disseminated through the country, the Canada thistle would receive a substantial check (55).

Seeds of the idea to use plant pathogens to kill weeds have lain dormant since their sowing. There has been an occasional sprout to indicate survival. Dormancy has not been due to infertile seed but to the lack of cultivation by plant pathologists. The potential for biological control of weeds with plant pathogens might be anticipated from the success that has been realized in controlling weeds with insects. Entomologists have proceeded through their dormancy period and have made biological control of weeds with insects a significant advance in weed control (30, 60, 61, 145). A number of past and present successes in controlling weeds with plant pathogens demonstrate the feasibility of this approach and point toward expanded activity.

All of us are aware of recent pressures to use means other than chemicals to control pests. Also, it is becoming more apparent in pest control that reliance on a single method of control is hazardous. Multiple methods of control give more assurance of overall success. Plant pathogens as weed control agents fit nicely into such a scheme. Effective plant pathogens would have at least three advantages over chemical herbicides. (*a*) They can be specific to the weed. (*b*) Residue and toxicity problems would be

[1] Published with the approval of the Director of the Arkansas Agricultural Experiment Station.

[2] The author should write a review of the assistance that he has received. It is hoped that those who have helped will recognize their contribution and know that it is appreciated. Special thanks are due R. E. Inman for his help and encouragement. Author's present address is Shade Tree and Ornamental Plants Laboratory, U.S.D.A., Delaware, Ohio.

reduced or eliminated altogether, and (*c*) there would be no accumulation of the herbicide in the soil and underground water.

Plant pathogens probably play an important role in nature in the reduction of weeds, but we know little about it because plant pathologists have given little attention to diseases of weeds. The potential of a plant pathogen to destroy a population of plants is obvious when we consider examples such as Dutch Elm Disease, Late Blight of Potato, and Chestnut Blight.

I shall discuss the meager literature making up the life blood of this infant field. R. P. True's dictum seems pertinent. Speaking of the use of microbial antagonists in the control of forest pathogens, True said, "This is also a fine area for the fruitful interplay of faith and skepticism. Some of us will naturally emphasize one, some the other, but let us not start out, at this time, to be judicial about a subject concerning which we know so little."

Since no type of disease or pathogen seems generally more destructive than others, there is no advantage in discussing weed diseases according to their causes or symptoms. Instead I shall proceed through the weed hosts that have been considered as possible victims of plant pathogens and attempt to expose areas that need research. Seedling diseases will be considered as a group because of their similarity in different weeds.

Control of Specific Weeds

Cactus.—The successful biological control of the prickly pear cactus in Australia serves as a salient example of the fruitful rewards of effective biological weed control. Plant diseases as well as insects played a role in the control of prickly pear in Australia and other countries. The relative importance of the pathogens is not clear since most of the intensive work was done with the insects rather than with the associated bacteria and fungi.

Success of the Australians in controlling the prickly pear resulted from well-organized and executed activities of the Commonwealth Prickly Pear Board (32–35) which involved extensive study as well as the introduction of a number of insects and associated microbes. One insect, *Cactoblastis cactorum,* proved to be far superior to others. There are also reports of the successful use of *C. cactorum* and presumably associated pathogens to control prickly pear in Hawaii (46), South Africa (30), and the West Indies (123).

Since *C. cactorum* has not been reared free of microorganisms, it has not been possible to assess the role of plant pathogens in the development of the insect on the host and the decline of prickly pear. Dodd (35) is of the opinion that plant pathogens are "given the opportunity of completing the work of eradication" following the invasion of *C. cactorum. Gloeosporium lunatum* E & E. and bacterial soft-rot organisms are considered the primary invaders following the attacks of *C. cactorum* (34). The diseases that they cause are restricted to the succulent upper segments of the prickly pear and Dodd feels that the lower segments and butts are destroyed by a "physiological breakdown" in which various fungi may be contributing factors.

The intimate association of plant pathogens with *C. cactorum* needs further investigation as this might explain the superior success with this insect and point to the selection of other associates for other weeds. In speaking of the use of pathogen vector relationships in weed control, Huffaker (61) states: "Perhaps the least explored but theoretically most promising approach is the combined use of disease organisms and insects." The possibility that plant pathogens have played an unsuspected role in successful weed control with insects needs investigation.

The association of insects with *Erwinia carnegieana* Standring, in causing the death of the giant cactus *Carnegiea giganteus* Englm., has been reported by Lightle et al. (75) and Boyle (14). In this case the cacti involved are considered aesthetically desirable and not weeds, but there is a similarity in the decline of these cacti to that of *Opuntia* spp. The bacterium which causes the decline of the giant cactus (92) is inoculated by an insect, *Cactobrosis fernaldialis*. Johnston & Hitchcock (64) found that a number of insects can introduce a rot-causing bacterium into *Opuntia* sp.

A number of fungi which cause diseases of *Opuntia* spp. have been described (23, 25, 34, 46, 48). A very destructive *Fusarium* disease of prickly pear (*Opuntia megacantha* Salm. Dick) has been found in Hawaii (23). Unfortunately this disease is much more virulent on the desirable red forms of the cactus than on the undesirable white forms. The red forms of the cactus serve as a forage plant in arid sections. Carpenter (23) reports that in the arid sections of Hawaii "the prickly pear, and the dew and misty rains are such important sources of water that cattle are reported to reach maturity without learning to drink."

Fullaway (46) reports an unsuccessful attempt to use *Fusarium oxysporum* Schlect on a large scale to control the white form of *O. megacantha* in Hawaii. Chock (25), has shown that *F. oxysporum* can effectively kill the white forms of *O. megacantha* and is easily inoculated into plants. He introduced the fungus into cactus in a number of unique ways such as by coating missiles with spores and propelling them with a sling shot or shot gun. Chock (25) found that with the white forms, "Disregarding age or size, all plants inoculated with this *Fusarium* eventually died to the ground."

The red and white cactus exemplifies a problem that runs throughout the field of biological control of weeds. The same plant may be a weed in one environment and an economically important plant in another. Unfortunately, plant pathogens cannot make a distinction between desirable and undesirable plants.

Mistletoes.—The dwarf mistletoes (*Arceuthobium* spp.) are major pathogens of conifers in western North America. There has been considerable interest in the biological control of these weeds with fungal parasites (11–13, 37, 38, 40, 41, 49, 50, 70, 85, 91, 120, 138–140). Wellman (137) and Wicker & Shaw (140) have recently reviewed this subject.

Three fungal parasites seem to hold some promise in the biological control of *Arceuthobium* spp. Ellis (41) described a fungus, *Septogloeum gillii* D. E. Ell., which he indicates might exercise a measure of biological control of dwarf mistletoes in locations. Mielke (86) was unsuccessful when he tried to establish epidemics of *S. gillii* in Northern Utah.

Wallrothiella arceuthobii (Peck) Sacc. is a fungus which may invade and completely destroy the seeds of *Arceuthobium* spp. Through the reduction in seed production, some biological control is accomplished (38). *Colletotrichum* blight of dwarf mistletoe (91) is indicated as having promise in biological control. Parmeter et al. (91) indicate that *Colletotrichum gloeosporioides* Penz. (*sensu* von Arx) has advantage over other pathogens described for biological control of *Arceuthobium* sp. in that it can be grown more readily in culture and the disease develops more rapidly on the host. The endophytic system of the host is invaded by *C. gloeosporioides* which does not seem to be the case with *S. gillii* and *W. arceuthobii.*

Wicker & Shaw (138) list the attributes considered necessary for a pathogen to be a successful biological agent against mistletoes. These are generally applicable in the selection of a pathogen to control weeds. These attributes are: distribution which coincides with that of the target host, ecologic amplitude sufficient to insure persistence within its host range, production of abundant inoculum (spores) for establishment of epiphytotics, high infectivity, high virulence, and an efficient mode of action for curtailing development of the target host.

A number of fungi have been reported on Loranthacae other than the dwarf mistletoes, but little attention has been given their possible roles in biological control. Gill & Hawksworth (50) list the nearly 100 fungi that have been reported on mistletoes.

Water weeds.—The recognition of water weed problems is increasing with increased impoundment, utilization, and need of water. These weeds interfere with water flow, navigation, waste disposal, and recreation; they also cause health hazards to humans by allowing pathogen-carrying insects to reproduce. The use of chemicals in the control of water weeds is particularly undesirable because of our inability to confine distribution of the chemicals.

Recent discovery by Safferman & Morris (116–119) of a viral disease of blue-green algae has stimulated work on the use of the virus, (LPP-1) to control undesirable algal "blooms" in sewage. The virus attacks algae belonging to the genera *Lyngbya, Plectonema,* and *Phormidium.* Diseased cells lyse and in culture plaques develop. *Plectonema boryanum* (Safferman & Morris) is the predominant species in algal blooms and it is very susceptible to the LPP-1 virus. Considerable information is now available on the infection process of this virus and its multiplication in the host (16, 51, 79, 121, 126–128); the findings of Safferman & Morris have been confirmed in this country, as well in Israel (89), in India (124), and perhaps in the Ukraine (109).

Jackson (63) has conducted preliminary tests on the use of the LPP-1 virus to control algal bloom in sewage disposal pools at the Brookside treatment plant in Fayetteville, New York. Five 1000-gallon pools were inoculated with 100 ml preparations of LPP-1 virus. Within a week there was a striking difference between the inoculated and uninoculated pools, with the virus causing a substantial reduction in the algal population. These preliminary findings may point to a new and inexpensive method to control eutrophication.

Safferman & Morris (119) and others (89) have indicated that there are different strains of the LPP-1 virus and that there are probably a number of blue-green algal viruses. The opportunity would seem to exist to find other viruses of algae, as well as other uses of such viruses in weed control.

Blue-green algae are a major problem on some of the large reservoirs and inland seas associated with major hydroelectric dams in the Soviet Union, where they clog waterways and kill fish. At the height of algal epidemics Goryushin & Chaplinskaya (5, 52) noticed clear patches where the algae were apparently absent or had been killed. They were able to isolate an undescribed virus from these patches that would kill algae. Goryushin & Chaplinskaya are in the process of selecting a strain of the virus that might be used in the biological control of blue-green algae and their progress should be followed.

In late 1966 and 1967, Eurasian water milfoil (*Myriophyllum spicatum* L.) decreased dramatically in abundance and distribution along the upper Chesapeake Bay. Bayley & Southwick (6) feel that the reduction in milfoil is due to a disease originally described by Elser (42) as "Northeast disease." From their preliminary studies, they think that this disease is caused by a virus since the pathogen is transmissible in tap water after it has been passed through a 0.2 μ filter. They have not yet demonstrated heat lability and particle reproduction through dilution of successive passages. Bayley & Southwick (6) have indicated that the future development of this disease is unpredictable at the moment. The opportunity of using this disease against this major water weed would seem hopeful.

A group in the Commonwealth Institute of Biological Control in India has been conducting research since 1962 on natural enemies of aquatic weeds. This work has been sponsored by the U. S. A. with PL-480 funds. Raj & Ponnappa have described a number of diseases of aquatic weeds in India (71, 94, 95–98, 100, 101). A thread blight of water hyacinth (*Eichhornia crassipes* (Mart Solma) caused by *Marasmiellus inoderma* (Berk.) Sing. shows promise in the biological control of this weed. Perhaps the Indian pathogens could be introduced into the U. S. A. and other countries to control certain aquatic weeds.

Disease epidemics in phytoplankton caused by aquatic fungi are apparently common as Canter (20–22) and Fott (44) have shown. Phytoplankton populations can be reduced substantially by the attacks of aquatic fungi especially chitrids. This seems worthy of much further exploration.

Weed trees.—With more intensive forest management, the removal of undesirable trees is important; also weed trees have been a constant headache in range management. The conversion of forested lands from one dominant species to another causes the established trees to become weeds by definition. This occurs repeatedly where there are attempts to convert hardwood forests into more profitable pine stands.

The conversion of predominantly oak stands to pine in certain areas has been found to be a profitable undertaking. However, presently known chemical and mechanical means of removing undesirable hardwoods are unpredictable and expensive. French & Schroeder (45) have effectively used the oak wilt fungus *Ceratocystis fagacearum* (Bretz) Hunt to convert marginal hardwood stands to pine in Minnesota. They compared inoculation of oaks with *C. fagacearum* to conventional chemical methods of control with 2,4,5-T. The biological control method was decidedly more efficient because of lower cost, greater ease of application, higher percentage of mortality with little or no resprouting, total lack of injury to other tree species, and no need to re-treat with the fungus as is necessary with 2,4,5-T. French & Schroeder (45) found very little spread of the fungus from the treated areas to adjacent oaks. The question that needs answering in situations like this is, "Just how much of a threat is made to adjacent areas by locally increasing the inoculum of a pathogen?"

The use of oak wilt as a tree eradicant will naturally meet with rather violent reactions. There is considerable controversy as to whether or not oak wilt is presently an epidemic disease. Some contend that 50 per cent of the oak population will be affected by this disease in 40 years (84) while others think that the disease is remaining static (143). The emotional reactions associated with approaches such as those of French & Schroeder should not override the real possibility that there may be safe and effective ways to use pathogens such as *C. fagacearum* for weed control.

Persimmon, *Diospyrus virginiana* L., is a major weed in pastures in some sections of the U.S.A. It is particularly difficult to kill with chemical herbicides, and resprouts profusely when attempts are made to remove it mechanically. Work at the University of Arkansas and the Noble Foundation in Ardmore, Oklahoma (4, 142, 144), has shown that persimmon can be controlled effectively and reasonably with persimmon wilt caused by *Cephalosporium diospyri* Crandall. A method has been developed whereby trees can be treated without causing any apparent hazard to untreated trees (144). This involves cutting all the persimmon in a pasture at the ground line and inoculating the stumps with a spore suspension of *C. diospyri.* The stumps develop sprouts which subsequently die, but no spores are produced as they normally are in standing trees which are naturally killed. If the assumption is correct that long distance spread is dependent upon spores, this method of treatment is not a threat to uninoculated persimmon, providing that all persimmons in the pasture are treated and killed.

Persimmon is of positive economic importance in that it is harvested in some areas for the manufacture of golf-club-heads and weaving shuttles.

Persimmon is also grown and developed in some areas for its fruit and it provides food for some wildlife. Therefore, the use of *C. diospyri* as a biological control of persimmon must be undertaken on a sound and appropriate biological base. Ranchers in Oklahoma consider persimmon such a pest that the state legislature has declared it a "noxious weed." This allows landowners to pursue this method of biological control without possible legal entanglements. It is interesting that some ranchers in Oklahoma 20 years ago carried "diseased wood" from areas of dying persimmon to healthy stands; thereby they were practicing biological control of persimmon before scientists recognized the disease they were using.

Weir (135) has described a dangerous weed tree that was introduced into Cuba from Senegal. This tree, *Dichrostachys nutans* (Pers.) Benth., or marabu, forms impenetrable forests on abandoned cane land where conditions for its growth are optimum. Weir (135) found two parasitic fungi associated with dead and dying trees. *Ganoderma pulverulentum* Murr. entered through wounds, causing a white, undifferentiated decay of the heartwood. The primary cause of the death of the trees was *Ustulina zonata* (Lev.) Sacc., a fungus with a wide host range in the tropics. Weir (135) held hope that study of the diseases of marabu would provide tools for the biological control of this weed.

The main obstacle to the use of plant pathogens in the control of weed trees is the same as that for cacti. Weed trees may be economically valuable in other ecological niches. However, the expense of removing weed trees by chemical and mechanical means points toward the desirability of a biological means of control.

Dodder.—The widespread nature of dodder and the close relationship with its host makes it particularly difficult to control with chemicals and it is therefore a prospect for biological control. Leach has described a disease of dodder (*Cuscuta epithymum* Murr. and *C. campestris* Yuncker) caused by *Colletotrichum destructivum* O'Gara (72). He suggests that this disease might be useful in the biological control of dodder and he has volunteered cultures to anyone interested in pursuing this project. He apparently has had no takers.

The Russians have successfully pursued the biological control of dodder with a fungus *Alternaria cuscutacidae* Rudak. They have developed techniques for the mass production of the fungus and the inoculation of dodder with it (74, 110–115). Dodder in certain alfalfa plantings has been effectively controlled by application of spores of *A. cuscutacidae.* Less success has been realized in sugar beet plantings (115).

Rudokov and others have tried to isolate the factors that influence the inoculation of dodder with *A. cuscutacidae.* Erratic results with this pathogen have prompted such studies. They found that some of the variability in their results could be related to the crop host, the time of application of the pathogen, environmental factors (particularly temperature and moisture), and variability in the pathogenicity of the organism.

One of the most enlightening considerations in these studies concerns the influence of associated organisms on the effectiveness of *A. cuscutacidae* as a weed eradicant. Rudakov (111) feels that species of *Cladosporium, Fusarium, Rhizoctonia, Trichoderma, Penicillium,* and bacteria improve the effectiveness of *A. cuscutacidae* by acting as secondary invaders to destroy the dodder. So it may be necessary to inoculate with certain secondary organisms as well as with the primary pathogen. Rudakov's initial attempts to do this have yielded inconsistent results. Rudakov et al. (113) have also been investigating other primary pathogens of dodder. They have found a species of *Curvularia* which appears more aggressive than *A. cuscutacidae.*

Although one of the arguments for biological control of weeds is to reduce the introduction of foreign chemicals into our environment, plant pathogens may present hazards themselves. Rudakov (111) had this in mind when he fed to cattle food laden with spores of *A. cuscutacidae.* Considerations like this must be made in the application of plant pathogens to food crops. Our increased awareness of the potential effects of mycotoxins makes such studies imperative.

The Russians have progressed with application of *A. cuscutacidae* to the point that they have set up "factories" for the mass production of inoculum. Application of this pathogen is increasing and it should be informative and helpful to follow future developments.

Burr.—Butler (19) provides a useful example of how a disease already present in a country can be used more effectively to kill a weed. Bathurst burr, *Xanthium spinosum* L., was found to be attacked by a destructive disease caused by *Colletotrichum xanthii* Halst., which was restricted to certain areas of New South Wales. This disease effectively reduced burr populations when temperature and moisture conditions were favorable for development of the disease.

Through field inoculations Butler showed that *C. xanthii* could be artificially established in previously uninfected areas. He also found that this pathogen persisted and caused destruction of the bathurst burr in subsequent years. Butler (19) distributed cultures of *C. xanthii* to landowners in areas with an average annual rainfall of 25 in or higher. It was recommended that inoculations be made in the fall, during or immediately following a period of rain. Most farmers reported successful establishment of the disease.

Crofton weed.—*Eupatorium adenophorum* spread very aggressively between 1940 and 1950 in the southeastern border districts of Australia. A gall-forming fly, *Procecidochares utilis* Stone, which was introduced from Hawaii in 1952, became established in the Crofton weed population and apparently checked its spread. Unfortunately certain parasitic Hymenoptera that were native to Australia began reducing the gall insect population. A fortunate development was the buildup of a leaf-spot disease of Crofton weed caused by *Cercospora eupatorii.* The introduction of this pathogen was coincidental with the introduction of *P. utilis* and according to Dodd

(36) it may have come into Australia on hairs of this fly. Adults of *P. utilis* have been found to carry spores of *C. eupatorii* and are capable of being vectors.

Since 1952 Crofton weed has not spread or increased. Dodd (36) attributes this to the combined effects of *P. utilis,* the *Cercospora* leaf spot, and a native Ceramybycid stem and root borer. It is interesting that *P. utilis* has been used in New Zealand to control *E. adenophorum,* and no evidence has been found of the involvement of the *Cercospora* leaf-spotting fungus (correspondence with J. M. Hay, Department of Scientific and Industrial Research, Nelson, New Zealand).

Diseases of miscellaneous weeds.—Most of the attention that has been given to diseases of weeds has been related to their role as reservoirs for pathogens of economic plants (1, 28, 39, 43, 57, 58, 88, 90, 103, 108, 134, 136, 141). A number of isolated diseases of weeds could be considered as biological control agents either in their normal environment or as introduced pathogens. Taubenhaus (129) reported *Phymatotrichum* root rot on winter and spring weeds in south central Texas and *Sclerotium rolfsii* Sacc. is known to attack weeds (28). There have been various diseases described on dock-leaved persicary (29), field bindweed (7), pigweed (10), dandelion (59, 66), crabgrass (132), *Portulaca* sp. (99), etc.

There have been a number of isolated reports of plant pathogens effective in weed control with little documentation of the work. Muenscher (86) reports the complete destruction of, ironically, the "live-for-ever" plant (*Sedum triphyllum*) by a fungus distributed by the New York State College of Agriculture of Ithaca. He states that, "The fungus disease was so successful that all the "live-for-ever" plants on the College farm were killed and the fungus disappeared before it had been isolated so that its identity could be determined." The New Jersey Agricultural Experiment Station, at the turn of the century, offered to send collections of Canada thistle (*Cirsium arvense* L. Scop.) infected by a rust *Puccinia obtegeus* LK Tul. to farmers for the purpose of inoculating healthy Canada thistles (55).

Isolated reports from different parts of the world can be found on the control of weeds with plant pathogens. Dr. A. J. Wapshere and his group in France are considering the use of *Puccinia chondrillina* as a control of *Chondrilla funces* (correspondence with R. E. Inman). In a Japanese weed book by Dr. Makoto Hanzawa (1910) (correspondence with Dr. Yasuo Kaschara, the Ohara Institute for Agricultural Biology, Kurashaki) it is stated that *Breea setosa* Kitam was controlled by a rust *P. obtegeus* (LK) Tul and that *Peronospora effusa* (Grev.) Rabh. inhibits remarkably the growth of the weed *Chenopodium album* L. Cockayne in New Zealand reported on the control of California thistle (*Coricus arvensis*) by the California thistle rust organism *P. obtegeus,* and described means of preparing inoculum for mass inoculations (26). Shadow in Germany (122) considered the enemies of a thistle (*C. arvense* L. Scop.) including *P. obtegeus* and

Verticillium albo-atrum Rke et Berth. as potential agents for biological control of this weed. In Scotland and Australia the control of braken ferns with plant diseases has been considered (2, 27, 53, 87). In South Africa, some moderate control of bramble (*Rubus fruticosus* L.) by a rust *Kuehneola albida* (Kuhn.) N. Magu. is reported (81, 133). Loos (78) in Ceylon has described a virus disease of *Emilia scabra* DC. In Sweden, Hofsten (59) unsuccessfully tried to select strains of *Phoma taraxaci* sufficiently virulent to control common dandelions (*Taraxacum* spp.) Brod tested the use of *Cercospora mercurialis* Passer to control *Mercurialis annua* L. in Germany (15). He found that the long incubation time of the fungus and the regenerative ability of affected weeds made this pathogen-weed combination ineffective for biological control.

Seedling diseases.—The sooner a weed is removed from competition with economic plants the better. Thus the use of seedling diseases (preemergent and postemergent) has considerable potential in the biological control of weeds. Observations on the effects of damping off and root rots of economic plants attest to the effectiveness of such diseases in preventing the establishment of plants.

Kiewnick (67, 68) has made a detailed study of the effects of soil pathogens on wild oat (*Avena fatua* L.) seed and has found that a large number of soil pathogens have a profound influence on seed viability. Seeds were killed directly by fungi and by fungal metabolites. Kiewnick concluded that the best approach toward using soil pathogens in weed control would be to create conditions which favor the pathogens rather than trying to infest the soil with pathogens. He found that through the addition of manure to the soil the life span of the wild oat seed was reduced. He thought that this was due to the creation of favorable conditions for pathogens of wild oat seed.

A pathogen can be a double threat in that it can cause diseases of the seedling and the mature plant. Butler (19) attributes the success of *C. xanthii* on burr to the destruction of both seedlings and mature plants. Such pathogens should be efficient weed killers.

A number of workers have suggested that seedling diseases of weeds may be useful in biological control (54, 67, 68, 80). Also, metabolites of microorganisms can cause destruction of weed seed and seedlings. Metabolites of such microorganisms should include some good preemergent herbicides if they were only fished out.

Increased understanding of the ecology of soil-inhabiting pathogens should prove useful in the control of certain weeds. The profound influence of soil microorganisms on the establishment of economic plants is becoming more apparent. Knowledge in this area might be utilized to create soil ecosystems which favor certain pathogens of weeds.

Present Projects

Many scattered attempts have been made to use plant pathogens to control weeds, but only recently have research programs had this as their major goal.

Dr. Robert E. Inman with the Stanford Research Institute is pioneering in the use of plant pathogens to control weeds. He is presently working in Rome through the sponsorship of the USDA Agricultural Research Service, Crops Research Division. Dr. Inman has developed an excellent philosophical discussion on the use of plant pathogens in weed control in the reports and proposals that he has presented to the ARS.

The first project undertaken by Dr. Inman was to establish the "feasibility" of using plant diseases to control specific weeds. Because of the limitation of time, "feasibility" was considered to be demonstrated "when pathogens of certain weed relatives in foreign countries are shown to be more pathogenic on U. S. weeds than on their foreign hosts." For example in the Sudan a *Helminthosporium* sp. which is a secondary parasite associated with *Cercospora* in lesions on *Setaria verticillata* (L.) Beauv., when inoculated on U. S. *Setaria* selections the fungus produced primary infections which were increased quantitatively and qualitatively by subsequent serial inoculations. Thus, Dr. Inman concludes that, "the potential of a given pathogen cannot be estimated on the basis of damage caused to its native host." There would then appear to be extensive untapped reservoirs of pathogens which could be used to control weeds.

Dr. Inman's next approach, after establishing "feasibility," was to introduce certain important U. S. weeds (under quarantine) into Italy and search out and inoculate them with various pathogens in Europe. From various tests, he is concentrating on the possibility of curly-dock control with *Uromyces rumicis* (Schiem.) Wint. The reports and proposals to ARS USDA (62) would be very useful to anyone planning to take this research approach. He has outlined many of his procedures and problems and hopefully will publish this information soon.

Dr. Inman has outlined some principles relative to phytopathogenic weed control that are instructive:

"(*a*) host resistance acts as the primary deterrent to biological control of weeds by restricting disease to insignificant levels; (*b*) natural weed populations may be expected to display a relatively high degree of natural resistance to most local disease organisms; (*c*) disease susceptibility is, therefore, the exception rather than the rule and high degrees of susceptibility are exceptional indeed; (*d*) most natural plant disease epidemics in the past have resulted from the accidental importation of foreign pathogens; (*e*) the fact that local plant populations develop resistance to local pathogens or insect pests does not preclude their susceptibility to forms from which they have been protected by natural barriers through the recent course of evolution; (*f*) the natural home of a pathogen does not necessarily represent the environment most conducive to its survival."

Dr. V. P. Rao and his associates with the Commonwealth Institute of Biological Control in Bangalore, India, have studied the natural enemies of witchweed, water hyacinth, and other aquatic weeds affecting waterways in India (102). Plant pathogens have been found and described on species of *Eichhornia* (100), *Glycosmis* (94), *Hygrophila* (71, 98), *Passiflora* (96), *Marsilea* (96), *Sphenoclea* (95), and *Striga* spp. (101). This work has not reached the stage where candidates to be used as weed eradicants have been chosen. If the fungi found in India are native, they may hold more promise as weed killers through their introduction into other countries or other areas of India.

Dr. G. H. Bridgmon and his students at the University of Wyoming have surveyed the diseases of pasture land weeds in Wyoming (18, 76, 77, 82, 83). They have concluded that diseases already play an important role in the suppression of weeds in pastures. They found species of *Fusarium* associated with diseased Canada thistles (*C. arvense*) and a *Bacillus* sp. that causes a marked reduction in seedling establishment, plant growth, and seed production of *C. arvense*. Further investigation with the fusarial disease of Canada thistle showed that an isolate of *Fusarium roseum* Lk. may have potential as a biological agent against certain selections of thistle and perhaps undesirable millets (76, 77).

Ecological Interaction of Weeds and Plant Pathogens

Weeds and plant pathogens are but two components of the ecosystem. They interact with each other and with other components in a variety of ways, concerning many of which we are probably not aware. For example, it may become possible to control human diseases with plant pathogens by using them to control water weeds, the breeding sites of vectors of certain human pathogens.

Among other interactions, plant pathogens can not only cause diseases of weeds, they can produce chemicals which influence the growth and reproduction of weeds, predispose weeds to the effects of other factors such as herbicides, break herbicides down in the soil, compete with favorable organisms in the weed rhizosphere, and influence weed nutrition through their saprophytic growth in the soil.

Weeds in turn can provide a food source for pathogens, act as alternate hosts for certain pathogens, provide a reservoir of inoculum for economically important plants, influence the growth and reproduction of pathogens in the soil, and influence the food and water relations of plant pathogens.

Because of the complex ecological entanglement of weeds and plant pathogens, it would be narrow to consider only the ability of plant pathogens to cause diseases of weeds. By considering the other interactions of weeds and pathogens we may be able to develop additional methods of weed control through manipulation of the ecosystem.

Native Versus Introduced Pathogens

It is generally assumed that introduced pathogens will be more effective than native pathogens in weed control. This would tend to discourage the investigation of native diseases of weeds as local weed control agents. However, there are numerous possible exceptions to this rule. Probably a number of already introduced pathogens hold promise. Some of these pathogens may, through natural barriers, poor vector relationships, etc., be restricted in their dissemination.

Within a country, numerous natural and artificial barriers restrict the dissemination of native pathogens. Therefore, there may be good candidates for the biological control of weeds lying in some unexplored ecological niche within a country. Also, it should be remembered that many of the more important weeds are introduced. In their establishment they may not yet have come in contact with a pathogen which could possibly eliminate them. Man's search and examination of native pathogens should, therefore, hold promise.

The environment and natural barriers of native pathogens are constantly changing. Most of this change is the result of man's activity. This is brought home many times when an insignificant disease of an economic plant becomes significant because of a change in cultural practices, varieties, etc. Therefore, the potential of native pathogens may not be realized until certain natural barriers are removed or certain cultural practices undertaken.

The use of introduced pathogens is perhaps the best approach toward finding "super pathogens" to kill weeds. The amount of resistance in a particular weed to a pathogen is related to the natural barriers between it and the pathogen, and the number and effectiveness of natural barriers are related to spatial separation.

Genetic Considerations

In selecting and applying plant pathogens to kill weeds we need to keep in mind the genetic lessons we have learned in studying diseases of economic plants. It would behoove us to remember that pathogens vary and that hosts vary. It is also inevitable that in the application of plant-pathogens to kill weeds we will cause the selective development of more resistant weeds. Weeds are already becoming resistant to certain herbicides (56).

That weeds are so plentiful and adaptable is held as evidence (69) that they possess a more elastic genetic base than most other plants. This could be held as discouraging in the selection of pathogens to kill weeds, since weeds would be expected more easily to develop resistance than other plants; and because of their free development and hybridization weeds would be expected to have more polygenic or horizontal resistance than cultivated plants.

If a weed exists in a large and accessible genetic pool of relatives, the opportunities for interspecific and introgressive hybridization would be favorable. On the other hand, if a weed exists in a somewhat homogenous genetic pool its variability is limited. It could be further conjectured that introduced weeds would have less of a genetic pool to draw on than endemic weeds. This would be a fortunate situation since many of our more destructive weeds are introduced; it would follow that they perhaps have less potential for developing resistance to plant pathogens. This may be one area where local pathogens may prove especially useful as control agents.

We should also keep in mind that weeds have come face to face with plant pathogens before. Resistance to various pathogens has had to evolve in weeds for them to have survived. So in the selection of appropriate pathogens to kill weeds we need to explore their present resistance to pathogens.

Increased knowledge of the genetics of plant pathogens should help in the development and selection of "super pathogens" to be used to kill weeds. It would appear that detailed genetic knowledge of the host and pathogens, such as we have with grains and rusts, would be necessary for the effective long-term use of a pathogen against a weed.

Interactions in Chemical Weed Control

It is recognized that plant pathogens produce an array of chemicals which affect plants in various ways. The possibility that these chemicals might have herbicidal activity has been investigated only slightly (7, 3). Also, the facts that plant pathogens can break down herbicides (8, 9), and that certain herbicides are fungicidal (24, 65) need to be considered when attempting to control weeds. Pathogens and herbicides may interact in other ways. The pathogen may predispose the plant to herbicide attack (3, 93) or the herbicide may affect the pathogenicity of a pathogen on a weed.

Shklyar & Khalimova (125) discuss the possibilities of finding selective herbicides as products of microbial metabolism. They have found that a compound produced by *Pseudomonas mycophaga* n. sp. is inhibitory to certain weeds and certain plant pathogens. Fulton et al. (47) and Templeton et al. (130, 131) have described a metabolite (tentoxin) of *Alternaria tenuis* Nees ex Cda. that inhibits chlorophyll production in a number of plants. This chemical holds some promise as a herbicide. Leben & Keitt (73) discuss the phytotoxicity of Antimycin A and its possible use as a herbicide.

Rodriguez-Kabana et al. (104–107) have found that certain herbicides affect the growth of certain plant pathogens and Altman & Ross (3) feel that plant pathogens are a possible factor in the unexpected damage from preplant herbicides in sugar beets. The interaction of herbicides and soil microorganisms has been reviewed by Bollen (8, 9).

Diaz (31) found a chemical produced by *Helminthosporium speciferum* (Bain) Nicot. that stimulates the germination of crabgrass. It is conceivable that such a chemical when used in combinations with a postemergence herbicide would increase its effectiveness. This illustrates potential manipulations of the chemical products of plant pathogens that might be made to combat weeds.

Some plant pathogens are effective defoliants. The possibility that chemical defoliants might be found among these organisms or that the organisms themselves might be used as defoliants seems worthy of investigation.

Commencement

I have chosen to write a commencement rather than a conclusion. To write a conclusion for this subject in its present state of growth seems premature. So, let us consider where we might go from here.

Where do we look for possible weed-pathogen combinations for biological control? The success that has been realized so far has not come from any one approach. Ideally, we should like to confront a weed having no resistance with a pathogen having high pathogenicity. It seems unlikely, however, that such combinations will be found where the weed and pathogen have evolved together for extended periods, since some sort of parasitic balance would have developed. This logic tends to discourage the exploration of endemic pathogens to control endemic weeds. However, it must be kept in mind that within countries there are natural barriers, newly introduced pathogens, newly introduced weeds, changing environments, etc., which might permit the discovery and use of endemic pathogens to control endemic weeds. Also, the genetic manipulation of endemic pathogens may permit the selection of "super pathogens."

The selection of pathogens with high virulence as weed eradicants seems like a good approach. Our increased knowledge of microbial genetics and increased ability to control the genetic makeup of plant pathogens should support progress in this area. It has been clearly demonstrated in certain fungi that virulence can be increased by the hybridization of certain isolates either through sexual or parasexual processes. Also, industrial microbiologists have been able to develop isolates of microorganisms for highly specific functions. So the way seems clear for the development of microbial herbicides through the selection of either highly virulent pathogenic isolates or isolates that produce certain herbicidal products.

The greatest hope in the development of effective pathogen-weed combinations would still seem to be the introduction of foreign pathogens to control established weeds. There is much we can learn here from the entomologists since they have so effectively used this approach in the biological control of both weeds and insects. There are always hazards associated with this approach. However, these are not insurmountable and should influence but not deter future work in this area.

Introducing pathogens from one country to another demands good international cooperation. Workers in different countries who have common interests in the control of weeds with pathogens should develop valuable information that they could exchange. After all, a rather innocuous pathogen on bramble in the U.S.A. may be just what is needed to stop the spread of brambles in South Africa. Therefore, development of international interest and cooperation on this subject could probably do more toward accelerating progress than anything else.

The establishment of weed nurseries throughout the world where the major weeds of various countries are represented should yield various sorts of information. These nurseries would have to be maintained under strict quarantine to prevent the escape of foreign weeds and other pests. Such nurseries could act as an automatic screen for potential pathogens in weed control. Also, the planting of weeds from different areas under different environments might allow the discovery of unique ecological information.

There is much basic ecological information to be gained from the manipulation of plant pathogens in various ecosystems. Wilson (145) has indicated that this has been a major contribution of the work with weed control by insects. He indicates that rather than drawing on previous ecological information, people in this field have contributed substantially to its development. Experiments in the control of weeds with plant pathogens, if properly designed, should permit valuable ecological evaluations. Plant pathologists have been interested primarily in the intimate ecological relations of a host and parasite and much less information is available on the independent ecology of pathogens and hosts. The possibility exists of better distinguishing certain physical and biological factors which affect disease development by studying and comparing pathogen-host combinations in different countries.

Are there perhaps certain types of diseases which would be more adaptable for biological weed control than others? Based on the few examples that we have, the answer to this question would be 'no.' Also, in assessing the most destructive diseases of economic plants we find representatives of all major types of diseases and major groupings of pathogens.

A goal which has perhaps discouraged development of the biological control of weeds with pathogens has been the requirement that the pathogen completely destroy the weed. This is many times an unrealistic and unnecessary goal. It must be realized that weeds are under severe competition in the ecosystem. In some instances a rather mild disease of a weed may give the advantage to other desirable plants and result in the displacement of the weed. Also the use of mild diseases in combination with chemical and mechanical means of control may hold promise.

Are there perhaps types of agriculture in which the biological control of weeds with pathogens would be more suitable? Huffaker (see 30) states: "Obviously the more simplified the human economy and ecology of an area the better are the chances of attempting biological control of weeds without operating at cross purposes with self-interests." This has been found to be the case where the biological control of persimmon with *C. diospyri* has been considered. In Oklahoma where grazing dominates the agricultural use of land this method has been promoted, whereas in Arkansas, where there is more diversified land use this method has been opposed.

More investigation of the combined effects of insects and diseases on weeds is needed. In the study of biological weed control cooperative studies between entomologists and plant pathologists should prove fruitful. Insect-disseminated pathogens can become efficient weed killers. Also, the combined effect of insect and pathogen damage on weeds has not been measured. The predisposition of plants to insect attack by diseases and *vice versa* needs consideration. All we need are some compatible entomologists and plant pathologists.

Consideration also needs to be given to multiple diseases on weeds. The important secondary effects of microorganisms on weed destruction have been demonstrated by Dodd (34) and Rudakov (111) but no attention has been given to the impact of multiple diseases. Synergistic effects such as those realized with certain virus combinations may be useful in weed eradication.

In attempts at establishing plant-disease epidemics in weed populations, much can be learned from a study of past disease epidemics of economic crops. The effective spread of Dutch elm disease can largely be explained by the efficiency of the vector-fungus relationship in this disease. The rapid development of chestnut blight has been explained largely by the prodigious production of inoculum. Although late blight of potato has rather specific environmental requirements for its development, the disease can develop rapidly under these conditions and much of its success has been attributed to the prodigious production of inoculum.

Various factors that promote plant-disease epidemics can be defined. The genetics of pathogenicity and resistance are basic for the development of epidemics. The underlying success of certain introduced pathogens can be found in their genetic advantage for pathogenicity.

That climate exerts a profound effect on disease development has been held as a discouraging factor in the use of plant pathogens in weed control. Even if the proper pathogen and weed are brought together, the disease is ineffective if proper weather conditions don't occur at the appropriate time. The argument is also made that if endemic diseases were going to kill the weed they would have done so already, and that factors responsible for limiting this activity are environmental conditions. Man does have less ability to manipulate the climate than any other part of the ecosystem, but this should not discourage the use of plant pathogens, but rather should encourage close attention to climatologic factors. Also, there are situations such as in greenhouses, etc., where climate can be controlled and perhaps managed in the use of pathogens to kill weeds.

The possibility that plant pathogens introduced to control weeds may attack plants of economic value needs careful consideration. The testing of foreign pathogens on a native host under native environmental conditions presents a problem. The establishment of areas with extreme quarantines will be necessary. Inman (62) has suggested the establishment of introductory centers on islands near mainlands. Here extensive tests could be made of the ability of the candidate organism to attack economic plants.

Unlike the application of chemicals, the process of pathogen introduction cannot be arbitrarily discontinued after it has been started. Once the pathogen is introduced into a favorable environment it can be expected to multiply and spread, although methods may be available to limit or control a pathogen once it is introduced. Conventional methods that are used in the control of plant diseases of economic plants might be applicable in certain situations. There is also the possibility that non-sporulating or weakly sporulating isolates could be selected for use, in order to prevent overland spread. Natural barriers that would limit the application of a pathogen may be utilized or created.

Since many plant pathogens are highly specific in their host preference, a good degree of safety can be obtained by the proper selection of plant pathogens for introduction. An example is the persimmon-wilt organism, *C. diospyri,* which apparently attacks only the persimmon. The persimmon is a member of the Ebony family in which there are no other economically important species in the U.S.A. Therefore, the use of the persimmon-wilt organisms to kill weed persimmon appears to pose no threat to other plants in the U.S.A. Unsuccessful attempts to inoculate various economic plants with *C. diospyri* have borne this out.

Much can be gained from studying the work of entomologists in the biological control of weeds. There are a number of reviews on this subject (30, 60, 61, 145). At the same time plant pathologists should not be overly influenced by these studies in making their own approaches. At this stage new and imaginative thinking should be encouraged.

Realization that nature is delicately "balanced" is perhaps never keener than when we are dealing with pests. The natural balance can be disturbed in numerous ways: introduction of a foreign organism, changed environment of native organism, introduction or elimination of hyperparasites of existing pests, changed cultural practices, selective destruction and distribution of organisms, etc. Man is recognized as the prime disturber of the biological balance, but he also has increasing power to manipulate biological entities. Two of the major competitors for man's food are weeds and plant pathogens. It is a challenge to pit these two forces against one another and tip the biological balance in man's favor!

LITERATURE CITED

1. Aamodt, O. S., Mallock, J. G. Smutty wheat caused by *Ustilago utriculosa* on dock-leaved persicary. *Can. J. Res.*, **7**, 578–92 (1932)
2. Alcock, N. L., Braid, K. W. The control of bracken. *Scot. Forestry J.*, **11**, 68–73 (1928)
3. Altman, J., Ross, M. Plant pathogens as a possible factor in unexpected preplant herbicide damage in sugarbeets. *Plant Disease Reptr.*, **51**, 86–88 (1967)
4. Anon. Persimmon control with disease shows promise. *The Oklahoma Farmer-Stockman,* 34 (1968)
5. Anon. Soviet virus claimed to clear green water. *New Scientist,* **36,** 677 (1967)
6. Bayley, S., Southwick, C. H. Milfoil disease in Chesapeake Bay. *Weed Sci. Soc. Abstr.*, 52 (1968)
7. Bever, W. M., Seely, C. I. A preliminary report on a fungus disease of the field bindweed, *Convolvulus arvensis. Phytopathology,* **30,** 774–79 (1940)
8. Bollen, W. B. Interactions between pesticides and soil microorganisms. *Ann. Rev. Microbiol.*, **15,** 69–92 (1961)
9. Bollen, W. B. Herbicides and soil microorganisms. *Western Weed Control Conf. Proc.*, Las Vegas, Nevada, 48–50 (1962)
10. Boosalis, M. G., Scharen, A. L. The susceptibility of pigweed to *Rhizoctonia solani* in irrigated fields of western Nebraska. *Plant Disease Reptr.*, **44,** 815–18 (1960)
11. Bourchier, R. J. *Septogloeum gillii* on lodgepole pine mistletoe. *Can. Dept. Agr., Forest Biol. Div.*, **10,** 3 (1954)
12. Bourchier, R. J. New host relationship for fungus parasites of the lodgepole pine mistletoe in Alberta. *Can. Dept. Agr., Forest Biol. Div.*, **11,** 2 (1955)
13. Bourchier, R. J. Parasites of dwarf mistletoes. *Can. Dept. Agr., Ann. Rept. Forest Insect Disease Surv.*, 111 pp. (1955)
14. Boyle, A. M. Further studies of the bacterial necrosis of the giant cactus. *Phytopathology,* **39,** 1029–52 (1949)
15. Brod, G. Studien uber *Cercospora mercurialis Passer* in Hinblich auf eine biologische Bekampfung des Schutt—Bingelkrautes (*Mercurialis annua* L.). *Phytopathol. Z.*, **24,** 431–32 (1955)
16. Brown, R. M., Smith, K. M., Walne, P. L. Replication cycle of the blue-green algal virus LPP-1. *Nature,* **212,** 729–30 (1966)
17. Bourn, W. S., Jenkins, B. *Rhizoctonia* disease on certain aquatic plants. *Botan. Gaz.*, **85**(4), 413–26 (1928)
18. Bruzloff, D. F., *Preliminary Investigations of the Stem Blight of* Asclepias speciosa *and its Causal Organism* (M.S. thesis, Univ. of Wyoming, Laramie, Wyo., 1952)
19. Butler, F. C. Anthracnose and seedling blight of Bathurst burr caused by *Colletotrichum xanthii* Halst. *Australia J. Agr. Res.*, **2,** 401–10 (1951)
20. Canter, H. M., Lund, J. W. G. Studies on plankton parasites. I. Fluctuations in the numbers of *Asterionella formosa* Hass. in relation to fungal epidemics. *New Phytologist,* **47,** 238–61 (1948)
21. Canter, H. M. Fungal parasites of the phytoplankton. I. *Ann. Botany (London),* **14,** 263–89 (1950)
22. Canter, H. M. Fungal parasites of the phytoplankton. II. *Ann. Botany (London),* **15,** 129–56 (1951)
23. Carpenter, C. W. Fusarium disease of the prickly pear. *Hawaiian Planters' Record,* **48,** 59–63 (1944)
24. Chappell, W. E., Miller, L. I. The effects of certain herbicides on plant pathogens. *Plant Disease Reptr.*, **40,** 52–56 (1956)
25. Chock, Q. C. Control of the prickly pear in Hawaii. Rept. of the Board of Commissioners of Agri. & Forestry of the Territory of Hawaii for the Biennial Review ended Dec. 31 (1944)
26. Cockayne, A. H. California thistle rust as a check on the spread of California thistle. *Intern. Rev. Sci. Pract. Agr.*, **7,** 451 (1916)
27. Conway, E. The bracken problem. *Outlook Agr.*, **2,** 158–67 (1959)
28. Cooley, J. S. Susceptibility of crop plants and weeds to *Sclerotium rolfsii. Phytopathology,* **28,** 594–95 (1938)
29. Cunningham, G. H. Natural control of weeds and insects by fungi.

New Zealand **Rept. Agr. Bull.** *132* (1927)

30. Debach, P., *Biological Control of Insect Pests and Weeds* (Reinhold, New York, 844 pp., 1964)
31. Diaz, C. A., *The Etiology of "Spring Dead Spot" of Bermuda Grass* (M.S. thesis, Univ. of Arkansas, Fayetteville, Ark., 1964)
32. Dodd, A. P. The biological control of prickly pear in Australia. *Australian Council Sci. Ind. Res. Bull.*, **34,** 1–44 (1927)
33. Dodd, A. P. The present position and future prospects in relation to the biological control of prickly pear. *Council Sci. Ind. Res. J.*, **6,** 8–13 (1933)
34. Dodd, A. P. *The Biological Campaign Against Prickly Pear* (Prickly Pear Board, Brisbane, 117 pp., 1940)
35. Dodd, A. P. The biological control of prickly-pear in Australia. *Monog. Biol.*, **8,** 565–77 (1959)
36. Dodd, A. P. Biological control of *Eupatorium adenophorum* in Queensland. *Australian J. Sci.*, **23,** 356–65 (1961)
37. Dowding, E. S. The vegetation of Alberta. III : The sandhill areas of central Alberta with particular reference to the ecology of *Arceuthobium americanum.* Nutt. *J. Ecol.*, **17,** 82–105 (1929)
38. Dowding, E. S. *Wallrothiella arceuthobii,* a parasite of the Jack-pine mistletoe. *Can. J. Res.*, **5,** 219–30 (1931)
39. Dykstra, T. P. Weeds as possible carriers of leaf roll and rugose mosaic of potato. *J. Agr. Res.*, **47,** 17–32 (1933)
40. Ellis, D. E. A fungus disease of *Arceuthobium. Phytopathology,* **29,** 995–96 (1939)
41. Ellis, D. E. Anthracnose of dwarf mistletoe caused by a new species of *Septogloeum. J. Elisha Mitchell Sci. Soc.*, **62,** 25–50 (1946)
42. Elser, H. J. Observations on diseases of watermilfoil and other aquatic plants, Maryland: 1962–1967. *Weed Sci. Soc. Abstr.*, 51 (1968)
43. Freitag, J. H., Severin, H. H. P. Insect transmission, host range, and properties of the crinkle-leaf strain of western-celery-mosaic virus. *Hilgardia,* **16,** 361–69 (1945)
44. Fott, B. *Phycidium scendesmi* spec. nova, a new chytrid destroying mass cultures of algae. *Z. Allgem. Mikrobiol.*, **7** (2), 97–102 (1967)
45. French, D. W., Schroeder, D. B. The oak wilt fungus, *Ceratocystis fagacearum* (Bretz) Hunt, as a selective silvicide. *Forest Sci.* (In press) (1969)
46. Fullaway, D. T. Biological control of cactus in Hawaii. *J. Econ. Entomol.*, **47,** 696–97 (1954)
47. Fulton, N. D., Bollenbacher, K., Templeton, G. E. A metabolite from *Alternaria tenuis* that inhibits chlorophyll production. *Phytopathology,* **55,** 49–51 (1965)
48. Gervasi, A. Su un fungo parasita di *Opuntia robusta* Wendl.: *Physalospora opuntiae robustae* n. sp. *Riv. patol. vegetale,* **31,** 3–12 (1941)
49. Gill, L. S. A new host for *Septogloeum gillii. Plant Disease Reptr.*, **36,** 300 (1952)
50. Gill, L. S., Hawksworth, F. G. The mistletoes, a literature review. *U. S. Dept. Agr., Tech. Bull. 1242,* 87 pp. (1961)
51. Goldstein, D. A., Bendet, I. J., Louffer, M. S., Smith, K. M. Some biological and physicochemical properties of blue-green algal virus LPP-1. *Virology,* **32,** 601–13 (1967)
52. Goryushin, V. A., Chaplinskaya, S. M. Existence of viruses of blue-green algae. *Mikrobiol. Zh. Akad. Nauk Ukr. RSR,* **28,** 94–97 (English summary) (1966)
53. Gregor, M. J. F. The possible utilisation of disease as a factor in bracken control. *Scot. Forestry J.*, **46,** 52–59 (1932)
54. Halsted, B. D. Fungi injurious to weed seedlings. *New Jersey Expt. Sta. Rept.*, 326–27 (1893)
55. Halsted, B. D. Weeds and their most common fungi. *New Jersey Expt. Sta. Rept.*, 379–81 (1894)
56. Harper, J. L. The evaluation of weeds in relation to the resistance to herbicides. *Brit. Weed Control Conf., 3rd, Blackpool, 1956, Proc.*, **1,** 179–88
57. Hein, A. Die Bedentung der Unkrauter fur die Epidemelogie pflanzlicher Virosen. *Deut.-Landwirtsch. Jahrb.*, **4,** 521–25 (1953)
58. Hein, A. Beitrage zur Kenntnis der Viruskrankeiten an Unkrautern I: Das Malva-virus. *Phytopathol. Z.*, **28,** 205–34 (1956)

59. Hofsten, C. G. von. *Studies on the genus* Taraxacum *Wigg. with special reference to the group Vulgaria DT in Scandinavia* (Swedish; Engl. summary, LTs Foerlag, Stockholm, 432 pp., 1954)
60. Huffaker, C. B. Fundamentals of biological control of weeds. *Hilgardia,* **27,** 101–57 (1957)
61. Huffaker, C. B. Principles of weed control. *Natl. Acad. Sci., Natl. Res. Council, Publ.* (In press)
62. Inman, R. E. Study of phytopathogens as weed-control agents. Final report by Stanford Res. Inst. for *U.S. Dept. Agr., ARS* (1967)
63. Jackson, D. F., *Interaction Between Algal Populations and Viruses in Model Pools—A Possible Control for Algal Blooms* (Presented at ASCE Ann. and Natl. Meeting on Water Resources Eng., Statler Hilton, New York, 1967)
64. Johnston, T. H., Hitchcock, L. A bacteriosis in prickly pear plants (*Opuntia* spp.). *Trans. Proc. Roy. Soc. S. Australia,* **47,** 162–64 (1923)
65. Kaufman, D. D. Effect of *s*-triazine and phenylurea herbicides on soil fungi in corn- and soybean-cropped soil. *Phytopathology,* **54,** 897 (1964)
66. Kassanis, B. A virus attacking lettuce and dandelion. *Nature,* 154–16 (1944)
67. Kiewnick, L. Untersuchungen uber den Einfluss der Samen-und Bodenmikroflora auf die Lebensdauer der Spelzfruchte des Flughafers (*Avena fatua* L.). I. Vorkommen, Artzusammensetzung und Eigenschaften der Mikroorganismen an Plughaferfruchten. *Weed Res.,* **3,** 322–32 (1963)
68. Kiewnick, L. Untersuchungen uber den Einfluss der Samen und Bodenmikroflora auf die Lebensdauer der Spelzfruchte des Flughafers (*Avena fatua* L.). II. Zum Einfluss der Mikroflora auf die Lebensdauer der Samen im Boden. *Weed Res.,* **4,** 31–43 (1964)
69. King, L. J. *Weeds of the World: Biology and Control* (Hill & Interscience, London, 526 pp., 1966)
70. Kuijt, J. Distribution of dwarf mistletoes (*Arceuthobium* spp.) and their fungus hyperparasites in northern Canada. *Can. Natl. Museum Bull.,* **186,** 134–48 (1963)
71. Laundon, G. F., Ponnappa, K. M. A new species of *Uredo* on *Hygrophila. Current Sci. (India),* **19,** 492–93 (1966)
72. Leach, C. M. A disease of dodder caused by the fungus *Colletotrichum destructivum. Plant Disease Reptr.,* **42,** 827–29 (1946)
73. Leben, C., Keitt, G. W. Phytotoxicity of Antimycin A. *Antibiot. Chemotherapy,* **6,** 191–93 (1956)
74. Levishko, A. P. V. Biologicheskii metod borby s povilikoy. Sel'skol *Skachoz Turkmensk. Laboratorii,* **1,** 68–70 (1962)
75. Lightle, P. C., Standring, E. T., Brown, J. G. A bacterial necrosis of the giant cactus. *Phytopathology,* **32,** 303–13 (1942)
76. Lloyd, E. H. *The Relationship of a* Fusarium *Fungus to Canada Thistle and Agronomic Crops* (M. S. thesis, Univ. Wyoming, Laramie, Wyo., 1964)
77. Lloyd, E. H., Bridgmon, G. H. A Fusarium species pathogenic on *Cirsium arvense. J. Colo. Wyo. Acad. Sci.,* **5,** 40 (1964)
78. Loos, C. A. A virus disease of *Emilia scabra. Trop. Agr. (Trinidad),* **97,** 18–21 (1941)
79. Luftig, G. R., Haselkorn, R. Morphology of blue-green algae and properties of its deoxyribonucleic acid. *J. Virol. (Kyoto),* **1,** 344–61 (1967)
80. Maguire, J. S. Laboratory germination of seeds of weedy and native plants. *Wash. Agr. Expt. Sta. Cir. 349,* 15 p. (1955)
81. Marais, J. P. Beware of the American bramble. *Farming S. Africa,* **36,** 44–47 (1961)
82. Maxfield, J. E. *A Survey of Plant Diseases as Potential Biological Control Agents for Undesirable Plants* (M. S. thesis, Univ. of Wyoming, Laramie, Wyo., 1962)
83. Maxfield, J. E., Bridgmon, G. H. A survey of diseases on some Wyoming pest plants. *J. Colo. Wyo. Acad. Sci.,* **5,** 29–30 (1962)
84. Merrill, W. The oak wilt epidemics in Pennsylvania and West Virginia: an analysis. *Phytopathology,* **57,** 1206–10 (1967)
85. Mielke, J. L. Infection experiments with *Septogloeum gillii,* a fungus parasitic on dwarf mistletoe. *J. Forestry,* **57,** 925–26 (1959)

86. Muenscher, W. C. *Weeds,* 2nd ed. (Macmillan, New York, 1955)
87. O'Brien, T. P., Robinson, B. D., Smith, L. W. Field and glasshouse trials on the control of bracken (*Pterdium aquilinum*) in Victoria. *Australian J. Expt. Agr. Animal Husbandry,* **3,** 92–97 (1963)
88. Oshima, N., Livingston, C. H., Harrison, M. D. Weeds as carriers of two potato pathogens in Colorado. *Plant Disease Reptr.,* **47,** 466–69 (1963)
89. Padan, E., Shilo, M., Kislev, N. Isolation of "Cyanophages" from freshwater ponds and their interaction with *Plectonema boryanum. Virology,* **32,** 234–46 (1967)
90. Padwick, C. W. Influence of wild and cultivated plants on the multiplication, survival, and spread of cereal foot-rooting fungi in the soil. *Can. J. Res.,* **21,** 575–89 (1935)
91. Parmeter, J. R., Hood, J. R., Scharpf, R. F. *Colletotrichum* blight of dwarf mistletoe. *Phytopathology,* **49,** 812–15 (1959)
92. Pasinetti, L., Buzzati-Traverso. A. Su alcune forme di cancrena dellecactacee dovute a nuoni micromiceti e ad un batterio. *Nuova giorn. botan. ital.,* **42,** 89–123 (1935)
93. Pinckard, J. A., Standifer, L. C. An apparent interaction between cotton herbicidal injury and seedling blight. *Plant Disease Reptr.,* **50,** 172–74 (1966)
94. Ponnappa, K. M. A new species of *Phyllostictina* on *Glycosmis* from Coorg. *Current Sci. (India),* **36,** 526–27 (1967)
95. Ponnappa, K. M. *Cercosporidium helleri* on *Sphenoclea zeylanica*—a new record for India. *Current Sci. (India),* **36,** 273 (1967)
96. Ponnappa, K. M. *Leptosphaerulina trifolii* on *Passiflora leschenaultii* and *Marsilea quadrifoliata*—two new host records for India. *Current Sci. (India),* **36,** 329–30 (1967)
97. Ponnappa, K. M. Some interesting fungi. I. Miscellaneous fungi. *Proc. Indian Acad. Sci.,* **66,** 266–72 (1967)
98. Ponnappa, K. M. Some interesting fungi. II. *Cercospora hygrophilae* sp. nov. and *Stenella plectroniae* sp. nov. *Proc. Indian Acad. Sci.,* **57,** 31–34 (1968)
99. Rader, W. E. *Helminthosporium portulacae,* a new pathogen of *Portulaca oleracea* L. *Mycologia,* **40,** 342–46 (1948)
100. Raj, T. R. N. Thread blight of water hyacinth. *Current Sci. (India),* **34,** 618–19 (1965)
101. Raj, T. R. N. Fungi occurring on witchweed in India. *Commonwealth Inst. Biol. Control Tech. Bull. No. 7,* 75–80 (1966)
102. Rao, V. P. Survey for natural enemies of witchweed and water hyacinth and other aquatic weeds affecting waterways in India. U. S. Pl-480 Project, Report for 1965
103. Riggs, R. D., Hamblen, M. L. Additional weed hosts of the *Heterodera glycines. Plant Disease Reptr.,* **50,** 15–16 (1966)
104. Rodriguez-Kabana, R., Curl, E. A., Funderburk, H. H., Jr. Effect of four herbicides on growth of *Rhizoctonia solani. Phytopathology,* **56,** 1332–33 (1966)
105. Rodriguez-Kabana, R., Curl, E. A., Funderburk, H. H., Jr. Effect of atrazine on growth response of *Sclerotium rolfsii* and *Trichoderma viride. Can. J. Microbiol.,* **13,** 1343–49 (1967)
106. Rodriguez-Kabana, R., Curl, E. A., Funderburk, H. H., Jr. Effect of paraquat on growth of *Sclerotium rolfsii* in liquid culture and soil. *Phytopathology,* **57,** 911–15 (1967)
107. Rodriguez-Kabana, R., Curl, E. A., Funderburk, H. H., Jr. Herbicides affect growth of root disease fungus. *Highlights Agr. Res.,* **14,** (1967)
108. Rogers, C. H. Cotton root-rot and weeds in native hay meadow of central Texas. *J. Am. Soc. Agron.,* **28,** 820–23 (1936)
109. Rubenchik, L. I., Bershova, O. I., Novikova, N. S., Koptyeva, Zh. P. Lysis of blue-green alga *Microcystis pulverea. Microbiol. J. (Kiev),* **2,** 88–91 (English summary) (1966)
110. Rudakov, O. L. Grib alternarii—vrag poviliki. *Sel'skoe Khoziaistvo Kirgizii SSR,* **6,** 42–43 (1960)
111. Rudakov, O. L. Gribnaya parazita poviliki (*Cuscuta lupuliformis* Kroch.) ivo proizvosdstvo primenniye. *Izdatelstov Akad. Nauk Kirgizii SSR, Frunze,* 67 pp. (1961)
112. Rudakov, O. L. Pervye rezultaty

biologicheskoy borby s povilikoy. *Zashchita Ractenii ot Vrediteli Boleznei,* **6,** 23–24 (1961)

113. Rudakov, O., Zuev, P. I., Al'Khovskaya, T. F. Novie gribniye parazity poviliki. *Sel'skoe Khoz. Kirg.,* **7,** 40 (1962)
114. Rudakov, O. L. Pervye rezultaty biologicheskoy borby S. povilikoj. *Zashchita Rast. ot Vreditelei i Boleznei,* **8,** 25–26 (1963)
115. Rudakov, O. L., Zuev, P. L., Al'Khovskaya, T. F. Proizvodstvennoe uspytanie biologicheskova metoda borby s povilikoy v kanstkom sveklosobkhoze. *Sb. Rabot Mikol. Algol Kirgizii SSR,* **63,** 27–44 (1963)
116. Safferman, R. S., Morris, M. E. Algal virus: isolation. *Science,* **140,** 679–80 (1963)
117. Safferman, R. S., Morris, M. E. Control of algae with viruses. *J. Am. Water Works Assoc.,* **56,** 1217–24 (1964)
118. Safferman, R. S., Morris, M. E. Growth characteristics of the blue-green algal virus LPP–1. *J. Bacteriol.,* **88,** 771–75 (1964)
119. Safferman, R. S., Morris, M. E. Observations on the occurrence, distribution and seasonal incidence of blue-green algal viruses. *Appl. Microbiol.,* **15,** 1219–22 (1967)
120. Scharpf, R. F. Cultural variation and pathogenicity of the Colletotrichum blight of dwarf mistletoe. *Phytopathology,* **54,** 905–6 (1964)
121. Schneider, I. R., Diener, T. O., Safferman, R. S. Blue-green algal virus LPP–1: Purification and partial characterization. *Science,* **114,** 1127 (1964)
122. Shadow, K. Feinde der Acker-Kratzdistel. *Biol. Schule,* **15,** 353–55 (1966)
123. Simmonds, F. J., Bennett, F. D. Biological control of *Opuntia* spp. by *Cactoblastis cactorum* in the Leeward Islands (West Indies). *Entomophaga,* **11,** 183–89 (1966)
124. Singh, R. N., Singh, P. K. Isolation of cyanophages from India. *Nature,* **216,** 1020–21 (1967)
125. Shklyar, M. S., Khalimova, L. A. Otokicheskom dejstvii baktepii roda Pseudomonas na semena ractenii *Uzbekskoj Biol. Zh.,* **5,** 21–25 (1962)
126. Smith, K. M., Brown, R. M., Jr., Goldstein, D. A., Walne, P. L. Culture methods for the blue-green alga *Plectonema boryanum* and its virus with an electron microscope study of virus-infected cells. *Virology,* **28,** 580–91 (1966)
127. Smith, K. M., Brown, R. M., Jr., Walne, P. L. Ultrastructure and time lapse studies on the replication cycle of the blue-green algal virus LPP–1. *Virology,* **31,** 329–37 (1967)
128. Smith, K. M., Brown, R. M., Jr., Walne, P. L., Goldstein, D. A. Electron microscopy of the infectious process of the blue green algal virus. *Virology,* **30,** 182–92 (1966)
129. Taubenhaus, J. J. *Phymatotrichum* root rot on winter and spring weeds of South Central Texas. *Am. J. Botany,* **23,** 167–68 (1936)
130. Templeton, G. E., Grable, C. I., Fulton, N. D., Bollenbacher, K. Factors affecting the amount and pattern of chlorosis caused by a metabolite of *Alternaria tenuis. Phytopathology,* **57,** 516–18 (1967)
131. Templeton, G. E., Meyer, W. L., Grable, C. I., Seigel, C. W. The chlorosis toxin from *Alternaria tenuis* is a cyclic-tetrapeptide. *Phytopathology,* **47,** 833 (1967)
132. Tullis, E. C. Covered smut of crabgrass. *Plant Disease Reptr.,* **46,** 64 (1962)
133. Wager, C. A. Can rust kill the bramble? *Farming S. Africa,* **22,** 831–32 (1947)
134. Walker, A. G. Rhizomatous grass weeds and *Ophiobolus graminis* Sacc. *Ann. Appl. Biol.,* **32,** 177–79 (1945)
135. Weir, J. R. The problem of *Dichrostachys nutans,* a weed tree, in Cuba with remarks on its pathology. *Phytopathology,* **17,** 137–46 (1927)
136. Wellman, F. L. Control of southern celery mosaic in Florida by removing weeds that serve as sources of mosaic infection. *Tech. Bull. U. S. Dept. Agr., No. 548,* 16 pp. (1937)
137. Wellman, F. L. Parasitism among neotropical phanerogams. *Ann. Rev. Phytopathol.,* **2,** 43–56 (1964)
138. Wicker, E. F., Shaw, C. G. Fungi which provide some local biological control of *Arceuthobium* spp. in the Pacific Northwest. *Phytopathology,* **52,** 757 (1962)
139. Wicker, E. F. Appraisal of biological

control of *Arceuthobium campylopodum* f. *campylopodum* by *Colletotrichum gloeosporioides*. *Plant Disease Reptr.*, **51**, 311–13 (1967)

140. Wicker, E. F., Shaw, C. G. Fungal parasites of dwarf mistletoes. *Mycologia*, **60**, 372–83 (1968)

141. Wilheim, S., Thomas, H. E. *Solanum sarachoides*, an important weed host to *Verticillium albo-atrum* *Phytopathology*, **42**, 519–20 (1952)

142. Wilson, C. L. Permission wilt—Friend or foe. *Ark. Farm Research*, **12**, 9 (1963)

143. Wilson, C. L., Tucker, M. C., Tiner, J. V. Oak wilt in Arkansas. *Plant Disease Reptr.*, **48**, 370–73 (1964)

144. Wilson, C. L. Consideration of the use of persimmon wilt as a silvicide for weed persimmons. *Plant Disease Reptr.*, **49**, 789–91 (1965)

145. Wilson, F. The biological control of weeds. *Ann. Rev. Entomol.*, **9**, 225–44 (1964)

OTHER REVIEWS OF INTEREST TO PLANT PATHOLOGISTS

Kornberg, A., Spudich, J. A., Nelson, D. L., Deutscher, M. P. Origin of proteins in sporulation. *Ann. Rev. Biochem.*, **37,** 51–78 (1968)

Hardy, R. W. F., Burns, R. C. Biological nitrogen fixation. *Ann. Rev. Biochem.*, **37,** 331–58 (1968)

Rothfield, L., Finkelstein, A. Membrane Biochemistry. *Ann. Rev. Biochem.*, **37,** 463–96 (1968)

Winteringham, F. P. W. Mechanisms of selective insecticidal action. *Ann. Rev. Entomol.*, **14,** 409–42 (1969)

Bartnicki-Garcia, S. Cell wall chemistry, morphogenesis, and taxonomy of fungi. *Ann. Rev. Microbiol.*, **22,** 87–108 (1968)

Anderson, E. S. The ecology of transferable drug resistance in the enterobacteria. *Ann. Rev. Microbiol.*, **22,** 131–80 (1968)

Wright, D. E. Toxins produced by fungi. *Ann. Rev. Microbiol.*, **22,** 269–82 (1968)

Branton, D. Membrane structure. *Ann. Rev. Plant Physiol.*, **20,** 209–38 (1969)

Pratt, H. K., Goeschl, J. D. Physiological roles in ethylene in plants. *Ann. Rev. Plant Physiol.*, **20,** 541–84 (1969)

Zelitch, I. Stomatal control. *Ann. Rev. Plant Physiol.*, **20,** 329–50 (1969)

Mazur, P. Freezing injury in plants. *Ann. Rev. Plant Physiol.*, **20,** 419–48 (1969)

Casida, J. E., Lykken, L. Metabolism of organic pesticide chemicals in higher plants. *Ann. Rev. Plant Physiol.*, **20,** 607–36 (1969)

AUTHOR INDEX

C

L

SUBJECT INDEX

Q

R

CUMULATIVE INDEXES

VOLUMES 3-7

INDEX OF CONTRIBUTING AUTHORS

INDEX OF CHAPTER TITLES

VOLUMES 3-7